# A MODERN APPROACH TO
# QUANTUM MECHANICS

# International Series in Pure and Applied Physics

**Bjorken and Drell:** *Relativistic Quantum Fields*
**Bjorken and Drell:** *Relative Quantum Mechanics*
**Fetter and Walecka:** *Quantum Theory of Many-Particle Systems*
**Fetter and Walecka:** *Theoretical Mechanics of Particles and Continua*
**Feynman and Hibbs:** *Quantum Mechanics and Path Integrals*
**Itzyksón and Zuber:** *Quantum Field Theory*
**Morse and Feshbach:** *Methods of Theoretical Physics*
**Park:** *Introduction to the Quantum Theory*
**Schiff:** *Quantum Mechanics*
**Stratton:** *Electromagnetic Theory*
**Tinkham:** *Group Theory and Quantum Mechanics*
**Townsend:** *A Modern Approach to Quantum Mechanics*
**Wang:** *Solid-State Electronics*

---

The late F. K. Richtmyer was Consulting Editor of the Series from its inception in 1929 to his death in 1939. Lee A. DuBridge was Consulting Editor from 1939 to 1946; and G. P. Harnwell from 1947 to 1954. Leonard I. Schiff served as consultant from 1954 until his death in 1971.

# A MODERN APPROACH TO QUANTUM MECHANICS

**John S. Townsend**
*Professor of Physics*
*Harvey Mudd College*
*Claremont, California*

**McGraw-Hill, Inc.**
New York   St. Louis   San Francisco   Auckland   Bogotá   Caracas
Lisbon   London   Madrid   Mexico   Milan   Montreal   New Delhi
Paris   San Juan   Singapore   Sydney   Tokyo   Toronto

This book was set in Times Roman by Publication Services.
The editors were Susan J. Tubb and John M. Morriss;
the production supervisor was Friederich W. Schulte.
The cover was designed by Rafael Hernandez.
Project supervision was done by Publication Services.
R. R. Donnelley & Sons Company was printer and binder.

**A MODERN APPROACH TO QUANTUM MECHANICS**

1 2 3 4 5 6 7 8 9 0 DOC DOC 9 0 9 8 7 6 5 4 3 2

ISBN 0-07-065119-1

Library of Congress Cataloging-in-Publication Data

Townsend, John S.
    A modern approach to quantum mechanics / John S. Townsend.
        p.      cm. — (International series in pure and applied physics)
    ISBN 0-07-065119-1
    1. Quantum theory.      I. Title.      II. Series.
    QC174.12.T69        1992                                    91-16732
    530.1′2—dc20

# ABOUT THE AUTHOR

**John S. Townsend** was born in Boston, Massachusetts, in 1946 and grew up in Chambersburg, Pennsylvania. He received his B.S. degree summa cum laude in physics from Duke University and his Ph.D. in physics from The Johns Hopkins University, where he held an NSF Graduate Fellowship. After postdoctoral positions at Johns Hopkins and the Stanford Linear Accelerator Center, he joined the physics department at Harvey Mudd College, the science and engineering college of The Claremont Colleges, where he has taught since 1975. Professor Townsend's primary research area is in theoretical particle physics. On a recent sabbatical leave, he was a Science Fellow at the Center for International Security and Arms Control at Stanford University. Professor Townsend is married, has two daughters, and enjoys playing tennis.

# CONTENTS

# 13 Scattering 368

# 14 Photons and Atoms 399

## Appendixes

## Index

# PREFACE

There have been two revolutions in the way we view the physical world in the twentieth century: relativity and quantum mechanics. In quantum mechanics the revolution has been both profound—requiring a dramatic revision in the structure of the laws of mechanics that govern the behavior of all particles, be they electrons or photons—and far-reaching in its impact—determining the stability of matter itself, shaping the interactions of particles on the atomic, nuclear, and particle physics level, and leading to macroscopic quantum effects ranging from lasers and superconductivity to neutron stars and radiation from black holes. Moreover, in a triumph for twentieth-century physics, special relativity and quantum mechanics have been joined together in the form of quantum field theory. Field theories such as quantum electrodynamics have been tested with an extremely high precision, with agreement between theory and experiment verified to better than nine significant figures. It should be emphasized that while our understanding of the laws of physics is continually evolving, always being subjected to experimental scrutiny, so far no confirmed discrepancy between theory and experiment for quantum mechanics has been detected.

This book is intended for an upper-division course in quantum mechanics. The most likely audience for the book consists of students who have completed a course in modern physics that includes an introduction to quantum mechanics in the form of one-dimensional wave mechanics. Such a modern physics course is likely to emphasize the historical development of the subject. Rather than continue with a similar approach in a second course, I have chosen to introduce the fundamentals of quantum mechanics through a detailed discussion of the physics of intrinsic spin. Such an approach has a number of significant advantages. First, students find starting a course with something "new" such as intrinsic spin both interesting and exciting, and they enjoy making connections with what they have seen before. Second, spin systems provide us with many beautiful but straightforward illustrations of the essential structure of quantum mechanics, a structure that is not obscured by the mathematics of wave mechanics. Quantum mechanics can be presented through concrete examples. I believe that most physicists learn through specific examples and then find it easy to generalize. By starting with spin, students are given plenty of time to assimilate this novel and striking material. I have found that they seem to learn this key introductory material easily and well—material that was often perceived to be difficult when I came to it midway through a course that began with wave mechanics. Third, when we do come to wave mechanics, students see that wave mechanics is only one aspect of quantum mechanics, not the fundamental core of the subject. They see at an early stage that wave mechanics and matrix mechanics are just different ways of calculating based on the same underlying quantum mechanics and that the approach they use depends on the particular problem they are addressing.

I have been inspired by two sources, an "introductory" treatment in Volume III of *The Feynman Lectures on Physics* and an advanced exposition in J. J. Sakurai's *Modern Quantum Mechanics*. Overall, I believe that wave mechanics is probably the best way to *introduce* students to quantum mechanics. Wave mechanics makes the largest overlap with what students know from classical physics and shows them the strange behavior of quantum mechanics in a familiar environment. This is probably why students find their first introduction to wave mechanics so stimulating. However, starting a second course with wave mechanics runs the risk of diminishing much of the excitement and enthusiasm for the entirely new way of viewing nature that is demanded by quantum mechanics. It becomes sort of old hat, material the student has seen before, repeated in more depth. It is, I believe, with the second exposure to quantum mechanics that something like Feynman's approach has its best chance to be effective. But to be effective, a quantum mechanics text needs to make lots of contact with the way most physicists think and calculate in quantum mechanics using the language of kets and operators. This is Sakurai's approach in his graduate-level textbook. In a sense, the approach that I am presenting here can be viewed as a superposition of these two approaches, but at the junior-senior level.

Chapter 1 introduces the concepts of a quantum state vector, complex probability amplitudes, and the probabilistic interpretation of quantum mechanics in the context of analyzing a number of Stern-Gerlach experiments carried out with spin-$\frac{1}{2}$ particles. By introducing ket vectors at the beginning, we have the framework for matrix mechanics and for thinking about states as having an existence quite apart from the way we happen to choose to represent them. Moreover, there is a natural role for operators; in Chapter 2 they rotate spin states so that the spin "points" in a different direction. I do not follow a postulatory approach, but rather I allow the basic physics of this spin system to drive the introduction of concepts such as Hermitian operators, eigenvalues, and eigenstates.

In Chapter 3 the commutation relations of the generators of rotations are determined from the behavior of ordinary vectors under rotations. Most of the material in this chapter is fairly conventional; what is not so conventional is the introduction of the operator techniques for determining the angular momentum eigenstates and eigenvalue spectrum and the derivation of the uncertainty relations from the commutation relations at such an early stage. Since so much of our initial discussion of quantum mechanics revolves around intrinsic spin, it is important for students to see how quantum mechanics can be used to determine from first principles the spin states that have been introduced in Chapters 1 and 2, without having to appeal only to experimental results.

Chapter 4 is devoted to time evolution of states. The natural operation in time development is to translate states forward in time. The Hamiltonian enters as the generator of time translations, and the states are shown to obey the Schrödinger equation. Most of the chapter is devoted to physical examples. In Chapter 5 another physical system, the spin-spin interaction of an electron and proton in the ground state of hydrogen, is used to introduce the spin states of two spin-$\frac{1}{2}$ particles. The total-spin-0 state serves as the basis for a discussion of the Einstein-Podolsky-Rosen paradox and the Bell inequalities.

The main theme of Chapter 6 is making contact with the usual formalism of wave mechanics. The special problems in dealing with states such as position states that have a spectrum in the continuum are analyzed. The momentum operator enters naturally as the generator of translations. Sections 6.8 through 6.10 include a general discussion with examples of solutions to the Schrödinger equation, which can serve as a review for students with a good background in one-dimensional wave mechanics.

Chapter 7 is devoted to the simple harmonic oscillator, which merits a chapter all its own. Although the material in Chapter 8 on path integrals may be skipped without affecting subsequent chapters (with the exception of Section 14.1, on the Aharonov-Bohm effect), I believe that path integrals should be discussed, if possible, since this formalism provides real insight into quantum dynamics. However, I have found it difficult to fit this material into our one-semester course, which is taken by all physics majors as well as some students majoring in other disciplines. Rather, I have chosen to postpone path integrals to a second course and then to insert the material in Chapter 8 before Chapter 14. Incidentally, the material on path integrals is the only part of the book that may require students to have had an upper-division mechanics course, one in which the principle of least action is discussed.

Chapters 9 through 13 cover fully three-dimensional problems, including the two-body problem, orbital angular momentum, central potentials, time-independent perturbations, identical particles, and scattering. An effort has been made to include as many physical examples as possible.

Although this is a textbook on nonrelativistic quantum mechanics, I have chosen to include a discussion of the quantized radiation field in the final chapter, Chapter 14. The use of ket and bra vectors from the beginning and the discussion of solutions to problems such as angular momentum and the harmonic oscillator in terms of abstract raising and lowering operators should have helped to prepare the student for the exciting jump to a quantized electromagnetic field. By quantizing this field, we can really understand the properties of photons, we can calculate the lifetimes for spontaneous emission from first principles, and we can understand why a laser works. By looking at higher-order processes such as photon-atom scattering, we can also see the essentials of Feynman diagrams. Although the atom is treated nonrelativistically, it is still possible to gain a sense of what quantum field theory is all about at this level without having to face the complications of the relativistic Dirac equation. For the instructor who wishes to cover time-dependent perturbation theory but does not have time for all of this chapter, Section 14.6 stands on its own.

Although SI units are the standard for undergraduate education in electricity and magnetism, I have chosen in this text to use Gaussian units, which are more commonly used to describe microscopic phenomena. However, with the possible exception of the last chapter, with its quantum treatment of the electromagnetic field, the choice of units has little impact. My own experience suggests that students who are generally at home with SI units are comfortable (as indicated in a number of footnotes throughout the text) replacing $e^2$ with $e^2/4\pi\varepsilon_0$ or ignoring the factor of $c$ in the Bohr magneton whenever they need to carry out numerical calculations. In addition, electromagnetic units are discussed in Appendix A.

A comprehensive solutions manual for the instructor is available from the publisher, upon request of the instructor.

Finally, some grateful acknowledgments are certainly in order. Students in my quantum mechanics classes have given me useful feedback as I tried out versions of this book in manuscript form. One of my colleagues at Harvey Mudd, Tom Helliwell, read the first draft of the manuscript and offered some valuable comments, as well as encouragement. Art Weldon of West Virginia University suggested a number of ways to improve the accuracy and effectiveness of the manuscript. I have also benefited from the comments of the following McGraw-Hill reviewers: William Dalton, St. Cloud State University; Michael Grady, SUNY–Fredonia; Richard Hazeltine, University of Texas at Austin; Jack Mochel, University of Illinois at Urbana-Champaign; and Jae Y. Park, North Carolina State University. The Pew Science Program provided support for two Harvey Mudd students, Doug Dunston and Doug Ridgway, who helped in the preparation of the text and the figures, respectively. Helen White helped in checking the galley proofs. And I especially want to thank my family for cheerfully letting me devote so much time to this project.

*John S. Townsend*

# A MODERN APPROACH TO QUANTUM MECHANICS

# CHAPTER

# 1

# STERN-GERLACH
# EXPERIMENTS

We begin our discussion of quantum mechanics with a conceptually simple experiment in which we measure the intrinsic spin angular momentum of an atom. This experiment was first carried out by O. Stern and W. Gerlach in 1922 using a beam of silver atoms. We will refer to the measuring apparatus as a Stern-Gerlach device. The results of experiments with a number of such devices are easy to describe but, as we shall see, nonetheless startling in their consequences.

## 1.1  THE ORIGINAL STERN-GERLACH EXPERIMENT

Before analyzing the experiment, we need to know something about the relationship between the intrinsic spin angular momentum of a particle and its corresponding magnetic moment. To the classical physicist, angular momentum is always orbital angular momentum, namely, $\mathbf{L} = \mathbf{r} \times \mathbf{p}$. Although the earth is said to have spin angular momentum $I\omega$ due to its rotation about its axis as well as orbital angular momentum due to its revolution about the sun, both types of angular momentum are just different forms of $\mathbf{L}$. The intrinsic spin angular momentum $\mathbf{S}$ of a microscopic particle is not at all of the same sort as orbital angular momentum, but it is real angular momentum nonetheless.

To get a feeling for the relationship that exists between the angular momentum of a charged particle and its corresponding magnetic moment, we first use a classical example and then point out some of its limitations. Consider a particle with charge $q$ moving in a circular orbit of radius $r$ with speed $v$. The magnetic moment $\mu$ is given by

$$\mu = \frac{IA}{c} = \left(\frac{q}{T}\right)\frac{\pi r^2}{c} = \frac{qvr}{2c} = \frac{q}{2mc}L \tag{1.1}$$

where $A$ is the area of the circle formed by the orbit, the current $I$ is the charge $q$ divided by the period $T = (2\pi r/v)$, and $L = mvr$ is the orbital angular momentum of the particle.[1] Since the magnetic moment and the orbital angular momentum are parallel or antiparallel depending on the sign of the charge $q$, we may express this relationship in the vector form

$$\boldsymbol{\mu} = \frac{q}{2mc}\mathbf{L} \qquad (1.2)$$

This relationship between $\mathbf{L}$ and $\boldsymbol{\mu}$ turns out to be generally true whenever the mass and charge coincide in space. One can obtain different constants of proportionality by adjusting the charge and mass distributions independently. For example, a solid spherical ball of mass $m$ rotating about an axis through its center with the charge $q$ distributed uniformly only on the surface of the ball has a constant of proportionality of $5q/6mc$.

When we come to intrinsic spin of a particle, we write

$$\boldsymbol{\mu} = \frac{gq}{2mc}\mathbf{S} \qquad (1.3)$$

where the value of the constant $g$ is experimentally determined to be $g = 2.00$ for an electron, $g = 5.58$ for a proton, or even $g = -3.82$ for a neutron.[2] One might be tempted to presume that $g$ is telling us about how the charge and mass are distributed for the different particles and that intrinsic spin angular momentum is just orbital angular momentum of the particle itself as it spins about its axis. We will see as we go along that such a simple classical picture of intrinsic spin is entirely untenable and that the intrinsic spin angular momentum we are discussing is a very different beast indeed. In fact, it appears that even a point particle in quantum mechanics may have intrinsic spin angular momentum.[3] Although there

---

[1] If you haven't seen them before, the Gaussian units we are using for electromagnetism may take a little getting used to. A comparison of SI and Gaussian units is given in Appendix A. In SI units the magnetic moment is just $IA$, so you can ignore the factor of $c$, the speed of light, in expressions such as (1.1) if you wish to convert to SI units.

[2] Each of these $g$ factors has its own experimental uncertainty. Recent measurements by R. S. Van Dyck, Jr., P. B. Schwinberg, and H. G. Dehmelt, *Phys. Rev. Lett.*, **59**, 26 (1987), have shown that $g/2$ for an electron is 1.0011596521884(43), where the factor of 43 reflects the uncertainty in the last two places. Relativistic quantum mechanics predicts that $g = 2$ for an electron. The deviations from this value can be accounted for by quantum field theory. The much larger deviations from $g = 2$ for the proton and neutron are due to the fact that these particles are not fundamental but are composed of charged constituents called quarks.

[3] It is amusing to note that in 1926 S. Goudsmit and G. Uhlenbeck as graduate students "discovered" the electron's spin from an analysis of atomic spectra. They wrote up their results for their advisor P. Ehrenfest, who then advised them to discuss the matter with H. Lorentz. When Lorentz showed them that a classical model of the electron required that the electron must be spinning at a speed on the surface approximately ten times the speed of light, they went to Ehrenfest to tell him of their foolishness. He informed them that he had already submitted their paper for publication and that they shouldn't worry since they were "both young enough to be able to afford a stupidity." *Physics Today*, June 1976, pp. 40–48.

are no classical arguments that we can give to justify (1.3), we can note that such a relationship between the magnetic moment and the intrinsic spin angular momentum is at least consistent with dimensional analysis. At this stage, you can think of $g$ as a dimensionless factor that has been inserted to make the magnitudes as well as the dimensions come out right.

Let's turn to the Stern-Gerlach experiment itself. Figure 1.1$a$ shows a schematic diagram of the apparatus. A collimated beam of silver atoms is produced by evaporating silver in a hot oven and selecting those atoms that pass through a series of narrow slits. The beam is then directed between the poles of a magnet. One of the pole pieces is flat; the other has a sharp tip. Such a magnet produces an inhomogeneous magnetic field, as shown in Fig. 1.1$b$. When a *neutral* atom with a magnetic moment $\boldsymbol{\mu}$ enters the magnetic field $\mathbf{B}$, it experiences a force $\mathbf{F} = \boldsymbol{\nabla}(\boldsymbol{\mu}\cdot\mathbf{B})$, since $-\boldsymbol{\mu}\cdot\mathbf{B}$ is the energy of interaction of a magnetic dipole with an external magnetic field. If we call the direction in which the inhomogeneous magnetic field is large the $z$ direction, we see that

$$F_z = \boldsymbol{\mu} \cdot \frac{\partial \mathbf{B}}{\partial z} \simeq \mu_z \frac{\partial B_z}{\partial z} \tag{1.4}$$

Notice that we have taken the magnetic field gradient $\partial B_z/\partial z$ in the figure to be negative, so that if $\mu_z$ is negative as well, then $F_z$ is positive and the atoms are deflected in the positive $z$ direction. Classically, $\mu_z = |\boldsymbol{\mu}|\cos\theta$, where $\theta$ is the angle that the magnetic moment $\boldsymbol{\mu}$ makes with the $z$ axis. Thus $\mu_z$ should take on a continuum of values ranging from $+\mu$ to $-\mu$. Since the atoms coming from the oven are not polarized with their magnetic moments pointing in a preferred direction, we should find a corresponding continuum of deflections. In the original Stern-Gerlach experiment, the silver atoms were detected by allowing them to

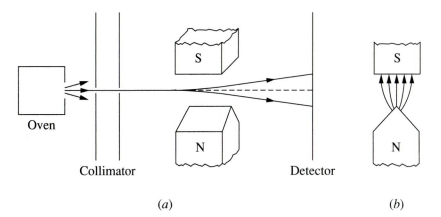

$(a)$ $\qquad\qquad\qquad\qquad\qquad\qquad\qquad$ $(b)$

**FIGURE 1.1**
($a$) A schematic diagram of the Stern-Gerlach experiment. ($b$) A cross-sectional view of the pole pieces of the magnet depicting the inhomogeneous magnetic field they produce.

build up to a visible deposit on a glass plate. Figure 1.2 shows the results of this original experiment. The surprising result is that $\mu_z$ takes on only *two* values, corresponding to the values $\pm\hbar/2$ for $S_z$. Numerically, $\hbar = h/2\pi = 1.055 \times 10^{-27} \text{erg} \cdot \text{s} = 6.582 \times 10^{-16} \text{eV} \cdot \text{s}$, where $h$ is Planck's constant.

Silver atoms are composed of 47 electrons and a nucleus. Atomic theory tells us the total orbital and total spin angular momentum of 46 of the electrons is equal to zero, and the 47th electron has zero orbital angular momentum. Moreover, as (1.3) shows, the nucleus makes a very small contribution to the magnetic moment of the atom because the mass of nucleus is so much larger than the mass of the electron. Therefore, the magnetic moment of the silver atom is effectively due to the magnetic moment of a single electron. Thus, in carrying out their experiment, Stern and Gerlach measured the component of the intrinsic spin angular momentum of an electron along the $z$ axis and found it to take on only two discrete values, $+\hbar/2$ and $-\hbar/2$, commonly called "spin up" and "spin

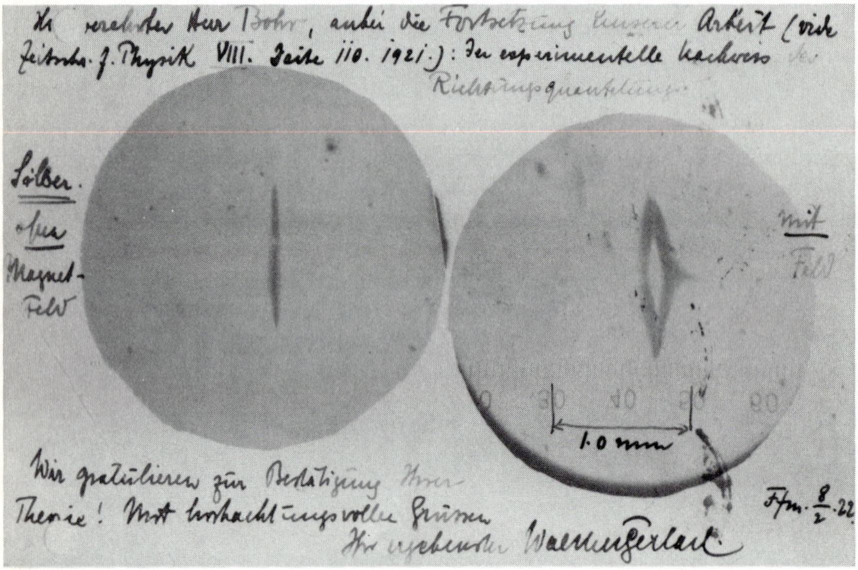

**FIGURE 1.2**
A postcard from Walther Gerlach to Niels Bohr, dated February 8, 1922. Note that the images on the postcard have been rotated by $90°$ relative to Fig. 1.1, where the collimating slit is horizontal. The left-hand image of the beam profile without the magnetic field shows the effect of the finite width of this collimating slit. The right-hand image shows the beam profile with the magnetic field. Only in the center of the apparatus is the magnitude of the magnetic field gradient sufficiently strong to cause splitting. The pattern is smeared because of the range of speeds of the atoms coming from the oven. Translation of the message: "My esteemed Herr Bohr, attached is the continuation of our work [vide *Zeitschr. f. Phys.* **8**, 110 (1921)]: the experimental proof of directional quantization. We congratulate you on the confirmation of your theory! With respectful greetings. Your most humble Walther Gerlach." *(Photograph copyright the Niels Bohr Archive. Used by permission.)*

down," respectively. Later, we will see that these values are characteristic of a spin-$\frac{1}{2}$ particle. Incidentally, we chose to make the bottom N pole piece of the Stern-Gerlach (SG) device the one with the sharp tip for a simple reason. With this configuration, $B_z$ decreases as $z$ increases, making $\partial B_z/\partial z$ negative. As we noted earlier, atoms with a negative $\mu_z$ are deflected upward in this field. Now an electron has charge $q = -e$ and from (1.3) with $g = 2$, $\mu_z = (-e/m_e c)S_z$. Thus a silver atom with $S_z = \hbar/2$, a spin-up atom, will conveniently be deflected upward.

## 1.2 FOUR EXPERIMENTS

Now that we have seen how the actual Stern-Gerlach experiment was done, let's turn our attention to four simple experiments that will tell us much about the structure of quantum mechanics. If you like, you can think of these experiments as thought experiments so that we needn't focus on any technical difficulties that might be faced in carrying them out.

### Experiment 1

Let us say a particle that exits an SGz device, one with its inhomogeneous magnetic field parallel to the $z$ axis, with $S_z = +\hbar/2$ is in the state $|+\mathbf{z}\rangle$. The symbol $|+\mathbf{z}\rangle$, known as a *ket vector*, is a convenient way of denoting this state. Suppose a beam of particles, *each* of which is in this state, enters another SGz device. We find that *all* the particles exit in the state $|+\mathbf{z}\rangle$; that is, the measurement of $S_z$ yields the value $+\hbar/2$ for each of the particles, as indicated in Fig. 1.3a.

### Experiment 2

Consider a beam of particles exiting the SGz device in the state $|+\mathbf{z}\rangle$, as in Experiment 1. We next send this beam into an SGx device, one with its inhomogeneous magnetic field oriented along the $x$ axis. We find that 50 percent of the particles exit the second device with $S_x = \hbar/2$ and are therefore in the state $|+\mathbf{x}\rangle$, while the other 50 percent exit with $S_x = -\hbar/2$ and are therefore in the state $|-\mathbf{x}\rangle$ (see Fig. 1.3b). For completeness, we also note that if we select the beam of particles exiting the initial SGz device in the state $|-\mathbf{z}\rangle$ instead of $|+\mathbf{z}\rangle$ and send this beam through the SGx device, we also find that 50 percent of the particles yield $\hbar/2$ for a measurement of $S_x$ and 50 percent yield $-\hbar/2$ for a measurement of $S_x$.

### Experiment 3

Let's add a third SG device to Experiment 2, but this time with its inhomogeneous magnetic field oriented along the $z$ axis (see Fig. 1.3c). If we send the beam of particles exiting the SGx device in the state $|+\mathbf{x}\rangle$ through the last SGz device, we find that 50 percent of the particles exit in the state $|+\mathbf{z}\rangle$ and 50 percent exit in the state $|-\mathbf{z}\rangle$. Initially, none of the particles entering the SGx device was in the state $|-\mathbf{z}\rangle$. Thus making the measurement of $S_x$ with the second device has modified the state of the system. We cannot think of the beam entering the last SGz device

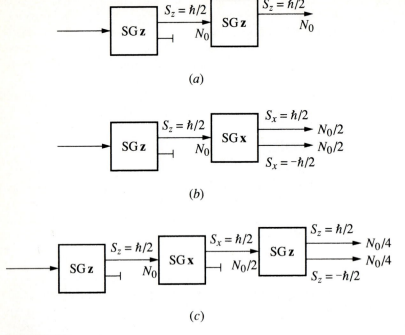

*(a)*

*(b)*

*(c)*

**FIGURE 1.3**
A block diagram of *(a)* Experiment 1, *(b)* Experiment 2, and *(c)* Experiment 3. $N_0$ is the number of particles in the beam exiting the first SG device with $S_z = \hbar/2$.

as comprised of particles with $S_z = \hbar/2$ *and* $S_x = \hbar/2$, as one might expect from the results of the measurements of the first two SG devices. This cannot account for the 50 percent of the beam that exits the last SGz device with $S_z = -\hbar/2$. We will see shortly that $S_z$ and $S_x$ are incompatible observables; namely, we cannot know both of them simultaneously. In the macroscopic world, on the other hand, it seems to be easy to create a state with two definite nonzero components of the angular momentum, as, for example, is the case for a spinning top whose angular momentum is oriented at $45°$ to both the $x$ and $z$ axes. This is an indication that the quantum world is fundamentally different from our everyday macroscopic experience. We will see this more clearly as we go on to consider the next Stern-Gerlach experiment.

## Experiment 4

In this experiment we use a modified SG device, introduced in thought experiments by Richard Feynman.[4] This SG device, shown in Fig. 1.4a, is comprised

---

[4] R. P. Feynman, R. B. Leighton, and M. Sands, *The Feynman Lectures on Physics*, Addison-Wesley, Reading, Mass., 1965, vol. 3, Chapter 5.

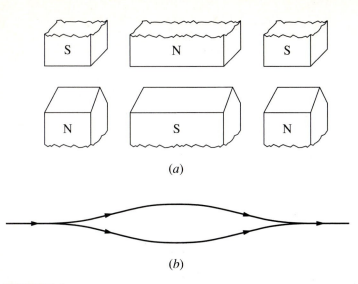

**FIGURE 1.4**
(*a*) The three magnets of a modified Stern-Gerlach device. (*b*) The paths that a spin-up and spin-down particle would follow in traversing the device.

of three high-gradient magnets, placed back to back, instead of a single magnet. The first magnet is a typical Stern-Gerlach magnet, followed by a second magnet with the same cross section as the first but twice as long and with the polarity opposite that of the first magnet. This second magnet pushes a particle with a magnetic moment in the opposite direction to the first magnet. Thus, in traversing the first half of the length of this magnet, the particle is decelerated and brought to rest in the transverse direction. In traversing the second half, the particle is accelerated back toward the axis. Although the third magnet is just like the first magnet, here it decelerates the particle so that the particle returns to the axis in the same state as it entered the first magnet. The net effect of the three magnets is to recombine the beams so that their condition upon exiting the third magnet is just like it was before entering the first magnet. Figure 1.4*b* indicates the paths that spin-up and spin-down particles would follow in this modified SG device.

You might think such a device serves no purpose, but we can use a modified SG device to make a measurement and select a particular spin state. For example, if the direction of the inhomogeneous magnetic field of the three magnets is along the $x$ axis, we can select a particle in the $|+\mathbf{x}\rangle$ spin state by blocking the path that a particle in the $|-\mathbf{x}\rangle$ spin state would take, as indicated in Fig. 1.5. Then all the particles exiting the modified three-magnet SGx device would be in the state $|+\mathbf{x}\rangle$. In fact, we can repeat Experiment 3 with the SGx device replaced by a modified SGx device. If the $|-\mathbf{x}\rangle$ state is filtered out by inserting a block in the lower path, we find, of course, exactly the same results as in Experiment 3; that is, when we measure with the last SGz device, we find 50 percent of the particles in the state $|+\mathbf{z}\rangle$ and 50 percent in the state $|-\mathbf{z}\rangle$. Similarly, if we filter out the

**FIGURE 1.5**
Selecting a spin-up state with a modified Stern-Gerlach device by blocking the spin-down state.

state $|+\mathbf{x}\rangle$ by inserting a block in the upper path, we also find 50 percent of the par-
ticles exiting the last SGz device in the state $|+\mathbf{z}\rangle$ and 50 percent in the state $|-\mathbf{z}\rangle$.

We are now ready for Experiment 4. As in Experiment 3, a beam of particles
in the state $|+\mathbf{z}\rangle$ from an initial SGz device enters an SGx device, but in this
experiment it is a modified SGx device in which we do *not* block one of the paths
and, therefore, do *not* make a measurement of $S_x$. We then send the beam from
this modified SGx device into another SGz device. As indicated in Fig. 1.6, we
find that 100 percent of the particles exit the last SGz device in the state $|+\mathbf{z}\rangle$,
just as if the modified SGx device were absent from the experiment and we were
repeating Experiment 1.

Before carrying out Experiment 4, it might seem obvious that 50 percent
of the particles passing through the modified SGx device are in the state $|+\mathbf{x}\rangle$
and 50 percent are in the state $|-\mathbf{x}\rangle$. But the results of Experiment 4 contradict
this assumption, since, if it were true, we would expect to find 50 percent of
the particles in the state $|+\mathbf{z}\rangle$ and 50 percent of the particles in the state $|-\mathbf{z}\rangle$
when the unfiltered beam exits the last SGz device. Our results are completely
incompatible with the hypothesis that the particles traversing the modified SGx
device have *either* $S_x = \hbar/2$ *or* $S_x = -\hbar/2$. Moreover, even if we carry out the
experiment with a beam of such low intensity that one particle at a time is passing
through the SG devices, we still find that each of the particles has $S_z = \hbar/2$ when
it leaves the last SGz device. Thus, the issue raised by this experiment cannot be
resolved by some funny business involving the interactions of the particles in the
beams as they pass through the modified SGx device.

So far, we have been able to describe the results of these Stern-Gerlach
experiments simply in terms of the percentage of particles exiting the SG devices
in a particular state because the experiments have been carried out on a beam
of particles, namely, on a large number of particles. For a *single* particle, it is

**FIGURE 1.6**
A block diagram of Experiment 4. Note that we cannot indicate the path followed through the three-
magnet modified SGx device since no measurement is carried out to select either a $|+\mathbf{x}\rangle$ or $|-\mathbf{x}\rangle$ spin
state.

generally *not* possible to predict with certainty the outcome of the measurement in advance. In Experiment 2, for example, *before* a measurement of $S_x$ on a particle in the state $|+\mathbf{z}\rangle$, all we can say is that there is a 50 percent *probability* of obtaining $S_x = \hbar/2$ and a 50 percent *probability* of obtaining $S_x = -\hbar/2$. However, probabilities alone do not permit us to understand Experiment 4. We cannot explain the results of this experiment by adding the probabilities that a particle passing through the modified SGx device is in the state $|+\mathbf{x}\rangle$ or in the state $|-\mathbf{x}\rangle$, since this fails to account for the differences when comparing the results of Experiment 3, in which 50 percent of the particles in the state $|+\mathbf{x}\rangle$ (or $|-\mathbf{x}\rangle$) yield $S_z = -\hbar/2$, with the results of Experiment 4, in which none of the particles has $S_z = -\hbar/2$ when exiting the last SGz device. Somehow in Experiment 4 we must eliminate the probability that the particle is in the state $|-\mathbf{z}\rangle$ when it enters the last SGz device. What we need is some sort of "interference" that can cancel out the $|-\mathbf{z}\rangle$ state. Such interference is common in the physics of waves, where two waves can interfere destructively to produce minima as well as constructively to produce maxima. With waves it isn't the intensities that interfere but rather the amplitudes. We will use the same terminology for our Stern-Gerlach experiments and introduce a *probability amplitude* that we will "square" to get the probability. If we don't observe which path is taken in the modified SGx device by inserting a block, or filter, we must add the amplitudes to take the two different paths corresponding to the $|+\mathbf{x}\rangle$ and $|-\mathbf{x}\rangle$ states. Even a single particle can have an amplitude to be in both states, to take both paths; when we add, or superpose, the amplitudes, we obtain an amplitude for the particle to be in the state $|+\mathbf{z}\rangle$ only.[5] In summary, when we *don't* make a measurement in the modified SG device, we must add the amplitudes, *not* the probabilities.

## 1.3 THE QUANTUM STATE VECTOR

In our description of the state of a particle in quantum mechanics, we have been using a new notation in which states, such as $|+\mathbf{z}\rangle$, are denoted by abstract vectors called ket vectors. Such a description includes as much information about the state of the particle as we are permitted in quantum mechanics. For example, the ket $|+\mathbf{x}\rangle$ is just a shorthand way of saying that the spin state of the particle is such that if we were to make a measurement of $S_x$, the intrinsic spin angular momentum in the $x$ direction, we would obtain the value $\hbar/2$. There are clearly other attributes that are required to give a complete description of the particle, such as the particle's position or momentum. However, for the time being we are concentrating on the

---

[5] In Section 2.3 we will discuss in more detail how this interference in Experiment 4 works. These results are reminiscent of the famous double-slit experiment, in which it seems logical to suppose that the particles go through one slit *or* the other, but the interference pattern on a distant screen is completely incompatible with this simple hypothesis. The double-slit experiment is discussed briefly in Section 6.7. If you are unfamiliar with this experiment from the perspective of quantum mechanics, an excellent discussion is given in *The Feynman Lectures on Physics*, vol. 3, Chapter 1.

spin degrees of freedom of the particle.[6] Later, in Chapter 6, we will see how to introduce other degrees of freedom in the description of the state of the particle.

Classical physics uses a different type of vector in its description of nature. Some of these ordinary vectors are more abstract than others. For example, consider the electric field $\mathbf{E}$, which is a useful but somewhat abstract vector. If there is an electric field present, we know that a test charge $q$ placed in the field will experience a force $\mathbf{F} = q\mathbf{E}$. Of course, even the force $\mathbf{F}$ will not be observed directly. We would probably allow the particle to be accelerated by the force, measure the acceleration, and then use Newton's law $\mathbf{F} = m\mathbf{a}$ to determine $\mathbf{F}$ and thence $\mathbf{E}$.

Let's suppose the electric field in the location where you are reading this book has a constant value, which you could determine in the way we have just outlined. How do you tell your friends about the value, both magnitude and direction, of $\mathbf{E}$? You might just point in the direction of $\mathbf{E}$ to show its direction. But what if your friends are not present and you want to write down $\mathbf{E}$ on a piece of paper? You would probably set up a coordinate system and choose basis vectors $\mathbf{i}$, $\mathbf{j}$, and $\mathbf{k}$ whose direction you could easily communicate. Using this coordinate system, you would denote the electric field as $\mathbf{E} = E_x\mathbf{i} + E_y\mathbf{j} + E_z\mathbf{k}$. In fact, we often use a shorthand notation in which we suppress the unit vectors and just say $\mathbf{E} = (E_x, E_y, E_z)$, although in the notation we will be using in our discussion of quantum mechanics, it would be better to denote this as $\mathbf{E} \rightarrow (E_x, E_y, E_z)$. How do we obtain the value for $E_x$, for example? We just project the electric field onto the $x$ axis. Formally, we take the dot product to find $E_x = \mathbf{i} \cdot \mathbf{E} = |\mathbf{E}|\cos\theta$, where $\theta$ is the angle the electric field $\mathbf{E}$ makes with the $x$ axis, as shown in Fig. 1.7.

Let's return to our discussion of quantum state vectors. If we send a spin-$\frac{1}{2}$ particle into an SGz device, we obtain only the values $\hbar/2$ and $-\hbar/2$, corresponding to the particle ending up in the state $|+\mathbf{z}\rangle$ or ending down in the state $|-\mathbf{z}\rangle$, respectively. These two states can be considered as vectors that form a basis for our abstract quantum mechanical vector space. If the particle is initially in the state $|+\mathbf{z}\rangle$, we have seen in Experiment 1 that there is zero amplitude for the particle to be in the state $|-\mathbf{z}\rangle$, which we denote by $\langle-\mathbf{z}|+\mathbf{z}\rangle = 0$. We can think of this as telling us that the vectors are orthogonal, the analogue of $\mathbf{i} \cdot \mathbf{j} = 0$ in our electric field example. Of course if we send a particle in the state $|+\mathbf{z}\rangle$ into an SGz device, we always find the particle in the state $|+\mathbf{z}\rangle$. In the language of quantum mechanical amplitudes this is clearly telling us that the amplitude $\langle+\mathbf{z}|+\mathbf{z}\rangle$ is nonzero. As we will see momentarily, it is convenient to require that our quantum

---

[6] The historical development of quantum mechanics initially focused on the more obvious degrees of freedom, such as a particle's position. In fact, Goudsmit was fond of relating how, when confronted with the need to introduce a new degree of freedom for the intrinsic spin of the electron in order to explain atomic spectra, he had to ask Uhlenbeck what was meant by the expression "degree of freedom."

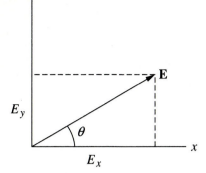

**FIGURE 1.7**

The $x$ and $y$ components of an electric field $\mathbf{E}$ making an angle $\theta$ with the $x$ axis can be obtained by taking the dot product of $\mathbf{E}$ with the unit vectors $\mathbf{i}$ and $\mathbf{j}$. For a classical vector such as $\mathbf{E}$, $E_x$ and $E_y$ can also be obtained by projecting $\mathbf{E}$ onto the $x$ and $y$ axes.

mechanical vectors be unit vectors and therefore satisfy $\langle +\mathbf{z}|+\mathbf{z}\rangle = 1$, just as $\mathbf{i} \cdot \mathbf{i} = 1$. We similarly require that $\langle -\mathbf{z}|-\mathbf{z}\rangle = 1$ as well.

Suppose the particle is in the state $|+\mathbf{x}\rangle$. From Experiment 3 we know that the particle has nonzero amplitudes, which we can call $c_+$ and $c_-$, to be in the states $|+\mathbf{z}\rangle$ and $|-\mathbf{z}\rangle$, respectively. We can express this state as $|+\mathbf{x}\rangle = c_+|+\mathbf{z}\rangle + c_-|-\mathbf{z}\rangle$, a linear combination of the states $|+\mathbf{z}\rangle$ and $|-\mathbf{z}\rangle$. In fact, it is convenient at this stage to consider an arbitrary spin state $|\psi\rangle$, which could be created by sending a beam of particles with intrinsic spin-$\frac{1}{2}$ through an SG device with its inhomogeneous magnetic field oriented in some arbitrary direction and selecting those particles that are deflected, for example, upward. In general, this state, like $|+\mathbf{x}\rangle$, will have nonzero amplitudes to yield both $\hbar/2$ and $-\hbar/2$ if a measurement of $S_z$ is made. Thus we will express this state $|\psi\rangle$ as

$$|\psi\rangle = c_+|+\mathbf{z}\rangle + c_-|-\mathbf{z}\rangle \qquad (1.5)$$

where the particular values for $c_+$ and $c_-$ depend on the orientation of the SG device. That an arbitrary state $|\psi\rangle$ can be expressed as a superposition of the states $|+\mathbf{z}\rangle$ and $|-\mathbf{z}\rangle$ means that these states form a complete set, just as the unit vectors $\mathbf{i}$, $\mathbf{j}$, and $\mathbf{k}$ form a complete set for expressing an electric field $\mathbf{E}$.

How can we formally determine the values of $c_+$ and $c_-$? In order to take the analogue of the dot product in our ordinary classical vector example, we need to introduce a new type of vector called a *bra vector*. For every ket $|\psi\rangle$ there corresponds a bra $\langle\psi|$. The fate of a bra such as $\langle\varphi|$ is to meet up with a ket $|\psi\rangle$ to form an amplitude, or inner product, $\langle\varphi|\psi\rangle$ in the form of a bracket—hence the name for bras and kets. The amplitude $\langle\varphi|\psi\rangle$ is the amplitude for a particle in the state $|\psi\rangle$ to be in the state $|\varphi\rangle$. From our earlier experiments we know that $\langle -\mathbf{z}|+\mathbf{z}\rangle = 0$, and similarly $\langle +\mathbf{z}|-\mathbf{z}\rangle = 0$, since a particle in the state $|-\mathbf{z}\rangle$, with $S_z = -\hbar/2$, has zero amplitude to be in the state $|+\mathbf{z}\rangle$, with $S_z = \hbar/2$. Thus from (1.5), we can deduce that

$$\langle +\mathbf{z}|\psi\rangle = c_+\langle +\mathbf{z}|+\mathbf{z}\rangle + c_-\langle +\mathbf{z}|-\mathbf{z}\rangle = c_+ \qquad (1.6a)$$

$$\langle -\mathbf{z}|\psi\rangle = c_+\langle -\mathbf{z}|+\mathbf{z}\rangle + c_-\langle -\mathbf{z}|-\mathbf{z}\rangle = c_- \qquad (1.6b)$$

or simply $c_\pm = \langle \pm \mathbf{z} | \psi \rangle$. This enables us to express (1.5) in the form

$$|\psi\rangle = \underbrace{\langle +\mathbf{z}|\psi\rangle}_{c_+}|+\mathbf{z}\rangle + \underbrace{\langle -\mathbf{z}|\psi\rangle}_{c_-}|-\mathbf{z}\rangle = |+\mathbf{z}\rangle\langle +\mathbf{z}|\psi\rangle + |-\mathbf{z}\rangle\langle -\mathbf{z}|\psi\rangle \qquad (1.7)$$

where in the last step we have positioned the amplitudes after the kets in a suggestive way. Note that the amplitudes $\langle +\mathbf{z}|\psi\rangle$ and $\langle -\mathbf{z}|\psi\rangle$, the brackets, are (complex) numbers, and thus the product of an amplitude times a ket vector is itself just a ket vector. It really doesn't matter whether we position the amplitude before or after the ket. Writing the ket vector $|\psi\rangle$ in the form (1.7) is analogous to expressing the electric field $\mathbf{E}$ in the form $\mathbf{E} = E_x\mathbf{i} + E_y\mathbf{j} + E_z\mathbf{k} = \mathbf{i}(\mathbf{i} \cdot \mathbf{E}) + \mathbf{j}(\mathbf{j} \cdot \mathbf{E}) + \mathbf{k}(\mathbf{k} \cdot \mathbf{E})$.

Since to each ket there corresponds a bra vector, we must be able to express $\langle \psi |$ in terms of $\langle +\mathbf{z}|$ and $\langle -\mathbf{z}|$ as

$$\langle \psi | = c'_+\langle +\mathbf{z}| + c'_-\langle -\mathbf{z}| \qquad (1.8)$$

Using the same technique as before, we see that

$$\langle \psi|+\mathbf{z}\rangle = c'_+\langle +\mathbf{z}|+\mathbf{z}\rangle + c'_-\langle -\mathbf{z}|+\mathbf{z}\rangle = c'_+ \qquad (1.9a)$$

$$\langle \psi|-\mathbf{z}\rangle = c'_+\langle +\mathbf{z}|-\mathbf{z}\rangle + c'_-\langle -\mathbf{z}|-\mathbf{z}\rangle = c'_- \qquad (1.9b)$$

Thus the bra corresponding to the ket in (1.7) is

$$\langle \psi | = \underbrace{\langle \psi|+\mathbf{z}\rangle}_{c'_+}\langle +\mathbf{z}| + \underbrace{\langle \psi|-\mathbf{z}\rangle}_{c'_-}\langle -\mathbf{z}| \qquad (1.10)$$

How are the amplitudes $\langle +\mathbf{z}|\psi\rangle$ and $\langle \psi|+\mathbf{z}\rangle$ related? Just as we require that $\langle +\mathbf{z}|+\mathbf{z}\rangle = 1$, we also require that $\langle \psi|\psi\rangle = 1$. We are demanding that all physical vectors in our abstract quantum mechanical vector space be unit vectors. As we will now see, this requirement is crucial to the probabilistic interpretation of quantum mechanics. If we use (1.7) and (1.10) to evaluate $\langle \psi|\psi\rangle$, we find

$$\langle \psi|\psi\rangle = \langle \psi|+\mathbf{z}\rangle\langle +\mathbf{z}|\psi\rangle + \langle \psi|-\mathbf{z}\rangle\langle -\mathbf{z}|\psi\rangle = 1 \qquad (1.11)$$

In Section 1.5 we will examine a final Stern-Gerlach experiment that will convince you that amplitudes such as $\langle +\mathbf{z}|\psi\rangle$ and $\langle -\mathbf{z}|\psi\rangle$ are in general complex numbers. The way to guarantee that equality (1.11) is satisfied for arbitrary $|\psi\rangle$'s is to have

$$\langle \psi|+\mathbf{z}\rangle = \langle +\mathbf{z}|\psi\rangle^* \quad \text{and} \quad \langle \psi|-\mathbf{z}\rangle = \langle -\mathbf{z}|\psi\rangle^* \qquad (1.12)$$

so that each of the terms in (1.11) is real. These results say that the amplitude for a particle in the state $|\psi\rangle$ to be in the states $|\pm\mathbf{z}\rangle$ is the complex conjugate of the amplitude for a particle in the states $|\pm\mathbf{z}\rangle$ to be in the state $|\psi\rangle$.

From (1.6) and (1.9), we see that $c'_+ = c_+^*$ and $c'_- = c_-^*$. Therefore, the bra corresponding to the ket (1.5) is

$$\langle \psi | = c_+^*\langle +\mathbf{z}| + c_-^*\langle -\mathbf{z}| \qquad (1.13)$$

The bra vector is generated from the ket vector by changing all the basis kets to their corresponding bras and by changing all amplitudes (complex numbers) to their complex conjugates.

With these results, we can express (1.11) as

$$\langle\psi|\psi\rangle = \langle+\mathbf{z}|\psi\rangle^*\langle+\mathbf{z}|\psi\rangle + \langle-\mathbf{z}|\psi\rangle^*\langle-\mathbf{z}|\psi\rangle$$
$$= c_+^* c_+ + c_-^* c_- = 1 \tag{1.14}$$

or

$$\langle\psi|\psi\rangle = |\langle+\mathbf{z}|\psi\rangle|^2 + |\langle-\mathbf{z}|\psi\rangle|^2 = 1 \tag{1.15}$$

where $|\langle+\mathbf{z}|\psi\rangle|^2 \equiv \langle+\mathbf{z}|\psi\rangle^*\langle+\mathbf{z}|\psi\rangle$ and $|\langle-\mathbf{z}|\psi\rangle|^2 \equiv \langle-\mathbf{z}|\psi\rangle^*\langle-\mathbf{z}|\psi\rangle$. We interpret $|\langle+\mathbf{z}|\psi\rangle|^2$ as the probability that a particle in the state $|\psi\rangle$ will be found to be in the state $|+\mathbf{z}\rangle$ if a measurement of $S_z$ is made with an SGz device and $|\langle-\mathbf{z}|\psi\rangle|^2$ as the probability that the particle will be found in the state $|-\mathbf{z}\rangle$. As (1.15) shows, the requirement that $\langle\psi|\psi\rangle = 1$ guarantees that the probability of finding the particle in either one state or the other sums to one, since there are only two results possible for a measurement of $S_z$ for a spin-$\frac{1}{2}$ particle.

The striking feature of (1.7) is that when both of the probability amplitudes $\langle+\mathbf{z}|\psi\rangle$ and $\langle-\mathbf{z}|\psi\rangle$ are nonzero, then a particle in the state $|\psi\rangle$ is really in a *superposition* of the states $|+\mathbf{z}\rangle$ and $|-\mathbf{z}\rangle$. There are probabilities of obtaining both $S_z = \hbar/2$ and $S_z = -\hbar/2$ if a measurement of $S_z$ is carried out. This is to be contrasted with classical mechanics, where for a particle *in a definite state* we do not expect measurements of, say, the orbital angular momentum of the particle at a particular time to yield two different values, such as $\mathbf{r}_1 \times \mathbf{p}_1$ and $\mathbf{r}_2 \times \mathbf{p}_2$.

## 1.4  ANALYSIS OF EXPERIMENT 3

As we noted earlier, Experiment 3 is telling us that a particle in the state $|+\mathbf{x}\rangle$ is in a superposition of the states $|+\mathbf{z}\rangle$ and $|-\mathbf{z}\rangle$: $|+\mathbf{x}\rangle = c_+|+\mathbf{z}\rangle + c_-|-\mathbf{z}\rangle$, since when we make measurements of $S_z$ with the last SGz device in the experiment, we have probabilities of obtaining both $\hbar/2$ and $-\hbar/2$. Because the probabilities are each 50 percent, we have

$$c_+^* c_+ = \langle+\mathbf{z}|+\mathbf{x}\rangle^*\langle+\mathbf{z}|+\mathbf{x}\rangle = |\langle+\mathbf{z}|+\mathbf{x}\rangle|^2 = \tfrac{1}{2} \tag{1.16a}$$

$$c_-^* c_- = \langle-\mathbf{z}|+\mathbf{x}\rangle^*\langle-\mathbf{z}|+\mathbf{x}\rangle = |\langle-\mathbf{z}|+\mathbf{x}\rangle|^2 = \tfrac{1}{2} \tag{1.16b}$$

The solution for $c_+$ and $c_-$ may be written as

$$c_+ = \frac{e^{i\delta_+}}{\sqrt{2}} \quad \text{and} \quad c_- = \frac{e^{i\delta_-}}{\sqrt{2}} \tag{1.17}$$

where $\delta_+$ and $\delta_-$ are real phases that allow for the possibility that $c_+$ and $c_-$ are complex. The ket for the state with $S_x = \hbar/2$ is then given by

$$|+\mathbf{x}\rangle = \frac{e^{i\delta_+}}{\sqrt{2}}|+\mathbf{z}\rangle + \frac{e^{i\delta_-}}{\sqrt{2}}|-\mathbf{z}\rangle \tag{1.18}$$

Notice that the probabilities (1.16) themselves do not give us any information about the values of these phases, since the phases cancel out when we calculate $c_+^* c_+$ and $c_-^* c_-$.

We can use the probabilities (1.16) to calculate the average value, or "expectation value," of $S_z$, which is the sum of each value obtained by a measurement of $S_z$ multiplied by the probability of obtaining that value:

$$\langle S_z \rangle = c_+^* c_+ \left( \frac{\hbar}{2} \right) + c_-^* c_- \left( -\frac{\hbar}{2} \right)$$

$$= \frac{1}{2} \left( \frac{\hbar}{2} \right) + \frac{1}{2} \left( -\frac{\hbar}{2} \right) = 0 \tag{1.19}$$

In this particular case, the expectation value doesn't coincide with any of the values that may be obtained by measuring $S_z$. An idealized set of data resulting from measurements of $S_z$ on a very large collection of particles, each with $S_x = \hbar/2$, is shown in Fig. 1.8. Clearly, there is an inherent uncertainty in the result of the measurements, since the measurements do not all yield the same value. We calculate this uncertainty by computing the standard deviation: we determine the average value of the data, take each data point, subtract the average value from it, square and average, and finally take the square root. Thus the square of the uncertainty is given by

$$\begin{aligned} (\Delta S_z)^2 &= \langle (S_z - \langle S_z \rangle)^2 \rangle \\ &= \langle S_z^2 - 2S_z \langle S_z \rangle + \langle S_z \rangle^2 \rangle \\ &= \langle S_z^2 \rangle - 2\langle S_z \rangle \langle S_z \rangle + \langle S_z \rangle^2 \\ &= \langle S_z^2 \rangle - \langle S_z \rangle^2 \end{aligned} \tag{1.20}$$

The expectation value $\langle S_z^2 \rangle$ is the sum of each value of $S_z^2$ multiplied by the probability of obtaining that value:

$$\langle S_z^2 \rangle = c_+^* c_+ \left( \frac{\hbar}{2} \right)^2 + c_-^* c_- \left( -\frac{\hbar}{2} \right)^2$$

$$= \frac{1}{2} \left( \frac{\hbar^2}{4} \right) + \frac{1}{2} \left( \frac{\hbar^2}{4} \right) = \frac{\hbar^2}{4} \tag{1.21}$$

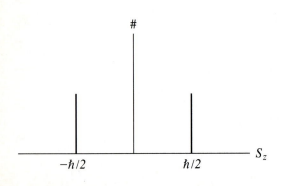

**FIGURE 1.8**
An idealized set of data resulting from measurements of $S_z$ on a collection of particles with $S_x = \hbar/2$.

Therefore, substituting (1.19) and (1.21) into (1.20), we find $\Delta S_z = \hbar/2$ for a particle in the state $|+\mathbf{x}\rangle$.

Of course, $\langle S_z \rangle = 0$ is not in disagreement with finding a single particle to be spin up if we make a measurement of $S_z$ on a particle in the state $|+\mathbf{x}\rangle$. To test predictions such as (1.19) requires a statistically significant sample. Suppose we make measurements of $S_z$ on 100 particles, each in the state $|+\mathbf{x}\rangle$, and find 55 of them to be spin up ($S_z = \hbar/2$) and 45 of them to be spin down ($S_z = -\hbar/2$). Should we be worried about a disagreement with the predictions of quantum mechanics? In general, if we make $N$ measurements, we should expect fluctuations that are on the order of $\sqrt{N}$. Thus with 100 measurements, deviations from $\langle S_z \rangle = 0$ on the order of 10 percent are reasonable. However, if we were to make $10^6$ measurements and find 550,000 particles spin up and 450,000 particles spin down, we should be concerned, since we should expect fluctuations of only about $\sqrt{N} = 1,000$, rather than the measured 50,000.

## 1.5 EXPERIMENT 5

We are now ready to consider the final Stern-Gerlach experiment of this chapter. In this experiment we replace the last SGz device of Experiment 3 with one that has its inhomogeneous magnetic field in the $y$ direction and thus make measurements of $S_y$ on particles exiting the SGx device in the state $|+\mathbf{x}\rangle$. From Experiment 3 we already know the results of this final experiment. We must find 50 percent of the particles with $S_y = \hbar/2$ and 50 percent of the particles with $S_y = -\hbar/2$, as indicated in Fig. 1.9. Although we are measuring $S_y$ instead of $S_z$ with the last SG device in this experiment, the percentage of the particles that go "up" and "down" must be the same for Experiment 3 and Experiment 5, since the axis that we called the $z$ axis in Experiment 3 could just as easily have been called the $y$ axis, either by us or by another observer viewing the experiment. In fact, this sort of argument tells us that if we were to replace the middle SGx device in Experiment 3 with an SGy device, we would still find that 50 percent of the particles have $S_z = \hbar/2$ and 50 percent have $S_z = -\hbar/2$ when exiting the last SGz device.

These simple results have important implications. Just as we are able to express the state $|+\mathbf{x}\rangle$ by (1.18), we can express the state $|+\mathbf{y}\rangle$ as a superposition of $|+\mathbf{z}\rangle$ and $|-\mathbf{z}\rangle$ in the form

$$|+\mathbf{y}\rangle = \frac{e^{i\gamma_+}}{\sqrt{2}}|+\mathbf{z}\rangle + \frac{e^{i\gamma_-}}{\sqrt{2}}|-\mathbf{z}\rangle = \frac{e^{i\gamma_+}}{\sqrt{2}}\left(|+\mathbf{z}\rangle + e^{i(\gamma_- -\gamma_+)}|-\mathbf{z}\rangle\right) \qquad (1.22)$$

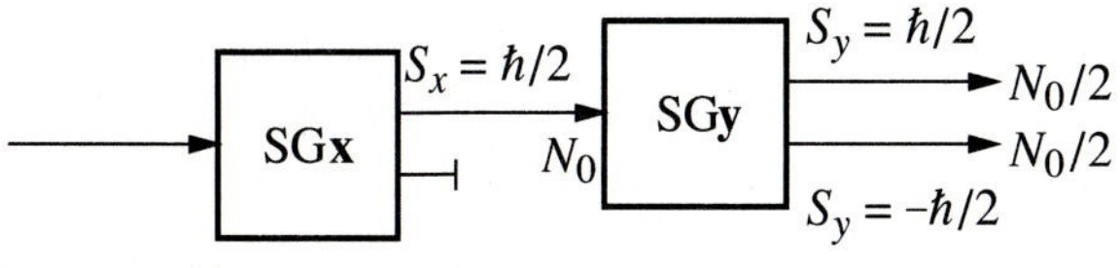

**FIGURE 1.9**
A block diagram of Experiment 5.

where we have written the complex numbers multiplying the kets $|+\mathbf{z}\rangle$ and $|-\mathbf{z}\rangle$ in such a way as to ensure that there is a 50 percent probability of obtaining $S_z = \hbar/2$ and a 50 percent probability of obtaining $S_z = -\hbar/2$. Note that in the last step we pulled out in front an overall phase factor $e^{i\gamma_+}$ for future computational convenience. Moreover, since in Experiment 5 there is a 50 percent probability of finding a particle with $S_y = \hbar/2$ when it exits the SGx device in the state $|+\mathbf{x}\rangle$, we must have

$$|\langle +\mathbf{y}|+\mathbf{x}\rangle|^2 = \tfrac{1}{2} \tag{1.23}$$

Now the bra corresponding to the ket (1.22) is

$$\langle +\mathbf{y}| = \frac{e^{-i\gamma_+}}{\sqrt{2}}\langle +\mathbf{z}| + \frac{e^{-i\gamma_-}}{\sqrt{2}}\langle -\mathbf{z}| = \frac{e^{-i\gamma_+}}{\sqrt{2}}\big(\langle +\mathbf{z}| + e^{-i(\gamma_- -\gamma_+)}\langle -\mathbf{z}|\big) \tag{1.24}$$

where we have replaced the complex numbers in (1.22) with their complex conjugates in going from (1.22) to (1.24). If we rewrite (1.18) by pulling out an overall phase factor:

$$|+\mathbf{x}\rangle = \frac{e^{i\delta_+}}{\sqrt{2}}\big(|+\mathbf{z}\rangle + e^{i(\delta_- -\delta_+)}|-\mathbf{z}\rangle\big) \tag{1.25}$$

then

$$\langle +\mathbf{y}|+\mathbf{x}\rangle = \frac{e^{i(\delta_+ -\gamma_+)}}{2}\big(\langle +\mathbf{z}| + e^{-i\gamma}\langle -\mathbf{z}|\big)\big(|+\mathbf{z}\rangle + e^{i\delta}|-\mathbf{z}\rangle\big)$$

$$= \frac{e^{i(\delta_+ -\gamma_+)}}{2}\big(1 + e^{i(\delta-\gamma)}\big) \tag{1.26}$$

where $\delta = \delta_- - \delta_+$ and $\gamma = \gamma_- - \gamma_+$ are the *relative* phases between the kets $|+\mathbf{z}\rangle$ and $|-\mathbf{z}\rangle$ for these two states, and we have used $\langle +\mathbf{z}|+\mathbf{z}\rangle = \langle -\mathbf{z}|-\mathbf{z}\rangle = 1$ and $\langle +\mathbf{z}|-\mathbf{z}\rangle = \langle -\mathbf{z}|+\mathbf{z}\rangle = 0$ in evaluating the amplitude. We finally calculate the probability:

$$|\langle +\mathbf{y}|+\mathbf{x}\rangle|^2 = \left[\frac{e^{i(\delta_+ -\gamma_+)}}{2}\big(1 + e^{i(\delta-\gamma)}\big)\right]\left[\frac{e^{-i(\delta_+ -\gamma_+)}}{2}\big(1 + e^{-i(\delta-\gamma)}\big)\right]$$

$$= \tfrac{1}{4}\big(1 + e^{i(\delta-\gamma)}\big)\big(1 + e^{-i(\delta-\gamma)}\big)$$

$$= \tfrac{1}{2}\big[1 + \cos(\delta - \gamma)\big] \tag{1.27}$$

Agreement with (1.23) requires $\delta - \gamma = \pm\pi/2$. The common convention, which we will see in Chapter 3, is to take $\delta = 0$. If in (1.22) and (1.25) we ignore the overall phases $\delta_+$ and $\gamma_+$, which appear in the amplitude (1.26) but do not enter into the calculation of the probability (1.27), we see that

$$|+\mathbf{x}\rangle = \frac{1}{\sqrt{2}}|+\mathbf{z}\rangle + \frac{1}{\sqrt{2}}|-\mathbf{z}\rangle \tag{1.28}$$

and

$$|+\mathbf{y}\rangle = \frac{1}{\sqrt{2}}|+\mathbf{z}\rangle + \frac{e^{i\pi/2}}{\sqrt{2}}|-\mathbf{z}\rangle = \frac{1}{\sqrt{2}}|+\mathbf{z}\rangle + \frac{i}{\sqrt{2}}|-\mathbf{z}\rangle \qquad (1.29)$$

where we have chosen $\gamma = \pi/2$. The choice $\gamma = -\pi/2$ yields the state

$$\frac{1}{\sqrt{2}}|+\mathbf{z}\rangle - \frac{i}{\sqrt{2}}|-\mathbf{z}\rangle = |-\mathbf{y}\rangle \qquad (1.30)$$

The reason for this ambiguity is that in discussing our series of Stern-Gerlach experiments we have not specified whether our coordinate system is right handed or left handed. The state we have called $|+\mathbf{y}\rangle$ is indeed the state with $S_y = \hbar/2$ in a right-handed coordinate system. The state we have called $|-\mathbf{y}\rangle$ is the state with $S_y = -\hbar/2$ in our right-handed coordinate system. Of course, this latter state, which is spin down along $y$, is spin up along $y$ in a left-handed coordinate system, as shown in Fig. 1.10. That is why we see both solutions appearing.[7]

These complications should not detract from the main message to be learned from Experiment 5. The simple fact is that (1.23) cannot be explained without a complex amplitude. The appearance of $i$'s such as the one in (1.29) is one of the key ingredients of a description of nature by quantum mechanics. Whereas in classical physics we often use complex numbers as an aid to do calculations, there they are not essential. The straightforward Stern-Gerlach experiments we have outlined in this chapter *demand* complex numbers for their explanation.

---

[7] We will see how to derive all of the results of this section from first principles in Chapter 3.

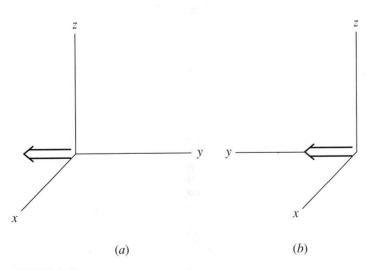

(a)                                    (b)

**FIGURE 1.10**
A state that is spin down along $y$ in the right-handed coordinate system shown in (a) is spin up along $y$ in the left-handed system shown in (b).

## 1.6  SUMMARY

The world of quantum mechanics is both strange and wonderful, in part because it is a world filled with surprises that so often run counter to our classical expectations. Yet as we go on, we will see the remarkable insight quantum mechanics gives us not just into microscopic phenomena but into the laws of classical mechanics as well. Since quantum mechanics subsumes classical mechanics, we cannot "derive" quantum mechanics from our classical, macroscopic experiences. Our strategy in this chapter has been to take a number of Stern-Gerlach experiments as our guide into this strange world of quantum behavior. From these experiments we can see many of the general features of quantum mechanics.

A quantum state is specified either by a ket vector $|\psi\rangle$ or a corresponding bra vector $\langle\psi|$. The complex numbers that we calculate in quantum mechanics result from a ket vector $|\psi\rangle$ meeting up with a bra vector $\langle\varphi|$, forming the bra(c)ket $\langle\varphi|\psi\rangle$, which we call the probability amplitude for a particle in the state $|\psi\rangle$ to be in the state $|\varphi\rangle$. The amplitude $\langle\psi|\varphi\rangle$ for a particle in the state $|\varphi\rangle$ to be in the state $|\psi\rangle$ is the complex conjugate of the amplitude for a particle in the state $|\psi\rangle$ to be in the state $|\varphi\rangle$:

$$\langle\psi|\varphi\rangle = \langle\varphi|\psi\rangle^* \tag{1.31}$$

The probability of finding a particle to be in the state $|\varphi\rangle$ when a measurement is made on a particle in the state $|\psi\rangle$ is given by $|\langle\varphi|\psi\rangle|^2$. Notice that the probability is unchanged if the ket $|\psi\rangle$ is multiplied by an *overall* phase factor $e^{i\delta}$: $|\psi\rangle \rightarrow e^{i\delta}|\psi\rangle$.

Although we have phrased our discussion so far solely in terms of the intrinsic spin angular momentum of a spin-$\frac{1}{2}$ particle, the structure that we see emerging has a broad level of applicability. Suppose that we are considering an observable $A$ for which the results of a measurement take on the discrete values $a_1, a_2, a_3, \ldots$. As we will see, angular momentum and energy are good examples of observables for which the results of measurements can be grouped in a discrete (although not necessarily finite) set. A general quantum state, expressed in the form of a ket vector $|\psi\rangle$, can be written as a superposition of the states $|a_1\rangle, |a_2\rangle, |a_3\rangle, \ldots$ that result if a measurement of $A$ yields $a_1, a_2, a_3, \ldots$, respectively:

$$|\psi\rangle = c_1|a_1\rangle + c_2|a_2\rangle + c_3|a_3\rangle + \cdots = \sum_n c_n|a_n\rangle \tag{1.32}$$

The corresponding bra vector is given by

$$\langle\psi| = c_1^*\langle a_1| + c_2^*\langle a_2| + c_3^*\langle a_3| + \cdots = \sum_n c_n^*\langle a_n| \tag{1.33}$$

The complex number

$$c_n = \langle a_n|\psi\rangle \tag{1.34}$$

is the amplitude to obtain $a_n$ if a measurement of $A$ is made for a particle in the state $|\psi\rangle$.[8]

Physically, we expect that

$$\langle a_i | a_j \rangle = 0 \qquad i \neq j \tag{1.35}$$

since if the particle is in a state for which the result of a measurement is $a_j$, there is zero amplitude of obtaining $a_i$ with $i \neq j$. The vectors $|a_i\rangle$ and $|a_j\rangle$ with $i \neq j$ are said to be orthogonal. The amplitude to obtain $a_i$ for a particle in the state $|a_i\rangle$ is taken to be one, that is,

$$\langle a_i | a_i \rangle = 1 \tag{1.36}$$

The vector $|a_i\rangle$ is then said to be normalized. Equations (1.35) and (1.36) can be nicely summarized by

$$\langle a_i | a_j \rangle = \delta_{ij} \tag{1.37}$$

where $\delta_{ij}$ is called the Kronecker delta and is defined by the relationship

$$\delta_{ij} = \begin{cases} 0 & i \neq j \\ 1 & i = j \end{cases} \tag{1.38}$$

We say that the set of vectors $|a_i\rangle$ form an orthonormal set of basis vectors. Equation (1.32) shows how an arbitrary vector $|\psi\rangle$ can be expressed in terms of this basis set. Thus the vectors $|a_i\rangle$ form a complete set.

Amplitudes such as (1.34) can be projected out of the ket $|\psi\rangle$ by taking the inner product of the ket $|\psi\rangle$ with the bra $\langle a_i |$:

$$\langle a_i | \psi \rangle = \sum_n c_n \langle a_i | a_n \rangle$$

$$= \sum_n c_n \delta_{in} = c_i \tag{1.39}$$

Thus the ket (1.32) can be written

$$|\psi\rangle = \sum_n |a_n\rangle \langle a_n | \psi \rangle \tag{1.40}$$

which is just a sum of ket vectors $|a_i\rangle$, each multiplied by the amplitude $\langle a_i | \psi \rangle$. Similarly, the amplitude $c_i^*$ can be projected out of the bra $\langle \psi |$ by taking the inner

---

[8] In this chapter we have used the shorthand notation $|S_z = \pm\hbar/2\rangle = |\pm\mathbf{z}\rangle$, $|S_x = \pm\hbar/2\rangle = |\pm\mathbf{x}\rangle$, and so on. Thus $\langle \pm\mathbf{z}|\psi\rangle$ are the amplitudes to obtain $S_z = \pm\hbar/2$ for a spin-$\frac{1}{2}$ particle in the state $|\psi\rangle$ if a measurement of $S_z$ is made.

product with the ket $|a_i\rangle$:

$$\langle\psi|a_i\rangle = \sum_n c_n^*\langle a_n|a_i\rangle$$

$$= \sum_n c_n^*\delta_{ni} = c_i^* \tag{1.41}$$

The bra (1.33) can thus be written as

$$\langle\psi| = \sum_n \langle\psi|a_n\rangle\langle a_n| \tag{1.42}$$

which is the sum of the bra vectors $\langle a_i|$, each multiplied by the amplitude $\langle\psi|a_i\rangle$.
The normalization requirement

$$\langle\psi|\psi\rangle = 1 \tag{1.43}$$

for a physical state $|\psi\rangle$ leads to

$$1 = \langle\psi|\psi\rangle = \left(\sum_i c_i^*\langle a_i|\right)\left(\sum_j c_j|a_j\rangle\right)$$

$$= \sum_i \sum_j c_i^* c_j\langle a_i|a_j\rangle$$

$$= \sum_i \sum_j c_i^* c_j\delta_{ij} = \sum_i |c_i|^2 \tag{1.44}$$

showing that the probabilities

$$|c_i|^2 = |\langle a_i|\psi\rangle|^2 \tag{1.45}$$

of obtaining the result $a_i$ if a measurement of $A$ is carried out sum to one. From these results it follows that the average value of the observable $A$ for a particle in the state $|\psi\rangle$ is given by

$$\langle A\rangle = \sum_n |c_n|^2 a_n \tag{1.46}$$

since the average value (expectation value) is the sum of the values obtained by the measurements weighted by the probabilities of obtaining those values. The uncertainty is given by

$$\Delta A = \left(\langle(A - \langle A\rangle)^2\rangle\right)^{1/2} = \left(\langle A^2\rangle - \langle A\rangle^2\right)^{1/2} \tag{1.47}$$

where

$$\langle A^2\rangle = \sum_n |c_n|^2 a_n^2 \tag{1.48}$$

Equations (1.46) and (1.48) illustrate the importance of *completeness,* that is, that any state can be expressed as a superposition of basis vectors, as in (1.32). Without this completeness, we would not know how to calculate the results of measurements for the observable $A$ for an arbitrary state.

One of the most striking features of the physical world is that if more than one of the $c_n$ in (1.32) is nonzero, then there are amplitudes to obtain different $a_n$ for a particle in a particular state $|\psi\rangle$. How should we interpret this result: Is the ket (1.32) telling us that the particle spends time in each of the states $|a_n\rangle$, and the probability $|\langle a_n|\psi\rangle|^2$ is just a reflection of how much time it spends in that particular state? Does this specification of the state as a superposition just reflect our lack of knowledge of which state the particle is really in? Is this why we must deal with probabilities? The answer to these questions is an emphatic no. Rather, (1.32) is to be read as a true superposition of the individual states $|a_n\rangle$, for if we parametrize the complex amplitudes in the form

$$\langle a_n|\psi\rangle = |\langle a_n|\psi\rangle|e^{i\delta_n} \tag{1.49}$$

where $|\langle a_n|\psi\rangle|$ is the magnitude, or modulus, of the amplitude and $\delta_n$ is the phase of the amplitude, the difference in phase (the relative phase) between the individual states in the superposition matters a great deal. As we have seen in our discussion of the spin-$\frac{1}{2}$ $|+\mathbf{x}\rangle$ and $|+\mathbf{y}\rangle$ kets, changing the relative phase between the kets $|+\mathbf{z}\rangle$ and $|-\mathbf{z}\rangle$ in such a superposition by $\pi/2$ changes a state with $S_x = \hbar/2$ into one with $S_y = \hbar/2$. Compare (1.28) and (1.29).[9] Thus the values of the relative phases in (1.32) dramatically affect how the states "add up," or how the amplitudes interfere with each other. Quantum mechanics is more than just a collection of probabilities. We live in a world in which the allowed states of a particle include *superpositions* of the states in which the particle possesses a definite attribute, such as the $z$ component of the particle's spin angular momentum, and thus by superposing such states we form states for which the particle does not have a definite value at all for such an attribute.

## PROBLEMS

**1.1.** Determine the field gradient of a 50-cm-long Stern-Gerlach magnet that would produce a 1 mm separation at the detector between spin-up and spin-down silver atoms that are emitted from an oven at $T = 1500$ K. Assume the detector (see Fig. 1.1) is located 50 cm from the magnet. *Note*: While the atoms in the oven have average kinetic energy $3kT/2$, the more energetic atoms strike the hole in the oven more frequently. Thus the *emitted* atoms have average kinetic energy $2kT$. The magnetic dipole moment of the silver atom is due to the intrinsic spin of a single electron. Appendix F gives the numerical value of the Bohr magneton, $e\hbar/2m_e c$, in a convenient form.

---

[9] This also shows that a spin-$\frac{1}{2}$ particle cannot have simultaneously a definite value for both the $x$ and $y$ components of its intrinsic spin angular momentum.

**1.2.** Show for a solid spherical ball of mass $m$ rotating about an axis through its center with a charge $q$ uniformly distributed on the surface of the ball that the magnetic moment $\boldsymbol{\mu}$ is related to the angular momentum $\mathbf{L}$ by the relation

$$\boldsymbol{\mu} = \frac{5q}{6mc}\mathbf{L}$$

*Reminder:* The factor of $c$ is a consequence of our using Gaussian units. If you work in SI units, just add the $c$ in by hand to compare with this result.

**1.3.** In Problem 3.2 we will see that the state of a spin-$\frac{1}{2}$ particle that is spin up along the axis whose direction is specified by the unit vector $\mathbf{n} = \sin\theta\cos\phi\,\mathbf{i} + \sin\theta\sin\phi\,\mathbf{j} + \cos\theta\,\mathbf{k}$, with $\theta$ and $\phi$ shown in Fig. 1.11, is given by

$$|+\mathbf{n}\rangle = \cos\frac{\theta}{2}|+\mathbf{z}\rangle + e^{i\phi}\sin\frac{\theta}{2}|-\mathbf{z}\rangle$$

(a) Verify that the state $|+\mathbf{n}\rangle$ reduces to the states $|+\mathbf{x}\rangle$ and $|+\mathbf{y}\rangle$ given in this chapter for the appropriate choice of the angles $\theta$ and $\phi$.

(b) Suppose that a measurement of $S_z$ is carried out on a particle in the state $|+\mathbf{n}\rangle$. What is the probability that the measurement yields (i) $\hbar/2$? (ii) $-\hbar/2$?

(c) Determine the uncertainty $\Delta S_z$ of your measurements.

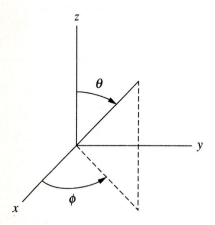

**FIGURE 1.11**
The angles $\theta$ and $\varphi$ specifying the orientation of an SGn device.

**1.4.** Repeat the calculations of Problem 1.3 (b) and (c) for measurements of $S_x$. *Hint:* Infer what the probability of obtaining $-\hbar/2$ for $S_x$ is from the probability of obtaining $\hbar/2$.

**1.5.** (a) What is the amplitude to find a particle that is in the state $|+\mathbf{n}\rangle$ (from Problem 1.3) with $S_y = \hbar/2$? What is the probability? Check your result by evaluating the probability for an appropriate choice of the angles $\theta$ and $\phi$.

(b) What is the amplitude to find a particle that is in the state $|+\mathbf{y}\rangle$ with $S_n = \hbar/2$? What is the probability?

**1.6.** Show that the state

$$|-\mathbf{n}\rangle = \sin\frac{\theta}{2}|+\mathbf{z}\rangle - e^{i\phi}\cos\frac{\theta}{2}|-\mathbf{z}\rangle$$

satisfies $\langle +\mathbf{n}|-\mathbf{n}\rangle = 0$, where the state $|+\mathbf{n}\rangle$ is given in Problem 1.3. Verify that $\langle -\mathbf{n}|-\mathbf{n}\rangle = 1$.

**1.7.** A beam of spin-$\frac{1}{2}$ particles is sent through a series of three Stern-Gerlach measuring devices, as illustrated in Fig. 1.12. The first SGz device transmits particles with $S_z = \hbar/2$ and filters out particles with $S_z = -\hbar/2$. The second device, an SGn device, transmits particles with $S_n = \hbar/2$ and filters out particles with $S_n = -\hbar/2$ where the axis **n** makes an angle $\theta$ in the $x$-$z$ plane with respect to the $z$ axis. Thus particles after passage through this SGn device are in the state $|+\mathbf{n}\rangle$ given in Problem 1.3 with the angle $\phi = 0$. A last SGz device transmits particles with $S_z = -\hbar/2$ and filters out particles with $S_z = \hbar/2$.

(a) What fraction of the particles transmitted through the first SGz device will survive the third measurement?

(b) How must the angle $\theta$ of the SGn device be oriented so as to maximize the number of particles that are transmitted by the final SGz device? What fraction of the particles survive the third measurement for this value of $\theta$?

(c) What fraction of the particles survive the last measurement if the SGn device is simply removed from the experiment?

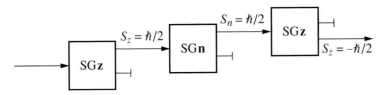

**FIGURE 1.12**

**1.8.** The state of a spin-$\frac{1}{2}$ particle is given by

$$|\psi\rangle = \frac{i}{\sqrt{3}}|+\mathbf{z}\rangle + \sqrt{\frac{2}{3}}|-\mathbf{z}\rangle$$

What are $\langle S_z \rangle$ and $\Delta S_z$ for this state? Suppose that an experiment is carried out on 100 particles, each of which is in this state. Make up a reasonable set of data for $S_z$ that could result from such an experiment. What if the measurements were carried out on 1,000 particles? What about 10,000?

**1.9.** Show that neither the probability of obtaining the result $a_i$ nor the expectation value $\langle A \rangle$ are affected by $|\psi\rangle \rightarrow e^{i\delta}|\psi\rangle$, that is, by an overall phase change for the state $|\psi\rangle$.

# CHAPTER

## 2

# ROTATION
# OF BASIS
# STATES
# AND MATRIX
# MECHANICS

In this chapter we will see that transforming a vector into a different vector in our quantum mechanical vector space requires an operator. We will also introduce a convenient shorthand notation in which we represent ket vectors by column vectors, bra vectors by row vectors, and operators by matrices. Our discussion will be primarily phrased in terms of the two-state spin-$\frac{1}{2}$ system introduced in Chapter 1, but we will also analyze another two-state system, the polarization of the electromagnetic field.

## 2.1  THE BEGINNINGS OF MATRIX MECHANICS

### Representing Kets and Bras

We have seen that we can express an arbitrary spin state $|\psi\rangle$ of a spin-$\frac{1}{2}$ particle as

$$|\psi\rangle = |+\mathbf{z}\rangle\langle+\mathbf{z}|\psi\rangle + |-\mathbf{z}\rangle\langle-\mathbf{z}|\psi\rangle = c_+|+\mathbf{z}\rangle + c_-|-\mathbf{z}\rangle \tag{2.1}$$

Such a spin state may, for example, be created by sending spin-$\frac{1}{2}$ particles through a Stern-Gerlach device with its magnetic field gradient oriented in some arbitrary direction. The complex numbers $c_\pm = \langle\pm\mathbf{z}|\psi\rangle$ tell us how our state $|\psi\rangle$ is oriented in our quantum mechanical vector space, that is, how much of $|\psi\rangle$ is projected onto each of the states $|+\mathbf{z}\rangle$ and $|-\mathbf{z}\rangle$.

One convenient way of representing $|\psi\rangle$ is just to keep track of these complex numbers. Just as we can avoid unit vectors in writing the classical electric field

$$\mathbf{E} = E_x\mathbf{i} + E_y\mathbf{j} + E_z\mathbf{k} \tag{2.2a}$$

by using the notation

$$\mathbf{E} \rightarrow (E_x, E_y, E_z) \tag{2.2b}$$

we can represent the ket (2.1) by the column vector

$$|\psi\rangle \xrightarrow[S_z \text{ basis}]{} \begin{pmatrix} \langle+\mathbf{z}|\psi\rangle \\ \langle-\mathbf{z}|\psi\rangle \end{pmatrix} = \begin{pmatrix} c_+ \\ c_- \end{pmatrix} \tag{2.3}$$

In this basis, the ket $|+\mathbf{z}\rangle$ is represented by the column vector

$$|+\mathbf{z}\rangle \xrightarrow[S_z \text{ basis}]{} \begin{pmatrix} \langle+\mathbf{z}|+\mathbf{z}\rangle \\ \langle-\mathbf{z}|+\mathbf{z}\rangle \end{pmatrix} = \begin{pmatrix} 1 \\ 0 \end{pmatrix} \tag{2.4}$$

and the ket $|-\mathbf{z}\rangle$ is represented by the column vector

$$|-\mathbf{z}\rangle \xrightarrow[S_z \text{ basis}]{} \begin{pmatrix} \langle+\mathbf{z}|-\mathbf{z}\rangle \\ \langle-\mathbf{z}|-\mathbf{z}\rangle \end{pmatrix} = \begin{pmatrix} 0 \\ 1 \end{pmatrix} \tag{2.5}$$

although the label under the arrow is really superfluous in (2.4) and (2.5) given the form of the column vectors on the right. Using (1.28), we can also write, for example,

$$|+\mathbf{x}\rangle \xrightarrow[S_z \text{ basis}]{} \begin{pmatrix} \langle+\mathbf{z}|+\mathbf{x}\rangle \\ \langle-\mathbf{z}|+\mathbf{x}\rangle \end{pmatrix} = \frac{1}{\sqrt{2}} \begin{pmatrix} 1 \\ 1 \end{pmatrix} \tag{2.6}$$

How do we represent bra vectors? We know that the bra vector corresponding to the ket vector (2.1) is

$$\langle\psi| = \langle\psi|+\mathbf{z}\rangle\langle+\mathbf{z}| + \langle\psi|-\mathbf{z}\rangle\langle-\mathbf{z}| = c_+^*\langle+\mathbf{z}| + c_-^*\langle-\mathbf{z}| \tag{2.7}$$

We can express

$$\langle\psi|\psi\rangle = \langle\psi|+\mathbf{z}\rangle\langle+\mathbf{z}|\psi\rangle + \langle\psi|-\mathbf{z}\rangle\langle-\mathbf{z}|\psi\rangle = 1 \tag{2.8}$$

conveniently as

$$\langle\psi|\psi\rangle = \underbrace{\left(\langle\psi|+\mathbf{z}\rangle, \langle\psi|-\mathbf{z}\rangle\right)}_{\text{bra vector}} \underbrace{\begin{pmatrix} \langle+\mathbf{z}|\psi\rangle \\ \langle-\mathbf{z}|\psi\rangle \end{pmatrix}}_{\text{ket vector}} = 1 \tag{2.9}$$

where we are using the usual rules of matrix multiplication for row and column vectors. This suggests that we represent the bra $\langle\psi|$ by the row vector

$$\langle\psi| \xrightarrow[S_z \text{ basis}]{} \left(\langle\psi| + \mathbf{z}\rangle, \langle\psi| - \mathbf{z}\rangle\right) \tag{2.10}$$

Since $\langle\psi|+\mathbf{z}\rangle = \langle+\mathbf{z}|\psi\rangle^*$ and $\langle\psi|-\mathbf{z}\rangle = \langle-\mathbf{z}|\psi\rangle^*$, (2.10) can also be expressed as

$$\langle\psi| \xrightarrow[S_z \text{ basis}]{} (\langle+\mathbf{z}|\psi\rangle^*, \langle-\mathbf{z}|\psi\rangle^*) = (c_+^*, c_-^*) \tag{2.11}$$

Comparing (2.11) with (2.3), we see that the row vector that *represents* the bra is the complex conjugate and transpose of the column vector that *represents* the corresponding ket vector. In this representation, an inner product such as (2.9) is carried out using the usual rules of matrix multiplication.

As an example, we may determine the representation for the ket $|-\mathbf{x}\rangle$ in the $S_z$ basis. We know from the Stern-Gerlach experiments that there is zero amplitude to obtain $S_x = -\hbar/2$ for a state with $S_x = \hbar/2$, that is, $\langle-\mathbf{x}|+\mathbf{x}\rangle = 0$. Making the amplitude $\langle-\mathbf{x}|+\mathbf{x}\rangle$ vanish requires that

$$|-\mathbf{x}\rangle \xrightarrow[S_z \text{ basis}]{} \frac{e^{i\delta}}{\sqrt{2}}\begin{pmatrix} 1 \\ -1 \end{pmatrix} \tag{2.12}$$

since then

$$\langle-\mathbf{x}|+\mathbf{x}\rangle = \frac{e^{-i\delta}}{\sqrt{2}}(1, -1)\frac{1}{\sqrt{2}}\begin{pmatrix} 1 \\ 1 \end{pmatrix} = 0 \tag{2.13}$$

Note that the $1/\sqrt{2}$ in front of the column vector in (2.12) has been chosen so that the ket $|-\mathbf{x}\rangle$ is properly normalized:

$$\langle-\mathbf{x}|-\mathbf{x}\rangle = \frac{e^{-i\delta}}{\sqrt{2}}(1, -1)\frac{e^{i\delta}}{\sqrt{2}}\begin{pmatrix} 1 \\ -1 \end{pmatrix} = 1 \tag{2.14}$$

The common convention is to choose the overall phase $\delta = 0$ so that

$$|-\mathbf{x}\rangle \xrightarrow[S_z \text{ basis}]{} \frac{1}{\sqrt{2}}\begin{pmatrix} 1 \\ -1 \end{pmatrix} \tag{2.15}$$

As another example, (1.29) indicates that the state with $S_y = \hbar/2$ is

$$|+\mathbf{y}\rangle = \frac{1}{\sqrt{2}}|+\mathbf{z}\rangle + \frac{i}{\sqrt{2}}|-\mathbf{z}\rangle \tag{2.16}$$

which may be represented in the $S_z$ basis by

$$|+\mathbf{y}\rangle \rightarrow \frac{1}{\sqrt{2}}\begin{pmatrix} 1 \\ i \end{pmatrix} \tag{2.17a}$$

The bra corresponding to this ket is represented in the same basis by

$$\langle+\mathbf{y}| \rightarrow \frac{1}{\sqrt{2}}(1, -i) \tag{2.17b}$$

Note the appearance of the $-i$ in this representation for the bra vector. Using these representations, we can check that

$$\langle +\mathbf{y}|+\mathbf{y}\rangle = \frac{1}{\sqrt{2}}(1, -i)\frac{1}{\sqrt{2}}\begin{pmatrix}1\\i\end{pmatrix} = 1 \qquad (2.18)$$

If we had used the row vector

$$\frac{1}{\sqrt{2}}(1, +i)$$

in evaluating the inner product, we would have obtained zero instead of one. Since $\langle -\mathbf{y}|+\mathbf{y}\rangle = 0$, this tells us that in the $S_z$ basis

$$\langle -\mathbf{y}| \rightarrow \frac{1}{\sqrt{2}}(1, +i) \qquad (2.19a)$$

and thus

$$|-\mathbf{y}\rangle \rightarrow \frac{1}{\sqrt{2}}\begin{pmatrix}1\\-i\end{pmatrix} \qquad (2.19b)$$

As a last example, we use matrix representations to calculate the probability that a spin-$\frac{1}{2}$ particle with $S_x = \hbar/2$ is found to have $S_y = \hbar/2$ when a measurement is carried out:

$$|\langle +\mathbf{y}|+\mathbf{x}\rangle|^2 = \left|\frac{1}{\sqrt{2}}(1, -i)\frac{1}{\sqrt{2}}\begin{pmatrix}1\\1\end{pmatrix}\right|^2$$

$$= \left|\frac{1-i}{2}\right|^2 = \frac{(1-i)}{2}\frac{(1+i)}{2} = \frac{1}{2} \qquad (2.20)$$

## Freedom of Representation

It is often convenient to use a number of different basis sets to express a particular state $|\psi\rangle$. Just as we can write the electric field in a particular coordinate system as (2.2), we could use a different coordinate system with unit vectors $\mathbf{i}'$, $\mathbf{j}'$, and $\mathbf{k}'$ to write the same electric field as

$$\mathbf{E} = E_{x'}\mathbf{i}' + E_{y'}\mathbf{j}' + E_{z'}\mathbf{k}' \qquad (2.21a)$$

or

$$\mathbf{E} \rightarrow (E_{x'}, E_{y'}, E_{z'}) \qquad (2.21b)$$

Of course, the electric field $\mathbf{E}$ hasn't changed. It still has the same magnitude and direction, but we have chosen a different set of unit vectors, or basis vectors, to express it. Similarly, we can take the quantum state $|\psi\rangle$ in (2.1) and write it in terms of the basis states $|+\mathbf{x}\rangle$ and $|-\mathbf{x}\rangle$ as

$$|\psi\rangle = |+\mathbf{x}\rangle\langle+\mathbf{x}|\psi\rangle + |-\mathbf{x}\rangle\langle-\mathbf{x}|\psi\rangle \tag{2.22}$$

which expresses the state as a superposition of the states with $S_x = \pm\hbar/2$ multiplied by the amplitudes for the particle to be in these states. We can then construct a column vector representing $|\psi\rangle$ in this basis using these amplitudes:

$$|\psi\rangle \xrightarrow[S_x \text{ basis}]{} \begin{pmatrix} \langle+\mathbf{x}|\psi\rangle \\ \langle-\mathbf{x}|\psi\rangle \end{pmatrix} \tag{2.23}$$

Thus the column vector representing the ket $|+\mathbf{x}\rangle$ is

$$|+\mathbf{x}\rangle \xrightarrow[S_x \text{ basis}]{} \begin{pmatrix} \langle+\mathbf{x}|+\mathbf{x}\rangle \\ \langle-\mathbf{x}|+\mathbf{x}\rangle \end{pmatrix} = \begin{pmatrix} 1 \\ 0 \end{pmatrix} \tag{2.24}$$

which is to be compared with the column vector (2.6). The ket $|+\mathbf{x}\rangle$ is the same state in the two cases; we have just written it out using the $S_z$ basis in the first case and the $S_x$ basis in the second case. Which basis we use is determined by what is convenient, such as what measurements we are going to perform on the state $|+\mathbf{x}\rangle$.

## 2.2 ROTATION OPERATORS

There is a nice physical way to transform the kets themselves from one basis set to another.[1] Recall that within classical physics a magnetic moment placed in a uniform magnetic field precesses about the direction of the field. When we discuss time evolution in Chapter 4, we will see that the interaction of the magnetic moment of a spin-$\frac{1}{2}$ particle with the magnetic field also causes the quantum spin state of the particle to rotate about the direction of the field as time progresses. In particular, if the magnetic field points in the $y$ direction and the particle is initially in the state $|+\mathbf{z}\rangle$, the spin will rotate in the $x$-$z$ plane. At some later time the particle will be in the state $|+\mathbf{x}\rangle$. With this example in mind, it is useful at this stage to introduce a rotation *operator* $\hat{R}(\frac{\pi}{2}\mathbf{j})$ that acts on the ket $|+\mathbf{z}\rangle$, a state that is spin up along the $z$ axis, and transforms it into the ket $|+\mathbf{x}\rangle$, a state that is spin up along the $x$ axis:

$$|+\mathbf{x}\rangle = \hat{R}\left(\tfrac{\pi}{2}\mathbf{j}\right)|+\mathbf{z}\rangle \tag{2.25}$$

Changing or transforming a ket in our vector space into a different ket requires an operator. To distinguish operators from ordinary numbers, we denote all operators with a hat.

---

[1] You may object to calling anything dealing directly with kets *physical* since ket vectors are abstract vectors specifying the quantum state of the system and involve, as we have seen, complex numbers.

What is the nature of the transformation effected by the operator $\hat{R}(\frac{\pi}{2}\mathbf{j})$? This operator just rotates the ket $|+\mathbf{z}\rangle$ by 90°, or $\pi/2$ radians, about the $y$ axis (indicated by the unit vector $\mathbf{j}$) in a counterclockwise direction as viewed from the positive $y$ axis, turning or rotating it into the ket $|+\mathbf{x}\rangle$, as indicated in Fig. 2.1$a$. The same rotation operator should rotate $|-\mathbf{z}\rangle$ into $|-\mathbf{x}\rangle$. In fact, since the most general state of a spin-$\frac{1}{2}$ particle may be expressed in the form of (2.1), the operator rotates this ket as well:

$$\hat{R}\left(\tfrac{\pi}{2}\mathbf{j}\right)|\psi\rangle = \hat{R}\left(\tfrac{\pi}{2}\mathbf{j}\right)\left(c_+|+\mathbf{z}\rangle + c_-|-\mathbf{z}\rangle\right)$$

$$= c_+\hat{R}\left(\tfrac{\pi}{2}\mathbf{j}\right)|+\mathbf{z}\rangle + c_-\hat{R}\left(\tfrac{\pi}{2}\mathbf{j}\right)|-\mathbf{z}\rangle$$

$$= c_+|+\mathbf{x}\rangle + c_-|-\mathbf{x}\rangle \tag{2.26}$$

Note that the operator acts on kets, not on the complex numbers.[2]

## The Adjoint Operator

What is the bra equation corresponding to the ket equation (2.25)? You may be tempted to guess that $\langle +\mathbf{x}| = \langle +\mathbf{z}|\hat{R}\left(\tfrac{\pi}{2}\mathbf{j}\right)$, but we can quickly see that this *cannot*

---

[2] An operator $\hat{A}$ satisfying

$$\hat{A}(a|\psi\rangle + b|\varphi\rangle) = a\hat{A}|\psi\rangle + b\hat{A}|\varphi\rangle$$

where $a$ and $b$ are complex numbers, is referred to as a *linear operator*.

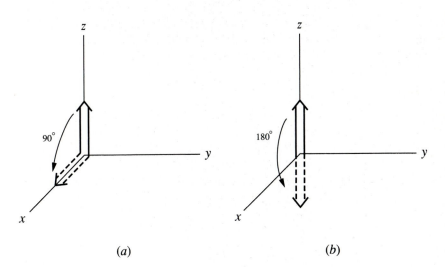

(a)                              (b)

**FIGURE 2.1**
Rotating $|+\mathbf{z}\rangle$ counterclockwise about the $y$ axis ($a$) by $\pi/2$ radians transforms the state into $|+\mathbf{x}\rangle$ and ($b$) by $\pi$ radians transforms the state into $|-\mathbf{z}\rangle$. The spin state of a spin-$\frac{1}{2}$ particle with a magnetic moment would rotate in the $x$-$z$ plane if the particle were placed in a magnetic field in the $y$ direction.

be correct, for if it were, we could calculate[3]

$$\langle +\mathbf{x}|+\mathbf{x}\rangle = \left[\langle +\mathbf{z}|\hat{R}\left(\tfrac{\pi}{2}\mathbf{j}\right)\right]\left[\hat{R}\left(\tfrac{\pi}{2}\mathbf{j}\right)|+\mathbf{z}\rangle\right] = \langle +\mathbf{z}|\hat{R}\left(\tfrac{\pi}{2}\mathbf{j}\right)\hat{R}\left(\tfrac{\pi}{2}\mathbf{j}\right)|+\mathbf{z}\rangle$$

We know that $\langle +\mathbf{x}|+\mathbf{x}\rangle = 1$, but since $\hat{R}\left(\tfrac{\pi}{2}\mathbf{j}\right)$ rotates by 90° around the $y$ axis, $\hat{R}\left(\tfrac{\pi}{2}\mathbf{j}\right)\hat{R}\left(\tfrac{\pi}{2}\mathbf{j}\right) = \hat{R}\left(\pi\mathbf{j}\right)$ performs a rotation of 180° about the $y$ axis. But $\hat{R}\left(\pi\mathbf{j}\right)|+\mathbf{z}\rangle = |-\mathbf{z}\rangle$ (see Fig. 2.1$b$), and since $\langle +\mathbf{z}|\hat{R}\left(\pi\mathbf{j}\right)|+\mathbf{z}\rangle = \langle +\mathbf{z}|-\mathbf{z}\rangle = 0$, we are left with a contradiction.

For the ket $|\psi\rangle = c_+|+\mathbf{z}\rangle + c_-|-\mathbf{z}\rangle$, the corresponding bra is $\langle\psi| = c_+^*\langle +\mathbf{z}| + c_-^*\langle -\mathbf{z}|$, with the complex numbers in the ket turning into their complex conjugates in the bra. Since we are dealing here with operators and not just complex numbers, we need an additional rule for determining the bra equation corresponding to a ket equation like (2.25) that involves an operator. We introduce a new operator $\hat{R}^\dagger$, called the *adjoint operator* of the operator $\hat{R}$, so that the bra equation corresponding to (2.25) is

$$\langle +\mathbf{x}| = \langle +\mathbf{z}|\hat{R}^\dagger\left(\tfrac{\pi}{2}\mathbf{j}\right) \tag{2.27}$$

We can then satisfy

$$1 = \langle +\mathbf{x}|+\mathbf{x}\rangle = \langle +\mathbf{z}|\hat{R}^\dagger\left(\tfrac{\pi}{2}\mathbf{j}\right)\hat{R}\left(\tfrac{\pi}{2}\mathbf{j}\right)|+\mathbf{z}\rangle = \langle +\mathbf{z}|+\mathbf{z}\rangle \tag{2.28}$$

if the adjoint operator $\hat{R}^\dagger$ is inverse of the operator $\hat{R}$. In particular, the adjoint operator $\hat{R}^\dagger(\tfrac{\pi}{2}\mathbf{j})$ is a rotation operator that can be viewed as operating to the right on the ket $\hat{R}(\tfrac{\pi}{2}\mathbf{j})|+\mathbf{z}\rangle$. If $\hat{R}(\tfrac{\pi}{2}\mathbf{j})$ rotates by 90° *counterclockwise*, then $\hat{R}^\dagger(\tfrac{\pi}{2}\mathbf{j})$ rotates by 90° *clockwise* so that $\hat{R}^\dagger(\tfrac{\pi}{2}\mathbf{j})\hat{R}(\tfrac{\pi}{2}\mathbf{j}) = 1$, and we are left with $\langle +\mathbf{z}|+\mathbf{z}\rangle = 1$.[4]

In general, an operator $\hat{U}$ satisfying $\hat{U}^\dagger\hat{U} = 1$ is called a *unitary* operator. Thus the rotation operator must be unitary in order that the amplitude for a state to be itself—that is, so that $\langle\psi|\psi\rangle = 1$—doesn't change under rotation. Otherwise, probability would not be conserved under rotation.

## The Generator of Rotations

Instead of performing rotations about the $y$ axis, let's rotate about the $z$ axis. If we rotate by 90° counterclockwise about the $z$ axis, we will, for example, turn $|+\mathbf{x}\rangle$ into $|+\mathbf{y}\rangle$, as indicated in Fig. 2.2$a$. Instead of carrying out this whole rotation initially, let us first focus on an infinitesimal rotation by an angle $d\phi$ about the $z$ axis, as shown in Fig. 2.2$b$. A useful way to express this infinitesimal rotation

---

[3] You can see why we position the operator to the right of the bra vector when we go to calculate an amplitude. Otherwise we would evaluate the inner product and the operator would be left alone with no vector to act on.

[4] As this example illustrates, the adjoint operator can act to the right on ket vectors as well to the left on bra vectors.

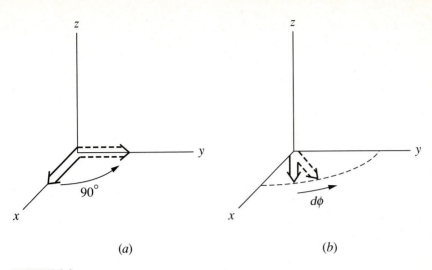

**FIGURE 2.2**
(a) Rotating $|+\mathbf{x}\rangle$ by $\pi/2$ radians counterclockwise about the $z$ axis transforms the state into $|+\mathbf{y}\rangle$.
(b) Rotation of a state by an infinitesimal angle $d\phi$ about the $z$ axis.

operator is in the form

$$\hat{R}(d\phi\,\mathbf{k}) = 1 - \frac{i}{\hbar}\hat{J}_z\,d\phi \qquad (2.29)$$

where we have introduced an operator $\hat{J}_z$ that "generates" rotations about the $z$ axis and moves us away from the identity element. Our form for $\hat{R}(d\phi\,\mathbf{k})$ clearly satisfies the requirement that $\hat{R}(d\phi\,\mathbf{k}) \to 1$ as $d\phi \to 0$. As we will see, the factor of $i$ and the factor of $\hbar$ have been introduced to bring out the physical significance of the operator $\hat{J}_z$. In particular, because the factor of $\hbar$ occurs in the denominator of the second term in (2.29), the operator $\hat{J}_z$ must have the dimensions of $\hbar$, namely, the dimensions of angular momentum. We will see that a convincing case can be made that we should identify this operator $\hat{J}_z$, the generator of rotations about the $z$ axis, with the $z$ component of the intrinsic spin angular momentum of the particle.

We first establish that $\hat{J}_z$ belongs to a special class of operators known as Hermitian operators. Physically, the operator $\hat{R}^{\dagger}(d\phi\,\mathbf{k})$ is the inverse of the rotation operator $\hat{R}(d\phi\,\mathbf{k})$. By taking the adjoint of (2.29), we can write this operator in the form

$$\hat{R}^{\dagger}(d\phi\,\mathbf{k}) = 1 + \frac{i}{\hbar}\hat{J}_z^{\dagger}\,d\phi \qquad (2.30)$$

where $\hat{J}_z^{\dagger}$ is the adjoint of the operator $\hat{J}_z$. Note that since the bra corresponding to the ket $c|\psi\rangle$ is $\langle\psi|c^*$, complex numbers get replaced by their complex conjugates when forming the adjoint operator. Thus $i \to -i$ in going from (2.29) to (2.30), which has the same effect as changing $d\phi$ to $-d\phi$, and therefore $\hat{R}^\dagger(d\phi\,\mathbf{k}) = \hat{R}(-d\phi\,\mathbf{k})$, provided $\hat{J}_z^\dagger = \hat{J}_z$. More formally, since the rotation operator $\hat{R}^\dagger(d\phi\,\mathbf{k})$ is the inverse of the rotation operator $\hat{R}(d\phi\,\mathbf{k})$, these operators must satisfy the condition

$$\hat{R}^\dagger(d\phi\,\mathbf{k})\hat{R}(d\phi\,\mathbf{k}) = \left(1 + \frac{i}{\hbar}\hat{J}_z^\dagger\,d\phi\right)\left(1 - \frac{i}{\hbar}\hat{J}_z\,d\phi\right)$$

$$= 1 + \frac{i}{\hbar}\left(\hat{J}_z^\dagger - \hat{J}_z\right)d\phi + O(d\phi^2) = 1 \qquad (2.31)$$

Since the angle $d\phi$ is infinitesimal, we can neglect the second-order terms in $d\phi$ and (2.31) will be satisfied only if $\hat{J}_z = \hat{J}_z^\dagger$. In general, an operator that is equal to its adjoint is called *self-adjoint* or *Hermitian*. Thus $\hat{J}_z$ must be a Hermitian operator. Hermitian operators have a number of nice properties that permit them to play major roles in quantum mechanics. After some specific examples, we will discuss some of these general properties in Section 2.8.[5]

One of the reasons that infinitesimal rotations are useful is that once we know how to perform an infinitesimal rotation about the $z$ axis by an angle $d\phi$, we can carry out a rotation by any finite angle $\phi$ by compounding an infinite number of infinitesimal rotations with

$$d\phi = \lim_{N\to\infty}\frac{\phi}{N}$$

The rotation operator $\hat{R}(\phi\mathbf{k})$ is then given by

$$\hat{R}(\phi\,\mathbf{k}) = \lim_{N\to\infty}\left[1 - \frac{i}{\hbar}\hat{J}_z\left(\frac{\phi}{N}\right)\right]^N = e^{-i\hat{J}_z\phi/\hbar} \qquad (2.32)$$

The last identity in (2.32) can be established by expanding both sides in a Taylor series and showing that they agree term by term (see Problem 2.1). In fact, a series expansion is really the only way to make sense of an expression such as an exponential of an operator.

## Eigenstates and Eigenvalues

What happens to a ket $|+\mathbf{z}\rangle$ if we rotate it about the $z$ axis—that is, what is $\hat{R}(\phi\mathbf{k})|+\mathbf{z}\rangle$? If you were to rotate a classical spinning top about its axis of rotation, it would still be in the same state with its angular momentum pointing in the same direction. Similarly, rotating a state of a spin-$\frac{1}{2}$ particle that is spin up along $z$

---

[5] Now you can see one reason for introducing the $i$ in the defining relation (2.29) for an infinitesimal rotation operator. Without it, the generator $\hat{J}_z$ would not have turned out to be Hermitian.

about the $z$ axis should still yield a state that is spin up along $z$, as illustrated in Fig. 2.3. In Chapter 1 we saw that the overall phase of a state does not enter into the calculation of probabilities, such as in (1.23). This turns out to be quite a general feature: two states that differ only in an *overall* phase are really the same state. We will now show that in order for $\hat{R}(\phi\mathbf{k})|+\mathbf{z}\rangle$ to differ from $|+\mathbf{z}\rangle$ only by an overall phase, it is necessary that

$$\hat{J}_z|+\mathbf{z}\rangle = (\text{constant})|+\mathbf{z}\rangle \tag{2.33}$$

In general, when an operator acting on a state yields a constant times the state, we call the state an *eigenstate* of the operator and the constant the corresponding *eigenvalue*.

First we will establish the eigenstate condition (2.33). If we expand the exponential in the rotation operator (2.32) in a Taylor series, we have

$$\hat{R}(\phi\mathbf{k})|+\mathbf{z}\rangle = \left[1 - \frac{i\phi\hat{J}_z}{\hbar} + \frac{1}{2!}\left(-\frac{i\phi\hat{J}_z}{\hbar}\right)^2 + \cdots\right]|+\mathbf{z}\rangle \tag{2.34}$$

If (2.33) is not satisfied and $\hat{J}_z|+\mathbf{z}\rangle$ is something other than a constant times $|+\mathbf{z}\rangle$, such as $|+\mathbf{x}\rangle$, the first two terms in the series will yield $|+\mathbf{z}\rangle$ plus a term involving $|+\mathbf{x}\rangle$, which would mean that $\hat{R}(\phi\mathbf{k})|+\mathbf{z}\rangle$ differs from $|+\mathbf{z}\rangle$ by other than a multiplicative constant. Note that other terms in the series cannot cancel this unwanted $|+\mathbf{x}\rangle$ term, since each term involving a different power of $\phi$ is linearly independent from the rest. Thus we deduce that the ket $|+\mathbf{z}\rangle$ must be an eigenstate, or eigenket, of the operator $\hat{J}_z$.

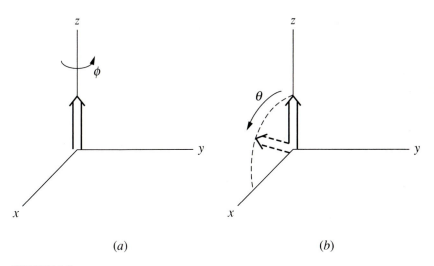

(a)                                                    (b)

**FIGURE 2.3**

($a$) Rotating $|+\mathbf{z}\rangle$ by angle $\phi$ about the $z$ axis with the operator $\hat{R}(\phi\mathbf{k})$ does not change the state, in contrast to action of the operator $\hat{R}(\theta\mathbf{j})$, which rotates $|+\mathbf{z}\rangle$ by angle $\theta$ about the $y$ axis, producing a different state, as indicated in ($b$).

Let's now turn our attention to the value of the constant, the eigenvalue, in (2.33). We will give a self-consistency argument to show that we will have agreement with the analysis of the Stern-Gerlach experiments in Chapter 1 provided

$$\hat{J}_z|\pm z\rangle = \pm \frac{\hbar}{2}|\pm z\rangle \tag{2.35}$$

This equation asserts that the eigenvalues for the spin-up and spin-down states are the values of $S_z$ that these states are observed to have in the Stern-Gerlach experiments.[6] First consider the spin-up state. If

$$\hat{J}_z|+z\rangle = \frac{\hbar}{2}|+z\rangle \tag{2.36a}$$

then

$$\hat{J}_z^2|+z\rangle = \hat{J}_z\frac{\hbar}{2}|+z\rangle = \frac{\hbar}{2}\hat{J}_z|+z\rangle = \left(\frac{\hbar}{2}\right)^2|+z\rangle \tag{2.36b}$$

and so on. From (2.34), we obtain

$$\hat{R}(\phi\mathbf{k})|+z\rangle = \left[1 - \frac{i\phi}{2} + \frac{1}{2!}\left(-\frac{i\phi}{2}\right)^2 + \cdots\right]|+z\rangle = e^{-i\phi/2}|+z\rangle \tag{2.37}$$

The state has picked up an overall phase, just as we would hope if the state is not to change. The value of the phase is determined by the eigenvalue in (2.36a).

In order to see why the eigenvalue should be $\hbar/2$, let's consider what happens if we rotate a spin-down state $|-z\rangle$ about the $z$ axis, that is, if we evaluate $\hat{R}(\phi\mathbf{k})|-z\rangle$. Just as before, we can argue that $|-z\rangle$ must be an eigenstate of $\hat{J}_z$. We can also argue that the eigenvalue for $|-z\rangle$ must be different from that for $|+z\rangle$. After all, if the eigenvalues were the same, applying the rotation operator $\hat{R}(\phi\mathbf{k})$ to the state

$$|+x\rangle = \frac{1}{\sqrt{2}}|+z\rangle + \frac{1}{\sqrt{2}}|-z\rangle \tag{2.38}$$

would not rotate the state, since $|+z\rangle$ and $|-z\rangle$ would each pick up the same phase factor, and the state in (2.38) would itself pick up just an overall phase. Therefore, it would still be the same state. But if we rotate the state $|+x\rangle$ by an angle $\phi$ in the $x$-$y$ plane, we expect the state to change. If we try

$$\hat{J}_z|-z\rangle = -\frac{\hbar}{2}|-z\rangle \tag{2.39}$$

---

[6] You can start to see why we introduced a factor of $1/\hbar$ in the defining relation (2.29) between the infinitesimal rotation operator and the generator of rotations.

for the eigenvalue equation for the spin-down state, we find

$$\hat{R}(\phi\mathbf{k})|-\mathbf{z}\rangle = \left[1 + \frac{i\phi}{2} + \frac{1}{2!}\left(\frac{i\phi}{2}\right)^2 + \cdots\right]|-\mathbf{z}\rangle = e^{i\phi/2}|-\mathbf{z}\rangle \qquad (2.40)$$

Using (2.37) and (2.40), we see that

$$\hat{R}(\phi\mathbf{k})|+\mathbf{x}\rangle = \frac{e^{-i\phi/2}}{\sqrt{2}}|+\mathbf{z}\rangle + \frac{e^{i\phi/2}}{\sqrt{2}}|-\mathbf{z}\rangle$$

$$= e^{-i\phi/2}\left(\frac{1}{\sqrt{2}}|+\mathbf{z}\rangle + \frac{e^{i\phi}}{\sqrt{2}}|-\mathbf{z}\rangle\right) \qquad (2.41)$$

which is clearly a different state from (2.38) for $\phi \neq 0$. In particular, with the choice $\phi = \pi/2$, we obtain

$$\hat{R}\left(\tfrac{\pi}{2}\mathbf{k}\right)|+\mathbf{x}\rangle = e^{-i\pi/4}\left(\frac{1}{\sqrt{2}}|+\mathbf{z}\rangle + \frac{e^{i\pi/2}}{\sqrt{2}}|-\mathbf{z}\rangle\right)$$

$$= e^{-i\pi/4}\left(\frac{1}{\sqrt{2}}|+\mathbf{z}\rangle + \frac{i}{\sqrt{2}}|-\mathbf{z}\rangle\right) = e^{-i\pi/4}|+\mathbf{y}\rangle \qquad (2.42)$$

where we have replaced the term in the brackets by the state $|+\mathbf{y}\rangle$ that we determined in (1.29). Since two states that differ only by an overall phase are the same state, we see that rotating the state $|+\mathbf{x}\rangle$ by 90° counterclockwise about the $z$ axis does generate the state $|+\mathbf{y}\rangle$ *when (2.35) holds.* Thus we are led to a striking conclusion: When the operator that generates rotations about the $z$ axis acts on the spin-up-along-$z$ and spin-down-along-$z$ states, it throws out a constant (the eigenvalue) times the state (the eigenstate); the eigenvalues for the two states are just the values of the $z$ component of the intrinsic spin angular momentum that characterize these states.

Finally, let us note something really perplexing about the effects of rotations on spin-$\frac{1}{2}$ particles: namely,

$$\hat{R}(2\pi\mathbf{k})|+\mathbf{z}\rangle = e^{-i\pi}|+\mathbf{z}\rangle = -|+\mathbf{z}\rangle \qquad (2.43a)$$

and

$$\hat{R}(2\pi\mathbf{k})|-\mathbf{z}\rangle = e^{i\pi}|-\mathbf{z}\rangle = -|-\mathbf{z}\rangle \qquad (2.43b)$$

Thus, if we rotate a spin-$\frac{1}{2}$ state by 360° and *end up right where we started,* we find that the state changes sign. Earlier we remarked that we could actually perform these rotations on our spin systems by inserting them in a magnetic field. When we come to time evolution in Chapter 4, we will see how this strange prediction (2.43) for spin-$\frac{1}{2}$ particles may be verified experimentally.

## 2.3 THE IDENTITY AND PROJECTION OPERATORS

In general, the operator $\hat{R}(\theta\mathbf{n})$ changes a ket into a different ket by rotating it by an angle $\theta$ around the axis specified by the unit vector $\mathbf{n}$. Most operators tend to *do* something when they act on ket vectors, but it is convenient to introduce an operator that acts on a ket vector and does nothing: the *identity operator.* Surprisingly, we will see that this operator is a powerful operator that will be very useful to us.

We have expressed the spin state $|\psi\rangle$ of a spin-$\frac{1}{2}$ particle in the $S_z$ basis as $|\psi\rangle = |+\mathbf{z}\rangle\langle+\mathbf{z}|\psi\rangle + |-\mathbf{z}\rangle\langle-\mathbf{z}|\psi\rangle$. We can think of the rather strange-looking object

$$|+\mathbf{z}\rangle\langle+\mathbf{z}| + |-\mathbf{z}\rangle\langle-\mathbf{z}| \tag{2.44}$$

as the identity operator. It is an operator because when it is applied to a ket, it yields another ket. Moreover, if we apply it to the ket $|\psi\rangle$, we obtain

$$\big(|+\mathbf{z}\rangle\langle+\mathbf{z}| + |-\mathbf{z}\rangle\langle-\mathbf{z}|\big)|\psi\rangle = |+\mathbf{z}\rangle\langle+\mathbf{z}|\psi\rangle + |-\mathbf{z}\rangle\langle-\mathbf{z}|\psi\rangle = |\psi\rangle \tag{2.45}$$

We earlier discussed a nice physical mechanism for inserting such an identity operator when we analyzed the effect of introducing a modified Stern-Gerlach device in Experiment 4 in Chapter 1. Here, since we are expressing an arbitrary state $|\psi\rangle$ in terms of the amplitudes to be in the states $|+\mathbf{z}\rangle$ and $|-\mathbf{z}\rangle$, we use a modified SG device with its magnetic field gradient oriented along the $z$ direction, as shown in Fig. 2.4a. The important point that we made in our discussion of the modified SG device was that because we do not make a measurement with such a device, the amplitudes to be in the states $|+\mathbf{z}\rangle$ and $|-\mathbf{z}\rangle$ combine together to yield the same state exiting as entering the device, just as if the device were absent. Hence, it is indeed an identity operator.

The identity operator (2.44) may be viewed as being composed of two operators called projection operators:

$$\hat{P}_+ = |+\mathbf{z}\rangle\langle+\mathbf{z}| \tag{2.46a}$$

and

$$\hat{P}_- = |-\mathbf{z}\rangle\langle-\mathbf{z}| \tag{2.46b}$$

They are called *projection operators* because

$$\hat{P}_+|\psi\rangle = |+\mathbf{z}\rangle\langle+\mathbf{z}|\psi\rangle \tag{2.47a}$$

projects out the component of the ket $|\psi\rangle$ along $|+\mathbf{z}\rangle$ and

$$\hat{P}_-|\psi\rangle = |-\mathbf{z}\rangle\langle-\mathbf{z}|\psi\rangle \tag{2.47b}$$

projects out the component of the ket $|\psi\rangle$ along $|-\mathbf{z}\rangle$.[7] That (2.44) is the identity

---

[7] Notice that the projection operator may be applied to a bra vector as well:

$$\langle\psi|\hat{P}_+ = \langle\psi|+\mathbf{z}\rangle\langle+\mathbf{z}| \qquad \langle\psi|\hat{P}_- = \langle\psi|-\mathbf{z}\rangle\langle-\mathbf{z}|$$

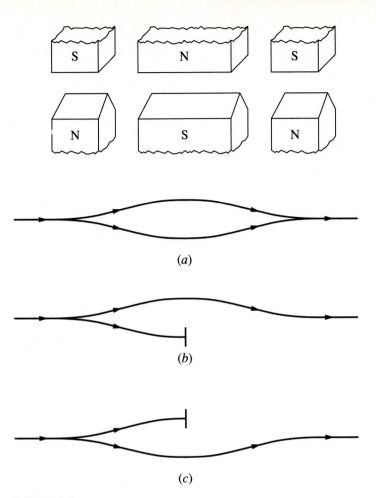

**FIGURE 2.4**
($a$) A modified Stern-Gerlach device serves as the identity operator. ($b$) Blocking the path that a spin-down particle follows produces the projection operator $\hat{P}_+$. ($c$) Blocking the path that a spin-up particle follows produces the projection operator $\hat{P}_-$.

operator may be expressed in terms of the projection operators as

$$\hat{P}_+ + \hat{P}_- = 1 \tag{2.48}$$

This relation is often referred to as a *completeness relation*. Projecting onto the two vectors corresponding to spin up and spin down are the only possibilities for a spin-$\frac{1}{2}$ particle. As (2.45) shows, (2.48) is equivalent to saying that an arbitrary state $|\psi\rangle$ can be expressed as a superposition of the two basis states $|+\mathbf{z}\rangle$ and $|-\mathbf{z}\rangle$.

Notice that if we apply the projection operator $\hat{P}_+$ to the basis states $|+\mathbf{z}\rangle$ and $|-\mathbf{z}\rangle$, we obtain

$$\hat{P}_+|+\mathbf{z}\rangle = |+\mathbf{z}\rangle\langle+\mathbf{z}|+\mathbf{z}\rangle = |+\mathbf{z}\rangle \tag{2.49a}$$

and

$$\hat{P}_+|-\mathbf{z}\rangle = |+\mathbf{z}\rangle\langle+\mathbf{z}|-\mathbf{z}\rangle = 0 \qquad (2.49b)$$

Thus $|+\mathbf{z}\rangle$ is an eigenstate of the projection operator $\hat{P}_+$ with eigenvalue 1, and $|-\mathbf{z}\rangle$ is an eigenstate of the projection operator $\hat{P}_+$ with eigenvalue 0. We can obtain a physical realization of the projection operator $\hat{P}_+$ from the modified SG device by blocking the path that would be taken by a particle in the state $|-\mathbf{z}\rangle$, that is, by blocking the lower path, as shown in Fig. 2.4b. Each particle in the state $|+\mathbf{z}\rangle$ entering the device exits the device. We can then say we have obtained the eigenvalue 1. Since none of the particles in the state $|-\mathbf{z}\rangle$ that enters the device also exits the device, we can say we have obtained the eigenvalue 0 in this case.

Similarly, we can create a physical realization of the projection operator $\hat{P}_-$ by blocking the upper path in the modified SG device, as shown in Fig. 2.4c. Then each particle in the state $|-\mathbf{z}\rangle$ that enters the device also exits the device:

$$\hat{P}_-|-\mathbf{z}\rangle = |-\mathbf{z}\rangle\langle-\mathbf{z}|-\mathbf{z}\rangle = |-\mathbf{z}\rangle \qquad (2.50a)$$

while none of the particles in the state $|+\mathbf{z}\rangle$ exits the device:

$$\hat{P}_-|+\mathbf{z}\rangle = |-\mathbf{z}\rangle\langle-\mathbf{z}|+\mathbf{z}\rangle = 0 \qquad (2.50b)$$

Hence the eigenvalues of $\hat{P}_-$ are 1 and 0 for the states $|-\mathbf{z}\rangle$ and $|+\mathbf{z}\rangle$, respectively.

Notice that each of the particles that has traversed one of the projection devices is certain to pass through a subsequent projection device of the same type:

$$\hat{P}_+^2 = \big(|+\mathbf{z}\rangle\langle+\mathbf{z}|\big)\big(|+\mathbf{z}\rangle\langle+\mathbf{z}|\big)$$
$$= |+\mathbf{z}\rangle\langle+\mathbf{z}|+\mathbf{z}\rangle\langle+\mathbf{z}| = |+\mathbf{z}\rangle\langle+\mathbf{z}| = \hat{P}_+ \qquad (2.51a)$$
$$\hat{P}_-^2 = \big(|-\mathbf{z}\rangle\langle-\mathbf{z}|\big)\big(|-\mathbf{z}\rangle\langle-\mathbf{z}|\big)$$
$$= |-\mathbf{z}\rangle\langle-\mathbf{z}|-\mathbf{z}\rangle\langle-\mathbf{z}| = |-\mathbf{z}\rangle\langle-\mathbf{z}| = \hat{P}_- \qquad (2.51b)$$

while a particle that passes a first projection device will surely fail to pass a subsequent projection device of the opposite type:

$$\hat{P}_+\hat{P}_- = \big(|+\mathbf{z}\rangle\langle+\mathbf{z}|\big)\big(|-\mathbf{z}\rangle\langle-\mathbf{z}|\big)$$
$$= |+\mathbf{z}\rangle\langle+\mathbf{z}|-\mathbf{z}\rangle\langle-\mathbf{z}| = 0 \qquad (2.52a)$$
$$\hat{P}_-\hat{P}_+ = \big(|-\mathbf{z}\rangle\langle-\mathbf{z}|\big)\big(|+\mathbf{z}\rangle\langle+\mathbf{z}|\big)$$
$$= |-\mathbf{z}\rangle\langle-\mathbf{z}|+\mathbf{z}\rangle\langle+\mathbf{z}| = 0 \qquad (2.52b)$$

These results are illustrated in Fig. 2.5.

Our discussion of the identity operator and the projection operators has arbitrarily been phrased in terms of the $S_z$ basis. We could as easily have expressed the same state $|\psi\rangle$ in terms of the $S_x$ basis as $|\psi\rangle = |+\mathbf{x}\rangle\langle+\mathbf{x}|\psi\rangle + |-\mathbf{x}\rangle\langle-\mathbf{x}|\psi\rangle$. Thus we can also express the identity operator as

$$|+\mathbf{x}\rangle\langle+\mathbf{x}| + |-\mathbf{x}\rangle\langle-\mathbf{x}| = 1 \qquad (2.53)$$

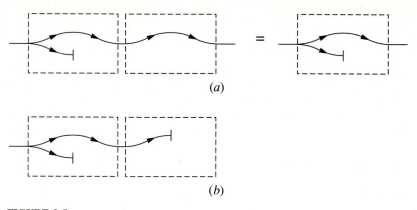

(a)

(b)

**FIGURE 2.5**
Physical realizations of (a) $\hat{P}_+^2 = \hat{P}_+$ and (b) $\hat{P}_- \hat{P}_+ = 0$.

and view it as being composed of projection operators onto the states $|+\mathbf{x}\rangle$ and $|-\mathbf{x}\rangle$.

Let's use this formalism to reexamine Experiment 4 of Chapter 1. In this experiment a particle in the state $|+\mathbf{z}\rangle$ passes through a modified SGx device and then enters an SGz device. Since the modified SGx device acts as an identity operator, the particle entering the last SGz device is still in the state $|+\mathbf{z}\rangle$ and thus the amplitude to find the particle in the state $|-\mathbf{z}\rangle$ vanishes: $\langle -\mathbf{z}|+\mathbf{z}\rangle = 0$. There is, however, another way to express this amplitude. We use the identity operator (2.53) to express the initial ket in terms of the amplitudes to be the states $|+\mathbf{x}\rangle$ and $|-\mathbf{x}\rangle$:

$$|+\mathbf{z}\rangle = |+\mathbf{x}\rangle\langle +\mathbf{x}|+\mathbf{z}\rangle + |-\mathbf{x}\rangle\langle -\mathbf{x}|+\mathbf{z}\rangle \qquad (2.54)$$

Then we have

$$\langle -\mathbf{z}|+\mathbf{z}\rangle = \langle -\mathbf{z}|+\mathbf{x}\rangle\langle +\mathbf{x}|+\mathbf{z}\rangle + \langle -\mathbf{z}|-\mathbf{x}\rangle\langle -\mathbf{x}|+\mathbf{z}\rangle \qquad (2.55)$$

Thus the amplitude for a particle with $S_z = \hbar/2$ to have $S_z = -\hbar/2$ has now been written as the sum of two amplitudes. We read each of these amplitudes from right to left. The first amplitude on the right-hand side is the amplitude for a particle with $S_z = \hbar/2$ to have $S_x = \hbar/2$ times the amplitude for a particle with $S_x = \hbar/2$ to have $S_z = -\hbar/2$. The second amplitude is the amplitude for a particle with $S_z = \hbar/2$ to have $S_x = -\hbar/2$ times the amplitude for a particle with $S_x = -\hbar/2$ to have $S_z = -\hbar/2$. Notice that we multiply the individual amplitudes together and then add the resulting two amplitudes with the $|+\mathbf{x}\rangle$ and $|-\mathbf{x}\rangle$ intermediate states together to determine the total amplitude.

We now calculate the probability:

$$\begin{aligned}
|\langle -\mathbf{z}|+\mathbf{z}\rangle|^2 =\ & |\langle -\mathbf{z}|+\mathbf{x}\rangle|^2|\langle +\mathbf{x}|+\mathbf{z}\rangle|^2 + |\langle -\mathbf{z}|-\mathbf{x}\rangle|^2|\langle -\mathbf{x}|+\mathbf{z}\rangle|^2 \\
& + \langle -\mathbf{z}|+\mathbf{x}\rangle\langle +\mathbf{x}|+\mathbf{z}\rangle\langle -\mathbf{z}|-\mathbf{x}\rangle^*\langle -\mathbf{x}|+\mathbf{z}\rangle^* \\
& + \langle -\mathbf{z}|+\mathbf{x}\rangle^*\langle +\mathbf{x}|+\mathbf{z}\rangle^*\langle -\mathbf{z}|-\mathbf{x}\rangle\langle -\mathbf{x}|+\mathbf{z}\rangle
\end{aligned} \qquad (2.56)$$

This looks like a pretty complicated way to calculate zero, but it is interesting to examine the significance of the four terms on the right-hand side. The first term is just the probability that a measurement of $S_x$ on the initial state yields $\hbar/2$ times the probability that a measurement of $S_z$ on a state with $S_x = \hbar/2$ yields $-\hbar/2$. The second term is the probability that a measurement of $S_x$ on the initial state yields $-\hbar/2$ times the probability that a measurement of $S_z$ on a state with $S_x = -\hbar/2$ yields $-\hbar/2$. These two terms, which sum to $\frac{1}{2}$, are just the terms we would have expected *if* we had made a measurement of $S_x$ with the modified SGx device. But we did not make a measurement and actually distinguish which path the particle followed in the modified SGx device.[8] Thus there are two additional terms in (2.56), *interference* terms, that arise because we added the amplitudes on the right-hand side together *before* squaring to get the probability. You can verify that these two interference terms do cancel the first two probabilities. These results are summarized in Fig. 2.6. In more general terms, if you do not make a measurement, you add the amplitudes to be in the different (indistinguishable) intermediate states, whereas if you do make a measurement that would permit you to distinguish among these states, you add the probabilities.

Finally, it is convenient to introduce the following shorthand notation. For a given two-dimensional basis, we can label our basis states by $|1\rangle$ and $|2\rangle$. We

---

[8] It should be emphasized that a measurement here means any physical interaction that would have permitted us *in principle* to distinguish which path is taken (such as arranging for the particle to leave a track in passing through the modified SG device), whether or not we actually choose to record this data.

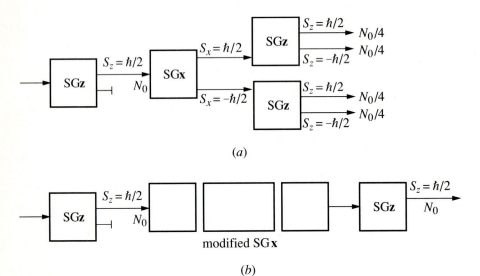

(a)

(b)

**FIGURE 2.6**
Block diagrams of experiments with SG devices in which (a) a measurement of $S_x$ is carried out, illustrating $|\langle -\mathbf{z}|-\mathbf{x}\rangle|^2|\langle -\mathbf{x}|+\mathbf{z}\rangle|^2 + |\langle -\mathbf{z}|+\mathbf{x}\rangle|^2|\langle +\mathbf{x}|+\mathbf{z}\rangle|^2 = \frac{1}{2}$; and (b) no measurement of $S_x$ is made, either by inserting a modified SGx device between the two SGz devices or by simply eliminating the SGx device pictured in (a), illustrating $|\langle -\mathbf{z}|-\mathbf{x}\rangle\langle -\mathbf{x}|+\mathbf{z}\rangle + \langle -\mathbf{z}|+\mathbf{x}\rangle\langle +\mathbf{x}|+\mathbf{z}\rangle|^2 = |\langle -\mathbf{z}|+\mathbf{z}\rangle|^2 = 0$.

can then express the identity operator as

$$\sum_i |i\rangle\langle i| = 1 \tag{2.57}$$

where the sum is from $i = 1$ to $i = 2$. The straightforward generalization of this relationship to larger dimensional bases will be very useful to us later.

## 2.4  MATRIX REPRESENTATIONS OF OPERATORS

In order to change, or transform, kets, operators are required. Although one can discuss concepts such as the adjoint operator abstractly in terms of its action on the bra vectors, it is helpful to construct matrix representations for operators, making concepts such as adjoint and Hermitian operators more concrete, as well as providing the framework for matrix mechanics. Equation (2.25) is a typical equation of the form

$$\hat{A}|\psi\rangle = |\varphi\rangle \tag{2.58}$$

where $\hat{A}$ is an operator and $|\psi\rangle$ and $|\varphi\rangle$ are, in general, different kets. We can also think of the eigenvalue equation (2.35) as being of this form with $|\varphi\rangle$ just a constant times $|\psi\rangle$. Just as we can express a quantum spin state $|\psi\rangle$ using the $S_z$ basis states by

$$|\psi\rangle = |+\mathbf{z}\rangle\langle+\mathbf{z}|\psi\rangle + |-\mathbf{z}\rangle\langle-\mathbf{z}|\psi\rangle \tag{2.59}$$

we can write a comparable expression for $|\varphi\rangle$:

$$|\varphi\rangle = |+\mathbf{z}\rangle\langle+\mathbf{z}|\varphi\rangle + |-\mathbf{z}\rangle\langle-\mathbf{z}|\varphi\rangle \tag{2.60}$$

Thus (2.58) becomes

$$\hat{A}(|+\mathbf{z}\rangle\langle+\mathbf{z}|\psi\rangle + |-\mathbf{z}\rangle\langle-\mathbf{z}|\psi\rangle) = |+\mathbf{z}\rangle\langle+\mathbf{z}|\varphi\rangle + |-\mathbf{z}\rangle\langle-\mathbf{z}|\varphi\rangle \tag{2.61}$$

In ordinary three-dimensional space, a vector equation such as $\mathbf{F} = m\mathbf{a}$ is really the three equations: $F_x = ma_x$, $F_y = ma_y$, and $F_z = ma_z$. We can formally obtain these three equations by taking the dot product of the vector equation with the basis vectors $\mathbf{i}$, $\mathbf{j}$, and $\mathbf{k}$; for example, $\mathbf{i}\cdot\mathbf{F} = \mathbf{i}\cdot m\mathbf{a}$ yields $F_x = ma_x$. Similarly, we can think of (2.61) as two equations that we obtain by projecting (2.61) onto our two basis states, that is, by taking the inner product of this equation with the bras $\langle+\mathbf{z}|$ and $\langle-\mathbf{z}|$:

$$\langle+\mathbf{z}|\hat{A}|+\mathbf{z}\rangle\langle+\mathbf{z}|\psi\rangle + \langle+\mathbf{z}|\hat{A}|-\mathbf{z}\rangle\langle-\mathbf{z}|\psi\rangle = \langle+\mathbf{z}|\varphi\rangle \tag{2.62a}$$

and

$$\langle-\mathbf{z}|\hat{A}|+\mathbf{z}\rangle\langle+\mathbf{z}|\psi\rangle + \langle-\mathbf{z}|\hat{A}|-\mathbf{z}\rangle\langle-\mathbf{z}|\psi\rangle = \langle-\mathbf{z}|\varphi\rangle \tag{2.62b}$$

These two equations can be conveniently cast in matrix form:

$$\begin{pmatrix} \langle+\mathbf{z}|\hat{A}|+\mathbf{z}\rangle & \langle+\mathbf{z}|\hat{A}|-\mathbf{z}\rangle \\ \langle-\mathbf{z}|\hat{A}|+\mathbf{z}\rangle & \langle-\mathbf{z}|\hat{A}|-\mathbf{z}\rangle \end{pmatrix} \begin{pmatrix} \langle+\mathbf{z}|\psi\rangle \\ \langle-\mathbf{z}|\psi\rangle \end{pmatrix} = \begin{pmatrix} \langle+\mathbf{z}|\varphi\rangle \\ \langle-\mathbf{z}|\varphi\rangle \end{pmatrix} \tag{2.63}$$

In the same way that we can *represent* a ket $|\psi\rangle$ in the $S_z$ basis by the column vector

$$|\psi\rangle \rightarrow \begin{pmatrix} \langle +\mathbf{z}|\psi\rangle \\ \langle -\mathbf{z}|\psi\rangle \end{pmatrix} \tag{2.64}$$

we can also *represent* the operator $\hat{A}$ in the $S_z$ basis by the $2 \times 2$ matrix in (2.63). Just as for states, we indicate a representation of an operator with an arrow:

$$\hat{A} \xrightarrow[S_z \text{ basis}]{} \begin{pmatrix} \langle +\mathbf{z}|\hat{A}|+\mathbf{z}\rangle & \langle +\mathbf{z}|\hat{A}|-\mathbf{z}\rangle \\ \langle -\mathbf{z}|\hat{A}|+\mathbf{z}\rangle & \langle -\mathbf{z}|\hat{A}|-\mathbf{z}\rangle \end{pmatrix} = \begin{pmatrix} A_{11} & A_{12} \\ A_{21} & A_{22} \end{pmatrix} \tag{2.65}$$

If we label our basis vectors by $|1\rangle$ and $|2\rangle$ for the states $|+\mathbf{z}\rangle$ and $|-\mathbf{z}\rangle$, respectively, we can express the matrix elements $A_{ij}$ in the convenient form

$$A_{ij} = \langle i|\hat{A}|j\rangle \tag{2.66}$$

where $i$ labels the rows and $j$ labels the columns of the matrix. Note that knowing the four matrix elements in (2.63) allows us to determine the action of the operator $\hat{A}$ on any state $|\psi\rangle$.

## Matrix Representations of the Projection Operators

As an example, the matrix representation of the projection operator $\hat{P}_+$ is given by

$$\hat{P}_+ \xrightarrow[S_z \text{ basis}]{} \begin{pmatrix} \langle +\mathbf{z}|\hat{P}_+|+\mathbf{z}\rangle & \langle +\mathbf{z}|\hat{P}_+|-\mathbf{z}\rangle \\ \langle -\mathbf{z}|\hat{P}_+|+\mathbf{z}\rangle & \langle -\mathbf{z}|\hat{P}_+|-\mathbf{z}\rangle \end{pmatrix} = \begin{pmatrix} 1 & 0 \\ 0 & 0 \end{pmatrix} \tag{2.67a}$$

where we have taken advantage of (2.50) in evaluating the matrix elements. Similarly, the matrix representation of the projection operator $\hat{P}_-$ is given by

$$\hat{P}_- \xrightarrow[S_z \text{ basis}]{} \begin{pmatrix} 0 & 0 \\ 0 & 1 \end{pmatrix} \tag{2.67b}$$

Thus, the completeness relation $\hat{P}_+ + \hat{P}_- = 1$ in matrix form becomes

$$\begin{pmatrix} 1 & 0 \\ 0 & 0 \end{pmatrix} + \begin{pmatrix} 0 & 0 \\ 0 & 1 \end{pmatrix} = \begin{pmatrix} 1 & 0 \\ 0 & 1 \end{pmatrix} = \mathbb{1} \tag{2.68}$$

where $\mathbb{1}$ is the identity matrix. The action of the projection operator $\hat{P}_+$ on the basis states is given by

$$\begin{pmatrix} 1 & 0 \\ 0 & 0 \end{pmatrix}\begin{pmatrix} 1 \\ 0 \end{pmatrix} = \begin{pmatrix} 1 \\ 0 \end{pmatrix} \tag{2.69a}$$

and

$$\begin{pmatrix} 1 & 0 \\ 0 & 0 \end{pmatrix}\begin{pmatrix} 0 \\ 1 \end{pmatrix} = \begin{pmatrix} 0 \\ 0 \end{pmatrix} \tag{2.69b}$$

in agreement with equations (2.49a) and (2.49b), respectively.

## Matrix Representation of $\hat{J}_z$

As another example, consider the operator $\hat{J}_z$, the generator of rotations about the $z$ axis. With the aid of (2.35), we can evaluate the matrix elements:

$$\hat{J}_z \xrightarrow[S_z \text{ basis}]{} \begin{pmatrix} \langle +z|\hat{J}_z|+z\rangle & \langle +z|\hat{J}_z|-z\rangle \\ \langle -z|\hat{J}_z|+z\rangle & \langle -z|\hat{J}_z|-z\rangle \end{pmatrix}$$

$$= \begin{pmatrix} (\hbar/2)\langle +z|+z\rangle & (-\hbar/2)\langle +z|-z\rangle \\ (\hbar/2)\langle -z|+z\rangle & (-\hbar/2)\langle -z|-z\rangle \end{pmatrix}$$

$$= \begin{pmatrix} \hbar/2 & 0 \\ 0 & -\hbar/2 \end{pmatrix} \tag{2.70}$$

The matrix is diagonal with the eigenvalues as the diagonal matrix elements because we are using the eigenstates of the operator as a basis and these eigenstates are orthogonal to each other. The eigenvalue equations $\hat{J}_z|+z\rangle = (\hbar/2)|+z\rangle$ and $\hat{J}_z|-z\rangle = (-\hbar/2)|-z\rangle$ may be expressed in matrix mechanics as

$$\begin{pmatrix} \hbar/2 & 0 \\ 0 & -\hbar/2 \end{pmatrix}\begin{pmatrix} 1 \\ 0 \end{pmatrix} = \frac{\hbar}{2}\begin{pmatrix} 1 \\ 0 \end{pmatrix} \tag{2.71}$$

and

$$\begin{pmatrix} \hbar/2 & 0 \\ 0 & -\hbar/2 \end{pmatrix}\begin{pmatrix} 0 \\ 1 \end{pmatrix} = -\frac{\hbar}{2}\begin{pmatrix} 0 \\ 1 \end{pmatrix} \tag{2.72}$$

respectively. Incidentally, we can write the matrix representation (2.70) in the form

$$\hat{J}_z \xrightarrow[S_z \text{ basis}]{} \begin{pmatrix} \hbar/2 & 0 \\ 0 & -\hbar/2 \end{pmatrix} = \frac{\hbar}{2}\begin{pmatrix} 1 & 0 \\ 0 & 0 \end{pmatrix} - \frac{\hbar}{2}\begin{pmatrix} 0 & 0 \\ 0 & 1 \end{pmatrix} \tag{2.73a}$$

which indicates that

$$\hat{J}_z = \frac{\hbar}{2}\hat{P}_+ - \frac{\hbar}{2}\hat{P}_- = \frac{\hbar}{2}|+z\rangle\langle +z| - \frac{\hbar}{2}|-z\rangle\langle -z| \tag{2.73b}$$

We could have also obtained this result directly in terms of bra and ket vectors by applying $\hat{J}_z$ to the identity operator (2.48).

## Matrix Elements of the Adjoint Operator

We next form the matrix representing the adjoint operator $\hat{A}^\dagger$. If an operator $\hat{A}$ acting on a ket $|\psi\rangle$ satisfies

$$\hat{A}|\psi\rangle = |\varphi\rangle \tag{2.74}$$

then, by definition,

$$\langle\psi|\hat{A}^{\dagger} = \langle\varphi| \tag{2.75}$$

(See Fig. 2.7.) If we take the inner product of (2.74) with the bra $\langle\chi|$, we have

$$\langle\chi|\hat{A}|\psi\rangle = \langle\chi|\varphi\rangle \tag{2.76}$$

while taking the inner product of (2.75) with the ket $|\chi\rangle$, we obtain

$$\langle\psi|\hat{A}^{\dagger}|\chi\rangle = \langle\varphi|\chi\rangle \tag{2.77}$$

Since $\langle\chi|\varphi\rangle = \langle\varphi|\chi\rangle^*$, we see that

$$\langle\psi|\hat{A}^{\dagger}|\chi\rangle = \langle\chi|\hat{A}|\psi\rangle^* \tag{2.78}$$

This straightforward but important result follows directly from our definition (2.75) of the adjoint operator. It can be used to tell us how the matrix representations of an operator and its adjoint are related. If we replace $|\psi\rangle$ and $|\chi\rangle$ with basis states such as $|+\mathbf{z}\rangle$ and $|-\mathbf{z}\rangle$, we obtain

$$\langle i|\hat{A}^{\dagger}|j\rangle = \langle j|\hat{A}|i\rangle^* \tag{2.79}$$

We denote this as

$$A_{ij}^{\dagger} = A_{ji}^* \tag{2.80}$$

which tells us that the matrix representing the operator $\hat{A}^{\dagger}$ is the transpose conjugate of the matrix representing $\hat{A}$. We can define the adjoint *matrix* $\mathbb{A}^{\dagger}$ as the transpose conjugate of the matrix $\mathbb{A}$.

We also find another important result. Since by definition a Hermitian operator $\hat{A}$ satisfies $\hat{A} = \hat{A}^{\dagger}$, then $\langle i|\hat{A}|j\rangle = \langle j|\hat{A}|i\rangle^*$, showing that *the matrix representation of a Hermitian operator equals its transpose conjugate matrix*. Our terminology for adjoint and Hermitian operators is consistent with the terminology used in linear algebra for their matrix representations. We can now see from the explicit matrix representations of the operators $\hat{P}_+$ in (2.67) and $\hat{J}_z$ in (2.70) that these are Hermitian operators, since the matrices are diagonal with real elements (the eigenvalues) on the diagonal. In Chapter 3 we will see examples of Hermitian operators with off-diagonal elements when we examine the matrix representations for $\hat{J}_x$ and $\hat{J}_y$ for spin-$\frac{1}{2}$ and spin-1 particles.

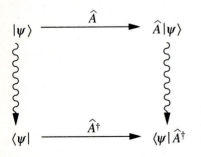

**FIGURE 2.7**
The adjoint operator $\hat{A}^{\dagger}$ of an operator $\hat{A}$ is defined by the correspondence between bras and kets.

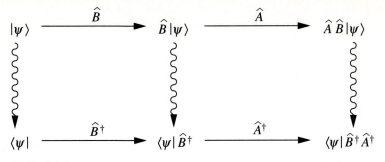

**FIGURE 2.8**
The adjoint of the product of operators is determined by the correspondence between bras and kets.

## The Product of Operators

We often must deal with situations where we have a product of operators, such as (2.51), which involves the product of two projection operators. Another way such a product of operators might arise is to perform two successive rotations on a state. To obtain the matrix representation of the product $\hat{A}\hat{B}$ of two operators, we first form the matrix element

$$\langle i|\hat{A}\hat{B}|j\rangle$$

If we insert the identity operator (2.57), we obtain

$$\langle i|\hat{A}\hat{B}|j\rangle = \langle i|\hat{A}\left(\sum_k |k\rangle\langle k|\right)\hat{B}|j\rangle = \sum_k \langle i|\hat{A}|k\rangle\langle k|\hat{B}|j\rangle = \sum_k A_{ik}B_{kj} \qquad (2.81)$$

which is the usual rule for the multiplication of the matrices representing $\hat{A}$ and $\hat{B}$.

As a simple illustration, the matrix representation of $\hat{P}_+^2 = \hat{P}_+$ is given by

$$\begin{pmatrix} 1 & 0 \\ 0 & 0 \end{pmatrix}\begin{pmatrix} 1 & 0 \\ 0 & 0 \end{pmatrix} = \begin{pmatrix} 1 & 0 \\ 0 & 0 \end{pmatrix} \qquad (2.82)$$

What is the adjoint operator for the product $\hat{A}\hat{B}$ of two operators? As Fig. 2.8 shows,

$$(\hat{A}\hat{B})^\dagger = \hat{B}^\dagger\hat{A}^\dagger \qquad (2.83)$$

## 2.5  CHANGING REPRESENTATIONS

The rotation operator $\hat{R}^\dagger$ can be used to rotate a ket $|\psi\rangle$ into a new ket $|\psi'\rangle$ in an active transformation:

$$|\psi'\rangle = \hat{R}^\dagger|\psi\rangle \qquad (2.84)$$

Recall that the rotation operator $\hat{R}^\dagger$ is just the inverse of the rotation operator $\hat{R}$, so if $\hat{R}$ rotates the state counterclockwise about the axis $\mathbf{n}$ by some angle $\theta$, then $\hat{R}^\dagger$ rotates the state clockwise about the axis $\mathbf{n}$ by the same angle $\theta$:

$$\hat{R}^\dagger(\theta\mathbf{n}) = \hat{R}(-\theta\mathbf{n}) \qquad (2.85)$$

We can form a representation for the ket $|\psi'\rangle$ in the $S_z$ basis, for example, in the usual way:

$$|\psi'\rangle \xrightarrow[S_z \text{ basis}]{} \begin{pmatrix} \langle+z|\hat{R}^\dagger|\psi\rangle \\ \langle-z|\hat{R}^\dagger|\psi\rangle \end{pmatrix} \tag{2.86}$$

There is, however, another way to view this transformation. Instead of the operator $\hat{R}^\dagger$ acting to the right on the ket, we can consider it as acting to the left on the bras. From our earlier discussion of the adjoint operator, we know that kets corresponding to the bras $\langle\pm z|\hat{R}^\dagger$ are $\hat{R}|\pm z\rangle$. Since $\hat{R}$ is the inverse of the operator $\hat{R}^\dagger$, we see that instead of $\hat{R}^\dagger$ rotating the state $|\psi\rangle$ into a new state $|\psi'\rangle$ as in (2.84), we may consider the operator $\hat{R}^\dagger$ in (2.86) to be performing the inverse rotation on the basis states that are used to form the representation.

Let's take a specific example using the $S_z$ and $S_x$ basis states to illustrate this in more detail. Recall that

$$|\pm x\rangle = \hat{R}(\tfrac{\pi}{2}\mathbf{j})|\pm z\rangle \tag{2.87a}$$

and

$$\langle\pm x| = \langle\pm z|\hat{R}^\dagger(\tfrac{\pi}{2}\mathbf{j}) \tag{2.87b}$$

Therefore, we will take the operator $\hat{R}^\dagger$ in (2.86) to be the specific rotation operator $\hat{R}^\dagger(\tfrac{\pi}{2}\mathbf{j})$ so that when it acts to the left on the bra vectors it transforms the $S_z$ basis to the $S_x$ basis according to (2.87b). But if $\hat{R}^\dagger(\tfrac{\pi}{2}\mathbf{j})$ acts to the right, it generates a new state

$$|\psi'\rangle = \hat{R}^\dagger(\tfrac{\pi}{2}\mathbf{j})|\psi\rangle \tag{2.88}$$

We can summarize our discussion in the following equation:

$$|\psi'\rangle \xrightarrow[S_z \text{ basis}]{} \begin{pmatrix} \langle+z|\psi'\rangle \\ \langle-z|\psi'\rangle \end{pmatrix} = \begin{pmatrix} \langle+z|\hat{R}^\dagger(\tfrac{\pi}{2}\mathbf{j})|\psi\rangle \\ \langle-z|\hat{R}^\dagger(\tfrac{\pi}{2}\mathbf{j})|\psi\rangle \end{pmatrix} = \begin{pmatrix} \langle+x|\psi\rangle \\ \langle-x|\psi\rangle \end{pmatrix} \xleftarrow[S_x \text{ basis}]{} |\psi\rangle \tag{2.89}$$

Read from the left, this equation gives the representation in the $S_z$ basis of the state $|\psi'\rangle$ that has been rotated by 90° clockwise around the $y$ axis, whereas read from the right, it shows the state $|\psi\rangle$ as being unaffected but the basis vectors being rotated in the opposite direction, by 90° counterclockwise around the $y$ axis. Both of these transformations lead to the same amplitudes, which we have combined into the column vector in (2.89). This alternative of rotating the basis states used to form a representation is often referred to as a *passive transformation* to distinguish it from an *active transformation* in which the state itself is rotated. A passive transformation is really just a rotation of our coordinate axes in our quantum mechanical vector space, as illustrated in Fig. 2.9.[9]

---

[9] If (2.43) did not seem sufficiently strange to you, try considering it from the perspective of a passive transformation. If we rotate our coordinate axes by 360° and end up with the same configuration of coordinate axes that we had originally, we find the state of a spin-$\tfrac{1}{2}$ particle has turned into the negative of itself.

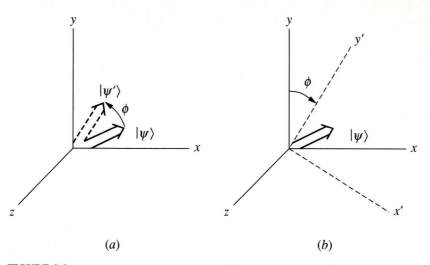

**FIGURE 2.9**
(a) Rotating a state by angle $\phi$ counterclockwise about an axis is equivalent to (b) rotating the coordinate axes by the same angle in the opposite direction, keeping the state fixed.

Equation (2.89) suggests a way to relate the column vector representing the ket $|\psi\rangle$ in one basis to the column vector representing the same ket $|\psi\rangle$ in another basis. If we start with the representation of the ket $|\psi\rangle$ in the $S_x$ basis and insert the identity operator, expressed in terms of $S_z$ basis states, between the bra and the ket vectors, we obtain

$$\begin{pmatrix}\langle+\mathbf{x}|\psi\rangle \\ \langle-\mathbf{x}|\psi\rangle\end{pmatrix} = \begin{pmatrix}\langle+\mathbf{x}|+\mathbf{z}\rangle & \langle+\mathbf{x}|-\mathbf{z}\rangle \\ \langle-\mathbf{x}|+\mathbf{z}\rangle & \langle-\mathbf{x}|-\mathbf{z}\rangle\end{pmatrix}\begin{pmatrix}\langle+\mathbf{z}|\psi\rangle \\ \langle-\mathbf{z}|\psi\rangle\end{pmatrix}$$

$$= \begin{pmatrix}\langle+\mathbf{z}|\hat{R}^\dagger(\frac{\pi}{2}\mathbf{j})|+\mathbf{z}\rangle & \langle+\mathbf{z}|\hat{R}^\dagger(\frac{\pi}{2}\mathbf{j})|-\mathbf{z}\rangle \\ \langle-\mathbf{z}|\hat{R}^\dagger(\frac{\pi}{2}\mathbf{j})|+\mathbf{z}\rangle & \langle-\mathbf{z}|\hat{R}^\dagger(\frac{\pi}{2}\mathbf{j})|-\mathbf{z}\rangle\end{pmatrix}\begin{pmatrix}\langle+\mathbf{z}|\psi\rangle \\ \langle-\mathbf{z}|\psi\rangle\end{pmatrix} \quad (2.90)$$

where the second line follows from (2.87b). We call the 2 × 2 matrix in (2.90) $\mathbb{S}^\dagger$, or more precisely in this specific example $\mathbb{S}^\dagger(\frac{\pi}{2}\mathbf{j})$, since it is really the matrix representation in the $S_z$ basis of the operator $\hat{R}^\dagger(\frac{\pi}{2}\mathbf{j})$ that rotates kets by 90° clockwise about the $y$ axis. Equation (2.90) transforms a given ket $|\psi\rangle$ in the $S_z$ basis into the $S_x$ basis.

We can transform from the $S_x$ basis to the $S_z$ basis in analogous fashion:

$$\begin{pmatrix}\langle+\mathbf{z}|\psi\rangle \\ \langle-\mathbf{z}|\psi\rangle\end{pmatrix} = \begin{pmatrix}\langle+\mathbf{z}|+\mathbf{x}\rangle & \langle+\mathbf{z}|-\mathbf{x}\rangle \\ \langle-\mathbf{z}|+\mathbf{x}\rangle & \langle-\mathbf{z}|-\mathbf{x}\rangle\end{pmatrix}\begin{pmatrix}\langle+\mathbf{x}|\psi\rangle \\ \langle-\mathbf{x}|\psi\rangle\end{pmatrix}$$

$$= \begin{pmatrix}\langle+\mathbf{z}|\hat{R}(\frac{\pi}{2}\mathbf{j})|+\mathbf{z}\rangle & \langle+\mathbf{z}|\hat{R}(\frac{\pi}{2}\mathbf{j})|-\mathbf{z}\rangle \\ \langle-\mathbf{z}|\hat{R}(\frac{\pi}{2}\mathbf{j})|+\mathbf{z}\rangle & \langle-\mathbf{z}|\hat{R}(\frac{\pi}{2}\mathbf{j})|-\mathbf{z}\rangle\end{pmatrix}\begin{pmatrix}\langle+\mathbf{x}|\psi\rangle \\ \langle-\mathbf{x}|\psi\rangle\end{pmatrix} \quad (2.91)$$

where in the first line we have inserted the identity operator, this time expressed in terms of the $S_x$ basis states. Also we have used (2.87a) to express the $2 \times 2$ matrix in the second line of the equation in terms of the matrix representation of the operator $\hat{R}(\frac{\pi}{2}\mathbf{j})$. Comparing the first lines of (2.90) and (2.91) reveals that the $2 \times 2$ matrix in (2.91) is the matrix $\mathbb{S}$, the adjoint matrix of the matrix $\mathbb{S}^{\dagger}$, since the matrix elements of $\mathbb{S}$ are simply obtained from the matrix elements of $\mathbb{S}^{\dagger}$ by taking the transpose conjugate. Also, a comparison of the second lines of (2.90) and (2.91) shows that the $2 \times 2$ matrix in (2.91) is the matrix representation of $\hat{R}(\frac{\pi}{2}\mathbf{j})$, while the $2 \times 2$ matrix in (2.90) is the matrix representation of $\hat{R}^{\dagger}(\frac{\pi}{2}\mathbf{j})$. Since the rotation operators are unitary, the matrices must satisfy

$$\mathbb{S}^{\dagger}\mathbb{S} = \mathbb{1} \tag{2.92}$$

which can also be verified by substituting equation (2.91) directly into equation (2.90).

We can now determine how the matrix representation of an operator in one basis is related to the matrix representation in some other basis. For example, the matrix representing an operator $\hat{A}$ in the $S_x$ basis is given by

$$\hat{A} \xrightarrow[S_x \text{ basis}]{} \begin{pmatrix} \langle +\mathbf{x}|\hat{A}|+\mathbf{x}\rangle & \langle +\mathbf{x}|\hat{A}|-\mathbf{x}\rangle \\ \langle -\mathbf{x}|\hat{A}|+\mathbf{x}\rangle & \langle -\mathbf{x}|\hat{A}|-\mathbf{x}\rangle \end{pmatrix} \tag{2.93}$$

A typical matrix element can be expressed as

$$\langle +\mathbf{x}|\hat{A}|-\mathbf{x}\rangle = \langle +\mathbf{z}|\hat{R}^{\dagger}(\tfrac{\pi}{2}\mathbf{j})\hat{A}\hat{R}(\tfrac{\pi}{2}\mathbf{j})|-\mathbf{z}\rangle$$

Inserting the identity element (2.44) before and after the operator $\hat{A}$ on the left-hand side or between each of the operators on the right-hand side [or using result (2.81) for the matrix representation of the product of operators] permits us to write

$$\hat{A} \xrightarrow[S_x \text{ basis}]{} \mathbb{S}^{\dagger}\mathbb{A}\mathbb{S} \tag{2.94}$$

where $\mathbb{A}$ is the matrix representation of $\hat{A}$ in the $S_z$ basis.[10]

---

[10] Equation (2.94)—or (2.90) and (2.91), for that matter—forms a good advertisement for the power of the identity operator. Rather than trying to remember such an equation, it is probably easier and safer to derive it whenever needed by starting with the matrix elements (or amplitudes) that you are trying to find and inserting the identity operator from the appropriate basis set in the appropriate place(s):

$$\begin{pmatrix} \langle +\mathbf{x}|\hat{A}|+\mathbf{x}\rangle & \langle +\mathbf{x}|\hat{A}|-\mathbf{x}\rangle \\ \langle -\mathbf{x}|\hat{A}|+\mathbf{x}\rangle & \langle -\mathbf{x}|\hat{A}|-\mathbf{x}\rangle \end{pmatrix}$$

$$= \begin{pmatrix} \langle +\mathbf{x}|+\mathbf{z}\rangle & \langle +\mathbf{x}|-\mathbf{z}\rangle \\ \langle -\mathbf{x}|+\mathbf{z}\rangle & \langle -\mathbf{x}|-\mathbf{z}\rangle \end{pmatrix} \begin{pmatrix} \langle +\mathbf{z}|\hat{A}|+\mathbf{z}\rangle & \langle +\mathbf{z}|\hat{A}|-\mathbf{z}\rangle \\ \langle -\mathbf{z}|\hat{A}|+\mathbf{z}\rangle & \langle -\mathbf{z}|\hat{A}|-\mathbf{z}\rangle \end{pmatrix} \begin{pmatrix} \langle +\mathbf{z}|+\mathbf{x}\rangle & \langle +\mathbf{z}|-\mathbf{x}\rangle \\ \langle -\mathbf{z}|+\mathbf{x}\rangle & \langle -\mathbf{z}|-\mathbf{x}\rangle \end{pmatrix}$$

Let's take the example of evaluating the matrix representation of $\hat{J}_z$ in the $S_x$ basis. Using (2.6) and (2.15) to evaluate the matrix $\mathbb{S}$ in (2.91), we find

$$\mathbb{S} = \frac{1}{\sqrt{2}}\begin{pmatrix} 1 & 1 \\ 1 & -1 \end{pmatrix} \tag{2.95}$$

Carrying out the matrix multiplication using the matrix representation of $\hat{J}_z$ in the $S_z$ basis from (2.70), we obtain

$$\hat{J}_z \xrightarrow[S_x \text{ basis}]{} \frac{1}{\sqrt{2}}\begin{pmatrix} 1 & 1 \\ 1 & -1 \end{pmatrix} \frac{\hbar}{2}\begin{pmatrix} 1 & 0 \\ 0 & -1 \end{pmatrix} \frac{1}{\sqrt{2}}\begin{pmatrix} 1 & 1 \\ 1 & -1 \end{pmatrix} = \frac{\hbar}{2}\begin{pmatrix} 0 & 1 \\ 1 & 0 \end{pmatrix} \tag{2.96}$$

Comparing (2.96) with (2.70), we see that the matrix representation of the operator is no longer diagonal, since we are not using the eigenstates of the operator as the basis.[11]

If we also take advantage of (2.90) to express the eigenstate $|+\mathbf{z}\rangle$ in the $S_x$ basis,

$$|+\mathbf{z}\rangle \xrightarrow[S_x \text{ basis}]{} \frac{1}{\sqrt{2}}\begin{pmatrix} 1 & 1 \\ 1 & -1 \end{pmatrix}\begin{pmatrix} 1 \\ 0 \end{pmatrix} = \frac{1}{\sqrt{2}}\begin{pmatrix} 1 \\ 1 \end{pmatrix} \tag{2.97}$$

we can express the eigenvalue equation $\hat{J}_z|+\mathbf{z}\rangle = (\hbar/2)|+\mathbf{z}\rangle$ in the $S_x$ basis:

$$\frac{\hbar}{2}\begin{pmatrix} 0 & 1 \\ 1 & 0 \end{pmatrix}\frac{1}{\sqrt{2}}\begin{pmatrix} 1 \\ 1 \end{pmatrix} = \frac{\hbar}{2}\frac{1}{\sqrt{2}}\begin{pmatrix} 1 \\ 1 \end{pmatrix} \tag{2.98}$$

Compare (2.98) with (2.71), where the same equation is written in the $S_z$ basis. Note that the eigenvalue equation is satisfied independently of the basis in which we choose to express it. This eigenvalue equation in its most basic form deals with operators and states, not with their representations, which we are free to choose in any way we want.

Before leaving this section, it is worth emphasizing again what we have learned. The S-matrices give us an easy way to transform both our states and our operators from one matrix representation to another. As the first line in both equations (2.90) and (2.91) shows, these S-matrices are composed of the amplitudes formed by taking the inner product of the basis kets of the representation we are transforming *from* with the basis bras of the representation we are transforming

---

[11] Alternatively, we could evaluate the matrix representation of $\hat{J}_z$ in the $S_x$ basis by expressing the basis states $|\pm\mathbf{x}\rangle$ in terms of $|\pm\mathbf{z}\rangle$ so that we can let $\hat{J}_z$ act on them directly. For example, the element in the first row, second column of (2.96) is given by

$$\langle+\mathbf{x}|\hat{J}_z|-\mathbf{x}\rangle = \tfrac{1}{2}\big(\langle+\mathbf{z}| + \langle-\mathbf{z}|\big)\,\hat{J}_z\big(|+\mathbf{z}\rangle - |-\mathbf{z}\rangle\big)$$

$$= \tfrac{1}{2}\big(\langle+\mathbf{z}| + \langle-\mathbf{z}|\big)\Big(\frac{\hbar}{2}|+\mathbf{z}\rangle - \Big(-\frac{\hbar}{2}\Big)|-\mathbf{z}\rangle\Big) = \frac{\hbar}{2}$$

*to.* It is often convenient, however, to return to the active viewpoint with which we started our discussion. Instead of the S-matrices transforming a given state from one basis to another, we can view the S-matrix as the matrix representation of the rotation operator that rotates the given state into a different state within a fixed representation. This will be our starting point in Chapter 3. As we have seen, an active rotation that transforms the state is just the inverse of the passive rotation that transforms the basis vectors used to form a particular representation.

## 2.6 EXPECTATION VALUES

It is interesting to see how we can use matrix mechanics to calculate expectation values of observables like the $z$ component of the angular momentum with which we have associated the operator $\hat{J}_z$. If a spin-$\frac{1}{2}$ particle is in the state

$$|\psi\rangle = |+\mathbf{z}\rangle\langle+\mathbf{z}|\psi\rangle + |-\mathbf{z}\rangle\langle-\mathbf{z}|\psi\rangle \tag{2.99}$$

then, as we saw in Section 1.4, the expectation value of $S_z$ is given by

$$\langle S_z\rangle = \left(\frac{\hbar}{2}\right)|\langle+\mathbf{z}|\psi\rangle|^2 + \left(-\frac{\hbar}{2}\right)|\langle-\mathbf{z}|\psi\rangle|^2 \tag{2.100}$$

That is, the average value of $S_z$ is the sum of the results $\hbar/2$ and $-\hbar/2$ of a measurement multiplied by the probability $|\langle+\mathbf{z}|\psi\rangle|^2$ and $|\langle-\mathbf{z}|\psi\rangle|^2$, respectively, of obtaining each result. We can express this expectation value in matrix mechanics as

$$\langle S_z\rangle = \left(\langle\psi|+\mathbf{z}\rangle, \langle\psi|-\mathbf{z}\rangle\right)\frac{\hbar}{2}\begin{pmatrix} 1 & 0 \\ 0 & -1 \end{pmatrix}\begin{pmatrix} \langle+\mathbf{z}|\psi\rangle \\ \langle-\mathbf{z}|\psi\rangle \end{pmatrix} \tag{2.101}$$

as can be verified by explicitly carrying out the matrix multiplication. The right-hand side of (2.101) is the representation in the $S_z$ basis of $\langle\psi|\hat{J}_z|\psi\rangle$. Thus, we can also express the expectation value in the form

$$\langle S_z\rangle = \langle\psi|\hat{J}_z|\psi\rangle \tag{2.102}$$

In the language of eigenstates and eigenvalues, the expectation value (2.100) is the sum of the eigenvalues with each weighted by the probability of obtaining that eigenvalue. The advantage of expressing the expectation value in the form (2.102) is that we needn't evaluate it in a representation in which the basis states are the eigenstates of the operator in question. For example, we could evaluate (2.102) in the $S_x$ basis by inserting the identity operator (2.53) between the bra vector and the operator and between the operator and the ket vector. Then we have

$$\langle S_z\rangle = \left(\langle\psi|+\mathbf{x}\rangle, \langle\psi|-\mathbf{x}\rangle\right)\begin{pmatrix} \langle+\mathbf{x}|\hat{J}_z|+\mathbf{x}\rangle & \langle+\mathbf{x}|\hat{J}_z|-\mathbf{x}\rangle \\ \langle-\mathbf{x}|\hat{J}_z|+\mathbf{x}\rangle & \langle-\mathbf{x}|\hat{J}_z|-\mathbf{x}\rangle \end{pmatrix}\begin{pmatrix} \langle+\mathbf{x}|\psi\rangle \\ \langle-\mathbf{x}|\psi\rangle \end{pmatrix} \tag{2.103}$$

You can verify that we can also go from (2.101) in the $S_z$ basis to (2.103) in the $S_x$ basis by inserting the identity operator $\mathsf{S}\mathsf{S}^\dagger$ before and after the 2 × 2 matrix in (2.101), provided we use the S-matrix (2.95) that transforms between these two basis sets.

As an example, let's return to (1.19), where we evaluated the expectation value of $S_z$ for the state $|+\mathbf{x}\rangle$. Substituting the column vector representation (2.6) for this ket in the $S_z$ basis into (2.101), we see that the expectation value may be written in matrix form as

$$\langle S_z \rangle = \frac{1}{\sqrt{2}} (1, \ 1) \frac{\hbar}{2} \begin{pmatrix} 1 & 0 \\ 0 & -1 \end{pmatrix} \frac{1}{\sqrt{2}} \begin{pmatrix} 1 \\ 1 \end{pmatrix} = 0 \qquad (2.104)$$

On the other hand, we can also evaluate this expectation value in the $S_x$ basis, in which case we use the column vector (2.24) to represent the ket $|+\mathbf{x}\rangle$ and the matrix (2.96) to represent the operator $\hat{J}_z$ in this basis. The expectation value is given by

$$\langle S_z \rangle = (1, \ 0) \frac{\hbar}{2} \begin{pmatrix} 0 & 1 \\ 1 & 0 \end{pmatrix} \begin{pmatrix} 1 \\ 0 \end{pmatrix} = 0 \qquad (2.105)$$

The results (2.104) and (2.105) of course agree. In (2.104) the matrix form for the operator is especially straightforward, while in (2.105) it is the representation for the state that is especially simple.

## 2.7 PHOTON POLARIZATION AND THE SPIN OF THE PHOTON

The previous discussion about representations of states and operators may seem somewhat mathematical in nature. The usefulness of this type of mathematics is just a reflection of the fundamental underlying linear-vector-space structure of quantum mechanics. We conclude this chapter by looking at how we can apply this formalism to another physical two-state system, the polarization of the electromagnetic field. Many polarization effects can be described by classical physics, unlike the physics of spin-$\frac{1}{2}$ particles, which is a purely quantum phenomenon. Nonetheless, analyzing polarization effects using quantum mechanics can help to illuminate the differences between classical and quantum physics and at the same time tell us something fundamental about the nature of the quantum of the electromagnetic field.

Instead of a beam of spin-$\frac{1}{2}$ atoms passing through a Stern-Gerlach device, we consider a beam of photons, traveling in the $z$ direction, passing through a linear polarizer. Those photons that pass through a polarizer with its transmission axis horizontal, that is, along the $x$ axis, are said to be in the state $|x\rangle$, and those photons that pass through a polarizer with its transmission axis vertical are said to be in the state $|y\rangle$.[12] These two polarization states form a basis and the

---

[12] These states are often referred to as $|x\rangle$ and $|y\rangle$. A different typeface is used to help distinguish these polarization states from position states, which will be introduced in Chapter 6.

basis states satisfy $\langle x|y \rangle = 0$, since a beam of photons that passes through a polarizer whose transmission axis is vertical will be completely absorbed by a polarizer whose transmission axis is horizontal. Thus none of the photons will be found to be in the state $|x\rangle$ if they are put into the state $|y\rangle$ by virtue of having passed through the initial polarizer (assuming that our polarizers function with 100 percent efficiency).

We can also create polarized photons by sending the beam through a polarizer whose transmission axis is aligned at some angle to our original $x$-$y$ axes. If the transmission axis is along the $x'$ axis or $y'$ axis shown in Fig. 2.10, the corresponding polarization states may be written as a superposition of the $|x\rangle$ and $|y\rangle$ polarization states as

$$|x'\rangle = |x\rangle\langle x|x'\rangle + |y\rangle\langle y|x'\rangle$$
$$|y'\rangle = |x\rangle\langle x|y'\rangle + |y\rangle\langle y|y'\rangle \qquad (2.106)$$

What are the amplitudes such as $\langle x|x'\rangle$, the amplitude for a photon linearly polarized along the $x'$ axis to be found with its polarization along the $x$ axis? A classical physicist asked to determine the intensity of light passing through a polarizer with its transmission axis along either the $x$ or the $y$ axis *after* it has passed through a polarizer with its transmission axis along $x'$, as pictured in Fig. 2.11, would calculate the component of the electric field along the $x$ or the $y$ axis and would square the *amplitude* of the field to determine the intensity passing through the second polarizer. If we denote the electric field after passage through the initial polarizer by $E_{x'}$, then the components of the field along the $x$ and $y$ axes are given by

$$E_x = E_{x'}\cos\phi \qquad E_y = E_{x'}\sin\phi$$

Thus the intensity of the light after passing through the second polarizer with its transmission axis along the $x$ or $y$ axis is proportional to $\cos^2\phi$ or $\sin^2\phi$, respectively. We can duplicate the classical results if we choose $\langle x|x'\rangle = \cos\phi$ and $\langle y|x'\rangle = \sin\phi$. Similarly, if the first polarizer has its transmission axis along the $y'$ axis and we denote the electric field after passage through this polarizer by

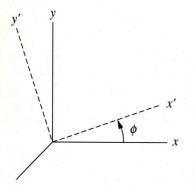

**FIGURE 2.10**
Two sets of transmission axes of a polarizer that may be used to create polarization states of photons traveling in the $z$ direction.

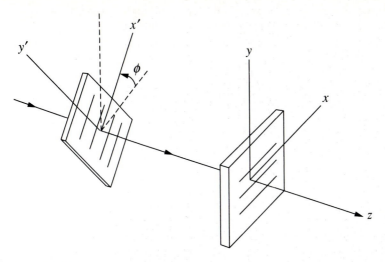

**FIGURE 2.11**
An $x'$ polarizer followed by an $x$ polarizer.

$E_{y'}$, then the components of the field along the $x$ and $y$ axes are given by

$$E_x = -E_{y'} \sin \phi \qquad E_y = E_{y'} \cos \phi$$

Again, we can duplicate the classical results if we choose $\langle y|y' \rangle = \cos \phi$ and $\langle x|y' \rangle = -\sin \phi$. Of course, the experiments outlined here alone do not give us any information about the phases of the amplitudes. However, since classical electromagnetic theory can account for interference phenomena such as the Young double-slit experiment, it is perhaps not too surprising that our conjectures about the amplitudes based on classical physics yield a valid quantum mechanical set, including phases:

$$|x'\rangle = \cos \phi |x\rangle + \sin \phi |y\rangle$$
$$|y'\rangle = -\sin \phi |x\rangle + \cos \phi |y\rangle \qquad (2.107)$$

Where do the quantum effects show up? Classical physics cannot account for the granular nature of the measurements, that a photomultiplier can detect photons coming in single lumps. Nor can it account for the inherently probabilistic nature of the measurements; we cannot do more than give a probability that a single photon in the state $|x'\rangle$ will pass through a polarizer with its transmission axis along $x$. For example, if the angle $\phi = 60°$, then a single photon after having passed through an $x'$ polarizer has a probability of $|\langle x|x'\rangle|^2 = \cos^2 60° = 0.25$ of passing through a second $x$ polarizer. Knowing the polarization state of the photon does not, in general, determine whether it will pass through a subsequent polarizer. All we can determine is the probability, much to the discomfiture of the classical physicist who would like to believe that such results should be completely determined if enough information is known about the state of the system. The classical and quantum predictions are, however, in complete accord when the intensity of the beams is high so that the number of photons is large.

We can use (2.107) to calculate the matrix $\mathbb{S}^\dagger$ that transforms from the $|x\rangle$-$|y\rangle$ basis to the $|x'\rangle$-$|y'\rangle$ basis:

$$\mathbb{S}^\dagger = \begin{pmatrix} \langle x'|x\rangle & \langle x'|y\rangle \\ \langle y'|x\rangle & \langle y'|y\rangle \end{pmatrix} = \begin{pmatrix} \cos\phi & \sin\phi \\ -\sin\phi & \cos\phi \end{pmatrix} \tag{2.108}$$

The matrix $\mathbb{S}$ that transforms from the $|x'\rangle$-$|y'\rangle$ basis to the $|x\rangle$-$|y\rangle$ basis is given by

$$\mathbb{S} = \begin{pmatrix} \langle x|x'\rangle & \langle x|y'\rangle \\ \langle y|x'\rangle & \langle y|y'\rangle \end{pmatrix} = \begin{pmatrix} \cos\phi & -\sin\phi \\ \sin\phi & \cos\phi \end{pmatrix} \tag{2.109}$$

You can check that these matrices satisfy $\mathbb{S}^\dagger\mathbb{S} = 1$. All the elements of the matrix $\mathbb{S}$ are real. In fact, it is an example of an orthogonal matrix familiar from classical physics for rotating a vector in the $x$-$y$ plane counterclockwise about the $z$ axis by an angle $\phi$. We can express $\mathbb{S}$ in terms of the rotation operator $\hat{R}(\phi\mathbf{k})$ that rotates the ket vectors themselves in this direction ($|x'\rangle = \hat{R}(\phi\mathbf{k})|x\rangle$ and $|y'\rangle = \hat{R}(\phi\mathbf{k})|y\rangle$):

$$\mathbb{S} = \begin{pmatrix} \langle x|\hat{R}(\phi\mathbf{k})|x\rangle & \langle x|\hat{R}(\phi\mathbf{k})|y\rangle \\ \langle y|\hat{R}(\phi\mathbf{k})|x\rangle & \langle y|\hat{R}(\phi\mathbf{k})|y\rangle \end{pmatrix} = \begin{pmatrix} \cos\phi & -\sin\phi \\ \sin\phi & \cos\phi \end{pmatrix} \tag{2.110}$$

There is another set of basis vectors that have a great deal of physical significance but cannot be obtained from $|x\rangle$-$|y\rangle$ basis by a simple rotation. We introduce

$$|R\rangle = \frac{1}{\sqrt{2}}(|x\rangle + i|y\rangle) \tag{2.111a}$$

$$|L\rangle = \frac{1}{\sqrt{2}}(|x\rangle - i|y\rangle) \tag{2.111b}$$

These states are referred to as right-circularly polarized and left-circularly polarized, respectively.

First, let's ask what the classical physicist would make of a right-circularly polarized electromagnetic plane wave of amplitude $E_0$ traveling in the $z$ direction,

$$\mathbf{E} = E_0\mathbf{i}e^{i(kz-\omega t)} + iE_0\mathbf{j}e^{i(kz-\omega t)} \tag{2.112a}$$

Of course, the classical physicist uses complex numbers only as a convenient way to express a wave. The physics is determined by the real part of (2.112a), or

$$\mathbf{E} = E_0\mathbf{i}\cos(kz - \omega t) - E_0\mathbf{j}\sin(kz - \omega t) \tag{2.112b}$$

The "extra" factor of $i$ in the $y$ component of $\mathbf{E}$ in (2.112a) here means that the $x$ and $y$ components of the electric field are 90° out of phase, as (2.112b) shows.

If we take $z = 0$ and examine the time dependence of the electromagnetic field, we see an **E** field that rotates in a circle as time progresses. If you curl your right hand in the direction of the changing **E**, your thumb points in the direction of propagation along the positive $z$ axis. The **E** field of the left-circularly polarized electromagnetic plane wave rotates in the opposite direction and thus would require you to curl your left hand in the direction of changing **E** to have your thumb point in the direction of propagation.

We can produce circularly polarized light by allowing linearly polarized light to fall on a birefringent crystal such as calcite that is cut so that the optic axis of the crystal lies in the $x$-$y$ plane. Light polarized parallel to the optic axis in a birefringent crystal has a different index of refraction than does light perpendicular to the optic axis. We can orient our coordinate axes so that the optic axis is along $x$ and the perpendicular axis is, of course, along $y$. Denoting the different indices of refraction by $n_x$ and $n_y$, we see from (2.112$a$) that light polarized parallel to the $x$ axis will pick up a phase $(n_x \omega/c)z$ in traversing a distance $z$ through the crystal. Similarly, light polarized parallel to the $y$ axis will gain a phase $(n_y \omega/c)z$. Thus a beam of linearly polarized light incident on such a crystal with its polarization axis inclined at $45°$ to the $x$ axis will have equal magnitudes for the $x$ and $y$ components of the electric field, as indicated in Fig. 2.12, and there will be a phase difference $[(n_x - n_y)\omega/c]z$ between these two components that grows as the light passes through a distance $z$ in the crystal. The crystal can be cut to a particular thickness, called a quarter-wave plate, so that the phase difference is $90°$ when the light of a particular wavelength exits the crystal, thus producing circularly polarized light.

What does the quantum physicist make of these circular polarization states (2.111)? Following the formalism of Section 2.2, it is instructive to ask how these states change under a rotation about the $z$ axis. If we consider a right-circularly polarized state that has been rotated by an angle $\phi$ counterclockwise about the $z$ axis, we see that it can be expressed as

$$|R'\rangle = \frac{1}{\sqrt{2}}(|x'\rangle + i|y'\rangle)$$

$$= \frac{1}{\sqrt{2}}\left[\cos\phi|x\rangle + \sin\phi|y\rangle + i(-\sin\phi|x\rangle + \cos\phi|y\rangle)\right]$$

$$= \frac{(\cos\phi - i\sin\phi)}{\sqrt{2}}(|x\rangle + i|y\rangle)$$

$$= e^{-i\phi}|R\rangle \tag{2.113}$$

Thus this state only picks up an overall phase factor when the state is rotated about the $z$ axis. Based on our experience with the behavior of spin-$\frac{1}{2}$ states under rotations, (2.113) indicates that the state is one with definite angular momentum

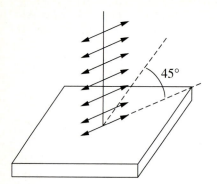

**FIGURE 2.12**
Plane-polarized light incident on a quarter-wave plate with its direction of polarization oriented at 45° to the optic axis will produce circularly polarized light.

in the $z$ direction. Since (2.32) shows that

$$|R'\rangle = \hat{R}(\phi\mathbf{k})|R\rangle = e^{-i\hat{J}_z\phi/\hbar}|R\rangle \qquad (2.114)$$

consistency with the preceding equation requires that

$$\hat{J}_z|R\rangle = \hbar|R\rangle \qquad (2.115)$$

Similarly, if we rotate the left-circularly polarized state by angle $\phi$ counterclockwise about the $z$ axis, we obtain

$$|L'\rangle = e^{i\phi}|L\rangle \qquad (2.116)$$

telling us that[13]

$$\hat{J}_z|L\rangle = -\hbar|L\rangle \qquad (2.117)$$

Thus the right- and left-circularly polarized states are eigenstates of $\hat{J}_z$, the operator that generates rotations about the $z$ axis, but with eigenvalues $\pm\hbar$, not the $\pm\hbar/2$ characteristic of a spin-$\frac{1}{2}$ particle. In Chapter 3 we will see that the eigenvalues of $\hat{J}_z$ for a spin-1 particle are $+\hbar$, 0, and $-\hbar$. Photons have intrinsic spin of 1 instead of $\frac{1}{2}$. The absence of the 0 eigenvalue for $\hat{J}_z$ for a photon turns out to be a special characteristic of a massless particle, which moves at speed $c$.

## 2.8  SUMMARY

In this chapter we have introduced operators in order to change a state into a different state. Since we are dealing here primarily with states of angular momentum, the natural operation is to rotate these states so that a state in which a component of the angular momentum has a definite value in a particular direction

---

[13] A particle with a positive (negative) projection of the intrinsic angular momentum along the direction of motion is said to have positive (negative) helicity. Photons thus come in two types, with both positive and negative helicity, corresponding to right- and left-circularly polarized light, respectively.

is rotated into a state in which the angular momentum has the same value in a different direction.[14] The operator that rotates states counterclockwise by angle $\phi$ about the $z$ axis is

$$\hat{R}(\phi\mathbf{k}) = e^{-i\hat{J}_z\phi/\hbar} \tag{2.118}$$

where the operator $\hat{J}_z$ is called the generator of rotations about the $z$ axis. In general, for an arbitrary operator $\hat{A}$, the bra corresponding to the ket

$$\hat{A}|\psi\rangle = |\varphi\rangle \tag{2.119a}$$

is

$$\langle\psi|\hat{A}^\dagger = \langle\varphi| \tag{2.119b}$$

where the dagger denotes the adjoint operator. Thus the rotated bra corresponding to the rotated ket

$$\hat{R}(\phi\mathbf{k})|\psi\rangle = e^{-i\hat{J}_z\phi/\hbar}|\psi\rangle \tag{2.120a}$$

is given by

$$\langle\psi|\hat{R}^\dagger(\phi\mathbf{k}) = \langle\psi|e^{i\hat{J}_z^\dagger\phi/\hbar} \tag{2.120b}$$

In order for probability to be conserved under rotation,

$$\langle\psi|\hat{R}^\dagger(\phi\mathbf{k})\hat{R}(\phi\mathbf{k})|\psi\rangle = \langle\psi|e^{i\hat{J}_z^\dagger\phi/\hbar}e^{-i\hat{J}_z\phi/\hbar}|\psi\rangle = \langle\psi|\psi\rangle \tag{2.121}$$

which requires that the generators of rotation be Hermitian:

$$\hat{J}_z^\dagger = \hat{J}_z \tag{2.122}$$

An operator like the rotation operator that satisfies $\hat{R}^\dagger\hat{R} = 1$ is called a unitary operator.

For a spin-$\frac{1}{2}$ particle, the spin-up-along-$z$ state $|+\mathbf{z}\rangle$ and spin-down-along-$z$ state $|-\mathbf{z}\rangle$ satisfy

$$\hat{J}_z|\pm\mathbf{z}\rangle = \pm\frac{\hbar}{2}|\pm\mathbf{z}\rangle \tag{2.123}$$

showing that when the generator of rotations about the $z$ axis acts on these states, the result is just the state itself multiplied by the value of $S_z$ that these states are observed to have when a measurement of the intrinsic spin angular momentum in the $z$ direction is carried out. Thus we can use a terminology in which we label the states $|\pm\mathbf{z}\rangle$ by $|S_z = \pm\hbar/2\rangle$, that is, we label the states by their values

---

[14] This way of describing a rotation of an angular momentum state may seem somewhat awkward, but in Chapter 3 we will see why we cannot say that the angular momentum simply points in a particular direction.

of $S_z$. Similarly, for example,

$$\hat{J}_x|\pm\mathbf{x}\rangle = \pm\frac{\hbar}{2}|\pm\mathbf{x}\rangle \tag{2.124}$$

where $\hat{J}_x$ is the generator of rotations about the $x$ axis. In Chapter 3 we will argue on more general grounds that we should identify the generator of rotations with the component of the angular momentum along the axis about which the rotation is taking place. In subsequent chapters we will see that the operator that generates displacements in space is the linear momentum operator and the operator that generates time translations (moves the state forward in time) is the energy operator. Thus we will see repeated a pattern in which a Hermitian operator $\hat{A}$ is associated with a physical observable and the result $a_n$ of a measurement for a particular state $|a_n\rangle$ satisfies

$$\hat{A}|a_n\rangle = a_n|a_n\rangle \tag{2.125}$$

Note that for a Hermitian, or self-adjoint, operator ($\hat{A} = \hat{A}^\dagger$), the bra equation corresponding to (2.125) is

$$\langle a_n|\hat{A} = \langle a_n|a_n^* \tag{2.126}$$

An equation in which an operator acting on a state yields a constant times the state is called an eigenvalue equation. In this case, the constant $a_n$ in (2.125) is called the eigenvalue and the state $|a_n\rangle$ [or $\langle a_n|$ in (2.126)] is called the eigenstate.

We will now show that the eigenvalues of a Hermitian operator are real. Taking the inner product of the eigenvalue equation (2.125) with the bra $\langle a_k|$, we obtain

$$\langle a_k|\hat{A}|a_n\rangle = a_n\langle a_k|a_n\rangle \tag{2.127}$$

Taking advantage of (2.126), this equation becomes

$$a_k^*\langle a_k|a_n\rangle = a_n\langle a_k|a_n\rangle \tag{2.128a}$$

or

$$(a_k^* - a_n)\langle a_k|a_n\rangle = 0 \tag{2.128b}$$

Note that if we take $k = n$, we find

$$(a_n^* - a_n)\langle a_n|a_n\rangle = 0 \tag{2.129}$$

and therefore the eigenvalues of a Hermitian operator are real ($a_n^* = a_n$), a necessary condition if these are to be the values that we obtain for a measurement. Moreover, (2.128b) shows that

$$\langle a_k|a_n\rangle = 0 \qquad a_k \neq a_n \tag{2.130}$$

as we argued in Chapter 1 must be true based on the fact that $\langle a_k|a_n\rangle$ is the amplitude to obtain $a_k$ for a particle in the state $|a_n\rangle$. This shows that the eigenstates of a Hermitian operator corresponding to distinct eigenvalues are orthogonal. Thus our association of Hermitian operators with observables such as angular momentum forms a nice, self-consistent physical picture.

We also see that we can express the expectation value $\langle A \rangle$ of the observable $A$ in terms of the operator $\hat{A}$ as

$$\langle A \rangle = \langle \psi | \hat{A} | \psi \rangle \tag{2.131}$$

For simplicity, let's consider the case where there are two eigenstates $|a_1\rangle$ and $|a_2\rangle$ with $a_1 \neq a_2$, as is the case for spin $\frac{1}{2}$. Since a general state can be written as

$$|\psi\rangle = c_1|a_1\rangle + c_2|a_2\rangle \tag{2.132}$$

then

$$
\begin{aligned}
\langle \psi | \hat{A} | \psi \rangle &= \left( c_1^* \langle a_1| + c_2^* \langle a_2| \right) \hat{A} \left( c_1 |a_1\rangle + c_2 |a_2\rangle \right) \\
&= \left( c_1^* \langle a_1| + c_2^* \langle a_2| \right) \left( c_1 a_1 |a_1\rangle + c_2 a_2 |a_2\rangle \right) \\
&= |c_1|^2 a_1 + |c_2|^2 a_2 \\
&= \langle A \rangle
\end{aligned}
\tag{2.133}
$$

where the last step follows since the penultimate line of (2.133) is just the sum of the eigenvalues weighted by the probability of obtaining each of those values, which is just what we mean by the expectation value.

Also note that, as in (1.39), (2.132) can be expressed in the form

$$|\psi\rangle = |a_1\rangle\langle a_1|\psi\rangle + |a_2\rangle\langle a_2|\psi\rangle \tag{2.134}$$

This suggests that we can write the identity operator in the form

$$|a_1\rangle\langle a_1| + |a_2\rangle\langle a_2| = 1 \tag{2.135}$$

which is also known as a completeness relation, because it is equivalent to saying that we can express an arbitrary state $|\psi\rangle$ as a superposition of the states $|a_1\rangle$ and $|a_2\rangle$, as shown in (2.134). The identity operator can be decomposed into projection operators

$$\hat{P}_1 = |a_1\rangle\langle a_1| \qquad \text{and} \qquad \hat{P}_2 = |a_2\rangle\langle a_2| \tag{2.136}$$

that project out of the state $|\psi\rangle$ the component of the vector in the direction of the eigenvector. For example,

$$\hat{P}_1|\psi\rangle = |a_1\rangle\langle a_1|\psi\rangle \tag{2.137}$$

If we insert the identity operator (2.135) between the ket and the bra in the amplitude $\langle \varphi | \psi \rangle$, we obtain

$$\langle \varphi | \psi \rangle = \langle \varphi | a_1 \rangle\langle a_1 | \psi \rangle + \langle \varphi | a_2 \rangle\langle a_2 | \psi \rangle \tag{2.138}$$

Thus, if a particle is in the state $|\psi\rangle$ and a measurement is carried out, the probability of finding the particle in the state $|\varphi\rangle$ can be written as

$$|\langle \varphi | \psi \rangle|^2 = |\langle \varphi | a_1 \rangle\langle a_1 | \psi \rangle + \langle \varphi | a_2 \rangle\langle a_2 | \psi \rangle|^2 \tag{2.139}$$

Note that the amplitudes $\langle \varphi | a_1 \rangle\langle a_1 | \psi \rangle$ and $\langle \varphi | a_2 \rangle\langle a_2 | \psi \rangle$ can interfere with each other. Equation (2.139) presumes that no measurement of the observable $A$ has actually taken place. If we were to actually insert a device that measured the observable $A$ for the state $|\psi\rangle$, we would then find the probability to obtain the state $|\varphi\rangle$ given by

$$|\langle\varphi|a_1\rangle|^2|\langle a_1|\psi\rangle|^2 + |\langle\varphi|a_2\rangle|^2|\langle a_2|\psi\rangle|^2 \tag{2.140}$$

which is just the sum of the probabilities of finding $|\psi\rangle$ in the states $|a_1\rangle$ and $|a_2\rangle$ times the probability that each of these states is in the state $|\varphi\rangle$. Equations (2.139) and (2.140) illustrate one of the fundamental principles of quantum mechanics: When we do not make a measurement that permits us to distinguish the intermediate states $|a_1\rangle$ and $|a_2\rangle$, we add the amplitudes and then square to get the probability, while if we do make a measurement that can distinguish which of the states $|a_1\rangle$ and $|a_2\rangle$ the particle is in, we add the individual probabilities, not the amplitudes. For a specific example, see the discussion at the end of Section 2.3.

A convenient shorthand notation is to use the eigenstates $|a_1\rangle$ and $|a_2\rangle$ as a basis and represent a ket such as (2.132) by a column vector

$$|\psi\rangle \xrightarrow[|a_1\rangle-|a_2\rangle \text{ basis}]{} \begin{pmatrix} c_1 \\ c_2 \end{pmatrix} = \begin{pmatrix} \langle a_1|\psi\rangle \\ \langle a_2|\psi\rangle \end{pmatrix} \tag{2.141}$$

a bra by a row vector

$$\langle\psi| \xrightarrow[\langle a_1|-\langle a_2| \text{ basis}]{} (c_1^*, c_2^*) = (\langle\psi|a_1\rangle, \langle\psi|a_2\rangle) \tag{2.142}$$

and an operator by a matrix

$$\hat{B} \xrightarrow[|a_1\rangle-|a_2\rangle \text{ basis}]{} \begin{pmatrix} \langle a_1|\hat{B}|a_1\rangle & \langle a_1|\hat{B}|a_2\rangle \\ \langle a_2|\hat{B}|a_1\rangle & \langle a_2|\hat{B}|a_2\rangle \end{pmatrix} \tag{2.143}$$

In this notation, an equation such as

$$\hat{B}|\psi\rangle = |\varphi\rangle \tag{2.144}$$

becomes

$$\begin{pmatrix} \langle a_1|\hat{B}|a_1\rangle & \langle a_1|\hat{B}|a_2\rangle \\ \langle a_2|\hat{B}|a_1\rangle & \langle a_2|\hat{B}|a_2\rangle \end{pmatrix} \begin{pmatrix} \langle a_1|\psi\rangle \\ \langle a_2|\psi\rangle \end{pmatrix} = \begin{pmatrix} \langle a_1|\varphi\rangle \\ \langle a_2|\varphi\rangle \end{pmatrix} \tag{2.145}$$

Knowing the matrix elements $\langle a_i|\hat{B}|a_j\rangle$ permits us to evaluate the action of the operator $\hat{B}$ on any state $|\psi\rangle$. As an example, we can use matrix mechanics to evaluate the expectation value of $B$ in the state $|\psi\rangle$:

$$\langle B \rangle = \langle\psi|\hat{B}|\psi\rangle = (\langle\psi|a_1\rangle, \langle\psi|a_2\rangle) \begin{pmatrix} \langle a_1|\hat{B}|a_1\rangle & \langle a_1|\hat{B}|a_2\rangle \\ \langle a_2|\hat{B}|a_1\rangle & \langle a_2|\hat{B}|a_2\rangle \end{pmatrix} \begin{pmatrix} \langle a_1|\psi\rangle \\ \langle a_2|\psi\rangle \end{pmatrix} \tag{2.146}$$

Finally, note that if basis states are the eigenstates of the operator, the matrix representation is diagonal with the eigenvalues forming the diagonal matrix elements:[15]

$$\hat{A} \xrightarrow[|a_1\rangle\text{-}|a_2\rangle \text{ basis}]{} \begin{pmatrix} a_1 & 0 \\ 0 & a_2 \end{pmatrix} \tag{2.147}$$

All of the results (2.132) through (2.147) can be extended in a straight-forward fashion to larger dimensional bases, as introduced in Section 1.6. For example, the identity operator is given by $\sum_n |a_n\rangle\langle a_n|$ in the more general case.

## PROBLEMS

**2.1.** Show that

$$\lim_{N \to \infty} \left(1 + \frac{x}{N}\right)^N = e^x$$

by comparing the Taylor series expansions for the two functions.

**2.2.** Use Dirac notation (the properties of kets, bras, and inner products) directly without explicitly using matrix representations to establish that the projection operator $\hat{P}_+$ is Hermitian. Use the fact that $\hat{P}_+^2 = \hat{P}_+$ to establish that the eigenvalues of the projection operator are 1 and 0.

**2.3.** Determine the matrix representation of the rotation operator $\hat{R}(\phi\mathbf{k})$ using the states $|+\mathbf{z}\rangle$ and $|-\mathbf{z}\rangle$ as a basis. Using your matrix representation, verify that the rotation operator is unitary, that is, that it satisfies $\hat{R}^\dagger(\phi\mathbf{k})\hat{R}(\phi\mathbf{k}) = 1$.

**2.4.** Determine the column vectors representing the states $|+\mathbf{x}\rangle$ and $|-\mathbf{x}\rangle$ using the states $|+\mathbf{y}\rangle$ and $|-\mathbf{y}\rangle$ as a basis.

**2.5.** What is the matrix representation of $\hat{J}_z$ using the states $|+\mathbf{y}\rangle$ and $|-\mathbf{y}\rangle$ as a basis? Use this representation to evaluate the expectation value of $S_z$ for a collection of particles each in the state $|-\mathbf{y}\rangle$.

**2.6.** Evaluate $\hat{R}(\theta\mathbf{j})|+\mathbf{z}\rangle$, where $\hat{R}(\theta\mathbf{j}) = e^{-i\hat{J}_y\theta/\hbar}$ is the operator that rotates kets counterclockwise by angle $\theta$ about the y axis. Show that $\hat{R}(\frac{\pi}{2}\mathbf{j})|+\mathbf{z}\rangle = |+\mathbf{x}\rangle$. *Suggestion:* Express the ket $|+\mathbf{z}\rangle$ as a superposition of the kets $|+\mathbf{y}\rangle$ and $|-\mathbf{y}\rangle$ and take advantage of the fact that $\hat{J}_y|\pm\mathbf{y}\rangle = (\pm\hbar/2)|\pm\mathbf{y}\rangle$; then switch back to the $|+\mathbf{z}\rangle$-$|-\mathbf{z}\rangle$ basis.

**2.7.** Work out the matrix representations of the projection operators $\hat{P}_+ = |+\mathbf{z}\rangle\langle+\mathbf{z}|$ and $\hat{P}_- = |-\mathbf{z}\rangle\langle-\mathbf{z}|$ using the states $|+\mathbf{y}\rangle$ and $|-\mathbf{y}\rangle$ of a spin-$\frac{1}{2}$ particle as a basis. Check that the results (2.51) and (2.52) are satisfied using these matrix representations.

**2.8.** A photon polarization state for a photon propagating in the z direction is given by

$$|\psi\rangle = \sqrt{\frac{2}{3}}|\mathbf{x}\rangle + \frac{i}{\sqrt{3}}|\mathbf{y}\rangle$$

---

[15] In general, there are an infinite number of sets of basis states that may be used to form representations in matrix mechanics. For example, in addition to the states $|\pm\mathbf{z}\rangle$, the states $|\pm\mathbf{x}\rangle$ can be used as a basis to represent states and operators for spin-$\frac{1}{2}$ particles. However, since $|\pm\mathbf{x}\rangle$ are not eigenstates of $\hat{J}_z$, the matrix representation of this operator using these states as a basis is not diagonal, as (2.96) shows.

(*a*) What is the probability that a photon in this state will pass through a perfect polarizer with its transmission axis oriented in the *y* direction?

(*b*) What is the probability that a photon in this state will pass through a perfect polarizer with its transmission axis *y'* making an angle $\phi$ with the *y* axis?

(*c*) A beam carrying *N* photons per second, each in the state $|\psi\rangle$, is totally absorbed by a black disk with its normal to the surface in the *z* direction. How large is the torque exerted on the disk? In which direction does the disk rotate? *Reminder:* The photon states $|R\rangle$ and $|L\rangle$ each carry a unit $\hbar$ of angular momentum parallel and antiparallel, respectively, to the direction of propagation of the photons.

(*d*) How would the result for each of these questions differ if the polarization state were

$$|\psi'\rangle = \sqrt{\frac{2}{3}}|x\rangle + \frac{1}{\sqrt{3}}|y\rangle$$

that is, the "*i*" in the state $|\psi\rangle$ is absent?

**2.9.** A system of *N* ideal linear polarizers is arranged in sequence, as shown in Fig. 2.13. The transmission axis of the first polarizer makes an angle of $\phi/N$ with the *y* axis. The transmission axis of every other polarizer makes an angle of $\phi/N$ with respect to the axis of the preceding one. Thus, the transmission axis of the final polarizer makes an angle $\phi$ with the *y* axis. A beam of *y*-polarized photons is incident on the first polarizer.

(*a*) What is the probability that an incident photon is transmitted by the array?

(*b*) Evaluate the probability of transmission in the limit of large *N*.

(*c*) Consider the special case with the angle $\phi = 90°$. Explain why your result is not in conflict with the fact that $\langle x|y\rangle = 0$.[16]

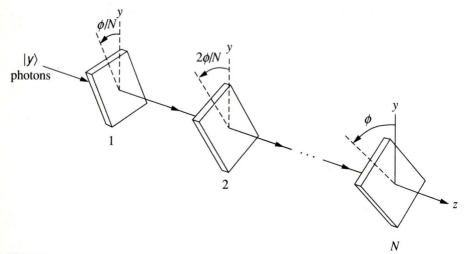

**FIGURE 2.13**

---

[16] A nice discussion of the quantum state using photon polarization states as a basis is given by A. P. French and E. F. Taylor, *An Introduction to Quantum Physics*, Norton, New York, 1978, Chapters 6 and 7. Problem 2.9 is adapted from this source.

**2.10.** (*a*) Determine a $2 \times 2$ matrix $\mathbb{S}$ that can be used to transform a column vector representing a photon polarization state using the linear polarization vectors $|x\rangle$ and $|y\rangle$ as a basis to one using the circular polarization vectors $|R\rangle$ and $|L\rangle$ as a basis.

(*b*) Using matrix multiplication, verify explicitly that the matrix $\mathbb{S}$ that you found in (*a*) is unitary.

**2.11.** Determine the matrix representation of the angular momentum operator $\hat{J}_z$ using both the circular polarization vectors $|R\rangle$ and $|L\rangle$ and the linear polarization vectors $|x\rangle$ and $|y\rangle$ as a basis. Check that the operator is Hermitian in both matrix representations.

**2.12.** Use both the matrix representations of the angular momentum operator $\hat{J}_z$ from Problem 2.11 to determine the expectation value of the angular momentum for the photon state $a|R\rangle + b|L\rangle$.

**2.13.** Use the matrix representation of the rotation operator $\hat{R}(\phi\mathbf{k})$ in the $|x\rangle$-$|y\rangle$ basis as given in (2.110) to establish that the photon circular polarization states (2.111), expressed as column vectors in the $|x\rangle$-$|y\rangle$ basis, are eigenstates of the rotation operator with the eigenvalues that appear in (2.113) and (2.116).

**2.14.** Construct projection operators out of bras and kets for $x$-polarized and $y$-polarized photons. Give physical examples of devices that can serve as these projection operators. Use (*a*) the properties of bras and kets and (*b*) the properties of the physical devices to show that the projection operators satisfy $\hat{P}_x^2 = \hat{P}_x$, $\hat{P}_y^2 = \hat{P}_y$, and $\hat{P}_x\hat{P}_y = \hat{P}_y\hat{P}_x = 0$.

**2.15.** Show that $\hat{J}_z = \hbar|R\rangle\langle R| - \hbar|L\rangle\langle L|$ for photons.

**2.16.** What is the probability that a right-circularly polarized photon will pass through a linear polarizer with its transmission axis along the $x'$ axis, which makes an angle $\phi$ with the $x$ axis?

**2.17.** Linearly polarized light of wavelength 5890 Å is incident normally on a birefringent crystal that has its optic axis parallel to the face of the crystal, along the $x$ axis. If the incident light is polarized at an angle of $45°$ to the $x$ and $y$ axes, what is the probability that the photons exiting a crystal of thickness 100.0 microns will be right-circularly polarized? The index of refraction for light of this wavelength polarized along $y$ (perpendicular to the optic axis) is 1.66 and the index of refraction for light polarized along $x$ (parallel to the optic axis) is 1.49.

**2.18.** A beam of linearly polarized light is incident on a quarter-wave plate with its direction of polarization oriented at $30°$ to the optic axis. Subsequently, the beam is absorbed by a black disk. Determine the rate angular momentum is transferred to the disk, assuming the beam carries $N$ photons per second.

**2.19.** (*a*) Show that if the states $|a_n\rangle$ form an orthonormal basis, so do the states $\hat{U}|a_n\rangle$, provided $\hat{U}$ is unitary.

(*b*) Show that the eigenvalues of a unitary operator can be written as $e^{i\theta}$.

# CHAPTER
# 3

# ANGULAR
# MOMENTUM

In this chapter we will see that the order in which we carry out rotations about different axes matters. Therefore, the operators that generate rotations about these different axes do not commute, leading to commutation relations that may be viewed as the defining relations for the angular momentum operators. We will use these commutation relations to determine the angular momentum eigenstates and eigenvalues. We will also see that the spin-$\frac{1}{2}$ states that have occupied much of our attention so far appear as a particular case of this general analysis of angular momentum in quantum mechanics.

## 3.1 ROTATIONS DO NOT COMMUTE AND NEITHER DO THE GENERATORS

Take your textbook and set up a convenient coordinate system centered on the book, as shown in Fig. 3.1. Rotate your text by $90°$ about the $x$ axis and then rotate it by $90°$ about the $y$ axis. Either note carefully the orientation of the text or, better still, borrow a copy of the text from a friend and perform the two rotations again, but this time first rotate about the $y$ axis by $90°$ and then about the $x$ axis by $90°$. The orientations of the two texts are different. Clearly, the order in which you carry out the rotations matters. We say that finite rotations about different axes do not commute.

In Section 2.7 we determined the matrix $\mathbb{S}$ that transforms a basis set of polarization states to another set that are related to the initial set by a rotation by angle $\phi$ counterclockwise about the $z$ axis. The matrix (2.109) is also the matrix that is used to rotate the components of an ordinary vector in the $x$-$y$ plane. Our familiarity with this example makes it a good one to use to analyze in more detail what happens when we make rotations about different axes. Rather than working directly with the actual operators that perform these rotations in our quantum mechanical vector space, we will initially work in a specific representation and

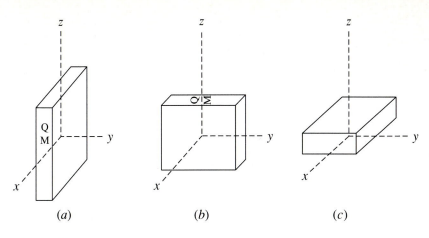

**FIGURE 3.1**

Noncommutativity of rotations. A book, shown in (a), is rotated in (b) by 90° around the x axis, then 90° about the y axis; in (c) the order of the rotations is reversed.

infer from the behavior that we see some fundamental properties about the operators themselves. The results we are interested in depend on the three-dimensional
structure of space and are properties that manifest themselves in all nontrivial
representations.

Let's consider an ordinary three-dimensional vector **A** and a vector **A′** that is
obtained by rotating **A** counterclockwise by an angle $\phi$ about the $z$ axis. How are
the components of **A** and **A′** related to each other? Denoting by $\theta$ the angle between the projection of **A** in the $x$-$y$ plane and the $x$ axis, as in Fig. 3.2a, we have

$$A'_x = \sqrt{A_x^2 + A_y^2}\,\cos(\phi + \theta) = \sqrt{A_x^2 + A_y^2}(\cos\phi\cos\theta - \sin\phi\sin\theta)$$

$$= A_x\cos\phi - A_y\sin\phi \tag{3.1a}$$

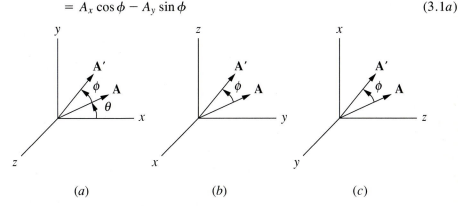

**FIGURE 3.2**

Rotating a vector **A** into a vector **A′** by angle $\phi$ counterclockwise about (a) the $z$ axis, (b) the $x$
axis, and (c) the $y$ axis. For simplicity, only the components of the vector in the plane perpendicular
to the axis of rotation are shown.

$$A'_y = \sqrt{A_x^2 + A_y^2} \sin(\theta + \phi) = \sqrt{A_x^2 + A_y^2}(\sin\phi\cos\theta + \sin\theta\cos\phi)$$

$$= A_x \sin\phi + A_y \cos\phi \tag{3.1b}$$

$$A'_z = A_z \tag{3.1c}$$

or, in matrix form,

$$\begin{pmatrix} A'_x \\ A'_y \\ A'_z \end{pmatrix} = \begin{pmatrix} \cos\phi & -\sin\phi & 0 \\ \sin\phi & \cos\phi & 0 \\ 0 & 0 & 1 \end{pmatrix} \begin{pmatrix} A_x \\ A_y \\ A_z \end{pmatrix} \tag{3.2}$$

Thus the matrix that rotates the vector by angle $\phi$ counterclockwise about the $z$ axis is given by

$$\mathbb{S}(\phi\mathbf{k}) = \begin{pmatrix} \cos\phi & -\sin\phi & 0 \\ \sin\phi & \cos\phi & 0 \\ 0 & 0 & 1 \end{pmatrix} \tag{3.3}$$

The $2 \times 2$ matrix in the upper left-hand corner is just the matrix (2.109). Because we are dealing here with a vector that has three components, the rotation matrix is a $3 \times 3$ matrix instead of the $2 \times 2$ matrix that we found for rotating polarization states. The additional elements in this matrix (3.3) simply show that the component of the vector in the $z$ direction is unaffected by a rotation about the $z$ axis.

We consider the special case where the angle is a small angle $\Delta\phi$ and retain terms in the Taylor series expansions for $\sin\Delta\phi$ and $\cos\Delta\phi$ through second order. It is necessary to work to at least this order to see the noncommutivity of the rotations. Thus

$$\mathbb{S}(\Delta\phi\mathbf{k}) = \begin{pmatrix} 1 - \Delta\phi^2/2 & -\Delta\phi & 0 \\ \Delta\phi & 1 - \Delta\phi^2/2 & 0 \\ 0 & 0 & 1 \end{pmatrix} \tag{3.4}$$

From Fig. 3.2$b$ we see that for a rotation about the $x$ axis by angle $\phi$, the matrix for the rotation can be obtained from the matrix (3.3) by letting $x \to y$, $y \to z$, and $z \to x$, that is, by a cyclic substitution. Therefore, the rotation matrix is

$$\mathbb{S}(\phi\mathbf{i}) = \begin{pmatrix} 1 & 0 & 0 \\ 0 & \cos\phi & -\sin\phi \\ 0 & \sin\phi & \cos\phi \end{pmatrix} \tag{3.5}$$

and consequently

$$\mathbb{S}(\Delta\phi\mathbf{i}) = \begin{pmatrix} 1 & 0 & 0 \\ 0 & 1 - \Delta\phi^2/2 & -\Delta\phi \\ 0 & \Delta\phi & 1 - \Delta\phi^2/2 \end{pmatrix} \tag{3.6}$$

Finally, we can obtain the matrix for a rotation about the $y$ axis from the matrix for a rotation about the $x$ axis by another cyclic substitution (see Fig. 3.2$c$). Thus

$$\mathbb{S}(\Delta\phi\mathbf{j}) = \begin{pmatrix} 1 - \Delta\phi^2/2 & 0 & \Delta\phi \\ 0 & 1 & 0 \\ -\Delta\phi & 0 & 1 - \Delta\phi^2/2 \end{pmatrix} \tag{3.7}$$

We now consider a rotation by $\Delta\phi$ about the $y$ axis followed by a rotation by the same angle about the $x$ axis. We subtract from it a rotation about the $x$ axis followed by a rotation about the $y$ axis. Multiplying the matrices (3.6) and (3.7), we obtain

$$\mathbb{S}(\Delta\phi\mathbf{i})\mathbb{S}(\Delta\phi\mathbf{j}) - \mathbb{S}(\Delta\phi\mathbf{j})\mathbb{S}(\Delta\phi\mathbf{i}) = \begin{pmatrix} 0 & -\Delta\phi^2 & 0 \\ \Delta\phi^2 & 0 & 0 \\ 0 & 0 & 0 \end{pmatrix}$$

$$= \mathbb{S}(\Delta\phi^2\mathbf{k}) - \mathbb{1} \tag{3.8}$$

where in the last step we have taken advantage of the explicit form of the matrix (3.4) when the rotation angle is $\Delta\phi^2$ and terms through order $\Delta\phi^2$ are retained.

From Section 2.5 we know that these S-matrices are the matrix representations of the rotation operators. For example, the matrix (3.3) is the representation of the rotation operator $\hat{R}(\phi\mathbf{k})$ in a particular basis.[1] Equation (3.8) shows that when we retain terms through second order in $\Delta\phi$, the operators themselves do not commute. Recall from (2.32) that the operator that rotates states by angle $\phi$ about the $z$ axis is

$$\hat{R}(\phi\mathbf{k}) = e^{-i\hat{J}_z\phi/\hbar} \tag{3.9}$$

where $\hat{J}_z$ is the generator of rotations. We can think of this as a special case of the more general rotation operator

$$\hat{R}(\phi\mathbf{n}) = e^{-i\hat{\mathbf{J}}\cdot\mathbf{n}\phi/\hbar} \tag{3.10}$$

that rotates states by angle $\phi$ about the axis defined by the unit vector $\mathbf{n}$. Thus the operators that rotate states by angle $\phi$ about the $x$ axis and the $y$ axis are given by

$$\hat{R}(\phi\mathbf{i}) = e^{-i\hat{J}_x\phi/\hbar} \quad \text{and} \quad \hat{R}(\phi\mathbf{j}) = e^{-i\hat{J}_y\phi/\hbar} \tag{3.11}$$

with generators $\hat{J}_x$ and $\hat{J}_y$, respectively. Thus, if we take the angle of rotation to be the small angle $\Delta\phi$ and expand the rotation operators through second order in $\Delta\phi$, (3.8) tells us that

---

[1] Although we have phrased our discussion so far in terms of how ordinary vectors change under rotations, we are effectively using spin-1 states like the ones we saw in Section 2.7 as a basis, but with three states instead of just the two states that are necessary to describe photon polarization. We argued in that section that the way the photon polarization states changed under rotation told us that photons are spin-1 particles. If photons traveling in the $z$ direction were to have a $|z\rangle$ polarization state as well as $|x\rangle$ and $|y\rangle$, this $|z\rangle$ polarization state would not be changed by performing a rotation about the $z$ axis, and the matrix representation of the rotation operator $\hat{R}(\phi\mathbf{k})$ using the $|x\rangle$, $|y\rangle$, and $|z\rangle$ states as a basis would look like (3.3) instead of (2.110). Later in this chapter we will see how spin-1 states do form a three-dimensional basis, when we determine the angular momentum spectrum. Again, particles like photons that move at $c$ require special treatment.

$$\left\{1 - \frac{i\hat{J}_x\Delta\phi}{\hbar} - \frac{1}{2}\left(\frac{\hat{J}_x\Delta\phi}{\hbar}\right)^2\right\}\left\{1 - \frac{i\hat{J}_y\Delta\phi}{\hbar} - \frac{1}{2}\left(\frac{\hat{J}_y\Delta\phi}{\hbar}\right)^2\right\}$$

$$- \left\{1 - \frac{i\hat{J}_y\Delta\phi}{\hbar} - \frac{1}{2}\left(\frac{\hat{J}_y\Delta\phi}{\hbar}\right)^2\right\}\left\{1 - \frac{i\hat{J}_x\Delta\phi}{\hbar} - \frac{1}{2}\left(\frac{\hat{J}_x\Delta\phi}{\hbar}\right)^2\right\}$$

$$= \left(1 - \frac{i\hat{J}_z\Delta\phi^2}{\hbar}\right) - 1 \tag{3.12}$$

The lowest-order nonvanishing terms involve $\Delta\phi^2$. Equating these terms, we obtain

$$\hat{J}_x\hat{J}_y - \hat{J}_y\hat{J}_x = i\hbar\hat{J}_z \tag{3.13}$$

or

$$\left[\hat{J}_x, \hat{J}_y\right] = i\hbar\hat{J}z \tag{3.14a}$$

where the left-hand side of the equation is called the *commutator* of the two operators $\hat{J}_x$ and $\hat{J}_y$. The commutator of two operators is just the product of the two operators subtracted from the product of the two operators with the order of the operators reversed. Notice how Planck's constant enters on the right-hand side of (3.14a).

If we were to repeat this whole procedure for rotations about the $y$ and $z$ axes and for rotations about the $z$ and $x$ axes, we would obtain two other commutation relations related to (3.14a) by the cyclic permutation $x \rightarrow y$, $y \rightarrow z$, and $z \rightarrow x$:

$$\left[\hat{J}_y, \hat{J}_z\right] = i\hbar\hat{J}_x \tag{3.14b}$$

and

$$\left[\hat{J}_z, \hat{J}_x\right] = i\hbar\hat{J}_y \tag{3.14c}$$

It would be difficult to overemphasize the importance of these commutation relations. In Section 3.3 we will see that they alone are sufficient to determine the eigenstates and the eigenvalues of the angular momentum operators. So far, our arguments to establish that these generators of rotations should be identified with the angular momentum operators are probably at best suggestive. The proof is in the results and the comparison with experiment.

Later we will see that the *orbital* angular momentum operators

$$\hat{\mathbf{L}} = \hat{\mathbf{r}} \times \hat{\mathbf{p}} \tag{3.15}$$

also obey these same commutation relations, that is, for example,

$$\left[\hat{L}_x, \hat{L}_y\right] = i\hbar\hat{L}_z \tag{3.16}$$

However, we have not introduced angular momentum operators through (3.15), but rather simply as the generators of rotations. Although this approach may seem more abstract and initially less physical, it is also more general and, in fact, essential. In Chapter 9 we will see that the eigenvalues of orbital angular momentum, as defined by (3.15), do *not* include the half-integral values that characterize spin-$\frac{1}{2}$ particles such as electrons, protons, neutrons, and neutrinos.

## 3.2 COMMUTING OPERATORS

The commutation relations of the generators of rotations show that the generators of rotations about different axes do not commute with each other. As we saw in Chapter 2, these generators are Hermitian operators. Before turning our attention toward solving the angular momentum eigenvalue problem, we need to ask what happens when two operators *do* commute. Consider two such linear Hermitian operators $\hat{A}$ and $\hat{B}$ that satisfy

$$[\hat{A}, \hat{B}] \equiv \hat{A}\hat{B} - \hat{B}\hat{A} = 0 \tag{3.17}$$

Suppose there exists only a single state $|a\rangle$ that is an eigenstate of $\hat{A}$ with eigenvalue $a$:

$$\hat{A}|a\rangle = a|a\rangle \tag{3.18}$$

If we apply the operator $\hat{B}$ to (3.18), we obtain

$$\hat{B}\hat{A}|a\rangle = \hat{B}a|a\rangle \tag{3.19}$$

On the left-hand side we take advantage of (3.17) and on the right-hand side we take advantage of the fact that $\hat{B}$ is a linear operator to write

$$\hat{A}\hat{B}|a\rangle = a\hat{B}|a\rangle \tag{3.20a}$$

or

$$\hat{A}(\hat{B}|a\rangle) = a(\hat{B}|a\rangle) \tag{3.20b}$$

where we have inserted the parentheses to isolate the state $\hat{B}|a\rangle$ on both sides. Equation (3.20) says that this state is an eigenstate of the operator $\hat{A}$ with eigenvalue $a$. Since we have presumed there is only one such state, we conclude that

$$\hat{B}|a\rangle = b|a\rangle \tag{3.21}$$

where $b$ is a constant, since if $|a\rangle$ satisfies (3.18), so does $b|a\rangle$ for any $b$. But (3.21) says that $|a\rangle$ is an eigenstate of $\hat{B}$ as well with eigenvalue $b$. Therefore, we can relabel the state $|a\rangle$ as $|a, b\rangle$ to show both of the eigenvalues and say that $\hat{A}$ and $\hat{B}$ have the eigenstate $|a, b\rangle$ in common. An example of a state that can be labeled by two eigenvalues is the state $|E, p\rangle$ of a *free* particle in one dimension, where $E$ is the energy and $p$ is the momentum of the particle.

   If there is more than one eigenstate of the operator $\hat{A}$ with eigenvalue $a$, we say that there is *degeneracy*. Our proof has established that each eigenstate of $\hat{A}$ is also an eigenstate of $\hat{B}$ for those states that are not degenerate. If there is

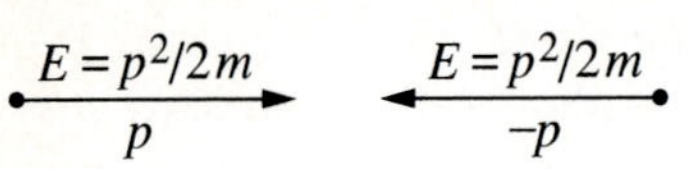

$$\xrightarrow{\quad E = p^2/2m \quad}$$
$$p$$

$$\xleftarrow{\quad E = p^2/2m \quad}$$
$$-p$$

**FIGURE 3.3**
A free particle with momentum $p$ has the same energy as one with momentum $-p$.

degeneracy, one can always find linear combinations of the degenerate eigenstates of $\hat{A}$ that are eigenstates of the Hermitian operator $\hat{B}$. Thus two Hermitian operators that commute have a complete set of eigenstates in common. This result follows from the fundamental spectral theorem of linear algebra. We will not prove it here, but we will have a number of opportunities in later chapters to verify that it holds in special cases. In fact, the example of the one-dimensional free particle can serve as an illustration, since for a particular energy $E = p^2/2m$ there is two-fold degeneracy: the states $|E, p\rangle$ and $|E, -p\rangle$ have the same energy but momenta $p$ and $-p$, respectively, corresponding to a particle moving to the right or the left (see Fig. 3.3). Note that you can certainly form states that are superpositions of the states $|E, p\rangle$ and $|E, -p\rangle$ (such as standing waves), so states with a definite energy need not have a definite momentum.

## 3.3 THE EIGENVALUES AND EIGENSTATES OF ANGULAR MOMENTUM

Although the commutation relations (3.14) show us that the generators of rotations about different axes do not commute with each other, the operator

$$\hat{\mathbf{J}}^2 = \hat{\mathbf{J}} \cdot \hat{\mathbf{J}} = \hat{J}_x^2 + \hat{J}_y^2 + \hat{J}_z^2 \tag{3.22}$$

does commute with each of the generators. In order to verify this, we choose $\hat{J}_z$, the generator of rotations about the $z$ axis, and use the identity (see Problem 3.1)

$$[\hat{A}, \hat{B}\hat{C}] = \hat{B}[\hat{A}, \hat{C}] + [\hat{A}, \hat{B}]\hat{C} \tag{3.23}$$

to obtain

$$[\hat{J}_z, \hat{J}_x^2 + \hat{J}_y^2 + \hat{J}_z^2] = [\hat{J}_z, \hat{J}_x^2] + [\hat{J}_z, \hat{J}_y^2]$$

$$= \hat{J}_x[\hat{J}_z, \hat{J}_x] + [\hat{J}_z, \hat{J}_x]\hat{J}_x + \hat{J}_y[\hat{J}_z, \hat{J}_y] + [\hat{J}_z, \hat{J}_y]\hat{J}_y$$

$$= i\hbar(\hat{J}_x\hat{J}_y + \hat{J}_y\hat{J}_x - \hat{J}_y\hat{J}_x - \hat{J}_x\hat{J}_y) = 0 \tag{3.24}$$

Because the operator $\hat{\mathbf{J}}^2$ commutes with $\hat{J}_z$, they have simultaneous eigenstates in common. We label the kets $|\lambda, m\rangle$, where

$$\hat{\mathbf{J}}^2|\lambda, m\rangle = \lambda\hbar^2|\lambda, m\rangle \tag{3.25a}$$

$$\hat{J}_z|\lambda, m\rangle = m\hbar|\lambda, m\rangle \tag{3.25b}$$

We have explicitly included the dimensions of the operators in the factors of $\hbar$ so that $\lambda$ and $m$ are dimensionless. Thus $|\lambda, m\rangle$ is a state for which a measurement of the $z$ component of the angular momentum yields the value $m\hbar$ and the magnitude squared of the angular momentum is $\lambda\hbar^2$.

We can see that $\lambda \geq 0$, as we would expect physically since $\lambda$ specifies the magnitude squared of the angular momentum in the state $|\lambda, m\rangle$. Consider

$$\langle \lambda, m|\hat{\mathbf{J}}^2|\lambda, m\rangle = \lambda\hbar^2\langle \lambda, m|\lambda, m\rangle \tag{3.26}$$

Like all physical states, the eigenstates satisfy $\langle \lambda, m|\lambda, m\rangle = 1$. A typical term in the left-hand side of (3.26) is of the form

$$\langle \lambda, m|\hat{J}_x^2|\lambda, m\rangle = \langle \psi|\psi\rangle \tag{3.27}$$

where we have defined $\hat{J}_x|\lambda, m\rangle = |\psi\rangle$, and $\langle \psi| = \langle \lambda, m|\hat{J}_x$ since $\hat{J}_x$ is Hermitian. Although the ket $|\psi\rangle$ is not normalized, we can always write it as $|\psi\rangle = c|\varphi\rangle$, where $c$ is a complex constant (that must have the dimensions of $\hbar$) and $|\varphi\rangle$ is a physical state satisfying $\langle \varphi|\varphi\rangle = 1$. In other words, the action of the operator $\hat{J}_x$ on a ket vector must yield another ket vector that belongs to the vector space.[2] Since $\langle \psi| = c^*\langle \varphi|$, we see that $\langle \psi|\psi\rangle = c^*c\langle \varphi|\varphi\rangle \geq 0$, where the equality would hold if $c = 0$. Our argument that (3.27) is positive semidefinite holds for each of the three pieces [see the form (3.22) of $\hat{\mathbf{J}}^2$] on the left-hand side of (3.26), and therefore $\lambda \geq 0$.

## An Example: Spin 1

To illustrate what we have discovered so far and suggest the next step, let's take the specific example involving the following three $3 \times 3$ matrices :

$$\hat{J}_x \rightarrow \frac{\hbar}{\sqrt{2}}\begin{pmatrix} 0 & 1 & 0 \\ 1 & 0 & 1 \\ 0 & 1 & 0 \end{pmatrix} \quad \hat{J}_y \rightarrow \frac{\hbar}{\sqrt{2}}\begin{pmatrix} 0 & -i & 0 \\ i & 0 & -i \\ 0 & i & 0 \end{pmatrix} \quad \hat{J}_z \rightarrow \hbar\begin{pmatrix} 1 & 0 & 0 \\ 0 & 0 & 0 \\ 0 & 0 & -1 \end{pmatrix}$$
$$\tag{3.28}$$

For now, don't worry about how we have obtained these matrices. Later in this chapter we will see how we can deduce the form of these matrices (see Problem 3.14). In the meantime, let's see what we can learn from the matrices themselves.

To begin, how can we be sure that these three matrices really *represent* angular momentum operators? Following our earlier discussion, it is sufficient to check (see Problem 3.13) that these matrices do indeed satisfy the commutation relations (3.14). We next calculate

$$\hat{\mathbf{J}}^2 = \hat{\mathbf{J}} \cdot \hat{\mathbf{J}} = \hat{J}_x^2 + \hat{J}_y^2 + \hat{J}_z^2 \rightarrow 2\hbar^2\begin{pmatrix} 1 & 0 & 0 \\ 0 & 1 & 0 \\ 0 & 0 & 1 \end{pmatrix} \tag{3.29}$$

---

[2] Because $\hat{J}_x$ is the generator of rotations about the $x$ axis, the ket $(1 - i\hat{J}_x \, d\phi/\hbar)|\lambda, m\rangle$ is just the ket that is produced by rotating the ket $|\lambda, m\rangle$ by angle $d\phi$ about the $x$ axis. Thus the ket $|\psi\rangle$ can be viewed as a linear combination of the rotated ket and the ket $|\lambda, m\rangle$, that is, a superposition of two physical states.

We see explicitly that $\hat{\mathbf{J}}^2$ is just a constant times the identity matrix and thus commutes with each of the components of $\hat{\mathbf{J}}$. The operator $\hat{J}_z$ is diagonal as well, suggesting that the matrix representations (3.28) are formed using the eigenstates of $\hat{J}_z$ as well as $\hat{\mathbf{J}}^2$ as a basis. The column vectors representing these eigenstates are given by[3]

$$\begin{pmatrix} 1 \\ 0 \\ 0 \end{pmatrix} \quad \begin{pmatrix} 0 \\ 1 \\ 0 \end{pmatrix} \quad \text{and} \quad \begin{pmatrix} 0 \\ 0 \\ 1 \end{pmatrix} \tag{3.30}$$

which have eigenvalues $\hbar$, $0$, and $-\hbar$, respectively, as can be verified by operating on them with the matrix representing $\hat{J}_z$. For example,

$$\hbar \begin{pmatrix} 1 & 0 & 0 \\ 0 & 0 & 0 \\ 0 & 0 & -1 \end{pmatrix} \begin{pmatrix} 1 \\ 0 \\ 0 \end{pmatrix} = \hbar \begin{pmatrix} 1 \\ 0 \\ 0 \end{pmatrix} \tag{3.31}$$

Similarly, we see that each of these states is an eigenstate of $\hat{\mathbf{J}}^2$ with eigenvalue $2\hbar^2$.

Since the matrix representations of $\hat{J}_x$ and $\hat{J}_y$ are not diagonal, the states (3.30) are not eigenstates of these operators. It is straightforward to evaluate the action of the operators $\hat{J}_x$ and $\hat{J}_y$ on the basis states. There is, however, a linear combination of these two operators, namely,

$$\hat{J}_x + i\hat{J}_y \rightarrow \sqrt{2}\hbar \begin{pmatrix} 0 & 1 & 0 \\ 0 & 0 & 1 \\ 0 & 0 & 0 \end{pmatrix} \tag{3.32}$$

whose action on the basis states exhibits an interesting pattern. Applying this operator to the basis states (3.30), we obtain

$$\sqrt{2}\hbar \begin{pmatrix} 0 & 1 & 0 \\ 0 & 0 & 1 \\ 0 & 0 & 0 \end{pmatrix} \begin{pmatrix} 0 \\ 0 \\ 1 \end{pmatrix} = \sqrt{2}\hbar \begin{pmatrix} 0 \\ 1 \\ 0 \end{pmatrix} \tag{3.33}$$

$$\sqrt{2}\hbar \begin{pmatrix} 0 & 1 & 0 \\ 0 & 0 & 1 \\ 0 & 0 & 0 \end{pmatrix} \begin{pmatrix} 0 \\ 1 \\ 0 \end{pmatrix} = \sqrt{2}\hbar \begin{pmatrix} 1 \\ 0 \\ 0 \end{pmatrix} \tag{3.34}$$

$$\sqrt{2}\hbar \begin{pmatrix} 0 & 1 & 0 \\ 0 & 0 & 1 \\ 0 & 0 & 0 \end{pmatrix} \begin{pmatrix} 1 \\ 0 \\ 0 \end{pmatrix} = \begin{pmatrix} 0 \\ 0 \\ 0 \end{pmatrix} \tag{3.35}$$

Thus, according to (3.33), the operator $\hat{J}_x + i\hat{J}_y$ acting on the state with eigenvalue $-\hbar$ for $\hat{J}_z$ turns it into a state with eigenvalue $0$, multiplied by $\sqrt{2}\hbar$. Similarly, as (3.34) shows, when the operator acts on the state with eigenvalue $0$ for $\hat{J}_z$, it turns it into a state with eigenvalue $\hbar$, multiplied by $\sqrt{2}\hbar$. This raising action terminates when the operator $\hat{J}_x + i\hat{J}_y$ acts on the state with eigenvalue $\hbar$, the maximum eigenvalue for $\hat{J}_z$. See (3.35). It can be similarly verified that the operator

---

[3] Compare these results with (2.70), (2.71), and (2.72) for a spin-$\frac{1}{2}$ particle.

$$\hat{J}_x - i\hat{J}_y \rightarrow \sqrt{2}\hbar \begin{pmatrix} 0 & 0 & 0 \\ 1 & 0 & 0 \\ 0 & 1 & 0 \end{pmatrix} \tag{3.36}$$

has a lowering action when it acts on the states with eigenvalues $\hbar$ and 0, turning them into states with eigenvalues 0 and $-\hbar$, respectively. In this case, the lowering action terminates when the operator (3.36) acts on the state with eigenvalue $-\hbar$, the lowest eigenvalue for $\hat{J}_z$.

## Raising and Lowering Operators

Let's return to our general analysis of angular momentum. The example suggests that it is convenient to introduce the two operators

$$\hat{J}_\pm = \hat{J}_x \pm i\hat{J}_y \tag{3.37}$$

in the general case. Notice that these are not Hermitian operators since

$$\hat{J}_+^\dagger = \hat{J}_x^\dagger + (-i)\hat{J}_y^\dagger = \hat{J}_x - i\hat{J}_y = \hat{J}_- \tag{3.38}$$

The utility of these operators derives from their commutation relations with $\hat{J}_z$:

$$\left[\hat{J}_z, \hat{J}_\pm\right] = \left[\hat{J}_z, \hat{J}_x \pm i\hat{J}_y\right] = i\hbar\hat{J}_y \pm i\left(-i\hbar\hat{J}_x\right) = \pm\hbar\hat{J}_\pm \tag{3.39}$$

To see the effect of $\hat{J}_+$ on the eigenstates, we evaluate $\hat{J}_z\hat{J}_+|\lambda, m\rangle$. We can use the commutation relation (3.39) to invert the order of the operators so that $\hat{J}_z$ can act directly on its eigenstate $|\lambda, m\rangle$. However, since the commutator between $\hat{J}_z$ and $\hat{J}_+$ is not zero but rather is proportional to the operator $\hat{J}_+$ itself, we pick up an additional contribution:

$$\begin{aligned} \hat{J}_z\hat{J}_+|\lambda, m\rangle &= \left(\hat{J}_+\hat{J}_z + \hbar\hat{J}_+\right)|\lambda, m\rangle \\ &= \left(\hat{J}_+ m\hbar + \hbar\hat{J}_+\right)|\lambda, m\rangle \\ &= (m + 1)\hbar\hat{J}_+|\lambda, m\rangle \end{aligned} \tag{3.40a}$$

Inserting some parentheses to help guide the eye:

$$\hat{J}_z\left(\hat{J}_+|\lambda, m\rangle\right) = (m + 1)\hbar\left(\hat{J}_+|\lambda, m\rangle\right) \tag{3.40b}$$

we see that $\hat{J}_+|\lambda, m\rangle$ is an eigenstate of $\hat{J}_z$ with eigenvalue $(m + 1)\hbar$. Hence $\hat{J}_+$ is referred to as a *raising operator*. The action of $\hat{J}_+$ on the state $|\lambda, m\rangle$ is to produce a new state with eigenvalue $(m + 1)\hbar$.

Also

$$\begin{aligned} \hat{J}_z\hat{J}_-|\lambda, m\rangle &= \left(\hat{J}_-\hat{J}_z - \hbar\hat{J}_-\right)|\lambda, m\rangle \\ &= \left(\hat{J}_- m\hbar - \hbar\hat{J}_-\right)|\lambda, m\rangle \\ &= (m - 1)\hbar\hat{J}_-|\lambda, m\rangle \end{aligned} \tag{3.41a}$$

Again, inserting some parentheses,

$$\hat{J}_z\left(\hat{J}_-|\lambda, m\rangle\right) = (m - 1)\hbar\left(\hat{J}_-|\lambda, m\rangle\right) \tag{3.41b}$$

showing that $\hat{J}_-|\lambda, m\rangle$ is an eigenstate of $\hat{J}_z$ with eigenvalue $(m - 1)\hbar$; hence $\hat{J}_-$ is a *lowering operator*. Notice that since $\hat{J}_+$ and $\hat{J}_-$ commute with $\hat{\mathbf{J}}^2$, the states $\hat{J}_\pm|\lambda, m\rangle$ are still eigenstates of the operator $\hat{\mathbf{J}}^2$ with eigenvalue $\lambda\hbar^2$:

$$\hat{\mathbf{J}}^2\left(\hat{J}_\pm|\lambda, m\rangle\right) = \hat{J}_\pm\hat{\mathbf{J}}^2|\lambda, m\rangle = \lambda\hbar^2\left(\hat{J}_\pm|\lambda, m\rangle\right) \tag{3.42}$$

## The Eigenvalue Spectrum

We now have enough information to determine the eigenvalues $\lambda$ and $m$, because there are bounds on how far we can raise or lower $m$. Physically (see Fig. 3.4), we expect that the square of the projection of the angular momentum on any axis should not exceed the magnitude of $\mathbf{J}^2$ and hence

$$m^2 \leq \lambda \tag{3.43}$$

Formally, since

$$\langle\lambda, m|(\hat{J}_x^2 + \hat{J}_y^2)|\lambda, m\rangle \geq 0 \tag{3.44}$$

we have

$$\langle\lambda, m|(\hat{\mathbf{J}}^2 - \hat{J}_z^2)|\lambda, m\rangle = (\lambda - m^2)\hbar^2\langle\lambda, m|\lambda, m\rangle \geq 0 \tag{3.45}$$

establishing (3.43).

Let's call the *maximum m* value $j$. Then we must have

$$\hat{J}_+|\lambda, j\rangle = 0 \tag{3.46}$$

since otherwise $\hat{J}_+$ would create a state $|\lambda, j + 1\rangle$, violating our assumption that $j$ is the maximum eigenvalue for $\hat{J}_z$.[4] Using

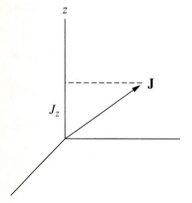

**FIGURE 3.4**
The projection of the angular momentum on the axis never exceeds the magnitude of the angular momentum. *Caution:* This is a classical picture; the angular momentum cannot point in any definite direction.

---

[4] Equation (3.35) demonstrates how this works for the special case of spin 1.

$$\hat{J}_-\hat{J}_+ = \left(\hat{J}_x - i\hat{J}_y\right)\left(\hat{J}_x + i\hat{J}_y\right)$$
$$= \hat{J}_x^2 + \hat{J}_y^2 + i\left[\hat{J}_x, \hat{J}_y\right]$$
$$= \hat{\mathbf{J}}^2 - \hat{J}_z^2 - \hbar\hat{J}_z \tag{3.47}$$

we see that

$$\hat{J}_-\hat{J}_+|\lambda, j\rangle = \left(\hat{\mathbf{J}}^2 - \hat{J}_z^2 - \hbar\hat{J}_z\right)|\lambda, j\rangle$$
$$= \left(\lambda - j^2 - j\right)\hbar^2|\lambda, j\rangle = 0 \tag{3.48}$$

or $\lambda = j(j + 1)$.

Similarly, if we call the *minimum* $m$ value $j'$, then

$$\hat{J}_-|\lambda, j'\rangle = 0 \tag{3.49}$$

and we find that

$$\hat{J}_+\hat{J}_-|\lambda, j'\rangle = \left(\hat{\mathbf{J}}^2 - \hat{J}_z^2 + \hbar\hat{J}_z\right)|\lambda, j'\rangle$$
$$= (\lambda - j'^2 + j')\hbar^2|\lambda, j'\rangle = 0 \tag{3.50}$$

In deriving this result, we have used

$$\hat{J}_+\hat{J}_- = \left(\hat{J}_x + i\hat{J}_y\right)\left(\hat{J}_x - i\hat{J}_y\right)$$
$$= \hat{J}_x^2 + \hat{J}_y^2 - i\left[\hat{J}_x, \hat{J}_y\right]$$
$$= \hat{\mathbf{J}}^2 - \hat{J}_z^2 + \hbar\hat{J}_z \tag{3.51}$$

Thus $\lambda = j'^2 - j'$. The solutions to the equation $j^2 + j = j'^2 - j'$, which results from setting these two values of $\lambda$ equal to each other, are $j' = -j$ and $j' = j + 1$. The second solution violates our assumption that the maximum $m$ value is $j$. Thus we find the minimum $m$ value is $-j$.

If we start at the $m = j$ state, the state with the maximum $m$, and apply the lowering operator a sufficient number of times, we must reach the state with $m = -j$, the state with the minimum $m$. If this were not the case, we would either reach a state with an $m$ value not equal to $-j$ for which (3.49) is satisfied or we would violate the bound on the $m$ values. But (3.49) determines uniquely the value of $j'$ to be $-j$. Since we lower an integral number of times, $j - j' = j - (-j) = 2j =$ an integer, and we deduce that the allowed values of $j$ are given by

$$j = 0, \tfrac{1}{2}, 1, \tfrac{3}{2}, 2, \ldots \tag{3.52}$$

As indicated in Fig. 3.5, the $m$ values for each $j$ run from $j$ to $-j$ in integral steps:

$$m = \underbrace{j, \ j - 1, j - 2, \ldots, -j + 1, -j}_{2j + 1 \text{ states}} \tag{3.53}$$

Given these results, we now change our notation slightly. It is conventional to denote a simultaneous eigenstate of $\hat{\mathbf{J}}^2$ and $\hat{J}_z$ by $|j, m\rangle$ instead of $|\lambda, m\rangle = |j(j + 1), m\rangle$. It is important to remember in this shorthand notation that

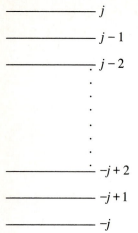

————————— $j$

————————— $j-1$

————————— $j-2$

.
.
.
.
.
.

————————— $-j+2$

————————— $-j+1$

————————— $-j$

**FIGURE 3.5**
The possible $m$ values for a fixed magnitude $\sqrt{j(j+1)}\hbar$ of the angular momentum.

$$\hat{\mathbf{J}}^2|j,m\rangle = j(j+1)\hbar^2|j,m\rangle \tag{3.54a}$$

as well as

$$\hat{J}_z|j,m\rangle = m\hbar|j,m\rangle \tag{3.54b}$$

   Let's examine a few of these states, for which the $m$ values are shown in Fig. 3.6.

1. The $j = 0$ state is denoted by $|0,0\rangle$. Since the magnitude of the angular momentum is zero for this state, it is not surprising that the projection of the angular momentum on the $z$ axis vanishes as well.

2. The $j = \frac{1}{2}$ states are given by $|\frac{1}{2}, \frac{1}{2}\rangle$ and $|\frac{1}{2}, -\frac{1}{2}\rangle$. Note that the eigenvalues of $\hat{J}_z$ for these states are $\hbar/2$ and $-\hbar/2$, respectively. These states are just the states $|+\mathbf{z}\rangle$ and $|-\mathbf{z}\rangle$ that have concerned us for much of Chapters 1 and 2. We now see the rationale for calling these states spin-$\frac{1}{2}$ states: the constant $j$ takes on the value $\frac{1}{2}$. However, the magnitude of the spin of the particle in these states is given by $[\frac{1}{2}(\frac{1}{2}+1)]^{1/2}\hbar = (3)^{1/2}\hbar/2$.

3. The angular momentum $j = 1$ states are denoted by $|1,1\rangle$, $|1,0\rangle$, and $|1,-1\rangle$. These spin-1 states are represented by the column vectors (3.30) in the example

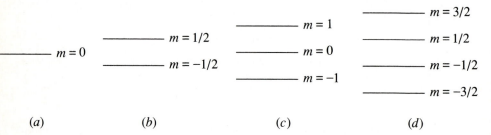

(a)                (b)                (c)                (d)

**FIGURE 3.6**
The $m$ values for (a) spin 0, (b) spin $\frac{1}{2}$, (c) spin 1, and (d) spin $\frac{3}{2}$.

of this section. The eigenvalues of $\hat{J}_z$ are $\hbar$, $0$, and $-\hbar$, which are the diagonal matrix elements of the matrix representing $\hat{J}_z$ in (3.28). The magnitude of the angular momentum for these states is given by $[1(1 + 1)]^{1/2}\hbar = (2)^{1/2}\hbar$.

4. There are four $j = \frac{3}{2}$ states: $|\frac{3}{2}, \frac{3}{2}\rangle$, $|\frac{3}{2}, \frac{1}{2}\rangle$, $|\frac{3}{2}, -\frac{1}{2}\rangle$, and $|\frac{3}{2}, -\frac{3}{2}\rangle$. The magnitude of the angular momentum is $[\frac{3}{2}(\frac{3}{2} + 1)]^{1/2}\hbar = (15)^{1/2}\hbar/2$.

As these examples illustrate, the magnitude $[j(j + 1)]^{1/2}\hbar$ of the angular momentum is always bigger than the maximum projection $j\hbar$ on the $z$ axis for any nonzero angular momentum. In Section 3.5 we will see how the uncertainty relations for angular momentum allow us to understand why the angular momentum does not line up along an axis.

## 3.4 THE MATRIX ELEMENTS OF THE RAISING AND LOWERING OPERATORS

We have seen in (3.40) and (3.42) that the action of the raising operator $\hat{J}_+$ on a state of angular momentum $j$ is to create a state with the same magnitude of the angular momentum but with the $z$ component increased by one unit of $\hbar$:

$$\hat{J}_+|j, m\rangle = c_+\hbar|j, m + 1\rangle \tag{3.55}$$

while the action of the lowering operator is

$$\hat{J}_-|j, m\rangle = c_-\hbar|j, m - 1\rangle \tag{3.56}$$

It is useful to determine the values of $c_+$ and $c_-$. Taking the inner product of the ket (3.55) with the corresponding bra and making use of (3.38), we obtain

$$\langle j, m|\hat{J}_-\hat{J}_+|j, m\rangle = c_+^*c_+\hbar^2\langle j, m + 1|j, m + 1\rangle \tag{3.57}$$

Substituting (3.47) for the operators, we find

$$\langle j, m|(\hat{\mathbf{J}}^2 - \hat{J}_z^2 - \hbar\hat{J}_z)|j, m\rangle = [j(j + 1) - m^2 - m]\hbar^2\langle j, m|j, m\rangle$$
$$= c_+^*c_+\hbar^2\langle j, m + 1|j, m + 1\rangle \tag{3.58}$$

Assuming the angular momentum states satisfy $\langle j, m|j, m\rangle = \langle j, m+1|j, m+1\rangle$, we can choose $c_+ = [j(j + 1) - m(m + 1)]^{1/2}$, or

$$\hat{J}_+|j, m\rangle = \sqrt{j(j + 1) - m(m + 1)}\,\hbar|j, m + 1\rangle \tag{3.59}$$

Note that when $m = j$, the square root factor vanishes and the raising action terminates, as it must. Similarly, we can establish that

$$\hat{J}_-|j, m\rangle = \sqrt{j(j + 1) - m(m - 1)}\,\hbar|j, m - 1\rangle \tag{3.60}$$

for which the square root factor vanishes when $m = -j$, as it must.

These results determine the matrix elements of the raising and lowering operators using the states $|j, m\rangle$ as a basis:

$$\langle j, m'|\hat{J}_+|j, m\rangle = \sqrt{j(j + 1) - m(m + 1)}\,\hbar\langle j, m'|j, m + 1\rangle$$
$$= \sqrt{j(j + 1) - m(m + 1)}\,\hbar\delta_{m',m+1} \tag{3.61}$$

and

$$\langle j, m'|\hat{J}_-|j, m\rangle = \sqrt{j(j + 1) - m(m - 1)}\, \hbar \langle j, m'|j, m - 1\rangle$$

$$= \sqrt{j(j + 1) - m(m - 1)}\, \hbar \delta_{m', m-1} \tag{3.62}$$

In obtaining these matrix elements, we have made use of $\langle j, m'|j, m\rangle = \delta_{m', m}$, since the amplitude to find a state having $J_z = m\hbar$ with $J_z = m'\hbar$, $m' \neq m$, is zero. The square root factors for $j = 1$ and $m = 1$, $0$, and $-1$ take on the values of either $\sqrt{2}$ or $0$ [compare these results with (3.33) through (3.35)]. In Section 3.6 we will see how useful the matrix elements (3.61) and (3.62) are for obtaining matrix representations of $\hat{J}_x$ and $\hat{J}_y$.

## 3.5   UNCERTAINTY RELATIONS AND ANGULAR MOMENTUM

In solving the angular momentum problem in Section 3.3, we took advantage of the commutation relation (3.24) to form simultaneous eigenstates of $\hat{\mathbf{J}}^2$ and $\hat{J}_z$. Since $[\hat{\mathbf{J}}^2, \hat{J}_x] = 0$ as well, we can also form simultaneous eigenstates of $\hat{\mathbf{J}}^2$ and $\hat{J}_x$. For the $j = \frac{1}{2}$ sector, the two eigenstates would be the states $|+\mathbf{x}\rangle$ and $|-\mathbf{x}\rangle$ that we discussed in the earlier chapters. We did not, however, try to form simultaneous eigenstates of $\hat{\mathbf{J}}^2$, $\hat{J}_x$, *and* $\hat{J}_z$. We now want to show that such simultaneous eigenstates are prohibited by the commutation relations of the angular momentum operators themselves, such as

$$[\hat{J}_x, \hat{J}_y] = i\hbar \hat{J}_z \tag{3.63}$$

This is why in Section 3.3 we chose *only one* of the components of $\hat{\mathbf{J}}$, together with the operator $\hat{\mathbf{J}}^2$, to label the eigenstates.

The commutation relation (3.63) is an example of two operators that do not commute and whose commutator can be expressed in the form

$$[\hat{A}, \hat{B}] = i\hat{C} \tag{3.64}$$

where $\hat{A}$, $\hat{B}$, and $\hat{C}$ are Hermitian operators. We will now demonstrate that a commutation relation of the form (3.64) implies a fundamental uncertainty relation. To derive the uncertainty relation, we use the Schwarz inequality

$$\langle \alpha|\alpha\rangle\langle \beta|\beta\rangle \geq |\langle \alpha|\beta\rangle|^2 \tag{3.65}$$

This is the analogue of the relation $(\mathbf{a} \cdot \mathbf{a})(\mathbf{b} \cdot \mathbf{b}) \geq (\mathbf{a} \cdot \mathbf{b})^2$, familiar from the ordinary real three-dimensional vector space. See Problem 3.7 for a derivation of (3.65).

We substitute

$$|\alpha\rangle = (\hat{A} - \langle A\rangle)|\psi\rangle \tag{3.66a}$$

$$|\beta\rangle = (\hat{B} - \langle B\rangle)|\psi\rangle \tag{3.66b}$$

into (3.65), where the expectation values

$$\langle A\rangle = \langle \psi|\hat{A}|\psi\rangle \tag{3.67a}$$

and

$$\langle B \rangle = \langle \psi | \hat{B} | \psi \rangle \tag{3.67b}$$

are real numbers because the operators are Hermitian. Notice that

$$\langle \alpha | \alpha \rangle = \langle \psi | (\hat{A} - \langle A \rangle)^2 | \psi \rangle = \Delta A^2 \tag{3.68a}$$

$$\langle \beta | \beta \rangle = \langle \psi | (\hat{B} - \langle B \rangle)^2 | \psi \rangle = \Delta B^2 \tag{3.68b}$$

where we have used the familiar definition of the uncertainty (see Section 1.4 or 1.6) and the fact that $\hat{A}$ and $\hat{B}$ are Hermitian operators. The right-hand side of the Schwarz inequality (3.65) for the states (3.66) becomes

$$\langle \alpha | \beta \rangle = \langle \psi | (\hat{A} - \langle A \rangle)(\hat{B} - \langle B \rangle) | \psi \rangle \tag{3.69}$$

For any operator $\hat{O}$, we may write

$$\hat{O} = \frac{\hat{O} + \hat{O}^\dagger}{2} + \frac{\hat{O} - \hat{O}^\dagger}{2} = \frac{\hat{F}}{2} + \frac{i\hat{G}}{2} \tag{3.70}$$

where $\hat{F} = \hat{O} + \hat{O}^\dagger$ and $\hat{G} = -i(\hat{O} - \hat{O}^\dagger)$ are Hermitian operators. If we take the operator $\hat{O}$ to be $(\hat{A} - \langle A \rangle)(\hat{B} - \langle B \rangle)$, we find

$$\hat{O} - \hat{O}^\dagger = [\hat{A}, \hat{B}] = i\hat{C} \tag{3.71}$$

and therefore $\hat{G} = \hat{C}$ in (3.70). Thus

$$|\langle \alpha | \beta \rangle|^2 = |\frac{1}{2} \langle \psi | \hat{F} | \psi \rangle + \frac{i}{2} \langle \psi | \hat{C} | \psi \rangle|^2$$

$$= \frac{|\langle \psi | \hat{F} | \psi \rangle|^2}{4} + \frac{|\langle \psi | \hat{C} | \psi \rangle|^2}{4} \geq \frac{|\langle C \rangle|^2}{4} \tag{3.72}$$

where we have made use of the fact that the expectation values of the Hermitian operators $\hat{F}$ and $\hat{C}$ are real. Combining (3.65), (3.68), and (3.72), we obtain

$$\Delta A^2 \Delta B^2 \geq \frac{|\langle C \rangle|^2}{4} \tag{3.73}$$

or simply

$$\Delta A \Delta B \geq \frac{|\langle C \rangle|}{2} \tag{3.74}$$

which is a very important result.

If we apply this uncertainty relation to the specific commutation relation (3.63), we find[5]

---

[5] In Chapter 6 we will see that the position and momentum operators satisfy

$$[\hat{x}, \hat{p}_x] = i\hbar$$

Thus (3.74) leads directly to the famous Heisenberg uncertainty relation $\Delta x \Delta p_x \geq \hbar/2$ as well.

$$\Delta J_x \Delta J_y \geq \frac{\hbar}{2}|\langle J_z\rangle| \qquad (3.75)$$

This uncertainty relation helps to explain a number of our earlier results. If a spin-$\frac{1}{2}$ particle is in a state with a definite value of $J_z$, $\langle J_z\rangle$ is either $\hbar/2$ or $-\hbar/2$, which is certainly nonzero. But (3.75) says that $\Delta J_x$ must then also be nonzero, and thus the particle cannot have a definite value of $J_x$ when it has a definite value of $J_z$. We now see why making a measurement of $S_z$ in the Stern-Gerlach experiments is bound to modify subsequent measurements of $S_x$. We cannot know both the $x$ and the $z$ components of the angular momentum of the particle with definite certainty. We can also see why in general the angular momentum doesn't line up along any axis: If the angular momentum were aligned completely along the $z$ axis, both the $x$ and $y$ components of the angular momentum would vanish. We would then know all three components of the angular momentum, in disagreement with the uncertainty relation (3.75), which requires that both $\Delta J_x$ and $\Delta J_y$ are nonzero in a state with a definite nonzero value of $J_z$. Thus the angular momentum never really "points" in any definite direction.

## 3.6   THE SPIN-$\frac{1}{2}$ EIGENVALUE PROBLEM

In this section we will see how we can use the results of this chapter to derive the spin states of a spin-$\frac{1}{2}$ particle that we deduced from the results of Stern-Gerlach experiments in Chapter 1. First we will make a small change in notation. It is customary in discussing angular momentum to call the angular momentum operators $\hat{J}_x$, $\hat{J}_y$, and $\hat{J}_z$ in general. We have introduced these operators as the generators of rotations. The commutation relations that we used in Section 3.3 depended only on the fact that rotations about different axes do not commute in a well-defined way. Our formulation is general enough to include all kinds of angular momentum, both intrinsic spin angular momentum and orbital angular momentum. That is one of the major virtues of introducing angular momentum in this way. In Chapter 9 we will see that for *orbital* angular momentum—angular momentum of the $\mathbf{r} \times \mathbf{p}$ type—only integral $j$'s are permitted. If our discussion of angular momentum is restricted to purely orbital angular momentum, it is conventional to denote the angular momentum operators by $\hat{L}_x$, $\hat{L}_y$, and $\hat{L}_z$. On the other hand, if our discussion is restricted to intrinsic spin angular momentum, it is customary to call the spin angular momentum operators $\hat{S}_x$, $\hat{S}_y$, and $\hat{S}_z$. Our discussion in Chapters 1 and 2 of the intrinsic spin angular momentum of particles like electrons and photons was restricted to angular momentum of the latter sort. Thus, we could return to Chapter 2, where we first introduced the generator of rotations about the $z$ axis, and relabel $\hat{J}_z$ to $\hat{S}_z$, because we were strictly concerned with rotating intrinsic spin states. In addition to renaming the operators for intrinsic spin, it is also common to relabel the basis states as $|s, m\rangle$, where

$$\hat{\mathbf{S}}^2|s, m\rangle = s(s + 1)\hbar^2|s, m\rangle \qquad (3.76a)$$

$$\hat{S}_z|s, m\rangle = m\hbar|s, m\rangle \qquad (3.76b)$$

For a spin-$\frac{1}{2}$ particle, $s = \frac{1}{2}$ and there are two spin states, $|\frac{1}{2}, \frac{1}{2}\rangle$ and $|\frac{1}{2}, -\frac{1}{2}\rangle$.

Before solving the eigenvalue problem for a spin-$\frac{1}{2}$ particle, it is useful to determine the matrix representations of the spin operators $\hat{S}_x$, $\hat{S}_y$, and $\hat{S}_z$. We will use as a basis the states $|\frac{1}{2}, \frac{1}{2}\rangle = |+\mathbf{z}\rangle$ and $|\frac{1}{2}, -\frac{1}{2}\rangle = |-\mathbf{z}\rangle$ that we found in Section 3.3. In fact, we already determined the matrix representation of $\hat{S}_z$ in this basis in Section 2.5. Of course, we were calling the operator $\hat{J}_z$ then. In agreement with (2.70) we have

$$\hat{S}_z \rightarrow \begin{pmatrix} \langle +\mathbf{z}|\hat{S}_z|+\mathbf{z}\rangle & \langle +\mathbf{z}|\hat{S}_z|-\mathbf{z}\rangle \\ \langle -\mathbf{z}|\hat{S}_z|+\mathbf{z}\rangle & \langle -\mathbf{z}|\hat{S}_z|-\mathbf{z}\rangle \end{pmatrix} = \frac{\hbar}{2}\begin{pmatrix} 1 & 0 \\ 0 & -1 \end{pmatrix} \tag{3.77}$$

in the $S_z$ basis.

In order to determine the matrix representations for $\hat{S}_x$ and $\hat{S}_y$, we start with the matrix representations of the raising and lowering operators $\hat{S}_+$ and $\hat{S}_-$, whose action on the basis states we already know. Forming the matrix representation in the $S_z$ basis for the raising operator using (3.61), we have

$$\hat{S}_+ \rightarrow \begin{pmatrix} \langle +\mathbf{z}|\hat{S}_+|+\mathbf{z}\rangle & \langle +\mathbf{z}|\hat{S}_+|-\mathbf{z}\rangle \\ \langle -\mathbf{z}|\hat{S}_+|+\mathbf{z}\rangle & \langle -\mathbf{z}|\hat{S}_+|-\mathbf{z}\rangle \end{pmatrix} = \hbar\begin{pmatrix} 0 & 1 \\ 0 & 0 \end{pmatrix} \tag{3.78}$$

reflecting the fact that

$$\hat{S}_+|+\mathbf{z}\rangle = \hat{S}_+|\tfrac{1}{2}, \tfrac{1}{2}\rangle = 0 \tag{3.79}$$

and

$$\hat{S}_+|-\mathbf{z}\rangle = \hat{S}_+|\tfrac{1}{2}, -\tfrac{1}{2}\rangle$$

$$= \sqrt{\tfrac{1}{2}(\tfrac{1}{2} + 1) - (-\tfrac{1}{2})(-\tfrac{1}{2} + 1)}\,\hbar|\tfrac{1}{2}, \tfrac{1}{2}\rangle$$

$$= \hbar|\tfrac{1}{2}, \tfrac{1}{2}\rangle = \hbar|+\mathbf{z}\rangle \tag{3.80}$$

Also, the matrix representation of the lowering operator in the $S_z$ basis can be obtained from (3.62):

$$\hat{S}_- \rightarrow \begin{pmatrix} \langle +\mathbf{z}|\hat{S}_-|+\mathbf{z}\rangle & \langle +\mathbf{z}|\hat{S}_-|-\mathbf{z}\rangle \\ \langle -\mathbf{z}|\hat{S}_-|+\mathbf{z}\rangle & \langle -\mathbf{z}|\hat{S}_-|-\mathbf{z}\rangle \end{pmatrix} = \hbar\begin{pmatrix} 0 & 0 \\ 1 & 0 \end{pmatrix} \tag{3.81}$$

reflecting the fact that

$$\hat{S}_-|-\mathbf{z}\rangle = \hat{S}_-|\tfrac{1}{2}, -\tfrac{1}{2}\rangle = 0 \tag{3.82}$$

and

$$\hat{S}_-|+\mathbf{z}\rangle = \hat{S}_-|\tfrac{1}{2}, \tfrac{1}{2}\rangle$$

$$= \sqrt{\tfrac{1}{2}(\tfrac{1}{2} + 1) - \tfrac{1}{2}(\tfrac{1}{2} - 1)}\,\hbar|\tfrac{1}{2}, \tfrac{1}{2}\rangle$$

$$= \hbar|\tfrac{1}{2}, -\tfrac{1}{2}\rangle = \hbar|-\mathbf{z}\rangle \tag{3.83}$$

As a check, note that since $\hat{S}_+^\dagger = \hat{S}_-$, we could also obtain (3.81) as the transpose, complex conjugate of the matrix (3.78). Recall (2.80).

With the matrix representations for $\hat{S}_+$ and $\hat{S}_-$, determining the matrix representations of $\hat{S}_x$ and $\hat{S}_y$ is straightforward. Since

$$\hat{S}_+ = \hat{S}_x + i\hat{S}_y \tag{3.84}$$

$$\hat{S}_- = \hat{S}_x - i\hat{S}_y \tag{3.85}$$

then

$$\hat{S}_x = \frac{\hat{S}_+ + \hat{S}_-}{2} \tag{3.86}$$

and

$$\hat{S}_y = \frac{\hat{S}_+ - \hat{S}_-}{2i} \tag{3.87}$$

Using the matrix representations (3.78) and (3.81) in the $S_z$ basis, we obtain

$$\hat{S}_x \rightarrow \frac{\hbar}{2}\begin{pmatrix} 0 & 1 \\ 1 & 0 \end{pmatrix} \tag{3.88}$$

and

$$\hat{S}_y \rightarrow \frac{\hbar}{2}\begin{pmatrix} 0 & -i \\ i & 0 \end{pmatrix} \tag{3.89}$$

The three $2 \times 2$ matrices in (3.88), (3.89), and (3.77) (without the factors of $\hbar/2$) are often referred to as Pauli spin matrices and are denoted by $\sigma_x$, $\sigma_y$, and $\sigma_z$, respectively. These three equations can then be expressed in the vector notation

$$\hat{\mathbf{S}} \rightarrow \frac{\hbar}{2}\boldsymbol{\sigma} \tag{3.90}$$

where $\hat{\mathbf{S}} = \hat{S}_x\mathbf{i} + \hat{S}_y\mathbf{j} + \hat{S}_z\mathbf{k}$ and $\boldsymbol{\sigma} = \sigma_x\mathbf{i} + \sigma_y\mathbf{j} + \sigma_z\mathbf{k}$.

We are now ready to find the eigenstates of $\hat{S}_x$ or $\hat{S}_y$. In fact, we can use the matrix representations (3.90) to determine the eigenstates of $\hat{S}_n = \hat{\mathbf{S}} \cdot \mathbf{n}$ and thus find the states that are spin up and spin down along an arbitrary axis specified by the unit vector $\mathbf{n}$. We will restrict our attention to the case where $\mathbf{n} = \cos\phi\,\mathbf{i} + \sin\phi\,\mathbf{j}$ lies in the $x$-$y$ plane, as indicated in Fig. 3.7. The choice $\phi = 0$ ($\phi = \pi/2$) will yield the eigenstates of $\hat{S}_x(\hat{S}_y)$ that we used extensively in Chapters 1 and 2. We will leave the more general case to the Problems (in particular, see Problem 3.2). We first express the eigenvalue equation in the form

$$\hat{S}_n|\mu\rangle = \mu\frac{\hbar}{2}|\mu\rangle \tag{3.91}$$

where, as we did earlier in our general discussion of angular momentum, we

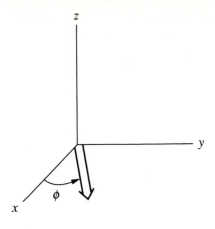

**FIGURE 3.7**
The spin-up-along-**n** state. $\mathbf{n} = \cos\phi\mathbf{i} + \sin\phi\mathbf{j}$.

have included a factor of $\hbar$ so that $\mu$ is dimensionless. The factor of $\frac{1}{2}$ in the eigenvalue has been included to make things turn out nicely. After all, we know the eigenvalues already. Since the eigenvalues of $\hat{S}_z$ are $\pm\hbar/2$ and since our choice of the $z$ axis is arbitrary, these must be the eigenvalues of $\hat{S}_n$ as well. Equation (3.91), however, does not presume particular eigenvalues, and we will see how solving the eigenvalue problem determines the allowed values of $\mu$.

As in (2.63), we obtain two equations that can be expressed in matrix form by taking the inner product of (3.91) with the two bra vectors $\langle+\mathbf{z}|$ and $\langle-\mathbf{z}|$:

$$\frac{\hbar}{2}\left[\begin{pmatrix} 0 & 1 \\ 1 & 0 \end{pmatrix}\cos\phi + \begin{pmatrix} 0 & -i \\ i & 0 \end{pmatrix}\sin\phi\right]\begin{pmatrix} \langle+\mathbf{z}|\mu\rangle \\ \langle-\mathbf{z}|\mu\rangle \end{pmatrix} = \mu\frac{\hbar}{2}\begin{pmatrix} \langle+\mathbf{z}|\mu\rangle \\ \langle-\mathbf{z}|\mu\rangle \end{pmatrix} \quad (3.92)$$

where the $2 \times 2$ matrix on the left-hand side is just the matrix representation of $\hat{S}_n = \hat{S}_x\cos\phi + \hat{S}_y\sin\phi$. Dividing out the common factor of $\hbar/2$, we can write this equation as

$$\begin{pmatrix} -\mu & e^{-i\phi} \\ e^{i\phi} & -\mu \end{pmatrix}\begin{pmatrix} \langle+\mathbf{z}|\mu\rangle \\ \langle-\mathbf{z}|\mu\rangle \end{pmatrix} = 0 \quad (3.93)$$

This is a homogeneous equation in the two unknowns $\langle+\mathbf{z}|\mu\rangle$ and $\langle-\mathbf{z}|\mu\rangle$. A nontrivial solution requires that the determinant of the coefficients vanishes. Otherwise, the $2 \times 2$ matrix in (3.93) has an inverse, and multiplying the equation by the inverse would leave just the column vector equal to zero, that is, the trivial solution. Thus

$$\begin{vmatrix} -\mu & e^{-i\phi} \\ e^{i\phi} & -\mu \end{vmatrix} = 0 \quad (3.94)$$

showing that $\mu^2 - e^{i\phi}e^{-i\phi} = \mu^2 - 1 = 0$, or $\mu = \pm 1$.

Now that we know the eigenvalues, we may determine the corresponding eigenstates. The state with $\mu = +1$ is an eigenstate of $\hat{S}_n$ with eigenvalue $\hbar/2$.

Thus, in our earlier notation, it is the state $|+\mathbf{n}\rangle$, and we can relabel it accordingly: $|\mu = 1\rangle = |+\mathbf{n}\rangle$. Substituting $\mu = +1$ into (3.93), we find that

$$\langle -\mathbf{z}|+\mathbf{n}\rangle = e^{i\phi}\langle +\mathbf{z}|+\mathbf{n}\rangle \tag{3.95}$$

The requirement that the state be normalized ($\langle +\mathbf{n}|+\mathbf{n}\rangle = 1$) is satisfied provided that

$$|\langle +\mathbf{z}|+\mathbf{n}\rangle|^2 + |\langle -\mathbf{z}|+\mathbf{n}\rangle|^2 = 1 \tag{3.96}$$

Substituting (3.95) into (3.96), we find

$$2|\langle +\mathbf{z}|+\mathbf{n}\rangle|^2 = 1 \tag{3.97}$$

Thus, up to an overall phase, we may choose $\langle +\mathbf{z}|+\mathbf{n}\rangle = 1/\sqrt{2}$, which with (3.95) shows that $\langle -\mathbf{z}|+\mathbf{n}\rangle = e^{i\phi}/\sqrt{2}$, or

$$|+\mathbf{n}\rangle = \frac{1}{\sqrt{2}}|+\mathbf{z}\rangle + \frac{e^{i\phi}}{\sqrt{2}}|-\mathbf{z}\rangle \tag{3.98}$$

Note how, up to an overall phase, this result agrees with (2.41), which we obtained by rotating the state $|+\mathbf{x}\rangle$ by an angle $\phi$ counterclockwise about the $z$ axis, namely, $|+\mathbf{n}\rangle = \hat{R}(\phi\mathbf{k})|+\mathbf{x}\rangle$.

The state with $\mu = -1$ is an eigenstate of $\hat{S}_n$ with eigenvalue $-\hbar/2$. We can thus relabel this state $|\mu = -1\rangle = |-\mathbf{n}\rangle$. If we substitute the value $\mu = -1$ into (3.93), we find that

$$\langle -\mathbf{z}|-\mathbf{n}\rangle = -e^{i\phi}\langle +\mathbf{z}|-\mathbf{n}\rangle \tag{3.99}$$

Satisfying

$$|\langle +\mathbf{z}|-\mathbf{n}\rangle|^2 + |\langle -\mathbf{z}|-\mathbf{n}\rangle|^2 = 1 \tag{3.100}$$

we obtain

$$|-\mathbf{n}\rangle = \frac{1}{\sqrt{2}}|+\mathbf{z}\rangle - \frac{e^{i\phi}}{\sqrt{2}}|-\mathbf{z}\rangle \tag{3.101}$$

These results are in agreement with our earlier forms for these states: setting $\phi = 0$ in (3.98) and (3.101) yields

$$|\pm\mathbf{x}\rangle = \frac{1}{\sqrt{2}}|+\mathbf{z}\rangle \pm \frac{1}{\sqrt{2}}|-\mathbf{z}\rangle \tag{3.102}$$

while setting $\phi = \pi/2$ yields

$$|\pm\mathbf{y}\rangle = \frac{1}{\sqrt{2}}|+\mathbf{z}\rangle \pm \frac{i}{\sqrt{2}}|-\mathbf{z}\rangle \tag{3.103}$$

However, in deriving (3.102) and (3.103) here, we have not had to appeal to the results from the Stern-Gerlach experiments. We have relied on only the commutation relations of the generators of rotations and their identification with the angular momentum operators. In a similar fashion, we can work out the spin eigenstates of a particle with arbitrary intrinsic spin $s$. In this latter case, because there are $2s + 1$ spin states for a particle with intrinsic spin $s$, the corresponding eigenvalue problem will involve $(2s + 1) \times (2s + 1)$ matrices. The procedure for determining the eigenstates and corresponding eigenvalues is the same as we have used in this section, but the algebra becomes more involved as the dimensionality of the matrices increases.

## 3.7   A STERN-GERLACH EXPERIMENT WITH SPIN-1 PARTICLES

Let's return to the sort of Stern-Gerlach experiments that we examined in Chapter 1, but this time let's perform one of these experiments with a neutral beam of spin-1 instead of spin-$\frac{1}{2}$ particles. Since the $z$ component of the angular momentum of a spin-1 particle can take on the three values $\hbar$, 0, and $-\hbar$, an unpolarized beam passing through an SGz device splits into three different beams, with the particles deflected upward, not deflected at all, or deflected downward, depending on the value of $S_z$ (see Fig. 3.8).

What happens if a beam of spin-1 particles passes through an SGy device? An unpolarized beam should split into three beams since $S_y$ can also take on the three values $\hbar$, 0, and $-\hbar$. If we follow this SGy device with an SGz device, we can ask, for example, what fraction of the particles with $S_y = \hbar$ will be found to have $S_z = \hbar$ when they exit the SGz device (see Fig. 3.9). Unlike the case of spin $\frac{1}{2}$, where it was "obvious" for two SG devices whose inhomogeneous magnetic fields were at right angles to each other that 50 percent of the particles would be spin up and 50 percent would be spin down when they exited the last SG device, here the answer is not so clear. In fact, you might try guessing how the particles will be distributed before going on. To answer this question, we need to calculate the amplitude to find a particle with $S_y = \hbar$ in a state with $S_z = \hbar$, that is, to calculate the amplitude $_z\langle 1, 1 | 1, 1 \rangle_y$, where we have put a subscript on the ket and bra indicating that they are eigenstates of $\hat{S}_y$ and $\hat{S}_z$, respectively. A natural way

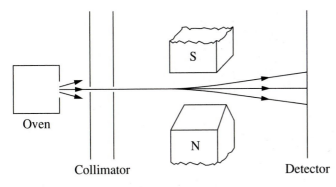

**FIGURE 3.8**
A schematic diagram indicating the paths that a spin-1 particle with $S_z$ equal to $\hbar$, 0, or $-\hbar$ would follow in a Stern-Gerlach device.

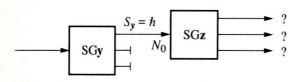

**FIGURE 3.9**

A block diagram for an experiment with spin-1 particles with two SG devices whose inhomogeneous magnetic fields are oriented at right angles to each other. What fraction of the particles exiting the SGy device with $S_y = \hbar$ exits the SGz device in each of the three channels?

to determine the amplitude $_z\langle 1, 1|1, 1\rangle_y$ is to determine the eigenstates of $\hat{S}_y$ for a spin-1 particle in the $S_z$ basis. We use the representation of $\hat{S}_y$ in the $\hat{S}_z$ basis from (3.28):

$$\hat{S}_y \xrightarrow[S_z \text{ basis}]{} \frac{\hbar}{\sqrt{2}} \begin{pmatrix} 0 & -i & 0 \\ i & 0 & -i \\ 0 & i & 0 \end{pmatrix} \tag{3.104}$$

The eigenvalue equation

$$\hat{S}_y|1, \mu\rangle_y = \mu\hbar|1, \mu\rangle_y \tag{3.105}$$

becomes the matrix equation

$$\frac{\hbar}{\sqrt{2}} \begin{pmatrix} 0 & -i & 0 \\ i & 0 & -i \\ 0 & i & 0 \end{pmatrix} \begin{pmatrix} a \\ b \\ c \end{pmatrix} = \mu\hbar \begin{pmatrix} a \\ b \\ c \end{pmatrix} \tag{3.106}$$

which can be expressed in the form

$$\begin{pmatrix} -\mu & -i/\sqrt{2} & 0 \\ i/\sqrt{2} & -\mu & -i/\sqrt{2} \\ 0 & i/\sqrt{2} & -\mu \end{pmatrix} \begin{pmatrix} a \\ b \\ c \end{pmatrix} = 0 \tag{3.107}$$

Note that we have represented the eigenstate by the column vector

$$|1, \mu\rangle_y \rightarrow \begin{pmatrix} _z\langle 1, 1|1, \mu\rangle_y \\ _z\langle 1, 0|1, \mu\rangle_y \\ _z\langle 1, -1|1, \mu\rangle_y \end{pmatrix} = \begin{pmatrix} a \\ b \\ c \end{pmatrix} \tag{3.108}$$

in the $S_z$ basis, where we have used $a$, $b$, and $c$ for the amplitudes for notational convenience. As we discussed in the preceding section, a nontrivial solution to (3.106) requires that the determinant of the coefficients in (3.107) must vanish:

$$\begin{vmatrix} -\mu & -i/\sqrt{2} & 0 \\ i/\sqrt{2} & -\mu & -i/\sqrt{2} \\ 0 & i/\sqrt{2} & -\mu \end{vmatrix} = 0 \tag{3.109}$$

showing that $-\mu(\mu^2 - \frac{1}{2}) + (i/\sqrt{2})(-i\mu/\sqrt{2}) = 0$, which can be written in the form $\mu(\mu^2 - 1) = 0$. Thus we see that the eigenvalues are indeed given by $\mu$ equals 1, 0, and $-1$, corresponding to eigenvalues $\hbar$, 0, and $-\hbar$ for $\hat{S}_y$, as expected. If we now, for example, substitute the eigenvalue $\mu = 1$ into (3.106), we obtain the equation

$$\frac{1}{\sqrt{2}}\begin{pmatrix} 0 & -i & 0 \\ i & 0 & -i \\ 0 & i & 0 \end{pmatrix}\begin{pmatrix} a \\ b \\ c \end{pmatrix} = \begin{pmatrix} a \\ b \\ c \end{pmatrix} \qquad (3.110)$$

indicating that for this eigenstate

$$-ib = \sqrt{2}\,a \qquad ia - ic = \sqrt{2}\,b \qquad \text{and} \qquad ib = \sqrt{2}\,c \qquad (3.111)$$

From the first and last of these equations we see that $c = -a$. Since $b = i\sqrt{2}\,a$, the column vector in the $S_z$ basis representing the eigenstate of $\hat{S}_y$ with eigenvalue $\hbar$ is given by

$$|1, 1\rangle_y \xrightarrow[S_z \text{ basis}]{} \begin{pmatrix} a \\ i\sqrt{2}\,a \\ -a \end{pmatrix} \qquad (3.112)$$

The requirement that the state be normalized is

$$\left(a^*, -i\sqrt{2}\,a^*, -a^*\right)\begin{pmatrix} a \\ i\sqrt{2}\,a \\ -a \end{pmatrix} = 4|a|^2 = 1 \qquad (3.113)$$

Thus, up to an overall phase, we can choose $a = \frac{1}{2}$, showing that

$$|1, 1\rangle_y \xrightarrow[S_z \text{ basis}]{} \frac{1}{2}\begin{pmatrix} 1 \\ i\sqrt{2} \\ -1 \end{pmatrix} \qquad (3.114)$$

or, expressed in terms of kets,

$$|1, 1\rangle_y = \frac{1}{2}|1, 1\rangle + i\frac{\sqrt{2}}{2}|1, 0\rangle - \frac{1}{2}|1, -1\rangle \qquad (3.115)$$

Note that we have not put subscripts on the kets on the right-hand side of (3.115) because, if there is no ambiguity, we will use the convention that without subscripts these are understood to be eigenkets of $\hat{S}_z$.

Based on our result, we can now ascertain how a beam of spin-1 particles exiting an SGy device in the state $|1, 1\rangle_y$, that is, with $S_y = \hbar$, will split when it passes through an SGz device. The probability of the particles exiting this SGz device with $S_z = \hbar$ is given by $|\langle 1, 1|1, 1\rangle_y|^2 = |\frac{1}{2}|^2 = \frac{1}{4}$; the probability of the particles exiting this SGz device with $S_z = 0$ is given by $|\langle 1, 0|1, 1\rangle_y|^2 = |i\sqrt{2}/2|^2 = \frac{1}{2}$; and the probability that the particles exit the SGz device with $S_z = -\hbar$ is given by $|\langle 1, -1|1, 1\rangle_y|^2 = |-\frac{1}{2}|^2 = \frac{1}{4}$. So when a beam of spin-1 "spin-up" particles from one SG device passes through another SG device whose inhomogeneous magnetic field is oriented at right angles to that of the initial device, 25 percent of the particles are deflected up, 50 percent of the particles are not deflected, and 25 percent of the particles are deflected down (see Fig. 3.10). This is to be compared with the 50 percent up and 50 percent down that we saw earlier for spin-$\frac{1}{2}$ particles in a similar experiment.

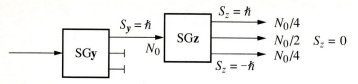

**FIGURE 3.10**
A block diagram showing the results of the Stern-Gerlach experiment with spin-1 particles.

## 3.8  SUMMARY

To a physicist, angular momentum along with linear momentum and energy constitute the "big three" space-time dynamical variables used to describe a system.[6] Angular momentum enters quantum mechanics in the form of three operators—$\hat{J}_x$, $\hat{J}_y$, and $\hat{J}_z$—that generate rotations of states about the $x$, $y$, and $z$ axes, respectively. Because finite rotations about different axes do not commute, the generators satisfy the commutation relations

$$\left[\hat{J}_x, \hat{J}_y\right] = i\hbar\hat{J}_z \qquad \left[\hat{J}_y, \hat{J}_z\right] = i\hbar\hat{J}_x \qquad \left[\hat{J}_z, \hat{J}_x\right] = i\hbar\hat{J}_y \qquad (3.116)$$

where the commutator of two operators $\hat{A}$ and $\hat{B}$ is defined by the relationship

$$\left[\hat{A}, \hat{B}\right] \equiv \hat{A}\hat{B} - \hat{B}\hat{A} \qquad (3.117)$$

Although the three generators $\hat{J}_x$, $\hat{J}_y$, and $\hat{J}_z$ do not commute with each other, they each commute with

$$\hat{\mathbf{J}}^2 = \hat{J}_x^2 + \hat{J}_y^2 + \hat{J}_z^2 \qquad (3.118)$$

Thus, we can find simultaneous eigenstates of $\hat{\mathbf{J}}^2$ and one of the components, for example, $\hat{J}_z$. These eigenstates are denoted by the kets $|j, m\rangle$, where

$$\hat{\mathbf{J}}^2|j, m\rangle = j(j + 1)\hbar^2|j, m\rangle \qquad (3.119a)$$

$$\hat{J}_z|j, m\rangle = m\hbar|j, m\rangle \qquad (3.119b)$$

Physically, we can see why $\hat{\mathbf{J}}^2$ and $\hat{J}_z$ commute, since the eigenvalue for $\hat{\mathbf{J}}^2$ specifies the magnitude of the angular momentum for the state and the *magnitude* of the angular momentum, like the length of any vector, is not affected by a rotation.

The linear combination of the generators

$$\hat{J}_+ = \hat{J}_x + i\hat{J}_y \qquad (3.120)$$

is a raising operator:

---

[6] Relativistically, we could term them the big two, grouping linear momentum and energy together as an energy-momentum four-vector. The importance of these variables arises primarily because of the conservation laws that exist for angular momentum, linear momentum, and energy. In Chapter 4 we will begin to see how these conservation laws arise. Intrinsic spin angular momentum plays an unusually important role, which we will see when we consider systems of identical particles in Chapter 12.

$$\hat{J}_+|j, m\rangle = \sqrt{j(j + 1) - m(m + 1)}\hbar|j, m + 1\rangle \qquad (3.121)$$

whereas $\hat{J}_- = \hat{J}_x - i\hat{J}_y$ is a lowering operator:

$$\hat{J}_-|j, m\rangle = \sqrt{j(j + 1) - m(m - 1)}\hbar|j, m - 1\rangle \qquad (3.122)$$

Since the magnitude of the projection of the angular momentum on an axis for a state must be less than the magnitude of the angular momentum itself, there are limits on how far you can raise or lower the $m$ values, which are sufficient to determine the allowed values of $j$ and $m$:

$$j = 0, \tfrac{1}{2}, 1, \tfrac{3}{2}, 2, \ldots \qquad (3.123)$$

and for any particular $j$, $m$ ranges from $+j$ to $-j$ in integral steps:

$$m = j, \; j - 1, \; j - 2, \ldots, -j + 1, \; -j \qquad (3.124)$$

The eigenstates of $\hat{J}_n = \hat{\mathbf{J}} \cdot \mathbf{n}$, the component of the angular momentum along an axis specified by the unit vector $\mathbf{n}$, can be determined by setting up the eigenvalue equation

$$\hat{J}_n|j, m\rangle_n = m\hbar|j, m\rangle_n \qquad (3.125)$$

using the eigenstates of $\hat{J}_z$ as a basis. Since for a particular $j$, there are $2j + 1$ different states $|j, m\rangle$, the eigenvalue equation (3.125) can be expressed as a matrix equation with the matrix representation of the operator $\hat{J}_n = \hat{\mathbf{J}} \cdot \mathbf{n} = \hat{J}_x n_x + \hat{J}_y n_y + \hat{J}_z n_z$ following directly from (3.119b), (3.120), (3.121), and (3.122). As an important example, the matrix representations for spin $\tfrac{1}{2}$ are given by

$$\hat{\mathbf{S}} \xrightarrow[S_z \text{ basis}]{} \frac{\hbar}{2}\boldsymbol{\sigma} \qquad (3.126)$$

with the Pauli matrices

$$\sigma_x = \begin{pmatrix} 0 & 1 \\ 1 & 0 \end{pmatrix} \quad \sigma_y = \begin{pmatrix} 0 & -i \\ i & 0 \end{pmatrix} \quad \text{and} \quad \sigma_z = \begin{pmatrix} 1 & 0 \\ 0 & -1 \end{pmatrix} \qquad (3.127)$$

In (3.126) we have labeled the angular momentum operators by $\hat{\mathbf{S}}$ instead of $\hat{\mathbf{J}}$, because when $j = \tfrac{1}{2}$ we know that we are dealing with intrinsic spin.

Finally, when two operators do not commute,

$$[\hat{A}, \hat{B}] = iC \qquad (3.128)$$

there is a fundamental uncertainty relation

$$\Delta A \Delta B \geq \frac{|\langle C \rangle|}{2} \qquad (3.129)$$

From this result follows uncertainty relations for angular momentum such as

$$\Delta J_x \Delta J_y \geq \frac{|\langle J_z \rangle|}{2} \qquad (3.130)$$

If the $z$ component of the angular momentum has a definite nonzero value, making the right-hand side of (3.130) nonzero, then we cannot specify either the $x$ or $y$ component of the angular momentum with certainty, because this would require the left-hand side of (3.130) to vanish, in contradiction to the inequality. This uncertainty relation is, of course, built into our results (3.123) and (3.124), which, like (3.130), follow directly from the commutation relations (3.116). Nonetheless, uncertainty relations such as (3.130) bring to the fore the sharp differences between the quantum and the classical worlds. In Chapter 6, we will see how (3.128) and (3.129) lead to the famous Heisenberg uncertainty relation $\Delta x \Delta p_x \geq \hbar/2$.

## PROBLEMS

**3.1.** Verify for the operators $\hat{A}$, $\hat{B}$, and $\hat{C}$ that
(a) $[\hat{A}, \hat{B} + \hat{C}] = [\hat{A}, \hat{B}] + [\hat{A}, \hat{C}]$
(b) $[\hat{A}, \hat{B}\hat{C}] = \hat{B}[\hat{A}, \hat{C}] + [\hat{A}, \hat{B}]\hat{C}$

**3.2.** Using the $|+\mathbf{z}\rangle$ and $|-\mathbf{z}\rangle$ states of a spin-$\frac{1}{2}$ particle as a basis, set up and solve as a problem in matrix mechanics the eigenvalue problem for $\hat{S}_n = \hat{\mathbf{S}} \cdot \mathbf{n}$, where the spin operator $\hat{\mathbf{S}} = \hat{S}_x \mathbf{i} + \hat{S}_y \mathbf{j} + \hat{S}_z \mathbf{k}$ and $\mathbf{n} = \sin\theta \cos\phi \mathbf{i} + \sin\theta \sin\phi \mathbf{j} + \cos\theta \mathbf{k}$. Show that the eigenstates may be written as

$$|+\mathbf{n}\rangle = \cos\frac{\theta}{2}|+\mathbf{z}\rangle + e^{i\phi}\sin\frac{\theta}{2}|-\mathbf{z}\rangle$$

$$|-\mathbf{n}\rangle = \sin\frac{\theta}{2}|+\mathbf{z}\rangle - e^{i\phi}\cos\frac{\theta}{2}|-\mathbf{z}\rangle$$

Rather than simply verifying that these are eigenstates by substituting into the eigenvalue equation, obtain these states by directly solving the eigenvalue problem, as in Section 3.6.

**3.3.** Show that the Pauli spin matrices satisfy $\sigma_i\sigma_j + \sigma_j\sigma_i = 2\delta_{ij}\mathbb{1}$, where $i$ and $j$ can take on the values 1, 2, and 3, with the understanding that $\sigma_1 = \sigma_x$, $\sigma_2 = \sigma_y$, and $\sigma_3 = \sigma_z$. Thus for $i = j$ show that $\sigma_x^2 = \sigma_y^2 = \sigma_z^2 = \mathbb{1}$, while for $i \neq j$ show that $\{\sigma_i, \sigma_j\} = 0$, where the curly brackets are called an *anticommutator*, which is defined by the relationship $\{\hat{A}, \hat{B}\} \equiv \hat{A}\hat{B} + \hat{B}\hat{A}$.

**3.4.** Verify that (a) $\boldsymbol{\sigma} \times \boldsymbol{\sigma} = 2i\boldsymbol{\sigma}$ and (b) $\boldsymbol{\sigma} \cdot \mathbf{a}\boldsymbol{\sigma} \cdot \mathbf{b} = \mathbf{a} \cdot \mathbf{b}\mathbb{1} + i\boldsymbol{\sigma} \cdot (\mathbf{a} \times \mathbf{b})$, where $\boldsymbol{\sigma} = \sigma_x\mathbf{i} + \sigma_y\mathbf{j} + \sigma_z\mathbf{k}$.

**3.5.** This problem demonstrates another way (also see Problem 3.2) to determine the eigenstates of $\hat{S}_n = \hat{\mathbf{S}} \cdot \mathbf{n}$. The operator

$$\hat{R}(\theta\mathbf{j}) = e^{-i\hat{S}_y\theta/\hbar}$$

rotates spin states by an angle $\theta$ counterclockwise about the $y$ axis.
(a) Show that this rotation operator can be expressed in the form

$$\hat{R}(\theta\mathbf{j}) = \cos\frac{\theta}{2} - \frac{2i}{\hbar}\hat{S}_y\sin\frac{\theta}{2}$$

*Suggestion:* Use the states $|+\mathbf{z}\rangle$ and $|-\mathbf{z}\rangle$ as a basis. Express the operator $\hat{R}(\theta\mathbf{j})$ in matrix form by expanding $\hat{R}$ in a Taylor series. Examine the explicit form for the matrices representing $\hat{S}_y^2$, $\hat{S}_y^3$, and so on.
(b) Apply $\hat{R}$ in matrix form to the state $|+\mathbf{z}\rangle$ to obtain the state $|+\mathbf{n}\rangle$ given in Problem 3.2 with $\phi = 0$, that is, rotated by angle $\theta$ in the $x$-$z$ plane.

**3.6.** Derive (3.60).

**3.7.** Derive the Schwarz inequality

$$\langle\alpha|\alpha\rangle\langle\beta|\beta\rangle \geq |\langle\alpha|\beta\rangle|^2$$

*Suggestion:* Use the fact that

$$\big(\langle\alpha|+\lambda^*\langle\beta|\big)\big(|\alpha\rangle + \lambda|\beta\rangle\big) \geq 0$$

and determine the value of $\lambda$ that minimizes the left-hand side of the equation.

**3.8.** Show that the operator $\hat{C}$ defined through $[\hat{A}, \hat{B}] = i\hat{C}$ is Hermitian, provided the operators $\hat{A}$ and $\hat{B}$ are Hermitian.

**3.9.** Calculate $\Delta S_x$ and $\Delta S_y$ for an eigenstate of $\hat{S}_z$ for a spin-$\frac{1}{2}$ particle. Check to see if the uncertainty relation $\Delta S_x \Delta S_y \geq \hbar|\langle S_z\rangle|/2$ is satisfied. Repeat your calculation for an eigenstate of $\hat{S}_x$.

**3.10.** Use the matrix representations of the spin-$\frac{1}{2}$ angular momentum operators $\hat{S}_x$, $\hat{S}_y$, and $\hat{S}_z$ in the $S_z$ basis to verify explicitly through matrix multiplication that

$$[\hat{S}_x, \hat{S}_y] = i\hbar\hat{S}_z$$

**3.11.** Determine the matrix representations of the spin-$\frac{1}{2}$ angular momentum operators $\hat{S}_x$, $\hat{S}_y$, and $\hat{S}_z$ using the eigenstates of $\hat{S}_y$ as a basis.

**3.12.** By examining their action on the basis states $|+\mathbf{z}\rangle$ and $|-\mathbf{z}\rangle$, verify for a spin-$\frac{1}{2}$ particle that (*a*)

$$\hat{S}_z = (\hbar/2)|+\mathbf{z}\rangle\langle+\mathbf{z}| - (\hbar/2)|-\mathbf{z}\rangle\langle-\mathbf{z}|$$

and (*b*) the raising and lowering operators may be expressed as

$$\hat{S}_+ = \hbar|+\mathbf{z}\rangle\langle-\mathbf{z}| \quad\text{and}\quad \hat{S}_- = \hbar|-\mathbf{z}\rangle\langle+\mathbf{z}|$$

**3.13.** Repeat Problem 3.10 using the matrix representations (3.28) for a spin-1 particle in the $J_z$ basis.

**3.14.** Use the spin-1 states $|1, 1\rangle$, $|1, 0\rangle$, and $|1, -1\rangle$ as a basis to form the matrix representations of the angular momentum operators and hence verify that the matrix representations (3.28) are correct.

**3.15.** Determine the eigenstates of $\hat{S}_x$ for a spin-1 particle in terms of the eigenstates $|1, 1\rangle$, $|1, 0\rangle$, and $|1, -1\rangle$ of $\hat{S}_z$.

**3.16.** A spin-1 particle exits an SGz device in a state with $S_z = \hbar$. The beam then enters an SGx device. What is the probability that the measurement of $S_x$ yields the value 0?

**3.17.** A spin-1 particle is in the state

$$|\psi\rangle \xrightarrow[S_z \text{ basis}]{} \frac{1}{\sqrt{14}}\begin{pmatrix} 1 \\ 2 \\ 3i \end{pmatrix}$$

(*a*) What are the probabilities that a measurement of $S_z$ will yield the values $\hbar$, 0, or $-\hbar$ for this state? What is $\langle S_z\rangle$?

(*b*) What is $\langle S_x\rangle$ for this state? *Suggestion:* Use matrix mechanics to evaluate the expectation value.

(*c*) What is the probability that a measurement of $S_x$ will yield the value $\hbar$ for this state?

**3.18.** Determine the eigenstates of $\hat{S}_n = \hat{\mathbf{S}} \cdot \mathbf{n}$ for a spin-1 particle, where the spin operator $\hat{\mathbf{S}} = \hat{S}_x\mathbf{i} + \hat{S}_y\mathbf{j} + \hat{S}_z\mathbf{k}$ and $\mathbf{n} = \sin\theta\cos\phi\,\mathbf{i} + \sin\theta\sin\phi\,\mathbf{j} + \cos\theta\,\mathbf{k}$. Use the matrix representation of the rotation operator in Problem 3.19 to check your result when $\phi = 0$.

**3.19.** Find the state with $S_n = \hbar$ of a spin-1 particle, where $\mathbf{n} = \sin\theta\mathbf{i} + \cos\theta\mathbf{k}$, by rotating a state with $S_z = \hbar$ by angle $\theta$ counterclockwise about the $y$ axis using the rotation operator $\hat{R}(\theta\mathbf{j}) = e^{-i\hat{S}_y\theta/\hbar}$. *Suggestion:* Use the matrix representation (3.104) for $\hat{S}_y$ in the $S_z$ basis and expand the rotation operator in a Taylor series. Work out the matrices through the one representing $\hat{S}_y^3$ in order to see the pattern and show that

$$\hat{R}(\theta\mathbf{j}) \xrightarrow[S_z \text{ basis}]{} \begin{pmatrix} \dfrac{1+\cos\theta}{2} & -\dfrac{\sin\theta}{\sqrt{2}} & \dfrac{1-\cos\theta}{2} \\ \dfrac{\sin\theta}{\sqrt{2}} & \cos\theta & -\dfrac{\sin\theta}{\sqrt{2}} \\ \dfrac{1-\cos\theta}{2} & \dfrac{\sin\theta}{\sqrt{2}} & \dfrac{1+\cos\theta}{2} \end{pmatrix}$$

**3.20.** A beam of spin-1 particles is sent through a series of three Stern-Gerlach measuring devices (Fig. 3.11). The first SGz device transmits particles with $S_z = \hbar$ and filters out particles with $S_z = 0$ and $S_z = -\hbar$. The second device, an SGn device, transmits particles with $S_n = \hbar$ and filters out particles with $S_n = 0$ and $S_n = -\hbar$, where the axis $\mathbf{n}$ makes an angle $\theta$ in the $x$-$z$ plane with respect to the $z$ axis. A last SGz device transmits particles with $S_z = -\hbar$ and filters out particles with $S_z = \hbar$ and $S_z = 0$.

(a) What fraction of the particles transmitted by the first SGz device will survive the third measurement? *Note:* The states with $S_n = \hbar$, $S_n = 0$, and $S_n = -\hbar$ in the $S_z$ basis follow directly from applying the rotation operator given in Problem 3.19 to states with $S_z = \hbar$, $S_z = 0$, and $S_z = -\hbar$, respectively.

(b) How must the angle $\theta$ of the SGn device be oriented so as to maximize the number of particles that are transmitted by the final SGz device? What fraction of the particles survive the third measurement for this value of $\theta$?

(c) What fraction of the particles survive the last measurement if the SGn device is removed from the experiment?

Repeat your calculation for parts (a), (b), and (c) if the last SGz device transmits particles with $S_z = 0$ only.

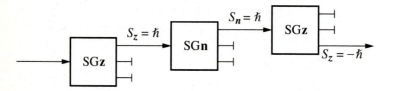

**FIGURE 3.11**

**3.21.** Introduce an angle $\theta$ defined by the relation $\cos\theta = J_z/|\mathbf{J}|$, reflecting the degree to which a particle's angular momentum lines up along the $z$ axis. What is the smallest value of $\theta$ for (a) a spin-$\frac{1}{2}$ particle, (b) a spin-1 particle, and (c) a macroscopic spinning top?

**3.22.** Show that if the two Hermitian operators $\hat{A}$ and $\hat{B}$ have a *complete* set of eigenstates in common, the operators commute.

# CHAPTER

# 4

# TIME
# EVOLUTION

Most of the interesting questions in physics, as in life, concern how things change with time. Just as we have introduced angular momentum operators to generate rotations, we will introduce an operator called the Hamiltonian to generate time translations of our quantum systems. After obtaining the fundamental equation of motion in quantum mechanics, the Schrödinger equation, we will examine the time evolution of a number of two-state systems, including spin precession and magnetic resonance of a spin-$\frac{1}{2}$ particle in an external magnetic field and the ammonia molecule.

## 4.1 THE HAMILTONIAN AND THE SCHRÖDINGER EQUATION

We begin our discussion of time development in quantum mechanics with the time-evolution operator $\hat{U}(t)$ that translates a ket vector forward in time:

$$\hat{U}(t)|\psi(0)\rangle = |\psi(t)\rangle \tag{4.1}$$

where $|\psi(0)\rangle$ is the initial state of the system at time $t = 0$ and $|\psi(t)\rangle$ is the state of the system at time $t$. In order to conserve probability,[1] time evolution should not affect the normalization of the state:

---

[1] In most applications of nonrelativistic quantum mechanics, the total probability of finding the particle doesn't vary in time. However, an electron could disappear, for example, by meeting up with its antiparticle, the positron, and being annihilated. Processes such as particle creation and annihilation require relativistic quantum field theory for their description.

$$\langle \psi(t)|\psi(t) \rangle = \langle \psi(0)|\hat{U}^\dagger(t)\hat{U}(t)|\psi(0) \rangle = \langle \psi(0)|\psi(0) \rangle = 1 \qquad (4.2)$$

which requires

$$\hat{U}^\dagger(t)\hat{U}(t) = 1 \qquad (4.3)$$

Thus the time-evolution operator must be unitary.

Just as we introduced the generator of rotations in (2.29) by considering an infinitesimal rotation, here we consider an infinitesimal time translation:

$$\hat{U}(dt) = 1 - \frac{i}{\hbar}\hat{H}\,dt \qquad (4.4)$$

where the operator $\hat{H}$ is the generator of time translations. Clearly, we need an operator in order to change the initial ket into a different ket at a later time. This is the role played by $\hat{H}$. Unitarity of the time-evolution operator dictates that $\hat{H}$ is a Hermitian operator (see Problem 4.1).

We can now show that $\hat{U}$ satisfies a first-order differential equation in time. Since

$$\hat{U}(t + dt) = \hat{U}(dt)\hat{U}(t) = \left(1 - \frac{i}{\hbar}\hat{H}\,dt\right)\hat{U}(t) \qquad (4.5)$$

then

$$\hat{U}(t + dt) - \hat{U}(t) = \left(-\frac{i}{\hbar}\hat{H}\,dt\right)\hat{U}(t) \qquad (4.6)$$

indicating that the time-evolution operator satisfies[2]

$$i\hbar\frac{d}{dt}\hat{U} = \hat{H}\hat{U}(t) \qquad (4.7)$$

We can also apply the operator equation (4.6) to the initial state $|\psi(0)\rangle$ to obtain

$$i\hbar\frac{d}{dt}|\psi(t)\rangle = \hat{H}|\psi(t)\rangle \qquad (4.8)$$

This equation, known as the *Schrödinger equation*, is the fundamental equation of motion that determines how states evolve in time in quantum mechanics. Schrödinger first proposed the equation in 1926, although not as an equation involving ket vectors but rather as a wave equation that follows from the position-space representation of (4.8), as we will see in Chapter 6.

---

[2] The derivative of an operator is defined in the usual way, that is,

$$\frac{d\hat{U}}{dt} = \lim_{\Delta t \to 0}\frac{\hat{U}(t + \Delta t) - \hat{U}(t)}{\Delta t}$$

If $\hat{H}$ is time independent, we can obtain a closed-form expression for $\hat{U}$ from a series of infinitesimal time translations:

$$\hat{U}(t) = \lim_{N \to \infty} \left[ 1 - \frac{i}{\hbar} \hat{H} \left( \frac{t}{N} \right) \right]^N = e^{-i\hat{H}t/\hbar} \qquad (4.9)$$

where we have taken advantage of Problem 2.1. Then

$$|\psi(t)\rangle = e^{-i\hat{H}t/\hbar} |\psi(0)\rangle \qquad (4.10)$$

Thus in order to solve the equation of motion in quantum mechanics when $\hat{H}$ is time independent, all we need is to know the initial state of the system $|\psi(0)\rangle$ *and* to be able to work out the action of the operator (4.9) on this state.

What is the physical significance of the operator $\hat{H}$? Like the generator of rotations, $\hat{H}$ is a Hermitian operator. From (4.4) we see that the dimensions of $\hat{H}$ are those of Planck's constant divided by time—namely, energy. In addition, when $\hat{H}$ itself is time independent, the expectation value of the observable to which the operator $\hat{H}$ corresponds is also independent of time:

$$\langle \psi(t) | \hat{H} | \psi(t) \rangle = \langle \psi(0) | \hat{U}^{\dagger}(t) \hat{H} \hat{U}(t) | \psi(0) \rangle = \langle \psi(0) | \hat{H} | \psi(0) \rangle \qquad (4.11)$$

since $\hat{H}$ commutes with $\hat{U}$.[3] All of these things suggest that we identify $\hat{H}$ as the energy operator, known as the *Hamiltonian*. Therefore

$$\langle E \rangle = \langle \psi | \hat{H} | \psi \rangle \qquad (4.12)$$

The eigenstates of the Hamiltonian, which are the energy eigenstates satisfying

$$\hat{H} | E \rangle = E | E \rangle \qquad (4.13)$$

play a special role in quantum mechanics. The action of the time-evolution operator $\hat{U}(t)$ on these states is easy to determine using the Taylor series for the exponential:

$$e^{-i\hat{H}t/\hbar} | E \rangle = \left[ 1 - \frac{i\hat{H}t}{\hbar} + \frac{1}{2!} \left( -\frac{i\hat{H}t}{\hbar} \right)^2 + \cdots \right] | E \rangle$$

$$= \left[ 1 - \frac{iEt}{\hbar} + \frac{1}{2!} \left( \frac{iEt}{\hbar} \right)^2 + \cdots \right] | E \rangle = e^{-iEt/\hbar} | E \rangle \qquad (4.14)$$

The operator $\hat{H}$ in the exponent can simply be replaced by the energy eigenvalue *when* the time-evolution operator acts on an eigenstate of the Hamiltonian. Thus if the initial state of the system is an energy eigenstate, $|\psi(0)\rangle = |E\rangle$, then

$$|\psi(t)\rangle = e^{-i\hat{H}t/\hbar} | E \rangle = e^{-iEt/\hbar} | E \rangle \qquad (4.15)$$

---

[3] To establish that $\hat{H}$ commutes with $\hat{U}$, use the Taylor-series expansion for $\hat{U}$, as in (4.14).

The state just picks up an overall phase as time progresses; thus, the physical state of the system does not change with time. We often call such an energy eigenstate a *stationary state* to emphasize this lack of time dependence.

You might worry that physics could turn out to be boring with a lot of emphasis on stationary states. However, if the initial state $|\psi(0)\rangle$ is a *superposition* of energy eigenstates with different energies, the *relative* phases between these energy eigenstates will change with time. Such a state is not a stationary state and the time-evolution operator will generate interesting time behavior. All we need to do to determine this time dependence is to express this initial state as a superposition of energy eigenstates, since we now know the action of the time-evolution operator on each of these states. We will see examples in Sections 4.3 and 4.5.

## 4.2 TIME DEPENDENCE OF EXPECTATION VALUES

The Schrödinger equation permits us to determine in general which variables exhibit time dependence for their expectation values. If we consider an observable $A$, then

$$\frac{d}{dt}\langle A\rangle = \frac{d}{dt}\langle\psi(t)|\hat{A}|\psi(t)\rangle$$

$$= \left(\frac{d}{dt}\langle\psi(t)|\right)\hat{A}|\psi(t)\rangle + \langle\psi(t)|\hat{A}\left(\frac{d}{dt}|\psi(t)\rangle\right) + \langle\psi(t)|\frac{\partial\hat{A}}{\partial t}|\psi(t)\rangle$$

$$= \left(\frac{1}{-i\hbar}\langle\psi(t)|\hat{H}\right)\hat{A}|\psi(t)\rangle + \langle\psi(t)|\hat{A}\left(\frac{1}{i\hbar}\hat{H}|\psi(t)\rangle\right) + \langle\psi(t)|\frac{\partial\hat{A}}{\partial t}|\psi(t)\rangle$$

$$= \frac{i}{\hbar}\langle\psi(t)|[\hat{H},\hat{A}]|\psi(t)\rangle + \langle\psi(t)|\frac{\partial\hat{A}}{\partial t}|\psi(t)\rangle \tag{4.16}$$

The appearance of the last term involving $\partial\hat{A}/\partial t$ in this equation allows for the possibility that the operator depends *explicitly* on time. Equation (4.16) shows that provided the operator corresponding to a variable does not have any explicit time dependence ($\partial\hat{A}/\partial t = 0$), the expectation value of that variable will be a constant of the motion whenever the operator commutes with the Hamiltonian.

What do we mean by explicit time dependence in the operator? Our examples in Sections 4.3 and 4.4 will probably illustrate this best. The Hamiltonian for a spin-$\frac{1}{2}$ particle in a constant magnetic field is given in (4.17). There is no explicit $t$ dependence in $\hat{H}$; therefore substituting $\hat{H}$ for the operator $\hat{A}$ in (4.16) indicates that energy is conserved, since $\hat{H}$ of course commutes with itself. However, if we examine the Hamiltonian (4.34) for a spin-$\frac{1}{2}$ particle in a time-dependent magnetic field, we see explicit time dependence within the Hamiltonian in the factor $\cos\omega t$. Such a Hamiltonian does not lead to an expectation value for the energy of the spin system that is independent of time because $\partial\hat{H}/\partial t \neq 0$. There is clearly an external system that is pumping electromagnetic energy into and out of the spin system.

## 4.3 PRECESSION OF A SPIN-$\frac{1}{2}$ PARTICLE IN A MAGNETIC FIELD

As our first example of quantum dynamics, let's consider the time evolution of the spin state of a spin-$\frac{1}{2}$ particle in a constant magnetic field. We will choose the $z$ axis to be in the direction of the magnetic field, $\mathbf{B} = B_0\mathbf{k}$, and take the charge of the spin-$\frac{1}{2}$ particle to be $q = -e$, that is, to have the same charge as an electron. The energy operator, or Hamiltonian, is given by

$$\hat{H} = -\hat{\boldsymbol{\mu}} \cdot \mathbf{B} = -\frac{gq}{2mc}\hat{\mathbf{S}} \cdot \mathbf{B} = \frac{ge}{2mc}\hat{S}_z B_0 = \omega_0 \hat{S}_z \tag{4.17}$$

where we have used (1.3) to relate the magnetic moment operator $\hat{\boldsymbol{\mu}}$ and the intrinsic spin operator $\hat{\mathbf{S}}$. We have also defined $\omega_0 = geB_0/2mc$. The eigenstates of $\hat{H}$ are the eigenstates of $\hat{S}_z$:

$$\hat{H}|+\mathbf{z}\rangle = \omega_0 \hat{S}_z |+\mathbf{z}\rangle = \frac{\hbar\omega_0}{2}|+\mathbf{z}\rangle = E_+|+\mathbf{z}\rangle \tag{4.18a}$$

$$\hat{H}|-\mathbf{z}\rangle = \omega_0 \hat{S}_z |-\mathbf{z}\rangle = -\frac{\hbar\omega_0}{2}|-\mathbf{z}\rangle = E_-|-\mathbf{z}\rangle \tag{4.18b}$$

where we have denoted the energy eigenvalues of the spin-up and spin-down states by $E_+$ and $E_-$, respectively.

What happens as time progresses? Since the Hamiltonian is time independent, we can take advantage of (4.9):

$$\hat{U}(t) = e^{-i\hat{H}t/\hbar} = e^{-i\omega_0\hat{S}_z t/\hbar} = e^{-i\hat{S}_z\phi/\hbar} = \hat{R}(\phi\mathbf{k}) \tag{4.19}$$

where in the last two steps we have expressed the time-development operator as the rotation operator that rotates states about the $z$ axis by angle $\phi = \omega_0 t$. Thus we see that placing the particle in a magnetic field in the $z$ direction rotates the spin of the particle about the $z$ axis as time progresses, with a period $T = 2\pi/\omega_0$. Using the terminology of classical physics, we say that the particle's spin is precessing about the $z$ axis, as depicted in Fig. 4.1. However, we should be careful not to carry over too completely the classical picture of a magnetic moment precessing in a magnetic field since in the quantum system the angular momentum—and hence the magnetic moment—of the particle cannot actually be pointing in a specific direction because of the uncertainty relations such as (3.75).

In order to see how we work out the details of quantum dynamics, let's take a specific example. With $\mathbf{B} = B_0\mathbf{k}$, we choose $|\psi(0)\rangle = |+\mathbf{x}\rangle$. The state $|+\mathbf{x}\rangle$ is a superposition of eigenstates of $\hat{S}_z$, and therefore from (4.18) it is a superposition of energy eigenstates with different energies. The state at time $t$ is given by

$$|\psi(t)\rangle = e^{-i\hat{H}t/\hbar}\left(\frac{1}{\sqrt{2}}|+\mathbf{z}\rangle + \frac{1}{\sqrt{2}}|-\mathbf{z}\rangle\right)$$

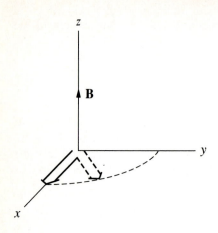

**FIGURE 4.1**
A spin-$\frac{1}{2}$ particle, initially in the state $|+\mathbf{x}\rangle$, precesses about the magnetic field, which points in the $z$ direction.

$$= \frac{e^{-iE_+t/\hbar}}{\sqrt{2}}|+\mathbf{z}\rangle + \frac{e^{-iE_-t/\hbar}}{\sqrt{2}}|-\mathbf{z}\rangle$$

$$= \frac{e^{-i\omega_0 t/2}}{\sqrt{2}}|+\mathbf{z}\rangle + \frac{e^{i\omega_0 t/2}}{\sqrt{2}}|-\mathbf{z}\rangle \tag{4.20}$$

This state does not simply pick up an overall phase as time progresses; it is not a stationary state. Equation (4.20) can also be written as

$$|\psi(t)\rangle = e^{-i\omega_0 t/2}\left(\frac{1}{\sqrt{2}}|+\mathbf{z}\rangle + \frac{e^{i\omega_0 t}}{\sqrt{2}}|-\mathbf{z}\rangle\right) \tag{4.21}$$

which is just an overall phase factor times the spin-up state $|+\mathbf{n}\rangle$ that we found in (3.98), provided we choose the azimuthal angle $\phi = \omega_0 t$.

Let's investigate how the probabilities of being in various spin states and the spin expectation values evolve in time. We use the expression (4.20) for $|\psi(t)\rangle$. Note that

$$|\langle+\mathbf{z}|\psi(t)\rangle|^2 = \left|\frac{e^{i\omega_0 t/2}}{\sqrt{2}}\right|^2 = \frac{1}{2} \tag{4.22a}$$

$$|\langle-\mathbf{z}|\psi(t)\rangle|^2 = \left|\frac{e^{-i\omega_0 t/2}}{\sqrt{2}}\right|^2 = \frac{1}{2} \tag{4.22b}$$

are independent of time, and therefore

$$\langle S_z\rangle = \frac{1}{2}\left(\frac{\hbar}{2}\right) + \frac{1}{2}\left(-\frac{\hbar}{2}\right) = 0 \tag{4.23}$$

is also a constant of the motion.

When we examine the components of the intrinsic spin in the $x$-$y$ plane, we do see explicit time dependence. Since

$$\langle +x|\psi(t)\rangle = \left(\frac{1}{\sqrt{2}}\langle +z| + \frac{1}{\sqrt{2}}\langle -z|\right)\left(\frac{e^{-i\omega_0 t/2}}{\sqrt{2}}|+z\rangle + \frac{e^{i\omega_0 t/2}}{\sqrt{2}}|-z\rangle\right)$$

$$= \frac{1}{\sqrt{2}}(1,\, 1)\frac{1}{\sqrt{2}}\left(\begin{array}{c} e^{-i\omega_0 t/2} \\ e^{i\omega_0 t/2} \end{array}\right) = \cos\frac{\omega_0 t}{2} \tag{4.24}$$

where in the second line we have used the matrix representations for the states in the $S_z$ basis, then

$$|\langle +x|\psi(t)\rangle|^2 = \cos^2\frac{\omega_0 t}{2} \tag{4.25}$$

As a check, note that the probability of the particle being spin up along the $x$ axis is one at time $t = 0$, as required by the initial condition. Similarly,

$$\langle -x|\psi(t)\rangle = \left(\frac{1}{\sqrt{2}}\langle +z| - \frac{1}{\sqrt{2}}\langle -z|\right)\left(\frac{e^{-i\omega_0 t/2}}{\sqrt{2}}|+z\rangle + \frac{e^{i\omega_0 t/2}}{\sqrt{2}}|-z\rangle\right)$$

$$= \frac{1}{\sqrt{2}}(1,\, -1)\frac{1}{\sqrt{2}}\left(\begin{array}{c} e^{-i\omega_0 t/2} \\ e^{i\omega_0 t/2} \end{array}\right) = -i\sin\frac{\omega_0 t}{2} \tag{4.26}$$

and

$$|\langle -x|\psi(t)\rangle|^2 = \sin^2\frac{\omega_0 t}{2} \tag{4.27}$$

The sum of the probabilities to be spin up or spin down along $x$ is one for all times, since these two states $|+x\rangle$ and $|-x\rangle$ form a complete set and probability is conserved. We can determine the average value of $S_x$ either as the sum of the eigenvalues multiplied by the probabilities of obtaining each of these eigenvalues,

$$\langle S_x\rangle = \cos^2\frac{\omega_0 t}{2}\left(\frac{\hbar}{2}\right) + \sin^2\frac{\omega_0 t}{2}\left(-\frac{\hbar}{2}\right) = \frac{\hbar}{2}\cos\omega_0 t \tag{4.28a}$$

or from

$$\langle S_x\rangle = \langle\psi(t)|\hat{S}_x|\psi(t)\rangle$$

$$= \frac{1}{\sqrt{2}}(e^{i\omega_0 t/2},\, e^{-i\omega_0 t/2})\frac{\hbar}{2}\left(\begin{array}{cc} 0 & 1 \\ 1 & 0 \end{array}\right)\frac{1}{\sqrt{2}}\left(\begin{array}{c} e^{-i\omega_0 t/2} \\ e^{i\omega_0 t/2} \end{array}\right)$$

$$= \frac{\hbar}{2}\cos\omega_0 t \tag{4.28b}$$

where we have used the representation for the bra and the ket vector and the operator in the $S_z$ basis.

A similar calculation yields

$$|\langle+\mathbf{y}|\psi(t)\rangle|^2 = \frac{1 + \sin\omega_0 t}{2} \tag{4.29a}$$

$$|\langle-\mathbf{y}|\psi(t)\rangle|^2 = \frac{1 - \sin\omega_0 t}{2} \tag{4.29b}$$

and

$$\langle S_y \rangle = \frac{\hbar}{2}\sin\omega_0 t \tag{4.30}$$

All of these results are consistent with the spin precessing counterclockwise around the $z$ axis with a period $T = 2\pi/\omega_0$, in agreement with our analysis using the explicit form (4.19) of the time-evolution operator as a rotation operator. If the charge $q$ of the particle is taken to be positive rather than negative, $\omega_0$ is negative, and the spin precesses in a clockwise direction.

Before going on to examine some examples of spin precession, it is worthwhile commenting on the time dependence of the expectation values (4.23), (4.28), and (4.30). First, note from (4.16) that

$$\frac{d}{dt}\langle S_z \rangle = \frac{i}{\hbar}\langle\psi|[\hat{H}, \hat{S}_z]|\psi\rangle \tag{4.31}$$

We can see from the explicit form of the Hamiltonian (4.17), which is just a constant multiple of $\hat{S}_z$, that $\hat{H}$ commutes with $\hat{S}_z$ and therefore $\langle S_z \rangle$ is time independent [as (4.23) shows]. It is interesting to consider this result from the perspective of rotational invariance. In particular, with the *external* magnetic field in the $z$ direction, rotations about the $z$ axis leave the spin Hamiltonian unchanged. Thus the generator $\hat{S}_z$ of these rotations must commute with $\hat{H}$, and consequently from (4.31) $\langle S_z \rangle$ is a constant of the motion. The advantage of thinking in terms of symmetry (a symmetry operation is one that leaves the system invariant) is that we can use symmetry to determine the constants of the motion *before* we actually carry out the calculations. We can also know in advance that $\langle S_x \rangle$ and $\langle S_y \rangle$ should vary with time. After all, since $\hat{S}_x$ and $\hat{S}_y$ generate rotations about the $x$ and $y$ axes, respectively, and the Hamiltonian is not invariant under rotations about these axes, $\hat{H}$ does not commute with these generators.

## The g Factor of the Muon

An interesting application of spin precession is the determination of the $g$ factor of the muon. The pion is a spin-0 particle that decays into a muon and a neutrino. The decay mode, for example, of the positively charged pion is $\pi^+ \rightarrow \mu^+ + \nu_\mu$, where the subscript on the neutrino indicates that it is a type of neutrino associated with the muon. Unlike photons, which are both right- and left-circularly polarized, neutrinos are purely left handed. For a spin-$\frac{1}{2}$ particle like the neutrino this means that the projection of the angular momentum along the direction of motion of

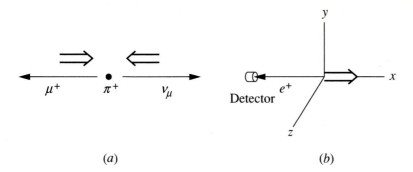

$(a)$ $(b)$

**FIGURE 4.2**
($a$) Conservation of linear and angular momentum requires that the decay of the spin-0 pion in its rest frame produces a left-handed $\mu^+$, since the $\nu_\mu$ is a purely left-handed particle. ($b$) The $\mu^+$ is brought to rest with its spin up along the $x$ axis and allowed to precess in a magnetic field in the $z$ direction. The positrons from the $\mu^+$ decay are emitted preferentially in the opposite direction to the spin of the $\mu^+$.

the neutrino is only $-\hbar/2$. There is no $+\hbar/2$ projection. Conservation of angular momentum in the decay of a pion at rest requires that the muon produced in this decay, which is also a spin-$\frac{1}{2}$ particle, be left handed as well (see Fig. 4.2). The muon is unstable and decays via $\mu^+ \rightarrow e^+ + \nu_e + \bar{\nu}_\mu$, with a lifetime of approximately 2.2 microseconds in the muon's restframe. As a consequence of the weak interactions responsible for the decay, the positron is preferentially emitted in a direction opposite to the spin direction of the muon, and therefore monitoring the decay of the muon gives us information about its spin orientation. If the muon is brought to rest, say in graphite, and placed in a magnetic field of magnitude $B_0$ along the $z$ direction with the initial spin state spin up along the $x$ axis as in our earlier discussion, the spin of the muon will precess. A detector located along the $x$ axis to detect the positrons that are produced in the decay should yield a counting rate proportional to (4.25) as the muon's spin precesses in the magnetic field. Figure 4.3 shows the data from a typical experiment that we can use to obtain a value for the $g$ factor (see Problem 4.7). The first measurements of this sort were carried out by Garwin et al.,[4] who found $g = 2.00 \pm 0.10$. The best experimental value for $g$, good to nine significant figures, comes from a spin-precession experiment carried out at CERN.[5] There is much interest in measuring the $g$ factor because its accurate determination can provide information about the strong and electro-weak interactions at short distances, as well as a detailed test of quantum electrodynamics.

---

[4] R. L. Garwin, L. M. Lederman, and M. Weinrich, *Phys. Rev.* **105**, 1415 (1957).

[5] This measurement [J. Bailey et al., *Nuc Phys.* **B150**, 1 (1979)] takes advantage of the fact that the difference between the frequency at which the muon circles in a constant magnetic field (its cyclotron frequency) and the frequency of spin precession for a muon initially polarized parallel or antiparallel to its direction of motion is proportional to $g - 2$.

Precession frequency = 807.5 kMz

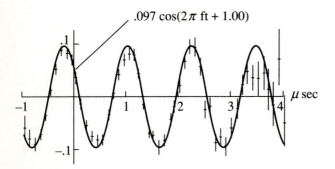

**FIGURE 4.3**
Data on the precession of a muon in a magnetic field of magnitude 60 gauss. [*From J. Sandweiss et al., Phys. Rev. Lett.* **30,** *1002 (1973).*]

## $2\pi$ Rotations of a Spin-$\frac{1}{2}$ Particle

As a second illustration of spin precession, let's consider a beautiful experiment that demonstrates that rotating a spin-$\frac{1}{2}$ particle through $2\pi$ radians causes the state of the particle to change sign, as shown in (2.43). At first thought, it might not seem feasible to test this prediction since the state of the particle picks up an overall phase as the result of such a rotation. However, as we saw in our discussion of Experiment 4 in Chapter 1, a single particle can have amplitudes to take two separate paths and how these amplitudes add up, or interfere, depends on their *relative* phases. Werner et al.[6] used neutrons as the spin-$\frac{1}{2}$ particles and constructed an interferometer of the type first developed for X rays. Their schematic of the interferometer is shown in Fig. 4.4a. A monoenergetic beam of thermal neutrons is split by Bragg reflection from a crystal of silicon into two beams at A, one of which traverses path ABD and the other path ACD. A silicon crystal is used to deflect the beams at B and C, as well as to recombine them at D. As in a typical interferometer, there will be constructive or destructive interference depending on the path difference between the the two legs ABD and ACD. The relative phase of the two beams can be altered, however, by allowing one of the beams to pass through a uniform magnetic field. As indicated by (4.21), there will be an additional phase difference of

$$\phi = \omega_0 T = \frac{g e B_0}{2 M c} T \tag{4.32}$$

introduced, where $M$ is the mass of the nucleon, $B_0$ is the magnitude of the uniform field on the path AC, and $T$ is the amount of time the beam spends in the

[6] S. A. Werner, R. Colella, A. W. Overhauser, and C. F. Eagen, *Phys. Rev. Lett.* **35,** 1053 (1975).

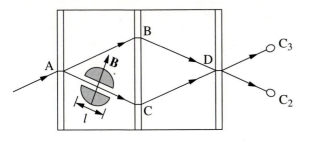

$(a)$

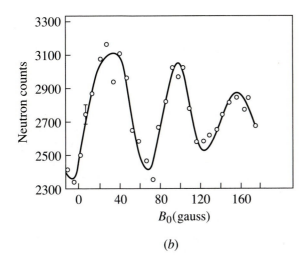

$B_0(\text{gauss})$

$(b)$

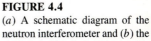

**FIGURE 4.4**
$(a)$ A schematic diagram of the neutron interferometer and $(b)$ the difference in counts between the counters $C_3$ and $C_2$ as a function of the magnetic field strength. [*From Werner et al.*, Phys. Rev. Lett. **35**, *1053 (1975)*.]

magnetic field.[7] In the experiment, the magnitude of the magnetic field strength could be varied between 0 and 500 gauss. The difference in $B_0$, which we call $\Delta B$, needed to produce successive maxima is determined by the requirement that

$$\frac{g\,e\Delta B}{2M\,c}\,T = 4\pi \tag{4.33}$$

---

[7] Three comments about this expression are in order. (1) Since a neutron is a neutral particle, it might seem strange for it to have a magnetic moment at all. That $g/2 = -1.91$ is an indication that the neutron is not itself a fundamental particle, but rather is composed of more fundamental charged constituents called quarks. (2) In nuclear physics, magnetic moments are generally expressed in terms of the nuclear magneton where the mass $M$ in (4.32) is really the mass of the proton. Since the mass of the proton differs from the mass of the neutron by less than 0.2 percent, we can ignore this distinction unless we are interested in results to this accuracy. (3) The time $T$ can be expressed as $T = lM/p$, where $p$ is the momentum of the neutron and $l$ is the path length in the magnetic field region. We can then use the de Broglie relation $p = h/\lambda$ [see (6.56)] to express this time in terms of the wavelength of the neutron. It is actually $\lambda$ that is determined when selecting the energy of the neutron beam using the techniques of crystal diffraction.

Notice that we have used the fact that a rotation by $4\pi$ radians is required to return the overall phase of spin-$\frac{1}{2}$ ket to its original value. As shown in Fig. 4.4b, Werner et al. found $\Delta B = 62 \pm 2$ gauss in their experiment. If rotating a ket by $2\pi$ radians were sufficient to keep the phase of the ket the same, the observed value of $\Delta B$ would have been one half as large as that found in the experiment. Thus the experimental results give an unambiguous confirmation of the unusual prediction (2.43) of quantum mechanics for spin-$\frac{1}{2}$ particles.

## 4.4 MAGNETIC RESONANCE

When a spin-$\frac{1}{2}$ particle precesses in a magnetic field in the $z$ direction, the probability of the particle being spin up or spin down along $z$ doesn't vary with time, as shown in (4.22). After all, the states $|+\mathbf{z}\rangle$ and $|-\mathbf{z}\rangle$ are stationary states of the Hamiltonian (4.17). However, if we alter the Hamiltonian by applying in addition an oscillating magnetic field transverse to the $z$ axis, we can induce transitions between these two states by properly adjusting the frequency of this transverse field. The energy difference $E_{+} - E_{-} = \hbar\omega_0$ can then be measured with high accuracy. This magnetic resonance gives us an excellent way of determining $\omega_0$. Initially, physicists used magnetic resonance techniques to make accurate determinations of $g$ factors and thus gain fundamental information about the nature of these particles. On the other hand, with known values for $g$, one can use the technique to make accurate determinations of the magnetic field $B_0$ in which the spin is precessing. For electrons or nuclei in atoms or molecules, this magnetic field is a combination of the known externally applied field and the local magnetic field at the site of the electron or nucleus. This local field provides valuable information about the nature of the bonds that electrons in the atom make with neighboring atoms in a molecule. More recently, magnetic resonance imaging (MRI) has become an important diagnostic tool in medicine.

The spin Hamiltonian for magnetic resonance is given by

$$\hat{H} = -\hat{\boldsymbol{\mu}} \cdot \mathbf{B} = -\frac{gq}{2mc}\hat{\mathbf{S}} \cdot \mathbf{B} = -\frac{gq}{2mc}\hat{\mathbf{S}} \cdot (B_1 \cos\omega t\mathbf{i} + B_0\mathbf{k}) \qquad (4.34)$$

where the magnetic field includes a constant magnetic field in the $z$ direction and an oscillating magnetic field in the $x$ direction. As we did for spin precession, we choose $q = -e$ and set $egB_0/2mc = \omega_0$. We also define $egB_1/2mc = \omega_1$. The Hamiltonian can now be written as

$$\hat{H} = \omega_0\hat{S}_z + \omega_1(\cos\omega t)\hat{S}_x \qquad (4.35)$$

This Hamiltonian is time dependent, so we cannot use the expression (4.9) for the time-evolution operator.[8]

---

[8] If we were to choose our total system to be sufficiently large, including, in this example, the energy of both the spin system *and* the elecromagnetic field, we would find that the total energy is conserved. Here we are treating the magnetic field as an external field acting on the quantum spin system.

To determine how spin states evolve in time, we return to the Schrödinger equation (4.8). Let's take the state of the particle at time $t = 0$ to be $|+\mathbf{z}\rangle$. We will work in the $S_z$ basis and express $|\psi(t)\rangle$ in this basis by

$$|\psi(t)\rangle \rightarrow \begin{pmatrix} a(t) \\ b(t) \end{pmatrix} \tag{4.36}$$

with the initial condition

$$|\psi(0)\rangle \rightarrow \begin{pmatrix} 1 \\ 0 \end{pmatrix} \tag{4.37}$$

In this basis, the time-development equation $\hat{H}|\psi(t)\rangle = i\hbar d|\psi(t)\rangle/dt$ is given by

$$\frac{\hbar}{2} \begin{pmatrix} \omega_0 & \omega_1 \cos\omega t \\ \omega_1 \cos\omega t & -\omega_0 \end{pmatrix} \begin{pmatrix} a(t) \\ b(t) \end{pmatrix} = i\hbar \begin{pmatrix} \dot{a}(t) \\ \dot{b}(t) \end{pmatrix} \tag{4.38}$$

where $\dot{a}(t) = da/dt$ and $\dot{b}(t) = db/dt$. This coupled set of first-order differential equations cannot be solved exactly. In practice, however, the transverse field $B_1$ is significantly weaker than the field $B_0$ in the $z$ direction and therefore the frequency $\omega_1$ is considerably smaller than $\omega_0$. We can take advantage of this fact to obtain an approximate solution to (4.38).

First, note that if $\omega_1 = 0$, the solution to (4.38) is

$$a(t) = a(0)e^{-i\omega_0 t/2} \qquad \text{and} \qquad b(t) = b(0)e^{i\omega_0 t/2} \tag{4.39}$$

in agreement with the time dependence of our earlier results (4.20). This suggests that we try writing

$$\begin{pmatrix} a(t) \\ b(t) \end{pmatrix} = \begin{pmatrix} c(t)e^{-i\omega_0 t/2} \\ d(t)e^{i\omega_0 t/2} \end{pmatrix} \tag{4.40}$$

where we expect that we have included the major part of the time dependence in the exponentials. If we substitute (4.40) into (4.38), we obtain

$$\begin{aligned} i\begin{pmatrix} \dot{c}(t) \\ \dot{d}(t) \end{pmatrix} &= \frac{\omega_1}{2}\cos\omega t \begin{pmatrix} d(t)e^{i\omega_0 t} \\ c(t)e^{-i\omega_0 t} \end{pmatrix} \\ &= \frac{\omega_1}{4}\begin{pmatrix} \left(e^{i(\omega_0+\omega)t} + e^{i(\omega_0-\omega)t}\right)d(t) \\ \left(e^{i(\omega-\omega_0)t} + e^{-i(\omega_0+\omega)t}\right)c(t) \end{pmatrix} \end{aligned} \tag{4.41}$$

Unless $\omega$ is chosen to be very near to $\omega_0$, both the exponentials in the second line of (4.41) are rapidly oscillating functions that when multiplied by a more slowly oscillating function such as $c(t)$ or $d(t)$, whose time scale is set by $\omega_1$, will cause the right-hand side of (4.41) to average to zero.[9] However, if $\omega$ is near $\omega_0$, the

---

[9] In a typical electron spin resonance (ESR) experiment in a field of $10^4$ gauss, $\omega_0 \sim 10^{11}$ Hz, while for nuclear magnetic resonance (NMR) with protons in a comparable field, $\omega_0 \sim 10^8$ Hz.

terms oscillating at $\omega_0 + \omega$ can be neglected with respect to those oscillating at $\omega_0 - \omega$, and these latter terms are now oscillating sufficiently slowly that $c$ and $d$ vary with time. Here we will solve for this time dependence when $\omega$ is equal to $\omega_0$, the resonant condition, and leave the more general case as a problem.

Setting $\omega = \omega_0$ and neglecting the exponentials oscillating at $2\omega_0$, we obtain

$$i\begin{pmatrix} \dot{c}(t) \\ \dot{d}(t) \end{pmatrix} = \frac{\omega_1}{4}\begin{pmatrix} d(t) \\ c(t) \end{pmatrix} \tag{4.42}$$

If we take the time derivative of these two coupled equations and then use (4.42) to eliminate the terms involving a single derivative, we obtain the uncoupled second-order differential equations

$$\begin{pmatrix} \ddot{c}(t) \\ \ddot{d}(t) \end{pmatrix} = -\left(\frac{\omega_1}{4}\right)^2 \begin{pmatrix} c(t) \\ d(t) \end{pmatrix} \tag{4.43}$$

The solution to (4.43) satisfying the initial condition $c(0) = 1$ and $d(0) = 0$ [see (4.42)] is $c(t) = \cos(\omega_1 t/4)$ and $d(t) = -i\sin(\omega_1 t/4)$. Thus the probability of finding the particle in the state $|-\mathbf{z}\rangle$ at time $t$ is given by

$$|\langle -\mathbf{z}|\psi(t)\rangle|^2 = b^*(t)b(t) = d^*(t)d(t) = \sin^2\frac{\omega_1 t}{4} \tag{4.44a}$$

for a spin-$\frac{1}{2}$ particle that initially resides in the state $|+\mathbf{z}\rangle$ at $t = 0$. Similarly, the probability of finding the particle in the state $|+\mathbf{z}\rangle$ is given by

$$|\langle +\mathbf{z}|\psi(t)\rangle|^2 = a^*(t)a(t) = c^*(t)c(t) = \cos^2\frac{\omega_1 t}{4} \tag{4.44b}$$

Of course, these two probabilities sum to one, since these two states form a complete set and probability is conserved in time. If a particle initially in the state $|+\mathbf{z}\rangle$ makes a transition to the state $|-\mathbf{z}\rangle$, the energy of the spin system is reduced by $E_+ - E_- = \hbar\omega_0$, assuming $\omega_0 > 0$. This energy is added to the electromagnetic energy of the oscillating field that is *stimulating* the transition. For $t$ between zero and $2\pi/\omega_1$, the probability of making a transition to the lower-energy state grows until $b^*(t)b(t) = 1$ and $a^*(t)a(t) = 0$. Then the particle is in the state $|-\mathbf{z}\rangle$. Next for $t$ between $2\pi/\omega_1$ and $4\pi/\omega_1$, the probability of being in the lower-energy state decreases and the probability of being in the higher-energy state grows as the system absorbs energy back from the electromagnetic field. This cycle of emission and absorption continues indefinitely (see Fig. 4.5).

As we noted earlier, there is a probability of inducing a transition between the two spin states even when the frequency $\omega$ is not equal to $\omega_0$. If the system is initially in the spin-up state, the probability of being in the lower-energy spin-down state at time $t$ is given by Rabi's formula (see Problem 4.9),

$$|\langle -\mathbf{z}|\psi(t)\rangle|^2 = \frac{\omega_1^2/4}{(\omega_0 - \omega)^2 + \omega_1^2/4}\sin^2\frac{\sqrt{(\omega_0 - \omega)^2 + \omega_1^2/4}}{2}t \tag{4.45}$$

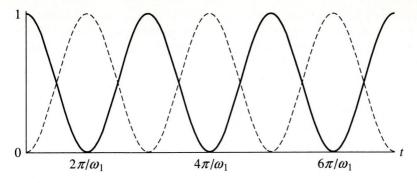

**FIGURE 4.5**
The probabilities $|\langle +\mathbf{z}|\psi(t)\rangle|^2$ (solid line) and $|\langle -\mathbf{z}|\psi(t)\rangle|^2$ (dashed line) for a spin-$\frac{1}{2}$ particle that is in the state $|+\mathbf{z}\rangle$ at $t = 0$ when the time-dependent magnetic field in the $x$ direction is tuned to the resonant frequency.

The maximum probability of transition is plotted as a function of $\omega$ in Fig. 4.6. Monitoring the losses and gains in energy to the oscillating field as a function of $\omega$ gives us a nice handle on whether the frequency of this field is indeed the resonant frequency of the spin system. Notice in (4.45) that making $B_1$ smaller makes $\omega_1$ smaller and the curve in Fig. 4.6 narrower, permitting a more accurate determination of $\omega_0$ .

In practice, the physical spin system consists of a large number of particles, either electrons or nuclei, that are in thermal equilibrium at some temperature $T$. The relative number of particles in the two energy states is given by the Boltzmann distribution, so slightly more of the particles are in the lower-energy state. There will be a net absorption of energy proportional to the *difference*

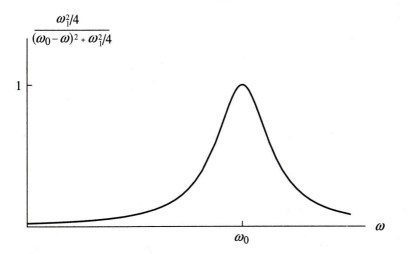

**FIGURE 4.6**
A sketch of the magnetic-resonance transition probability as a function of the frequency $\omega$ of the time-dependent magnetic field.

in populations of the two levels, since the magnetic field induces transitions in both directions. Of course, if we just sit at the resonant frequency, the populations will equalize quickly and there will be no more absorption. Thus, in practice, it is necessary to move the system away from resonance, often by varying slightly the field $B_0$, thus permitting thermal equilibrium to be reestablished. In the case of nuclear magnetic resonance, the nuclear magnetic moments are located at the center of the atoms, surrounded by electrons, and are relatively isolated thermally from their surroundings. Therefore, it can be difficult to get the nuclear spins to "relax" back to thermal equilibrium, even when the resonance condition no longer persists. In this case thermal contact can be increased by doping the sample with paramagnetic ions.

## 4.5 THE AMMONIA MOLECULE AND THE AMMONIA MASER

As our last example in this chapter of a two-state system, we consider the ammonia molecule.[10] At first glance, the ammonia molecule does not seem a promising hunting ground for a two-state system. After all, $NH_3$ is a complicated system of four nuclei and ten electrons interacting with each other to form bonds between the atoms, making the stable state of the molecule a pyramid with three hydrogen atoms forming the base and a nitrogen atom at the apex (see Fig. 4.7). Here we won't worry about all of this internal dynamics, nor will we concern ourselves with how the molecule as a whole is rotating or translating. Rather, we will take the molecule to be in a fixed state as regards all of these degrees of freedom and focus on the location of the nitrogen atom; namely, is the nitrogen atom above or below the plane formed by the three hydrogen atoms? The existence of a reasonably well-defined location for the nitrogen atom indicates that there is a potential well in which the nitrogen atom finds it energetically advantageous

---

[10] Our discussion of the ammonia molecule as a two-state system is inspired by the treatment in vol. 3 of *The Feynman Lectures on Physics*.

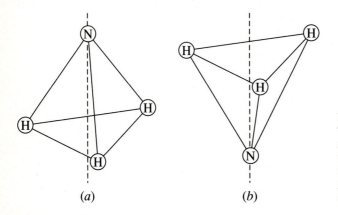

(a)  (b)

**FIGURE 4.7**
The two states of the ammonia molecule with (a) the nitrogen atom above the plane in state $|1\rangle$ and (b) the nitrogen atom below the plane in state $|2\rangle$.

to reside. However, the geometry of the molecule tells us that if there is a potential well above the plane, there must be a similar well below the plane. Which state does the nitrogen atom choose? Nature likes to find the lowest-energy state, so we are led to solve the energy eigenvalue problem to determine the allowed states and energies of the system.

We introduce two kets:

$$|1\rangle = |N \text{ above the plane}\rangle \quad \text{and} \quad |2\rangle = |N \text{ below the plane}\rangle \quad (4.46)$$

and construct the matrix representation of the Hamiltonian using these two states, depicted in Fig. 4.7, as basis states. The symmetry of the two physical configurations suggests that the expectation value of the Hamiltonian in these states, an energy that we denote by $E_0$, should be the same for the two states. Thus

$$\langle 1|\hat{H}|1\rangle = \langle 2|\hat{H}|2\rangle = E_0 \quad (4.47)$$

where $\hat{H}$ is the Hamiltonian of the system. What about the off-diagonal matrix elements? If we look back to our discussion of time evolution of the spin system in magnetic resonance, we see that when we set the off-diagonal matrix elements of the Hamiltonian in (4.38) equal to zero, the spin-up and spin-down states were stationary states; if the system were in one of these states initially, it remained in that state forever, as (4.39) shows. For the ammonia molecule, the vanishing of the off-diagonal matrix elements, such as $\langle 2|\hat{H}|1\rangle$, would mean that a molecule initially, for example, in the state $|1\rangle$, with the $N$ atom above the plane, would remain in that state. Now, if the potential barrier between the two wells were infinitely high, there would be no chance that a nitrogen atom above the plane in state $|1\rangle$ would be found below the plane in state $|2\rangle$. However, although the barrier formed by the three hydrogen atoms is large, it is not infinite, and there is a small amplitude for a nitrogen atom to tunnel between the two states. This means that the off-diagonal matrix element $\langle 2|\hat{H}|1\rangle$ is nonzero. We will take its value to be $-A$. Thus in the $|1\rangle$-$|2\rangle$ basis

$$\hat{H} \rightarrow \begin{pmatrix} \langle 1|\hat{H}|1\rangle & \langle 1|\hat{H}|2\rangle \\ \langle 2|\hat{H}|1\rangle & \langle 2|\hat{H}|2\rangle \end{pmatrix} = \begin{pmatrix} E_0 & -A \\ -A & E_0 \end{pmatrix} \quad (4.48)$$

where $A$ is a positive constant. We will see that this sign for $A$ is required to get the correct disposition of the energy levels. Note that if, as we have presumed, the off-diagonal matrix elements are real, Hermiticity of $\hat{H}$, as well as the symmetry of the situation, requires that they be equal. In principle, if we were really adept at carrying out quantum mechanics calculations for molecules, we would be able to calculate the value of $A$ from first principles. We think we understand all the physics of the electromagnetic interactions responsible for holding the molecule together, but $NH_3$ is composed of a large number of particles and no one is able to work out all the details. We can think of (4.48) as a phenomenological Hamiltonian where the value for a constant such as $A$ must be determined experimentally.

We are now ready to determine the energy eigenstates and eigenvalues of $\hat{H}$. The energy eigenvalue equation

$$\hat{H}|\psi\rangle = E|\psi\rangle \quad (4.49)$$

in the $|1\rangle$-$|2\rangle$ basis is given by

$$\begin{pmatrix} E_0 & -A \\ -A & E_0 \end{pmatrix} \begin{pmatrix} \langle 1|\psi\rangle \\ \langle 2|\psi\rangle \end{pmatrix} = E \begin{pmatrix} \langle 1|\psi\rangle \\ \langle 2|\psi\rangle \end{pmatrix} \qquad (4.50)$$

The eigenvalues are determined by requiring

$$\begin{vmatrix} E_0 - E & -A \\ -A & E_0 - E \end{vmatrix} = 0 \qquad (4.51)$$

which yields $E = E_0 \pm A$. We denote the energy eigenstate with energy $E_I = E_0 - A$ by $|I\rangle$. Substituting the eigenvalue into (4.50) shows that $\langle 1|I\rangle = \langle 2|I\rangle$, so that we may write[11]

$$|I\rangle = \frac{1}{\sqrt{2}}|1\rangle + \frac{1}{\sqrt{2}}|2\rangle \qquad (4.52)$$

Energy eigenstate $|II\rangle$ with energy $E_{II} = E_0 + A$ satisfies $\langle 1|II\rangle = -\langle 2|II\rangle$ and thus may be written as

$$|II\rangle = \frac{1}{\sqrt{2}}|1\rangle - \frac{1}{\sqrt{2}}|2\rangle \qquad (4.53)$$

The existence of tunneling between the states $|1\rangle$ and $|2\rangle$ has split the energy states of the molecule into two states with different energies, one with energy $E_0 - A$ and the other with energy $E_0 + A$, as shown in Fig. 4.8. The wavelength of the electromagnetic radiation emitted when the molecule makes a transition between these two energy states is observed to be $1\frac{1}{4}$ cm, corresponding to an energy separation $E_{II} - E_I = h\nu = hc/\lambda$ of $10^{-4}$ eV. This small energy separation is to be compared with a typical spacing of atomic energy levels that is on the order of electron volts, requiring optical or uv photons to excite the atom. Molecules also have vibrational and rotational energy levels, but these modes are excited by photons in the infrared or far infrared, respectively. Exciting an ammonia molecule from state $|I\rangle$ to state $|II\rangle$ requires electromagnetic radiation of an even longer wavelength, in the microwave part of the spectrum. The smallness of this energy difference $E_{II} - E_I = 2A$ is a reflection of the smallness of the amplitude for tunneling from state $|1\rangle$ to $|2\rangle$.

Notice that neither in energy eigenstate $|I\rangle$ nor $|II\rangle$ is the nitrogen atom located above or below the plane formed by the three hydrogen atoms. Under the transformation $|1\rangle \leftrightarrow |2\rangle$ that flips the position of the nitrogen atom, the state $|I\rangle$ is symmetric, that is, $|I\rangle \to |I\rangle$, while the state $|II\rangle$ is antisymmetric, that is, $|II\rangle \to -|II\rangle$. We can, however, localize the nitrogen atom above the plane, for example, by superposing the energy eigenstates:

---

[11] In the normalization of the state, we have neglected the nonzero amplitude $\langle 2|1\rangle$ because of its small magnitude.

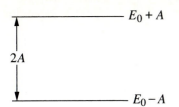

$$E_0 + A$$

$$2A$$

$$E_0 - A$$

**FIGURE 4.8**
The two energy levels of the ammonia molecule.

$$|1\rangle = \frac{1}{\sqrt{2}}|I\rangle + \frac{1}{\sqrt{2}}|II\rangle \tag{4.54}$$

If $|\psi(0)\rangle = |1\rangle$, then

$$|\psi(t)\rangle = e^{-i\hat{H}t/\hbar}\left(\frac{1}{\sqrt{2}}|I\rangle + \frac{1}{\sqrt{2}}|II\rangle\right)$$

$$= \frac{e^{-i(E_0-A)t/\hbar}}{\sqrt{2}}|I\rangle + \frac{e^{-i(E_0+A)t/\hbar}}{\sqrt{2}}|II\rangle$$

$$= e^{-i(E_0-A)t/\hbar}\left(\frac{1}{\sqrt{2}}|I\rangle + \frac{e^{-2iAt/\hbar}}{\sqrt{2}}|II\rangle\right) \tag{4.55}$$

where in the last step we have pulled an overall phase factor out in front of the ket. Since the initial state of the molecule is a superposition of energy states with different energies, the molecule is not in a stationary state. We see that the *relative* phase between the two energy eigenstates changes with time, and thus the state of the molecule is really varying in time. The motion is periodic with a period $T$ determined from $2AT/\hbar = 2\pi$. What is nature of the motion? When $t = T/2$, the relative phase is $\pi$ and

$$|\psi(T/2)\rangle = (\text{overall phase})\left(\frac{1}{\sqrt{2}}|I\rangle - \frac{1}{\sqrt{2}}|II\rangle\right) = (\text{overall phase})\,|2\rangle \tag{4.56}$$

The nitrogen atom is located below the plane. Thus the nitrogen atom oscillates back and forth above and below the plane with a frequency $\nu = 1/T = A\pi/\hbar = 2A/h = 24$ GHz, which is the same as the frequency of the electromagnetic radiation emitted when the molecule makes a transition between states $|II\rangle$ and $|I\rangle$.

## The Molecule in a Static External Electric Field

Since the valence electrons in the ammonia molecule tend to reside somewhat closer to the nitrogen atom, the nitrogen atom is somewhat negative and the hydrogen atoms are somewhat positive. Thus the molecule has an *electric* dipole moment $\boldsymbol{\mu}_e$ directed away from the nitrogen atom toward the plane formed by

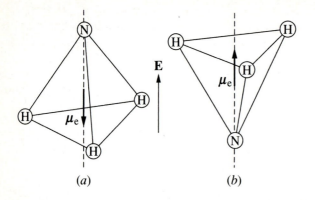

**FIGURE 4.9**
The electric dipole moment $\boldsymbol{\mu}_e$ of the ammonia molecule in (a) state $|1\rangle$ and (b) state $|2\rangle$. In the presence of an external electric field **E**, the two states acquire different energies, as indicated in (4.57).

the hydrogen atoms. Just as the *magnetic* dipole moment associated with its spin angular momentum allowed us to interact with a spin-$\frac{1}{2}$ particle in Stern-Gerlach or spin-precession experiments by inserting it in a magnetic field, we can interact with the ammonia molecule by placing it in an external electric field **E**, as indicated in Fig. 4.9. There is an energy of interaction with the electric field of the form $-\boldsymbol{\mu}_e \cdot \mathbf{E}$ that will differ depending on whether the nitrogen atom is above the plane in state $|1\rangle$ or below the plane in state $|2\rangle$. The presence of this electric field modifies the matrix representation of the Hamiltonian in the $|1\rangle$-$|2\rangle$ basis:[12]

$$\hat{H} \rightarrow \begin{pmatrix} \langle 1|\hat{H}|1\rangle & \langle 1|\hat{H}|2\rangle \\ \langle 2|\hat{H}|1\rangle & \langle 2|\hat{H}|2\rangle \end{pmatrix} = \begin{pmatrix} E_0 + \mu_e|\mathbf{E}| & -A \\ -A & E_0 - \mu_e|\mathbf{E}| \end{pmatrix} \qquad (4.57)$$

where we assume the external field is sufficiently weak that it does not affect the amplitude for the nitrogen atom to tunnel through the barrier. The eigenvalues are determined by the requirement that

$$\begin{vmatrix} E_0 + \mu_e|\mathbf{E}| - E & -A \\ -A & E_0 - \mu_e|\mathbf{E}| - E \end{vmatrix} = 0 \qquad (4.58a)$$

or

$$E = E_0 \pm \sqrt{(\mu_e|\mathbf{E}|)^2 + A^2} \qquad (4.58b)$$

See Fig. 4.10. Most external electric fields satisfy $\mu_e|\mathbf{E}| \ll A$, so we can expand the square root in a Taylor series or a binomial expansion to obtain

$$E \cong E_0 \pm A \pm \frac{1}{2}\frac{(\mu_e|\mathbf{E}|)^2}{A} \qquad (4.59)$$

As in the Stern-Gerlach experiments where we used an inhomogeneous magnetic field to make measurements of the intrinsic spin and select spin-up

---

[12] It is customary to use $\boldsymbol{\mu}_e$ for the electric dipole moment to avoid confusion with the symbol for momentum. We also use $|\mathbf{E}|$ for the magnitude of the electric field to avoid confusion with the symbol for energy.

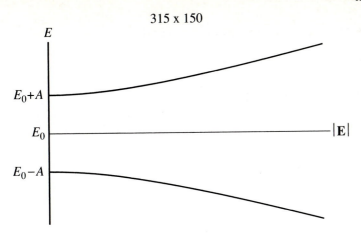

**FIGURE 4.10**
The energy levels of the ammonia molecule in an external electric field.

and spin-down states, here we use an inhomogeneous electric field to separate $NH_3$ molecules into those in states $|I\rangle$ and $|II\rangle$. If we call the direction in which the electric field increases the $z$ direction, then the force in that direction is given by

$$F_z \cong -\frac{\partial}{\partial z}\left[\pm\frac{(\mu_e|\mathbf{E}|)^2}{2A}\right] \tag{4.60}$$

Notice that the minus sign in (4.59) corresponds to the state with energy $E_0 - A$ in the absence of the external electric field. Hence a molecule in state $|I\rangle$ will be deflected in the positive $z$ direction, while a molecule in the state $|II\rangle$ will be deflected in the negative $z$ direction, as shown in Fig. 4.11. Because of the small value of $A$ in the denominator in (4.60), it is relatively easy to separate a beam of ammonia molecules in, for example, a gas jet by sending them through a region in which there is a large gradient in the electric field.

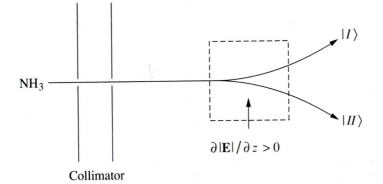

**FIGURE 4.11**
A beam of ammonia molecules passing through a region in which there is a strong electric field gradient separates into two beams, one with the molecules in state $|I\rangle$ and the other with the molecules in state $|II\rangle$.

## The Molecule in a Time-Dependent Electric Field

We are now ready to induce transitions between states $|I\rangle$ and $|II\rangle$ by applying a time-dependent electric field of the form $\mathbf{E} = \mathbf{E}_0 \cos \omega t$. There will be a resonant absorption or emission of electromagnetic energy, provided that $\hbar\omega$ is equal to the energy difference $2A$ between the two states. This sounds similar to the magnetic resonance effects that we treated in the previous section, and in fact the mathematics describing the two problems is essentially identical. To see this, consider the Hamiltonian in the $|1\rangle$-$|2\rangle$ basis as given in (4.57) with a time-dependent electric field. If we transform to the $|I\rangle$-$|II\rangle$ basis, we obtain (see Problem 4.10)

$$\hat{H} \rightarrow \begin{pmatrix} \langle I|\hat{H}|I\rangle & \langle I|\hat{H}|II\rangle \\ \langle II|\hat{H}|I\rangle & \langle II|\hat{H}|II\rangle \end{pmatrix} = \begin{pmatrix} E_0 - A & \mu_e|\mathbf{E}_0|\cos \omega t \\ \mu_e|\mathbf{E}_0|\cos \omega t & E_0 + A \end{pmatrix} \quad (4.61)$$

Comparing this matrix with that for the Hamiltonian in (4.38) of a spin-$\frac{1}{2}$ particle in an oscillating transverse magnetic field, we see that it is possible to draw a one-to-one correspondence between each term in the two matrices: $E_+ = \hbar\omega_0/2 \rightarrow E_0 + A$, $E_- = -\hbar\omega_0/2 \rightarrow E_0 - A$, and $\hbar\omega_1/2 \rightarrow \mu_e|\mathbf{E}_0|$. Thus one can follow the steps leading to the probability of making a transition between the spin-up and spin-down states in (4.44) and apply them to this new problem to obtain the probability of making a transition between states $|I\rangle$ and $|II\rangle$. Therefore, at resonance the probability of finding the ammonia molecule in state $|I\rangle$ for a molecule initially in the state $|II\rangle$ at time $t = 0$ is

$$|\langle I|\psi(t)\rangle|^2 = \sin^2 \frac{\mu_e|\mathbf{E}_0|t}{2\hbar} \quad (4.62)$$

We can combine the results of this section and the preceding one to provide a description of a simple ammonia MASER (Microwave Amplification by Stimulated Emission of Radiation). First we use an inhomogeneous electric field to select a beam of ammonia molecules that are in the upper-energy state $|II\rangle$; then we send this beam into a microwave cavity whose resonant frequency is tuned to 24 GHz, the resonant frequency of the ammonia molecule. If the molecules spend a time $T$ in the cavity such that $\mu_e|\mathbf{E}_0|T/2\hbar = \pi/2$, then according to (4.62) they will all make transitions from state $|II\rangle$ to state $|I\rangle$. The molecular energy released in this transition is fed into the cavity, where it can be used as microwave radiation.[13]

---

[13] The key element missing from our discussion of the MASER is the coherent nature of the radiation that it produces. So far we have treated the electromagnetic field as a classical field and have not taken into account its quantum properties, that is, that it is really composed of photons. We will examine this issue in more detail in Chapter 14.

## 4.6   THE ENERGY-TIME UNCERTAINTY RELATION

As our last topic on time evolution, let's consider the energy-time uncertainty relation

$$\Delta E \Delta t \gtrsim \frac{\hbar}{2} \tag{4.63}$$

This uncertainty relation is somewhat of a misnomer; unlike our previous uncertainty relations such as (3.74), only $\Delta E$ in (4.63) is a legitimate uncertainty. It reflects the spread in energy characterizing a particular state. To see the meaning of $\Delta t$, consider an example. Let's return to the ammonia molecule that is initially in state $|1\rangle$, with the nitrogen atom above the plane. As (4.55) shows, this state is not an energy eigenstate but a superposition of two energy eigenstates with different energies. The uncertainty in the energy of a molecule in this state is given by

$$
\begin{aligned}
\Delta E &= \left( \langle E^2 \rangle - \langle E \rangle^2 \right)^{1/2} \\
&= \left\{ \tfrac{1}{2}(E_0 + A)^2 + \tfrac{1}{2}(E_0 - A)^2 - \left[ \tfrac{1}{2}(E_0 + A) + \tfrac{1}{2}(E_0 - A) \right]^2 \right\}^{1/2} \\
&= A
\end{aligned}
\tag{4.64}
$$

We can express the time evolution of the state (4.55) in terms of the uncertainty $\Delta E$ as

$$|\psi(t)\rangle = (\text{overall phase}) \left( \frac{1}{\sqrt{2}} |I\rangle + \frac{e^{-2i \Delta E t/\hbar}}{\sqrt{2}} |II\rangle \right) \tag{4.65}$$

How long do we have to wait before the state of the molecule changes? The answer to this question is the quantity we call $\Delta t$. To be sure the state (4.65) has changed, we need to be sure the relative phase between the energy eigenstates $|I\rangle$ and $|II\rangle$ has changed significantly from its value of zero at $t = 0$ to something of order unity. This requires that the time interval $\Delta t$ satisfy $2\Delta E \Delta t/\hbar \gtrsim 1$, which is simply (4.63).[14] In (4.55) we saw that the time required for the nitrogen atom to appear below the plane in state $|2\rangle$ is determined by the requirement that the relative phase change by $\pi$. Thus the time interval $\Delta t$ determined by (4.63) is roughly one-third of the time required for the nitrogen atom to oscillate from above to below the plane.

---

[14] We have taken the lower limit in (4.63) as an *approximate* equality. In our example we have somewhat arbitrarily chosen to say that the system has changed when the phase in (4.65) reaches one. Since $\Delta t$ is an evolutionary time interval and not an uncertainty, there is no clear definition of $\Delta t$, as there is for an uncertainty such as $\Delta E$ in (4.64). A more general derivation of (4.63) and corresponding specification of $\Delta t$ is given in Problem 4.15.

Notice that if, instead of being in a superposition of energy eigenstates with different energies, the state of the molecule had been an energy eigenstate, there would be a definite value for the energy of the molecule, and hence $\Delta E = 0$. But in this case, the ket would only pick up an overall phase as time evolved, and the time interval $\Delta t$ required for the state to change would be infinite. An energy eigenstate is really a stationary state.

Our discussion and example should make clear that $\Delta t$ is not an uncertainty at all. Time in nonrelativistic quantum mechanics is just a *parameter* and not a dynamical variable like energy, angular momentum, position, or momentum with which there may be an uncertainty depending on the state of the system. When we discuss the state of the system at time $t$, there is no inherent limit on how accurately we can specify this time.

In the example we chose a particular initial state $|\psi\rangle$ and then examined the length of the *evolutionary time* $\Delta t$ for that state to change. Now that we understand the meaning of the uncertainty relation (4.63), we can turn this around slightly. An atom (or an ammonia molecule) in an excited-energy state will not remain in this state indefinitely, even if undisturbed by any outside influence. It will decay to lower-energy states with some lifetime $\tau$. In Chapter 14 we will see how to calculate the lifetime for excited states of the hydrogen atom using the Hamiltonian arising from the interactions of charged particles with the electromagnetic field. Thus an excited state is not a stationary state, and the lifetime $\tau$ sets a natural evolutionary time for that state. Therefore, from (4.63) there must be an uncertainty in the energy of the excited state given by $\Delta E \sim \hbar/\tau$. Photons emitted in this transition will have not a definite energy but rather a spread in energies. This is the origin of the natural linewidth (see Problem 4.16).

## 4.7 SUMMARY

Time development is where much of the action occurs in quantum mechanics. To move states forward in time, we introduce a time-evolution operator $\hat{U}(t)$ so that

$$|\psi(t)\rangle = \hat{U}(t)|\psi(0)\rangle \tag{4.66}$$

In order for probability to be conserved as time evolves,

$$\langle\psi(t)|\psi(t)\rangle = \langle\psi(0)|\hat{U}^\dagger(t)\hat{U}(t)|\psi(0)\rangle$$
$$= \langle\psi(0)|\psi(0)\rangle \tag{4.67}$$

and consequently the operator $\hat{U}(t)$ must be unitary:

$$\hat{U}^\dagger(t)\hat{U}(t) = 1 \tag{4.68}$$

The Hamiltonian $\hat{H}$, the energy operator, enters as the generator of time translations through the infinitesimal time-evolution operator:

$$\hat{U}(dt) = 1 - \frac{i}{\hbar}\hat{H}\,dt \tag{4.69}$$

The unitarity requirement (4.68) then dictates that the Hamiltonian is Hermitian. The time-evolution operator obeys the differential equation

$$\hat{H}\hat{U}(t) = i\hbar\frac{d}{dt}\hat{U}(t) \qquad (4.70)$$

leading to the Schrödinger equation:

$$\hat{H}|\psi(t)\rangle = i\hbar\frac{d}{dt}|\psi(t)\rangle \qquad (4.71)$$

A particularly useful solution to (4.70) occurs when the Hamiltonian is independent of time, in which case the time-development operator is given by

$$\hat{U}(t) = e^{-i\hat{H}t/\hbar} \qquad (4.72)$$

The action of the time-development operator (4.72) on an energy eigenstate $|E\rangle$ is given by

$$e^{-i\hat{H}t/\hbar}|E\rangle = e^{-iEt/\hbar}|E\rangle \qquad (4.73)$$

showing that a single energy eigenstate just picks up an overall phase as time evolves and is therefore a stationary state. Time evolution for a state $|\psi\rangle$ that can be expressed as a superposition of energy eigenstates as

$$|\psi(0)\rangle = \sum_n |E_n\rangle\langle E_n|\psi(0)\rangle \qquad (4.74)$$

is given by

$$|\psi(t)\rangle = e^{-i\hat{H}t/\hbar}\sum_n |E_n\rangle\langle E_n|\psi(0)\rangle$$

$$= \sum_n e^{-iE_nt/\hbar}|E_n\rangle\langle E_n|\psi(0)\rangle \qquad (4.75)$$

When the superposition (4.74) involves states with different energies, the relative phase between the energy eigenstates changes with time. The time $\Delta t$ (the evolutionary time) necessary for the system to change with time in this case satisfies

$$\Delta E \Delta t \geq \frac{\hbar}{2} \qquad (4.76)$$

where $\Delta E$ is the usual uncertainty in energy for the state $|\psi\rangle$.

Expectation values satisfy

$$\frac{d}{dt}\langle A\rangle = \frac{i}{\hbar}\langle\psi(t)|[\hat{H},\hat{A}]|\psi(t)\rangle + \langle\psi(t)|\frac{\partial\hat{A}}{\partial t}|\psi(t)\rangle \qquad (4.77)$$

which tells us that observables that do not explicitly depend on time will be constants of the motion when they commute with the Hamiltonian.

Although this chapter is devoted to time evolution, the similarity between the operators that generate rotations [see (3.10)] and the operator that generates time translations [see (4.72)] is striking. Or compare the form for an infinitesimal rotation operator $\hat{R}(d\phi\,\mathbf{n}) = 1 - i\hat{J}_n d\phi/\hbar$ for rotations by angle $d\phi$ about the axis specified by the unit vector $\mathbf{n}$ with the infinitesimal time translation operator (4.69). We can actually tie the rotation operator and the time-evolution operator together with a common thread—namely, symmetry. A symmetry operation is one that leaves the physical system unchanged or invariant. For example, if the Hamiltonian is invariant under rotations about an axis, the generator of rotations about that axis must commute with the Hamiltonian. But (4.77) then tells us that the component of the angular momentum along this axis is conserved, since its expectation value doesn't vary in time. Also, if the Hamiltonian is invariant under time translations, which simply means that $\hat{H}$ is independent of time, then of course energy is conserved. We will have more to say about symmetry, especially in Chapter 9, but this is our first indication of the important connection between symmetries of a physical system and conservation laws.

## PROBLEMS

**4.1.** Show that unitarity of the infinitesimal time-evolution operator (4.4) requires that the Hamiltonian $\hat{H}$ be Hermitian.

**4.2.** Show that if the Hamiltonian depends on time *and* $\left[\hat{H}(t_1), \hat{H}(t_2)\right] = 0$, the time-development operator is given by

$$\hat{U}(t) = \exp\left[-\frac{i}{\hbar}\int_0^t dt'\,\hat{H}(t')\right]$$

**4.3.** Use (4.16) to verify that expectation value of an observable $A$ does not change with time if the system is in an energy eigenstate (a stationary state) and $\hat{A}$ does not depend explicitly on time.

**4.4.** A beam of spin-$\frac{1}{2}$ particles with speed $v_0$ passes through a series of two SGz devices. The first SGz device transmits particles with $S_z = \hbar/2$ and filters out particles with $S_z = -\hbar/2$. The second SGz device transmits particles with $S_z = -\hbar/2$ and filters out particles with $S_z = \hbar/2$. Between the two devices is a region of length $l_0$ in which there is a uniform magnetic field $B_0$ pointing in the $x$ direction. Determine the smallest value of $l_0$ such that exactly 25 percent of the particles transmitted by the first SGz device are transmitted by the second device. Express your result in terms of $\omega_0 = egB_0/2mc$ and $v_0$.

**4.5.** A beam of spin-$\frac{1}{2}$ particles in the $|+\mathbf{z}\rangle$ state enters a uniform magnetic field $B_0$ in the $x$-$z$ plane oriented at an angle $\theta$ with respect to the $z$ axis. At time $T$ later, the particles enter an SGy device. What is the probability the particles will be found with $S_y = \hbar/2$? Check your result by evaluating the special cases $\theta = 0$ and $\theta = \pi/2$.

**4.6.** Verify that the expectation values (4.23), (4.28), and (4.30) for a spin-$\frac{1}{2}$ particle precessing in a uniform magnetic field $B_0$ in the z direction satisfy (4.16).

**4.7.** Use the data given in Fig. 4.3 to determine the $g$ factor of the muon.

**4.8.** A spin-$\frac{1}{2}$ particle, initially in a state with $S_n = \hbar/2$ with $\mathbf{n} = \sin\theta\mathbf{i} + \cos\theta\mathbf{k}$, is in a constant magnetic field $B_0$ in the $z$ direction. Determine the state of the particle at time $t$ and determine how $\langle S_x\rangle$, $\langle S_y\rangle$, and $\langle S_z\rangle$ vary with time.

**4.9.** Derive Rabi's formula (4.45).

**4.10.** Express the Hamiltonian (4.57) for the ammonia molecule in the $|I\rangle$-$|II\rangle$ basis to obtain (4.61). Assume the electric field $\mathbf{E} = \mathbf{E}_0\cos\omega t$. Compare this Hamiltonian with that for a spin-$\frac{1}{2}$ particle in a time-dependent magnetic field that appears in (4.38) and deduce the form for the probability of finding the molecule in state $|I\rangle$ at time $t$ if it is initially placed in the state $|II\rangle$; that is, what is the analogue of Rabi's formula (4.45) for the ammonia molecule?

**4.11.** A spin-1 particle with a magnetic moment $\boldsymbol{\mu} = (gq/2mc)\mathbf{S}$ is situated in a magnetic field $\mathbf{B} = B_0\mathbf{k}$ in the $z$ direction. At time $t = 0$ the particle is in a state with $S_y = \hbar$ [see (3.115)]. Determine the state of the particle at time $t$. Calculate how the expectation values $\langle S_x \rangle$, $\langle S_y \rangle$, and $\langle S_z \rangle$ vary in time.

**4.12.** A particle with intrinsic spin one is placed in a constant external magnetic field $B_0$ in the $x$ direction. The initial spin state of the particle is $|\psi(0)\rangle = |1, 1\rangle$, that is, a state with $S_z = \hbar$. Take the spin Hamiltonian to be

$$\hat{H} = \omega_0 \hat{S}_x$$

and determine the probability that the particle is in the state $|1, -1\rangle$ at time $t$. *Suggestion:* If you haven't already done so, you should first work out Problem 3.15 to determine the eigenstates of $\hat{S}_x$ for a spin-1 particle in terms of the eigenstates of $\hat{S}_z$.

**4.13.** Let

$$\begin{pmatrix} E_0 & 0 & A \\ 0 & E_1 & 0 \\ A & 0 & E_0 \end{pmatrix}$$

be the matix repesentation of the Hamiltonian for a three-state system with basis states $|1\rangle$, $|2\rangle$, and $|3\rangle$.
(a) If the state of the system at time $t = 0$ is $|\psi(0)\rangle = |2\rangle$, what is $|\psi(t)\rangle$?
(b) If the state of the system at time $t = 0$ is $|\psi(0)\rangle = |3\rangle$, what is $|\psi(t)\rangle$?

**4.14.** The matrix representation of the Hamiltonian for a photon propagating *along* the optic axis (taken to be the $z$ axis) of a quartz crystal using the linear polarization states $|x\rangle$ and $|y\rangle$ as a basis is given by

$$\hat{H} \xrightarrow[|x\rangle\text{-}|y\rangle \text{ basis}]{} \begin{pmatrix} 0 & -iE_0 \\ iE_0 & 0 \end{pmatrix}$$

(a) What are the eigenstates and eigenvalues of the Hamiltonian?
(b) A photon enters the crystal linearly polarized in the $x$ direction, that is, $|\psi(0)\rangle = |x\rangle$. What is the $|\psi(t)\rangle$, the state of the photon at time $t$? Express your answer in the $|x\rangle$-$|y\rangle$ basis. What is happening to the polarization of the photon as it travels through the crystal?

**4.15.** Consider any observable $A$ associated with the state of the system in quantum mechanics. Show that there is an uncertainty relation of the form

$$\Delta E \left( \frac{\Delta A}{|d\langle A\rangle/dt|} \right) \geq \frac{\hbar}{2}$$

provided the operator $\hat{A}$ does not depend explicitly on time. The quantity $\Delta A/|d\langle A\rangle/dt|$ is a time that we may call $\Delta t$. What is the physical significance of $\Delta t$?

**4.16.** The lifetime of hydrogen in the $2p$ state to decay to the $1s$ ground state is $1.6 \times 10^{-9}$s [see (14.181)]. Estimate the uncertainty $\Delta E$ in energy of this excited state. What is the corresponding linewidth in angstroms?

# CHAPTER
# 5

# A SYSTEM
# OF TWO
# SPIN-1/2
# PARTICLES

Let's turn our attention to systems containing two spin-$\frac{1}{2}$ particles. For definiteness, we focus initially on the spin-spin interaction of an electron and a proton in the ground state of hydrogen, which leads to hyperfine splitting of this energy level. We will see that the energy eigenstates are also eigenstates of total spin angular momentum with total-spin zero and total-spin one. The spin-0 state serves as the foundation for a discussion of the famous Einstein-Podolsky-Rosen paradox and the Bell inequalities. The experimental tests of the predictions of quantum mechanics on two-particle systems such as the spin-0 state have profound implications for our understanding of the nature of reality.

## 5.1 THE BASIS STATES FOR A SYSTEM OF TWO SPIN-$\frac{1}{2}$ PARTICLES

What are the spin basis states for a system of two spin-$\frac{1}{2}$ particles, such as an electron and a proton? A "natural" basis set is to label the states by the value of $S_z$ for each of the particles:

$$|+\mathbf{z}, +\mathbf{z}\rangle \qquad |+\mathbf{z}, -\mathbf{z}\rangle \qquad |-\mathbf{z}, +\mathbf{z}\rangle \qquad |-\mathbf{z}, -\mathbf{z}\rangle \qquad (5.1)$$

where the first element in these kets indicates the spin state of one of the particles (particle 1) and the second element indicates the spin state of the other particle (particle 2). In the notation of Chapter 3

$$|\pm\mathbf{z}, \pm\mathbf{z}\rangle = |s_1 = \tfrac{1}{2}, \ m_1 = \pm\tfrac{1}{2}, \ s_2 = \tfrac{1}{2}, \ m_2 = \pm\tfrac{1}{2}\rangle \qquad (5.2)$$

Another basis set we could choose—albeit one that looks somewhat less appealing—is to label one of the particles by its value of $S_x$ and the other by its value of $S_z$:

$$|+\mathbf{x}, +\mathbf{z}\rangle \qquad |+\mathbf{x}, -\mathbf{z}\rangle \qquad |-\mathbf{x}, +\mathbf{z}\rangle \qquad |-\mathbf{x}, -\mathbf{z}\rangle \qquad (5.3)$$

In fact, since these are both complete basis sets, we must be able to superpose the kets in (5.1) to obtain those in (5.3). For example,

$$|+\mathbf{x}, +\mathbf{z}\rangle = \frac{1}{\sqrt{2}}|+\mathbf{z}, +\mathbf{z}\rangle + \frac{1}{\sqrt{2}}|-\mathbf{z}, +\mathbf{z}\rangle \qquad (5.4)$$

Another way to transform from the basis set (5.1) to the basis (5.3) is to rotate the spin of the first particle, leaving the second fixed. Rotating the spin state of particle 1 by $\pi/2$ radians counterclockwise about the $y$ axis transforms, for example, the states $|\pm\mathbf{z}, +\mathbf{z}\rangle$ into the states $|\pm\mathbf{x}, +\mathbf{z}\rangle$. We denote the three generators of rotations for particle 1, the angular momentum operators, by $\hat{S}_{1x}$, $\hat{S}_{1y}$, and $\hat{S}_{1z}$, or, in vector form, $\hat{\mathbf{S}}_1$, where

$$\hat{\mathbf{S}}_1 = \hat{S}_{1x}\mathbf{i} + \hat{S}_{1y}\mathbf{j} + \hat{S}_{1z}\mathbf{k} \qquad (5.5)$$

Similarly, we can rotate the spin state of particle 2 with the three generators $\hat{S}_{2x}$, $\hat{S}_{2y}$, and $\hat{S}_{2z}$, or, in vector form, $\hat{\mathbf{S}}_2$. Since we can rotate the spin state of particle 1 independently of the spin state of particle 2, the generators of rotations for the two particles must commute:

$$[\hat{\mathbf{S}}_1, \hat{\mathbf{S}}_2] = 0 \qquad (5.6)$$

You might be concerned that if the spins interact with each other, rotating the spin state of particle 1 must affect the spin state of particle 2. That, however, is a different matter from determining the possible basis states that can be used to describe such a two-particle system. In the next section we will examine which linear combinations of the basis states (5.1) are eigenstates of the Hamiltonian when the spins do interact in a specified way. Just as in our analysis of the ammonia molecule where we selected a "natural" basis set with kets $|1\rangle$ and $|2\rangle$ that did not turn out to be eigenstates of the Hamiltonian, here too we cannot know a priori which combinations of the basis states will be coupled together as eigenstates of the Hamiltonian. Our choice of basis states does not make any presumption about how, or even whether, the particles interact.

It is worth noting that there is a useful way to express the basis states of two spin-$\frac{1}{2}$ particles in terms of single-particle spin states. As a specific example, we denote a state in which particle 1 has $S_{1z} = \hbar/2$ and particle 2 has $S_{2z} = -\hbar/2$ by the ket

$$|+\mathbf{z}, -\mathbf{z}\rangle = |+\mathbf{z}\rangle_1 \otimes |-\mathbf{z}\rangle_2 \qquad (5.7)$$

The kets on the right-hand side of (5.7) are the usual single-particle spin states, but two of them have been combined together in what is referred to as a *direct product* of the individual ket vectors, forming a two-particle state. We have inserted subscripts on the individual kets to emphasize which ket refers to the state of which particle. The symbol $\otimes$ emphasizes the fact that a special type of multiplication is taking place when we combine two vectors from *different* vector spaces. Until now, if we put two vectors together, it was always either in the form of an inner product, or amplitude, such as $\langle +\mathbf{z}|+\mathbf{x}\rangle$, or in the form of an outer product, such as appears in the projection operator $|+\mathbf{z}\rangle\langle +\mathbf{z}|$. Moreover, the two vectors were always vectors specifying a state of the same single particle; that is, they were from the same vector space. The right-hand side of (5.7) just expresses in a natural way that we can specify the basis states of a two-particle system with each of the particles in particular single-particle states. Actually, we can simplify our notation and dispense with the direct-product symbol $\otimes$ altogether, since there is really no other way to interpret the right-hand side of (5.7) except as the direct product of the two vector spaces. Thus the four basis states (5.1) of the two-particle system can also be expressed by a direct product of single-particle states as

$$|1\rangle = |+\mathbf{z}, +\mathbf{z}\rangle = |+\mathbf{z}\rangle_1|+\mathbf{z}\rangle_2 \qquad |2\rangle = |+\mathbf{z}, -\mathbf{z}\rangle = |+\mathbf{z}\rangle_1|-\mathbf{z}\rangle_2$$

$$|3\rangle = |-\mathbf{z}, +\mathbf{z}\rangle = |-\mathbf{z}\rangle_1|+\mathbf{z}\rangle_2 \qquad |4\rangle = |-\mathbf{z}, -\mathbf{z}\rangle = |-\mathbf{z}\rangle_1|-\mathbf{z}\rangle_2 \quad (5.8)$$

where we have ordered the states from $|1\rangle$ through $|4\rangle$ for notational convenience when we use these states as basis states for matrix representations in the next section.

## 5.2  THE HYPERFINE SPLITTING OF THE GROUND STATE OF HYDROGEN

We are ready to analyze the spin-spin interaction of the electron and the proton in hydrogen. Of course, the electron and proton interact predominantly through the Coulomb interaction $V(r) = -e^2/r$, which is independent of the spins of the particles. In Chapter 10 we will see that the energy eigenvalues of the Hamiltonian with this potential energy are given by $E_n = -13.6$ eV/$n^2$, where $n$ is a positive integer. In addition, there are relativistic corrections, due to effects such as spin-orbit coupling, that lead to a fine structure on these energy levels that does depend on the spin state of the electron. We will discuss this fine structure in detail in Chapter 11. There is, however, another interaction that involves the intrinsic spins of *both* the electron and the proton. Since the proton has a magnetic moment, the proton is a source of a magnetic field. The magnetic moment of the electron interacts with this magnetic field, generating an interaction energy proportional to the magnetic moments of both particles and thus, from (1.3), proportional to the intrinsic spins of both of the particles. Because the mass of the proton is roughly 2000 times larger than that of the electron, the magnetic moment of the proton is roughly 2000 times smaller than that of the electron and the overall scale of this spin-spin interaction turns out to be even smaller than the fine structure—hence the name hyperfine interaction.

The complete form of the spin-spin Hamiltonian follows directly from Maxwell's equations. It involves, of course, not just the magnetic moments of the particles but also the distance separating the particles. Fortunately, if we restrict our analysis to the ground state of hydrogen, a state with zero orbital angular momentum, the spin-spin Hamiltonian can be expressed in the simple form

$$\hat{H} = \frac{2A}{\hbar^2}\hat{\mathbf{S}}_1 \cdot \hat{\mathbf{S}}_2 \qquad (5.9)$$

where $\hat{\mathbf{S}}_1$ is the angular momentum operator of the electron and $\hat{\mathbf{S}}_2$ is the angular momentum operator of the proton. The factor of $\hbar^2$ in the denominator guarantees that the constant $A$ has the dimensions of energy. We will determine the value for $A$, which turns out to be positive, from experiment. In our analysis of the ammonia molecule, where there was a term in the Hamiltonian that we also denoted by $A$, this was essentially the best we could do; here, calculating $A$ is fairly straightforward, because the hydrogen atom is essentially a two-body problem with well-understood electromagnetic interactions.[1]

We are now ready to determine the energy eigenvalues and corresponding eigenstates of the Hamiltonian (5.9). In order to construct the $4 \times 4$ matrix representation of the Hamiltonian using the basis states (5.8), it is convenient to use the operator identity

$$2\hat{\mathbf{S}}_1 \cdot \hat{\mathbf{S}}_2 = 2\hat{S}_{1x}\hat{S}_{2x} + 2\hat{S}_{1y}\hat{S}_{2y} + 2\hat{S}_{1z}\hat{S}_{2z}$$
$$= \hat{S}_{1+}\hat{S}_{2-} + \hat{S}_{1-}\hat{S}_{2+} + 2\hat{S}_{1z}\hat{S}_{2z} \qquad (5.10)$$

where the first line reflects the definition of the ordinary dot product, albeit involving operators, while the second line follows from the definition of the raising and lowering operators for the two particles:

$$\hat{S}_{1+} = \hat{S}_{1x} + i\hat{S}_{1y} \qquad \hat{S}_{1-} = \hat{S}_{1x} - i\hat{S}_{1y} \qquad (5.11a)$$

$$\hat{S}_{2+} = \hat{S}_{2x} + i\hat{S}_{2y} \qquad \hat{S}_{2-} = \hat{S}_{2x} - i\hat{S}_{2y} \qquad (5.11b)$$

The expression (5.10) is useful since it permits us to evaluate the action of the Hamiltonian on each of the basis states. For example, a typical diagonal matrix element is

$$\langle 1|\hat{H}|1\rangle = \frac{A}{\hbar^2}\langle +\mathbf{z}, +\mathbf{z}|(\hat{S}_{1+}\hat{S}_{2-} + \hat{S}_{1-}\hat{S}_{2+} + 2\hat{S}_{1z}\hat{S}_{2z})|+\mathbf{z}, +\mathbf{z}\rangle$$

$$= \frac{A}{\hbar^2}\langle +\mathbf{z}, +\mathbf{z}|2\hat{S}_{1z}\hat{S}_{2z}|+\mathbf{z}, +\mathbf{z}\rangle = \frac{A}{2} \qquad (5.12)$$

Note that the raising and lowering operators change the basis state and therefore cannot contribute to a diagonal matrix element (in this case both the raising operators yield zero when they act to the right on the ket). A nonvanishing off-

---

[1] See, for example, S. Gasiorowicz, *Quantum Physics*, Wiley, New York, 1974, Chapter 17.

diagonal matrix element such as the element in the third row and second column is

$$\langle 3|\hat{H}|2\rangle = \frac{A}{\hbar^2}\langle -\mathbf{z}, +\mathbf{z}|(\hat{S}_{1+}\hat{S}_{2-} + \hat{S}_{1-}\hat{S}_{2+} + 2\hat{S}_{1z}\hat{S}_{2z})|+\mathbf{z}, -\mathbf{z}\rangle$$

$$= \frac{A}{\hbar^2}\langle -\mathbf{z}, +\mathbf{z}|\hat{S}_{1-}\hat{S}_{2+}|+\mathbf{z}, -\mathbf{z}\rangle = A \tag{5.13}$$

where in the second line we have retained the only operator term that can give a nonzero contribution to the matrix element.

Working out the remaining matrix elements, we find that the matrix representation of the Hamiltonian in the basis (5.8) is given by

$$\hat{H} \rightarrow \begin{pmatrix} A/2 & 0 & 0 & 0 \\ 0 & -A/2 & A & 0 \\ 0 & A & -A/2 & 0 \\ 0 & 0 & 0 & A/2 \end{pmatrix} \tag{5.14}$$

The energy eigenvalue equation $\hat{H}|\psi\rangle = E|\psi\rangle$ in this basis is

$$\begin{pmatrix} A/2 & 0 & 0 & 0 \\ 0 & -A/2 & A & 0 \\ 0 & A & -A/2 & 0 \\ 0 & 0 & 0 & A/2 \end{pmatrix} \begin{pmatrix} \langle 1|\psi\rangle \\ \langle 2|\psi\rangle \\ \langle 3|\psi\rangle \\ \langle 4|\psi\rangle \end{pmatrix} = E \begin{pmatrix} \langle 1|\psi\rangle \\ \langle 2|\psi\rangle \\ \langle 3|\psi\rangle \\ \langle 4|\psi\rangle \end{pmatrix} \tag{5.15}$$

The energy eigenvalues are determined by the requirement that the determinant of the coefficients vanishes:

$$\begin{vmatrix} A/2 - E & 0 & 0 & 0 \\ 0 & -A/2 - E & A & 0 \\ 0 & A & -A/2 - E & 0 \\ 0 & 0 & 0 & A/2 - E \end{vmatrix} = 0 \tag{5.16}$$

which yields $(A/2 - E)^2[(E + A/2)^2 - A^2] = 0$. Thus three of the eigenvalues are $E = A/2$ and one of them is $E = -3A/2$, as indicated in Fig. 5.1. If we substitute these energies into (5.15), we obtain the three column vectors

$$\begin{pmatrix} 1 \\ 0 \\ 0 \\ 0 \end{pmatrix} \quad \frac{1}{\sqrt{2}}\begin{pmatrix} 0 \\ 1 \\ 1 \\ 0 \end{pmatrix} \quad \text{and} \quad \begin{pmatrix} 0 \\ 0 \\ 0 \\ 1 \end{pmatrix} \tag{5.17}$$

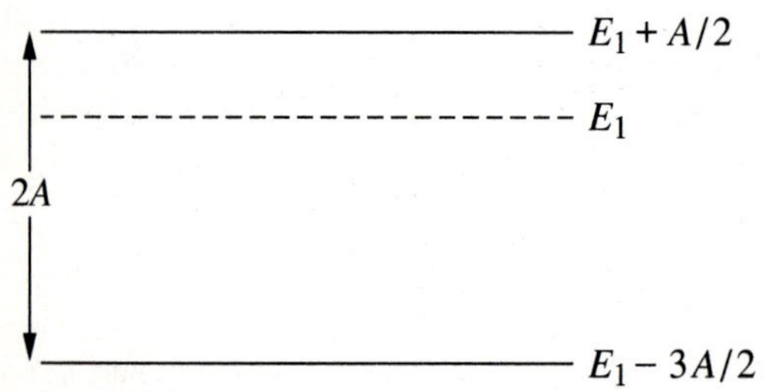

$E_1 + A/2$

$E_1$

$2A$

$E_1 - 3A/2$

**FIGURE 5.1**
The hyperfine splitting of the ground-state energy level of hydrogen. The energy $E_1$ is the energy of the ground state excluding the hyperfine interaction.

which represent the normalized eigenstates

$$|+\mathbf{z}, +\mathbf{z}\rangle \qquad (5.18a)$$

$$\frac{1}{\sqrt{2}}|+\mathbf{z}, -\mathbf{z}\rangle + \frac{1}{\sqrt{2}}|-\mathbf{z}, +\mathbf{z}\rangle \qquad (5.18b)$$

$$|-\mathbf{z}, -\mathbf{z}\rangle \qquad (5.18c)$$

with $E = A/2$ and the column vector

$$\frac{1}{\sqrt{2}}\begin{pmatrix} 0 \\ 1 \\ -1 \\ 0 \end{pmatrix} \qquad (5.19)$$

which represents the eigenstate

$$\frac{1}{\sqrt{2}}|+\mathbf{z}, -\mathbf{z}\rangle - \frac{1}{\sqrt{2}}|-\mathbf{z}, +\mathbf{z}\rangle \qquad (5.20)$$

with $E = -3A/2$. Thus there is a single two-particle spin state for the ground state, while the excited state is three-fold degenerate.

A photon emitted or absorbed in making a transition between these two energy levels must have a frequency $\nu$ determined by $h\nu = 2A$. For hydrogen this frequency is approximately 1420 MHz, corresponding to a wavelength $\lambda$ of about 21 cm, which is in the microwave part of the spectrum. The frequency has actually been measured to one part in $10^{13} - \nu = 1,420,405,751.768 \pm 0.001$ Hz—making it the most accurately known physical quantity.[2] The technique responsible for this unusual achievement is our old friend the maser. In the hydrogen maser, a beam of hydrogen atoms in the upper-energy state is selected by using a Stern-Gerlach device. The beam then enters a microwave cavity tuned to the resonant frequency. Because of the very long lifetime of a hydrogen atom in the upper-energy state,[3] the natural linewidth is especially narrow and consequently the spectral purity is especially high, permitting such an accurate determination of the resonant frequency. Incidentally, the theoretical value for the hyperfine splitting has "only" been calculated to 1 part in $10^6$, leaving considerable room for theoretical improvement.

Finally, it should be noted that the 21-cm line of hydrogen provides us with an extremely useful tool for discovering the density distribution and velocities of

---

[2] This measurement was first carried out by S. B. Crampton, D. Kleppner, and N. F. Ramsey, *Phys. Rev. Lett.* **11**, 338 (1963), who merely obtained $\nu = 1,420,405,751.800 \pm 0.028$ Hz.

[3] This is a magnetic dipole transition, not the more common electric dipole transition, which generally leads to a substantially shorter lifetime, as in $NH_3$. We will discuss these types of transitions in Chapter 14.

atomic hydrogen in interstellar space. The intensity of the radiation received by a radio-frequency antenna tuned to 1420 MHz is a measure of the concentration of the gas, while the Doppler frequency shifts of the radiation provide a measure of the velocity of the gas.

## 5.3  THE ADDITION OF ANGULAR MOMENTA FOR TWO SPIN-$\frac{1}{2}$ PARTICLES

In solving the energy eigenvalue problem, we have determined the eigenstates of the operator $2\hat{\mathbf{S}}_1 \cdot \hat{\mathbf{S}}_2$:

$$2\hat{\mathbf{S}}_1 \cdot \hat{\mathbf{S}}_2 \begin{bmatrix} |+\mathbf{z}, +\mathbf{z}\rangle \\ \frac{1}{\sqrt{2}}|+\mathbf{z}, -\mathbf{z}\rangle + \frac{1}{\sqrt{2}}|-\mathbf{z}, +\mathbf{z}\rangle \\ |-\mathbf{z}, -\mathbf{z}\rangle \end{bmatrix} = \frac{\hbar^2}{2} \begin{bmatrix} |+\mathbf{z}, +\mathbf{z}\rangle \\ \frac{1}{\sqrt{2}}|+\mathbf{z}, -\mathbf{z}\rangle + \frac{1}{\sqrt{2}}|-\mathbf{z}, +\mathbf{z}\rangle \\ |-\mathbf{z}, -\mathbf{z}\rangle \end{bmatrix}$$

(5.21)

and

$$2\hat{\mathbf{S}}_1 \cdot \hat{\mathbf{S}}_2 \left( \frac{1}{\sqrt{2}}|+\mathbf{z}, -\mathbf{z}\rangle - \frac{1}{\sqrt{2}}|-\mathbf{z}, +\mathbf{z}\rangle \right) = -\frac{3\hbar^2}{2} \left( \frac{1}{\sqrt{2}}|+\mathbf{z}, -\mathbf{z}\rangle - \frac{1}{\sqrt{2}}|-\mathbf{z}, +\mathbf{z}\rangle \right)$$

(5.22)

These eigenstates have a much deeper significance than has been apparent from just our discussion of the hyperfine splitting in hydrogen. To see this significance, first consider the infinitesimal rotation operator for a *system* of two spin-$\frac{1}{2}$ particles. In order to rotate a two-particle spin ket by angle $d\theta$ about the axis specified by the unit vector $\mathbf{n}$, we must rotate the spin state of each of the particles by the angle $d\theta$ about this axis. See Fig. 5.2. Thus the infinitesimal rotation operator for the system is given by

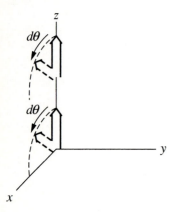

**FIGURE 5.2**
A schematic diagram showing the rotation of the spins of two spin-up-along-$z$ particles by angle $d\theta$ about the $y$ axis. Note that the generator of rotations $\hat{\mathbf{S}}$ rotates the spins but not the positions of the two particles.

$$\hat{R}(d\theta\,\mathbf{n}) = 1 - \frac{i}{\hbar}\hat{\mathbf{S}}\cdot\mathbf{n}\,d\theta$$

$$= \left(1 - \frac{i}{\hbar}\hat{\mathbf{S}}_1\cdot\mathbf{n}\,d\theta\right)\otimes\left(1 - \frac{i}{\hbar}\hat{\mathbf{S}}_2\cdot\mathbf{n}\,d\theta\right) \tag{5.23}$$

where in the first line we have introduced a new vector operator $\hat{\mathbf{S}}$ whose three components are the generators of rotations for the two-particle spin system, and in the second line we have expressed this system-rotation operator in the direct-product space in terms of the rotation operators for the individual particles, to first order in $d\theta$. Since the components of $\hat{\mathbf{S}}$ generate rotations, these components must satisfy the usual commutation relations (3.14) of angular momentum. We call $\mathbf{S}$ the total spin angular momentum. As (5.23) shows, the total spin angular momentum operator $\hat{\mathbf{S}}$ is related to the individual spin angular momentum operators in just the way we would expect:

$$\hat{\mathbf{S}} = \hat{\mathbf{S}}_1\otimes 1 + 1\otimes\hat{\mathbf{S}}_2 \tag{5.24}$$

or, more simply, $\hat{\mathbf{S}} = \hat{\mathbf{S}}_1 + \hat{\mathbf{S}}_2$, where, if we are operating in the direct-product space, the operator $\hat{\mathbf{S}}_1$ is understood to include the identity operator in the vector space of particle 2, and so on. The total spin angular momentum is just the sum of the individual spin angular momenta. You can also verify directly, using (5.6) and the commutation relations of the individual spin operators, that the components of the total spin operator (5.24) satisfy the usual commutation relations of angular momentum, such as

$$\left[\hat{S}_{1x} + \hat{S}_{2x}, \hat{S}_{1y} + \hat{S}_{2y}\right] = i\hbar\left(\hat{S}_{1z} + \hat{S}_{2z}\right) \tag{5.25}$$

Solving the angular momentum problem for total spin means finding the two-particle eigenstates of

$$\hat{S}^2 = \left(\hat{\mathbf{S}}_1 + \hat{\mathbf{S}}_2\right)^2 = \hat{S}_1^2 + \hat{S}_2^2 + 2\hat{\mathbf{S}}_1\cdot\hat{\mathbf{S}}_2 \tag{5.26a}$$

and

$$\hat{S}_z = \hat{S}_{1z} + \hat{S}_{2z} \tag{5.26b}$$

From our general analysis of angular momentum in Chapter 3, which was based solely on the fact that the angular momentum operators obey the commutation relations (3.14), we know that we can express these eigenstates of total spin in the form

$$\hat{S}^2|s, m\rangle = s(s + 1)\hbar^2|s, m\rangle \tag{5.27a}$$

$$\hat{S}_z|s, m\rangle = m\hbar|s, m\rangle \tag{5.27b}$$

Since

$$\hat{S}_1^2|\pm\mathbf{z}, \pm\mathbf{z}\rangle = \tfrac{1}{2}\left(\tfrac{1}{2} + 1\right)\hbar^2|\pm\mathbf{z}, \pm\mathbf{z}\rangle \tag{5.28a}$$

and

$$\hat{S}_2^2|\pm\mathbf{z}, \pm\mathbf{z}\rangle = \tfrac{1}{2}(\tfrac{1}{2} + 1)\hbar^2|\pm\mathbf{z}, \pm\mathbf{z}\rangle \tag{5.28b}$$

we see that the eigenstates of $2\hat{\mathbf{S}}_1 \cdot \hat{\mathbf{S}}_2$ are eigenstates of $\hat{\mathbf{S}}^2$ as well. Using the eigenvalue $\tfrac{1}{2}\hbar^2$ for $2\hat{\mathbf{S}}_1 \cdot \hat{\mathbf{S}}_2$ from (5.21), we find that each of the three states in (5.21) is an eigenstate of $\hat{\mathbf{S}}^2 = \hat{\mathbf{S}}_1^2 + \hat{\mathbf{S}}_2^2 + 2\hat{\mathbf{S}}_1 \cdot \hat{\mathbf{S}}_2$ with the eigenvalue $[\tfrac{1}{2}(\tfrac{1}{2} + 1) + \tfrac{1}{2}(\tfrac{1}{2} + 1) + \tfrac{1}{2}]\hbar^2 = 2\hbar^2$, or, in the notation of (5.27a), they have $s = 1$ and are spin-1 states. The eigenvalue of the single state in (5.22) is $[\tfrac{1}{2}(\tfrac{1}{2} + 1) + \tfrac{1}{2}(\tfrac{1}{2} + 1) - \tfrac{3}{2}]\hbar^2 = 0$, and thus it is an $s = 0$ state. In fact, each of the states in (5.21) and (5.22) is also an eigenstate of the $z$ component of total spin. For example,

$$\hat{S}_z|+\mathbf{z}, +\mathbf{z}\rangle = (\hat{S}_{1z} + \hat{S}_{2z})|+\mathbf{z}, +\mathbf{z}\rangle = \left(\frac{\hbar}{2} + \frac{\hbar}{2}\right)|+\mathbf{z}, +\mathbf{z}\rangle = \hbar|+\mathbf{z}, +\mathbf{z}\rangle \tag{5.29}$$

Thus, using the $|s, m\rangle$ notation for total spin, we find

$$|1, 1\rangle = |+\mathbf{z}, +\mathbf{z}\rangle \tag{5.30a}$$

$$|1, 0\rangle = \frac{1}{\sqrt{2}}|+\mathbf{z}, -\mathbf{z}\rangle + \frac{1}{\sqrt{2}}|-\mathbf{z}, +\mathbf{z}\rangle \tag{5.30b}$$

$$|1, -1\rangle = |-\mathbf{z}, -\mathbf{z}\rangle \tag{5.30c}$$

and

$$|0, 0\rangle = \frac{1}{\sqrt{2}}|+\mathbf{z}, -\mathbf{z}\rangle - \frac{1}{\sqrt{2}}|-\mathbf{z}, +\mathbf{z}\rangle \tag{5.31}$$

Thus we have learned how to "add" the spins of two spin-$\tfrac{1}{2}$ particles to make states of total spin.

It is worth noting here that there is another way to see which linear combinations of the basis states (5.8) are eigenstates of total spin. Since

$$[\hat{\mathbf{S}}^2, \hat{S}_z] = 0 \tag{5.32}$$

these two operators have eigenstates in common. Because the basis states $|+\mathbf{z}, +\mathbf{z}\rangle$ and $|-\mathbf{z}, -\mathbf{z}\rangle$ are eigenstates of $\hat{S}_z$ with eigenvalues $\hbar$ and $-\hbar$, respectively—and they are the only basis states with these eigenvalues for the $z$ component of the total spin—they must be eigenstates of $\hat{\mathbf{S}}^2$ as well. As we have seen, they are spin-1 states. On the other hand, there are two basis states, $|+\mathbf{z}, -\mathbf{z}\rangle$ and $|-\mathbf{z}, +\mathbf{z}\rangle$, that are eigenstates of $\hat{S}_z$ with eigenvalue 0:

$$\hat{S}_z|+\mathbf{z}, -\mathbf{z}\rangle = (\hat{S}_{1z} + \hat{S}_{2z})|+\mathbf{z}, -\mathbf{z}\rangle = \left(\frac{\hbar}{2} - \frac{\hbar}{2}\right)|+\mathbf{z}, -\mathbf{z}\rangle = 0 \tag{5.33a}$$

$$\hat{S}_z|-\mathbf{z}, +\mathbf{z}\rangle = (\hat{S}_{1z} + \hat{S}_{2z})|-\mathbf{z}, +\mathbf{z}\rangle = \left(-\frac{\hbar}{2} + \frac{\hbar}{2}\right)|-\mathbf{z}, +\mathbf{z}\rangle = 0 \tag{5.33b}$$

For spin 1, the allowed $m$ values are 1, 0, and $-1$, so a linear combination of the states $|+\mathbf{z}, -\mathbf{z}\rangle$ and $|-\mathbf{z}, +\mathbf{z}\rangle$ must be the missing $m = 0$ state. We can obtain this state by applying the lowering operator

$$\hat{S}_- = \hat{S}_{1-} + \hat{S}_{2-} \qquad (5.34)$$

to the state $|1, 1\rangle$ or applying the raising operator

$$\hat{S}_+ = \hat{S}_{1+} + \hat{S}_{2+} \qquad (5.35)$$

to the state $|1, -1\rangle$. For example,

$$\begin{aligned}
\hat{S}_-|1, 1\rangle &= (\hat{S}_{1-} + \hat{S}_{2-})|+\mathbf{z}, +\mathbf{z}\rangle \\
&= \hbar(|-\mathbf{z}, +\mathbf{z}\rangle + |+\mathbf{z}, -\mathbf{z}\rangle) \\
&= \sqrt{2}\hbar |1, 0\rangle \qquad (5.36)
\end{aligned}$$

where the last step follows from (3.60) for the total-spin state. Dividing through by the factor of $\sqrt{2}\hbar$ leads to a correctly normalized expression for the state $|1, 0\rangle$. The other total-spin state, the $|0, 0\rangle$ state, can be determined by finding the linear combination of the states $|+\mathbf{z}, -\mathbf{z}\rangle$ and $|-\mathbf{z}, +\mathbf{z}\rangle$ that is orthogonal to the $|1, 0\rangle$ state. Satisfying the condition $\langle 0, 0|1, 0\rangle = 0$ leads to (5.31), up to an arbitrary overall phase.[4]

In terms of total-spin states, the spin-spin interaction in hydrogen splits the ground-state energy level into two levels, with the *triplet* of spin-1 states forming the upper energy level and the *singlet* spin-0 state forming the true ground state. The magnitude of the hyperfine splitting in hydrogen is roughly $5.9 \times 10^{-6}$ eV, which is to be compared with the typical spacing between energy levels that is on the order of electron volts. The magnitude of the splitting is indeed quite small in this case.

An interesting example of spin-spin interaction where the magnitude of the interaction is much larger than that between the electron and the proton in hydrogen occurs in the strong nuclear interaction that binds quarks and antiquarks (both spin-$\frac{1}{2}$ particles) together to form mesons. In particular, a $u$ quark and a $\bar{d}$ antiquark bind together to form a $\pi^+$, a spin-0 particle. The rest-mass energy of the $\pi^+$ is roughly 140 MeV. Changing the total-spin state of the $u$-$\bar{d}$ system from a singlet spin-0 state to the triplet of spin-1 states generates a different particle, the spin-1 $\rho^+$, which has a rest-mass energy of roughly 770 MeV. Thus the energy cost of reorienting the spins of the constituent quarks is a hefty 630 MeV.

## Discussion of the Spin-0 and Spin-1 States

Before concluding this section, some discussion of these important spin-0 and spin-1 states is in order. Two of our initial basis states in (5.8), namely, the $|+\mathbf{z}, +\mathbf{z}\rangle$ and $|-\mathbf{z}, -\mathbf{z}\rangle$ states, cannot be spin-0 states because the projection of the total spin on the $z$ axis is nonzero for each of these states; the individual

---

[4] The results we have obtained are, in fact, a special case of a more general result: adding angular momentum $j_1$ to angular momentum $j_2$ generates states of total angular momentum $j$, where $j$ takes on values ranging from $j_1 + j_2$ to $|j_1 - j_2|$ in integral steps. See Appendix B for a way to generate these states using angular momentum raising and lowering operators.

spins are either both up or both down, respectively. The $\pm\hbar$ values for total $S_z$ for these states are, of course, consistent with their being spin-1 states. We can say that the other two basis states, $|+\mathbf{z}, -\mathbf{z}\rangle$ and $|-\mathbf{z}, +\mathbf{z}\rangle$, consist of states in which the spins of the individual particles are opposite to each other, one spin up and the other spin down in each case. Nonetheless, having the spins oppositely directed can produce states with total spin of one as well as zero, depending on the linear combination (5.30b) or (5.31) one chooses. Clearly, the relative phase between the state $|+\mathbf{z}, -\mathbf{z}\rangle$ and the state $|-\mathbf{z}, +\mathbf{z}\rangle$ in the superposition is of crucial importance.

To see the effect of this phase even more clearly, let's express the states $|0, 0\rangle$ and $|1, 0\rangle$ in terms of the $S_x$ basis states for each of the individual particles. The states $|0, 0\rangle$ and $|1, 0\rangle$ are, of course, still eigenstates of $\hat{S}_z = \hat{S}_{1z} + \hat{S}_{2z}$. From (3.102) we know that for a single spin-$\frac{1}{2}$ particle

$$|\pm\mathbf{z}\rangle = \frac{1}{\sqrt{2}}|+\mathbf{x}\rangle \pm \frac{1}{\sqrt{2}}|-\mathbf{x}\rangle \tag{5.37}$$

Using this result, we can express the two-particle total-spin states $|0, 0\rangle$ and $|1, 0\rangle$ as

$$
\begin{aligned}
|1, 0\rangle &= \frac{1}{\sqrt{2}}|+\mathbf{z}, -\mathbf{z}\rangle + \frac{1}{\sqrt{2}}|-\mathbf{z}, +\mathbf{z}\rangle \\
&= \frac{1}{\sqrt{2}}|+\mathbf{z}\rangle_1|-\mathbf{z}\rangle_2 + \frac{1}{\sqrt{2}}|-\mathbf{z}\rangle_1|+\mathbf{z}\rangle_2 \\
&= \frac{1}{\sqrt{2}}\left[\frac{1}{\sqrt{2}}(|+\mathbf{x}\rangle_1 + |-\mathbf{x}\rangle_1)\right]\left[\frac{1}{\sqrt{2}}(|+\mathbf{x}\rangle_2 - |-\mathbf{x}\rangle_2)\right] \\
&\quad + \frac{1}{\sqrt{2}}\left[\frac{1}{\sqrt{2}}(|+\mathbf{x}\rangle_1 - |-\mathbf{x}\rangle_1)\right]\left[\frac{1}{\sqrt{2}}(|+\mathbf{x}\rangle_2 + |-\mathbf{x}\rangle_2)\right] \\
&= \frac{1}{\sqrt{2}}|+\mathbf{x}\rangle_1|+\mathbf{x}\rangle_2 - \frac{1}{\sqrt{2}}|-\mathbf{x}\rangle_1|-\mathbf{x}\rangle_2 \\
&= \frac{1}{\sqrt{2}}|+\mathbf{x}, +\mathbf{x}\rangle - \frac{1}{\sqrt{2}}|-\mathbf{x}, -\mathbf{x}\rangle
\end{aligned}
\tag{5.38}
$$

and, in a similar fashion,

$$
\begin{aligned}
|0, 0\rangle &= \frac{1}{\sqrt{2}}|+\mathbf{z}, -\mathbf{z}\rangle - \frac{1}{\sqrt{2}}|-\mathbf{z}, +\mathbf{z}\rangle \\
&= -\left(\frac{1}{\sqrt{2}}|+\mathbf{x}, -\mathbf{x}\rangle - \frac{1}{\sqrt{2}}|-\mathbf{x}, +\mathbf{x}\rangle\right)
\end{aligned}
\tag{5.39}
$$

If we make measurements of both $S_{1x}$ and $S_{2x}$ on the individual spin-$\frac{1}{2}$ particles in the state $|1, 0\rangle$, we find them with a 50 percent probability both spin up or both spin down along the $x$ axis, reflecting the fact that this is really a spin-1 state. On the other hand, if we make measurements of both $S_{1x}$ and $S_{2x}$ on the particles in the state $|0, 0\rangle$, we always find the spins oppositely aligned, one spin up and the other spin down, but this time along the $x$ axis instead of the $z$ axis. In fact, if we measure the components of the spin of the individual particles along an arbitrary axis for the $|0, 0\rangle$ state, this opposite alignment of the individual spins must be maintained (see Problem 5.3), as would be expected for a state with total spin equal to zero.

## 5.4   THE EINSTEIN-PODOLSKY-ROSEN PARADOX

Consider a spin-0 particle at rest that decays into two spin-$\frac{1}{2}$ particles.[5] In order to conserve linear momentum, the two particles emitted in the decay must move in opposite directions. In order to conserve angular momentum, the spin state of the two-particle system must be $|0, 0\rangle$, assuming there is zero relative orbital angular momentum. Two experimentalists, A (Al) and B (Bert), set up Stern-Gerlach measuring devices along this line of flight, as depicted in Fig. 5.3. Each observer is prepared to make measurements of the intrinsic spin of the particles as they pass through their respective SG devices.

---

[5] We are viewing this as a thought experiment. An actual example of a spinless particle decaying into two spin-$\frac{1}{2}$ particles is the rare decay mode of the $\eta$ meson into a $\mu^{+}$-$\mu^{-}$ pair. However, in order to measure the spin with an SG device in our thought experiment, we need to presume that the particles emitted in the decay are neutral.

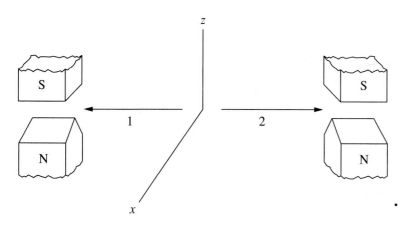

**FIGURE 5.3**
A schematic of the EPR experiment in which A measures the spin of particle 1 and B measures the spin of particle 2.

What can we say about the results of their measurements? Let's call the particle observed by A to be particle 1 and the one observed by B to be particle 2. We call the line of flight of the two particles the $y$ axis. If A and B both decide to make measurements of $S_z$ on their respective particles and A obtains a value of $S_{1z} = \hbar/2$, B must obtain a value of $S_{2z} = -\hbar/2$. From the expression (5.39) for the $|0, 0\rangle$ state, we see that there is a 50 percent probability to find the system in this $|+\mathbf{z}, -\mathbf{z}\rangle$ state. Similarly, if A obtains $S_{1z} = -\hbar/2$, B must obtain $S_{2z} = \hbar/2$. Again, there is a 50 percent probability to find the system in the state $|-\mathbf{z}, +\mathbf{z}\rangle$. The striking thing about these results is that if A measures first, A's measurement has *instantaneously* fixed or determined the value that B can obtain, even though the two particles may be completely noninteracting and A and B may be separated by light-years.

Although we have argued that the relative phase between the basis states $|+\mathbf{z}, -\mathbf{z}\rangle$ and $|-\mathbf{z}, +\mathbf{z}\rangle$ in the $|0, 0\rangle$ state matters, if you still tend to think in classical terms that a $|0, 0\rangle$ state is just a 50–50 mix of these two states, this instantaneous determination of B's result by A's measurement doesn't seem so strange. Imagine that you hold in your hand two colored balls that are indentical in feel but one is green and the other is red. You separate the balls without looking and put one in each hand. If you look at the ball in your left hand and find it is red, you have immediately determined that the ball in your right hand is green, even before you open your right hand. You would, of course, presume that the ball in your right hand was green all along, whether or not you had opened your left hand to check the color of that ball. This cannot, however, be an adequate explanation of what is going on in the spin system. The reason is that A and/or B can choose to measure a different component of the spin of the particles. Suppose that both A and B decide to measure $S_x$ instead of $S_z$. As (5.39) indicates, if A obtains the value $S_{1x} = \hbar/2$, then B must obtain $S_{2x} = -\hbar/2$. Similarly, if A obtains $S_{1x} = -\hbar/2$, then B must obtain $S_{2x} = \hbar/2$. Here again, the results of their measurements are completely correlated and measurements by A can determine the results of measurements by B. As we have seen in Chapter 3, however, the spin-$\frac{1}{2}$ particles cannot have definite values for both $S_z$ and $S_x$. The state of particle 2 as it travels toward B's SG device, for example, cannot be an eigenstate of both $\hat{S}_{2z}$ *and* $\hat{S}_{2x}$. In our example of the colored balls, it would be similar to the balls having two other colors such as blue and yellow, as well as red and green. Finding one of the balls to be yellow demands that the other is blue, just as finding one of the balls to be red demands that the other ball be green. However, a single ball cannot simultaneously have two colors and be, for example, both red and yellow. So what color is the ball in your right hand before you look?[6]

---

[6] Note that if A chooses to measure $S_z$ and B chooses to measure $S_x$, the results of their measurements will be completely uncorrelated. If A, for example, obtains $S_{1z} = \hbar/2$, then since

$$|\langle +\mathbf{z}, +\mathbf{x}|0, 0\rangle|^2 = |\langle +\mathbf{z}, -\mathbf{x}|0, 0\rangle|^2 = \tfrac{1}{4}$$

B has equal probabilities of obtaining $S_{2x} = \hbar/2$ and $S_{2x} = -\hbar/2$.

The idea that particles do not necessarily have definite attributes has been implicit in our discussion of quantum mechanics from the beginning. A single spin-$\frac{1}{2}$ particle in the state $|+\mathbf{x}\rangle$ does not have a definite value for $S_z$. Before a measurement is carried out, we can only give the probabilities of obtaining $S_z = \hbar/2$ or $S_z = -\hbar/2$; the particle has amplitudes to be in both the state $|+\mathbf{z}\rangle$ and the state $|-\mathbf{z}\rangle$. Once a measurement of $S_z$ is made, however, this uncertainty in the value of $S_z$ for the particle disappears; the particle is then in a state with a definite value of $S_z$. The new feature that is raised by our discussion of the two-particle system is that a measurement carried out on one of the particles can immediately determine the state of the other particle, even if the two particles are widely separated at the time of the measurement. This is the straightforward result of applying quantum mechanics to a two-particle system. A measurement of $S_{1z}$ unambiguously selects either the $|+\mathbf{z}, -\mathbf{z}\rangle$ or the $|-\mathbf{z}, +\mathbf{z}\rangle$ state. A measurement on part of the system in the form of a measurement on one of the particles in this two-particle system is really a measurement on the system as a whole.

Not everyone has been happy with this state of affairs. In particular, Albert Einstein never liked the idea that a single particle could be in a state in which the particle did not have a definite attribute, be it spin or position. In his view, this meant that physical properties did not have an objective reality independent of their being observed. For Einstein there was a more reasonable position. Although the results of measurements carried out on a single particle are in complete accord with quantum mechanics, these results do not of themselves demand that a particular particle does not have a definite attribute before the measurement is made. As we discussed in Section 1.4, testing the predictions of quantum mechanics requires measurements on a collection of particles, each of which is presumed to be in the same state. Thus Einstein could believe that 50 percent of the particles in the state $|+\mathbf{x}\rangle$ also had $S_z = \hbar/2$ and that 50 percent had $S_z = -\hbar/2$ but that we are unable to discriminate between these two types of particles, as if the attribute that would allow us to distinguish the particles was hidden from us—hence a hidden-variable theory of quantum mechanics.

In order to show how unsatisfactory the conventional interpretation of quantum mechanics really was, Einstein, Podolsky, and Rosen devised the ingenious thought-experiment on a two-particle system of the type that we have been describing in this section.[7] It is one thing to have a definite attribute for a particle dependent on having made a measurement of that attribute on the particle, but it is even more unusual to have that attribute determined by making a measurement on another particle altogether. To Einstein this was completely unacceptable: "But on one supposition we should, in my opinion, absolutely hold fast. The real factual

---

[7] A. Einstein, B. Podolsky, and N. Rosen, *Phys. Rev.* **47**, 777 (1935). The particular experiment described in their paper involved measurements of two different noncommuting variables, position and momentum, instead of two components of the intrinsic spin such as $S_z$ and $S_x$. The generalization of their argument to spin-$\frac{1}{2}$ particles was initially made by D. Bohm, *Quantum Theory*, Prentice-Hall, Englewood Cliffs, N.J., 1951, pp. 614–619.

situation of the system $S_2$ is independent of what is done with the system $S_1$, which is spatially separated from the former."[8] Because the conventional interpretation of quantum mechanics, which we have used in analyzing measurements by A and B on this two-particle system, is so completely at odds with what Einstein termed any "reasonable definition of the nature of reality," which includes this locality principle, the issue raised in their 1935 paper is generally referred to as the Einstein-Podolsky-Rosen paradox.

## 5.5 A NONQUANTUM MODEL AND THE BELL INEQUALITIES

Until 1964 it was believed that one could always construct a hidden-variable theory that would give all the same results as quantum mechanics. In that year, however, John S. Bell pointed out that alternative theories based on Einstein's locality principle actually yield a testable inequality that differs from the predictions of quantum mechanics.[9] As you might guess from our earlier discussion about measurements of $S_z$ for a particle in the state $|+x\rangle$, this disagreement cannot be observed in measurements on a single particle. Rather, it is a prediction about correlations that are observed in measurements made on a two-particle system such as the two spin-$\frac{1}{2}$ particles in a singlet spin state.

Let us first see how we can construct a local theory in which particles have their own independent attributes that can account for all the results of measuring $S_z$ or $S_x$ on a system of two particles in the $|0, 0\rangle$ state. As the particles travel outward toward the SG devices, there is no way to know in advance what the orientation of these devices will be. In fact, A and B may alter the orientation of their respective SG devices while the particles are in flight. The "local realist" wants each of the particles to possess its own definite attributes with no inherent uncertainty. Thus each particle must carry with it all the information, or instructions, necessary to tell the SG device what to yield if a measurement of $S_z$ or $S_x$ for that particle is made. For example, a *single* particle such as particle 1 may be of the type $\{+z, +x\}$, indicating that A obtains $\hbar/2$ for a measurement of $S_{1z}$ *or* $\hbar/2$ for a measurement of $S_{1x}$. Note that we are inventing a new $\{\ \}$ notation to provide a nonquantum description of the state of the particle. In this model, particle 1 is presumed to have definite values of $S_{1z}$ and $S_{1x}$, which is completely at odds with our earlier analysis of the allowed angular momentum states of a particle in quantum mechanics. However, in order to avoid obvious disagreements with experiments such as the Stern-Gerlach experiments of Chapter 1, we are not suggesting that A can simultaneously measure $S_{1z}$ *and* $S_{1x}$ for this particle. A's decision to measure $S_{1x}$, for example, on a particle of the type $\{+z, +x\}$ means

---

[8] A. Einstein, in P. A. Schilpp, ed., *Albert Einstein, Philosopher-Scientist*, Tudor, New York, 1949, p. 85.

[9] J. S. Bell, *Physics* **1**, 195 (1964).

that A forgoes the chance to measure $S_{1z}$ on this type of particle. The value of $S_{1z}$ of the particle is essentially hidden from us. In fact, making a measurement of $S_{1x}$ and obtaining $\hbar/2$ must alter the state of the particle. After this measurement of $S_{1x}$ on a collection of particles of the type $\{+\mathbf{z}, +\mathbf{x}\}$, 50 percent of the particles would now be of the type $\{+\mathbf{z}, +\mathbf{x}\}$ and 50 percent would be of the type $\{-\mathbf{z}, +\mathbf{x}\}$. In this way, the local realist can reproduce the results of Experiment 3 in Chapter 1 on a single particle.

Conservation of angular momentum for two particles in a spin-0 state requires that particle 2 be of the type $\{-\mathbf{z}, -\mathbf{x}\}$ if particle 1 is of the type $\{+\mathbf{z}, +\mathbf{x}\}$. Let us assume that four distinct groups of the two particles are produced in the decay of a collection of spin-0 particles:

$$
\begin{array}{cc}
\text{Particle 1} & \text{Particle 2} \\
\{+\mathbf{z}, +\mathbf{x}\} & \{-\mathbf{z}, -\mathbf{x}\} \\
\{+\mathbf{z}, -\mathbf{x}\} & \{-\mathbf{z}, +\mathbf{x}\} \\
\{-\mathbf{z}, +\mathbf{x}\} & \{+\mathbf{z}, -\mathbf{x}\} \\
\{-\mathbf{z}, -\mathbf{x}\} & \{+\mathbf{z}, +\mathbf{x}\}
\end{array}
\tag{5.40}
$$

and that each of these distinct groups of particles is produced in equal numbers. If A and B both make measurements of $S_z$ or both make measurements of $S_x$ on their respective particles, the results are consistent with conservation of angular momentum (and the predictions of quantum mechanics) since they always find the spin components of their particles pointing in opposite directions. In addition, if A, for example, makes measurements of $S_{1z}$ and obtains the value $\hbar/2$ and B makes measurements of $S_{2x}$, 50 percent of B's measurements will yield $\hbar/2$ and 50 percent will yield $-\hbar/2$, since 50 percent of B's particle must be of the type $\{-\mathbf{z}, -\mathbf{x}\}$ and 50 percent must be of the type $\{-\mathbf{z}, +\mathbf{x}\}$. Thus this simple, nonquantum model in which each of the particles in the two-particle system has definite attributes is able to reproduce the results of quantum mechanics. Moreover, in this model the results that B obtains are completely predetermined by the type of particle entering B's detector, independent of what A chooses to measure. This makes the local realist happy.

We now want to show that this simple model cannot reproduce all the results of quantum mechanics in a somewhat more complicated experiment in which A and B agree to make measurements of the spin along one of three, in general, nonorthogonal, coplanar directions specified by the vectors $\mathbf{a}$, $\mathbf{b}$, and $\mathbf{c}$. Each of the particles must now belong to a definite type such as $\{+\mathbf{a}, -\mathbf{b}, +\mathbf{c}\}$, for which a measurement by A or B on a particle of this type would yield $\hbar/2$ if the SG device is oriented along the direction specified by $\mathbf{a}$ or $\mathbf{c}$, but would yield $-\hbar/2$ if the SG device is oriented along the direction specified by $\mathbf{b}$. Again, in order to conserve angular momentum, if particle 1 is of the type $\{+\mathbf{a}, -\mathbf{b}, +\mathbf{c}\}$, then particle 2 must be of the type $\{-\mathbf{a}, +\mathbf{b}, -\mathbf{c}\}$, so that if A finds particle 1 to have its spin up or down along some axis, B finds particle 2 to have its spin oppositely directed along the same axis. There are now eight different groups that the two particles emitted in the decay of a spin-0 particle may reside in:

| Population | Particle 1 | Particle 2 | |
|---|---|---|---|
| $N_1$ | $\{+\mathbf{a}, +\mathbf{b}, +\mathbf{c}\}$ | $\{-\mathbf{a}, -\mathbf{b}, -\mathbf{c}\}$ | |
| $N_2$ | $\{+\mathbf{a}, +\mathbf{b}, -\mathbf{c}\}$ | $\{-\mathbf{a}, -\mathbf{b}, +\mathbf{c}\}$ | |
| $N_3$ | $\{+\mathbf{a}, -\mathbf{b}, +\mathbf{c}\}$ | $\{-\mathbf{a}, +\mathbf{b}, -\mathbf{c}\}$ | |
| $N_4$ | $\{+\mathbf{a}, -\mathbf{b}, -\mathbf{c}\}$ | $\{-\mathbf{a}, +\mathbf{b}, +\mathbf{c}\}$ | |
| $N_5$ | $\{-\mathbf{a}, +\mathbf{b}, +\mathbf{c}\}$ | $\{+\mathbf{a}, -\mathbf{b}, -\mathbf{c}\}$ | |
| $N_6$ | $\{-\mathbf{a}, +\mathbf{b}, -\mathbf{c}\}$ | $\{+\mathbf{a}, -\mathbf{b}, +\mathbf{c}\}$ | |
| $N_7$ | $\{-\mathbf{a}, -\mathbf{b}, +\mathbf{c}\}$ | $\{+\mathbf{a}, +\mathbf{b}, -\mathbf{c}\}$ | |
| $N_8$ | $\{-\mathbf{a}, -\mathbf{b}, -\mathbf{c}\}$ | $\{+\mathbf{a}, +\mathbf{b}, +\mathbf{c}\}$ | (5.41) |

First, let's consider an experiment in which A and B orient their SG devices at random along the axes $\mathbf{a}$, $\mathbf{b}$, and $\mathbf{c}$, making measurements of the spin of the particle along these axes.[10] Let's examine the correlations in their data for those cases in which their SG devices are oriented along *different* axes. In particular, let's see what fraction of their measurements yield values for the spins of the two particles that have opposite signs, such as would be the case, for example, if A finds particle 1 to have $S_{1a} = \hbar/2$ and B finds particle 2 to have $S_{2c} = -\hbar/2$. Clearly, all measurements made on particles in populations $N_1$ and $N_8$ will yield opposite signs for the spins of the two particles. On the other hand, for population $N_2$, when A finds $S_{1a} = \hbar/2$, B's measurement yields the result $S_{2b} = -\hbar/2$ (with the opposite sign) if B's SG device is oriented along $\mathbf{b}$, but if instead B's SG device is oriented along the $\mathbf{c}$ axis, B obtains $S_{2c} = \hbar/2$ (with the same sign). Similarly, if A's SG device is oriented along the $\mathbf{b}$ axis, A finds $S_{1b} = \hbar/2$ while B finds $S_{2a} = -\hbar/2$ or $S_{2c} = \hbar/2$, depending on whether B's SG device is oriented along $\mathbf{a}$ or $\mathbf{c}$, respectively. Finally, still for population $N_2$, if A's SG device is oriented along the $\mathbf{c}$ axis, A obtains $S_{1c} = -\hbar/2$, while B finds $S_{2a} = -\hbar/2$ or $S_{2b} = -\hbar/2$. Thus, overall for population $N_2$, $\frac{2}{6} = \frac{1}{3}$ of the measurements yield results with opposite signs when the SG devices are oriented along different axes. This ratio holds for all the populations $N_2$ through $N_7$. Since measurements on populations $N_1$ and $N_8$ always yield results with opposite signs, independent of the orientation of the SG devices, *at least one-third* of the measurements [in fact, $\frac{1}{3}\left(\frac{3}{4}\right) + \left(\frac{1}{4}\right) = \frac{1}{2}$ of the measurements if all eight populations occur with equal frequency] will find the particles with opposite signs for their spins when the two experimentalists orient their SG devices along different axes.

Although this result seems straightforward enough, we can quickly see that it is in complete disagreement with the predictions of quantum mechanics, at least for certain orientations of the axes $\mathbf{a}$, $\mathbf{b}$, and $\mathbf{c}$. We express the $|0, 0\rangle$ state as

---

[10] This thought experiment was suggested by N. D. Mermin, *Am. J. Phys.* **49**, 940 (1981). See also his discussion in *Physics Today*, April 1985. Our derivation of the Bell inequality (5.54) follows that given by J. J. Sakurai, *Modern Quantum Mechanics*, Benjamin-Cummings, Menlo Park, Calif., 1985.

$$|0, 0\rangle = \frac{1}{\sqrt{2}}|+\mathbf{a}, -\mathbf{a}\rangle - \frac{1}{\sqrt{2}}|-\mathbf{a}, +\mathbf{a}\rangle \tag{5.42}$$

The amplitude to find particle 1 with $S_{1a} = -\hbar/2$ and particle 2 with $S_{2b} = \hbar/2$ is given by

$$\langle -\mathbf{a}, +\mathbf{b}|0, 0\rangle = \frac{1}{\sqrt{2}}\langle -\mathbf{a}, +\mathbf{b}|+\mathbf{a}, -\mathbf{a}\rangle - \frac{1}{\sqrt{2}}\langle -\mathbf{a}, +\mathbf{b}|-\mathbf{a}, +\mathbf{a}\rangle$$

$$= -\frac{1}{\sqrt{2}}\langle -\mathbf{a}, +\mathbf{b}|-\mathbf{a}, +\mathbf{a}\rangle = \frac{1}{\sqrt{2}}\big(_1\langle -\mathbf{a}|-\mathbf{a}\rangle_1\big)\big(_2\langle +\mathbf{b}|+\mathbf{a}\rangle_2\big)$$

$$= -\frac{1}{\sqrt{2}}\langle +\mathbf{b}|+\mathbf{a}\rangle \tag{5.43}$$

where we have expressed the two-particle state in terms of a direct product of single-particle states to evaluate the amplitude in terms of single-particle amplitudes. We have also dropped the subscripts on the last amplitude, which involves only a single particle. From our earlier work (see Problem 3.2), we know that

$$|+\mathbf{n}\rangle = \cos\frac{\theta}{2}|+\mathbf{z}\rangle + e^{i\phi}\sin\frac{\theta}{2}|-\mathbf{z}\rangle \tag{5.44}$$

Thus $\langle +\mathbf{z}|+\mathbf{n}\rangle = \cos(\theta/2)$, where $\theta$ is the angle $\mathbf{n}$ makes with the $z$ axis. Therefore,

$$\langle +\mathbf{b}|+\mathbf{a}\rangle = \cos\frac{\theta_{ab}}{2} \tag{5.45}$$

where $\theta_{ab}$ is the angle between the $\mathbf{a}$ and the $\mathbf{b}$ axis, as shown in Fig. 5.4. The

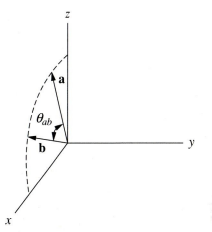

**FIGURE 5.4**
Two axes $\mathbf{a}$ and $\mathbf{b}$ used for measuring the spin.

quantum mechanical prediction for the probability of finding the particles in the state $|-\mathbf{a}, +\mathbf{b}\rangle$ is

$$|\langle +\mathbf{a}, -\mathbf{b}|0, 0\rangle|^2 = \frac{1}{2} \cos^2 \frac{\theta_{ab}}{2} \tag{5.46}$$

Similarly,

$$|\langle -\mathbf{a}, +\mathbf{b}|0, 0\rangle|^2 = \frac{1}{2} \cos^2 \frac{\theta_{ab}}{2} \tag{5.47}$$

Thus the total probability that A and B obtain opposite signs for the spin when they make measurements with A's SG device oriented along $\mathbf{a}$ and B's SG device oriented along $\mathbf{b}$ is given by

$$|\langle +\mathbf{a}, -\mathbf{b}|0, 0\rangle|^2 + |\langle -\mathbf{a}, +\mathbf{b}|0, 0\rangle|^2 = \cos^2 \frac{\theta_{ab}}{2} \tag{5.48}$$

Clearly the probability would be the same if A's SG device is oriented along $\mathbf{b}$ and B's device is oriented along $\mathbf{a}$. Now let's choose the axes $\mathbf{a}$, $\mathbf{b}$, and $\mathbf{c}$, as shown in Fig. 5.5. With the angle $\theta_{ab} = 120°$, the probability (5.48) is simply $\frac{1}{4}$. But the probability

$$|\langle +\mathbf{a}, -\mathbf{c}|0, 0\rangle|^2 + |\langle -\mathbf{a}, +\mathbf{c}|0, 0\rangle|^2 = \cos^2 \frac{\theta_{ac}}{2} \tag{5.49}$$

is also equal to $\frac{1}{4}$ since the angle $\theta_{ac} = 120°$. In fact, since the angle $\theta_{bc} = 120°$ as well, quantum mechanics predicts for the particular orientation of the axes shown in Fig. 5.5 that exactly one-quarter of the measurements will yield values with opposite signs for the spins along different axes, in direct disagreement with the model in which each of the particles possesses definite attributes, where at least one-third of the measurements yield values with opposite signs. Thus it should be possible to test which is right—quantum mechanics or a model in which the particles possess definite attributes—by performing an experiment.

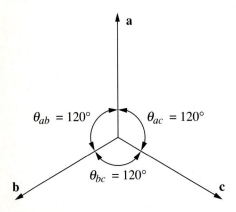

**FIGURE 5.5**
One orientation of the axes $\mathbf{a}$, $\mathbf{b}$, and $\mathbf{c}$ that leads to disagreement between quantum mechanics and a model in which the particles possess definite attributes.

Interestingly, we can extract a variety of inequalities from the supposition that the particles can be grouped into populations of the type (5.41), inequalities that may be easier to test in practice than the experiment that we have just described. Without having to specify the relative populations of the different groups (5.41), we may quickly see, for example, that certain inequalities such as

$$N_3 + N_4 \leq (N_2 + N_4) + (N_3 + N_7) \tag{5.50}$$

must hold. But

$$\frac{N_3 + N_4}{\sum_i N_i} = P(+\mathbf{a}; +\mathbf{b}) \tag{5.51}$$

is the probability that a measurement by A yields $S_{1a} = \hbar/2$ for particle 1 *and* a measurement by B yields $S_{2b} = \hbar/2$ for particle 2. Only populations $N_3$ and $N_4$ contain particle types satisfying both these conditions. Similarly,

$$\frac{N_2 + N_4}{\sum_i N_i} = P(+\mathbf{a}; +\mathbf{c}) \tag{5.52}$$

is the probability that a measurement by A yields $S_{1a} = \hbar/2$ for particle 1 *and* a measurement by B yields $S_{2c} = \hbar/2$. Also

$$\frac{N_3 + N_7}{\sum_i N_i} = P(+\mathbf{c}; +\mathbf{b}) \tag{5.53}$$

Thus the inequality (5.50) may be expressed as

$$P(+\mathbf{a}; +\mathbf{b}) \leq P(+\mathbf{a}; +\mathbf{c}) + P(+\mathbf{c}; +\mathbf{b}) \tag{5.54}$$

which is known as a Bell's inequality. In order to test this inequality, A and B just make measurements to determine the three probabilities. First A's SG device is oriented along $\mathbf{a}$, while B's is fixed along $\mathbf{b}$, and measurements are made to determine $P(+\mathbf{a}; +\mathbf{b})$. A and B then go on to measure $P(+\mathbf{a}; +\mathbf{c})$ and $P(+\mathbf{c}; +\mathbf{b})$.

The inequality (5.54) is in a form that is easy to compare with the predictions of quantum mechanics. In particular, since

$$\langle +\mathbf{a}, +\mathbf{b}|0, 0 \rangle = \frac{1}{\sqrt{2}} \langle +\mathbf{a}, +\mathbf{b}|+\mathbf{a}, -\mathbf{a} \rangle - \frac{1}{\sqrt{2}} \langle +\mathbf{a}, +\mathbf{b}|-\mathbf{a}, +\mathbf{a} \rangle$$

$$= \frac{1}{\sqrt{2}} \langle +\mathbf{b}|-\mathbf{a} \rangle \tag{5.55}$$

and

$$\langle +\mathbf{b}|-\mathbf{a} \rangle = \sin \frac{\theta_{ab}}{2} \tag{5.56}$$

the prediction of quantum mechanics for the probability is given by

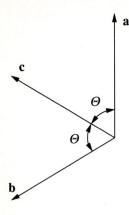

**FIGURE 5.6**
An orientation of the axes **a**, **b**, and **c** where **c** bisects the angle between **a** and **b**.

$$P(+\mathbf{a}; +\mathbf{b}) = |\langle +\mathbf{a}, +\mathbf{b}|0, 0\rangle|^2 = \frac{1}{2} \sin^2 \frac{\theta_{ab}}{2} \tag{5.57}$$

Note that if $\mathbf{b} = \mathbf{a}$, then $\theta_{ab} = 0$ and $P(+\mathbf{a}; +\mathbf{a}) = 0$, as it must for two particles in a state with total-spin 0. Also, if $\mathbf{b} = -\mathbf{a}$, then $\theta_{ab} = \pi$ and $P(+\mathbf{a}; -\mathbf{a}) = \frac{1}{2}$, again the usual result for a total-spin-0 state. If we generalize the result (5.57) to the other two terms in the Bell's inequality (5.54), we obtain

$$\sin^2 \frac{\theta_{ab}}{2} \leq \sin^2 \frac{\theta_{ac}}{2} + \sin^2 \frac{\theta_{cb}}{2} \tag{5.58}$$

As in our earlier discussion, this inequality is violated for certain orientations of **a**, **b**, and **c**. To see the disagreement in a particular case and to make the algebra easy, let's take the special case where **c** bisects the angle $\theta_{ab}$, as shown in Fig. 5.6. If we call $\theta_{ab} = 2\Theta$, then $\theta_{ac} = \theta_{cb} = \Theta$, and the inequality (5.58) becomes

$$\sin^2 \Theta \leq 2 \sin^2 \frac{\Theta}{2} \tag{5.59}$$

In particular, let $\Theta = \pi/3 = 60°$ as a specific example; we then obtain

$$\tfrac{3}{4} \leq \tfrac{1}{2} \tag{5.60}$$

again a marked disagreement between the predictions of quantum mechanics and those of a local, realistic theory. In fact, this particular choice of angles is the same as in our earlier discussion. Just let $\mathbf{c} \to -\mathbf{c}$ to go from Fig. 5.5 to Fig. 5.6. Then spin-down along **c** is spin-up along $-\mathbf{c}$. As Fig. 5.7 shows, (5.58) is violated for all angles $\Theta$ satisfying $0 < \Theta < \pi/2$. Thus it should be possible to test the predictions of quantum mechanics by observing the correlations in the spins of the two particles for a variety of angles. Based on our earlier discussion, if quantum mechanics is correct and Bell's inequality is violated, no local hidden-variable theory can be valid.

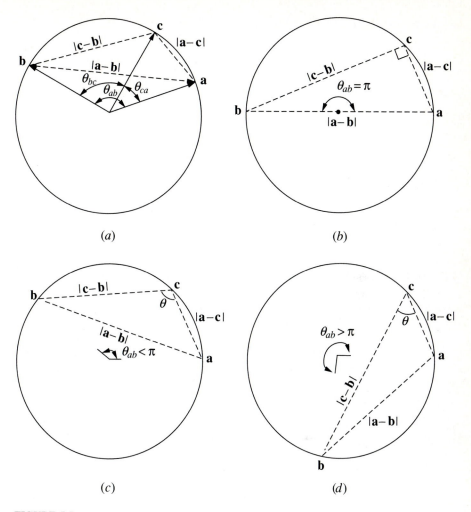

**FIGURE 5.7**

(a) The unit vectors **a**, **b**, and **c** specifying the orientation of three SG devices for measuring the spins of the two spin-$\frac{1}{2}$ particles emitted in a total-spin-0 state. Each of the SG devices has its measurement axis transverse to the direction of flight of the two particles, and therefore the unit vectors all lie in a plane with their tips on a circle. Note that the square of the length of the vector pointing between **a** and **b** is given by $|\mathbf{a} - \mathbf{b}|^2 = \mathbf{a}^2 + \mathbf{b}^2 - 2\mathbf{a} \cdot \mathbf{b} = 2(1 - \cos\theta_{ab}) = 4\sin^2\theta_{ab}/2$. Similarly, $|\mathbf{a} - \mathbf{c}|^2 = 4\sin^2\theta_{ac}/2$ and $|\mathbf{b} - \mathbf{c}|^2 = 4\sin^2\theta_{bc}/2$. Thus, expressed in terms of these lengths, the inequality (5.58) becomes $|\mathbf{a} - \mathbf{b}|^2 \leq |\mathbf{a} - \mathbf{c}|^2 + |\mathbf{c} - \mathbf{b}|^2$. (b) The angle $\theta_{ab}$ is taken to be $\pi$, in which case the triangle formed by $|\mathbf{a} - \mathbf{b}|$, $|\mathbf{a} - \mathbf{c}|$, and $|\mathbf{c} - \mathbf{b}|$ is a right triangle and therefore $|\mathbf{a} - \mathbf{b}|^2 = |\mathbf{a} - \mathbf{c}|^2 + |\mathbf{c} - \mathbf{b}|^2$. Note that in (b), (c), and (d), the vectors **a**, **b**, and **c** are not actually shown, but you can see their direction by noting the points where they intersect the unit circle. (c) The angle $\theta_{ab} < \pi$ and since the angle $\theta > \pi/2$, $|\mathbf{a} - \mathbf{b}|^2 > |\mathbf{a} - \mathbf{c}|^2 + |\mathbf{c} - \mathbf{b}|^2$ and the Bell inequality (5.58) is violated. (d) The angle $\theta_{ab} > \pi$, making $\theta < \pi/2$ and $|\mathbf{a} - \mathbf{b}|^2 < |\mathbf{a} - \mathbf{c}|^2 + |\mathbf{c} - \mathbf{b}|^2$, in accord with the Bell inequality.

## Experimental Tests and Implications

Bell's results have inspired a number of experiments. With the exception of one experiment that measured the spin orientation of protons in a singlet state, these experiments have all been carried out on the polarization state of pairs of photons rather than on spin-$\frac{1}{2}$ particles. A suitably correlated pair of photons is produced, for example, in the annihilation of positronium in its ground state (see Problem 5.6), but these photons are, of course, quite energetic gamma rays, for which there are no really efficient polarization filters. Suitable optical photons are produced in the cascade decays of atoms such as Ca or Hg excited by laser pumping in which the transition is of the form

$$(J = 0) \xrightarrow{\gamma} (J = 1) \xrightarrow{\gamma} (J = 0)$$

and the photons are emitted essentially back to back, with the atom itself not carrying very much momentum. The correlations are between measurements of the linear polarization for each of the photons. The most precise experiments of this type have been carried out by Aspect et al.[11] In one case the Bell inequality was violated by more than nine standard deviations. On the other hand, the agreement with the predictions of quantum mechanics is excellent, as shown in Fig. 5.8.

These results do not make the local realist happy. One of the disturbing features of these results to the local realist (and, perhaps, to you too) is understanding how A's measurement on particle 1 can instantaneously fix the result of B's measurement on particle 2 when the two measuring devices may be separated by arbitrarily large distances. In the experiments of Aspect et al., the separation between these devices was as large as 13 m. Although we do not have any mechanism in mind for how the setting of A's measuring device could influence B's device, in these experiments the devices are left in particular settings for extended periods of time. Maybe B's device "knows" about the settings of A's device in ways we don't understand. In order to eliminate the possibility of any influence,

[11] A. Aspect, P. Grangier, and G. Roger, *Phys. Rev. Lett.* **49**, 91 (1982).

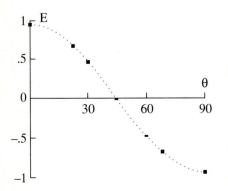

**FIGURE 5.8**
Correlation of polarizations as a function of the relative angle of the polarimeters. The indicated errors are $\pm 2$ standard deviations. The dotted curve is not a fit to the data, but the quantum mechanical predictions for the actual experiment. For ideal polarizers, the curve would reach the values $\pm 1$. *[From A. Aspect, P. Grangier, and G. Roger, Phys. Rev. Lett.* **49**, *91 (1982).]*

Aspect et al. have carried out one experiment in which the choice of analyzer setting was changed so rapidly that A's decision on what to measure could not have influenced B's result unless the information about the choice of setting was transmitted between A and B with a speed faster than the speed of light.[12] Even in this case the quantum mechanical correlations between the measurements persisted. Strange as these correlations may seem, they do not permit the possibility of faster-than-light communication. In the spin system, for example, 50 percent of B's measurements of $S_z$ yield $S_{2z} = \hbar/2$ and 50 percent yield $S_{2z} = -\hbar/2$ whether or not A has made a measurement and no matter what the orientation of A's SG device (see Problem 5.4). It is only when A and B compare their data after the experiment that they find a complete correlation between their results when they both oriented their SG devices in the same direction.

So where does all this leave us? Certainly with a sense of wonder about the way the physical world operates. It is hard to guess how Einstein would have responded to the recent experimental results. As we have noted, he believed particles should have definite attributes, or properties, independent of whether or not these properties were actually measured. As A. Pais recounts: "We often discussed his notions on objective reality. I recall that during one walk Einstein suddenly stopped, turned to me and asked whether I really believed that the moon exists only when I look at it."[13] In the microscopic world, the answer appears to be yes.

## 5.6  SUMMARY

In this chapter we have examined the quantum states of two particles. We use a number of different notations to specify a two-particle state. A general state $|\psi\rangle$ can be expressed in terms of the single-particle states by

$$|\psi\rangle = \sum_{i,j} c_{ij} |a_i\rangle_1 \otimes |b_j\rangle_2 \tag{5.61}$$

where in this form we are not presuming that the states $|a_i\rangle$ for particle 1 are necessarily the same as the states $|b_j\rangle$ for particle 2. The symbol $\otimes$ indicates a special product, called a direct product, of the kets from the two different vector spaces. We often dispense with the direct-product symbol and simply write

$$|\psi\rangle = \sum_{i,j} c_{ij} |a_i\rangle_1 |b_j\rangle_2 \tag{5.62}$$

or

$$|\psi\rangle = \sum_{i,j} c_{ij} |a_i, b_j\rangle = \sum_{i,j} |a_i, b_j\rangle\langle a_i, b_j|\psi\rangle \tag{5.63}$$

[12]  A. Aspect, J. Dalibard, and G. Roger, *Phys. Rev. Lett.* **49,** 1804 (1982).
[13]  A. Pais, *Rev. Mod. Phys.* **51,** 863 (1979).

where we understand the ket $|a_i, b_j\rangle$ to be a two-particle ket with particle 1 in the state $|a_i\rangle$ and particle 2 in the state $|b_j\rangle$.

Because our discussion of quantum mechanics so far has emphasized the spin degrees of freedom of a spin-$\frac{1}{2}$ particle, we have focused our attention in this chapter on the spin states of a system of two spin-$\frac{1}{2}$ particles. In particular, we have discovered that the eigenstates of total spin angular momentum $\hat{\mathbf{S}} = \hat{\mathbf{S}}_1 + \hat{\mathbf{S}}_2$ are given by

$$|1, 1\rangle = |+\mathbf{z}, +\mathbf{z}\rangle \tag{5.64a}$$

$$|1, 0\rangle = \frac{1}{\sqrt{2}}|+\mathbf{z}, -\mathbf{z}\rangle + \frac{1}{\sqrt{2}}|-\mathbf{z}, +\mathbf{z}\rangle \tag{5.64b}$$

$$|1, -1\rangle = |-\mathbf{z}, -\mathbf{z}\rangle \tag{5.64c}$$

and

$$|0, 0\rangle = \frac{1}{\sqrt{2}}|+\mathbf{z}, -\mathbf{z}\rangle - \frac{1}{\sqrt{2}}|-\mathbf{z}, +\mathbf{z}\rangle \tag{5.65}$$

where the labels for the kets on the left-hand side are just the total angular momentum states

$$\hat{\mathbf{S}}^2|s, m\rangle = s(s + 1)\hbar^2|s, m\rangle \tag{5.66a}$$

$$\hat{S}_z|s, m\rangle = m\hbar|s, m\rangle \tag{5.66b}$$

with $\hat{\mathbf{S}}$ the total spin. As for any angular momentum, we can use one of the components and the square of the magnitude of the total spin to label the total-spin states. The states in (5.64) form a triplet of total-spin-1 states, since $s = 1$ for these three states ($m = 1, 0, -1$), while the state (5.65) is a singlet $s = 0$ state. Notice in particular that the spin-1 states are unchanged (symmetric) if you exchange the spins of each of the particles, while the spin-0 state changes sign (antisymmetric) under exchange of the spins of the two particles. These spin states will play a pivotal role in our discussion of the possible states of identical particles in Chapter 12.

Although states such as (5.64) and (5.65) are natural extensions of our specification of a quantum state for a single particle to that for two particles, experiments carried out on states such as these tell us much about the physical "nature of reality." The total-spin-0 state has been our laboratory for an investigation of the correlations that exist between the measurements of the individual spins of the two particles in such a state. These correlations have been shown to be in experimental disagreement with those that would occur if the particles were each to possess definite attributes. In fact, in measurements on a two-particle spin state, measurements on one of the particles can fix the result of a measurement of the *other* particle, even though the particles may be separated by arbitrarily large distances *and* neither of the particles possesses a definite attribute before the measurement. Although this may seem paradoxical, it is a natural outcome of applying quantum mechanics to a two-particle system.

## PROBLEMS

**5.1.** Take the spin Hamiltonian for the hydrogen atom in an external magnetic field $B_0$ in the $z$ direction to be

$$\hat{H} = \frac{2A}{\hbar^2}\hat{\mathbf{S}}_1 \cdot \hat{\mathbf{S}}_2 + \omega_0 \hat{S}_{1z}$$

where $\omega_0 = g e B_0/2mc$, with $m$ the mass of the electron. The contribution $-\hat{\boldsymbol{\mu}}_2 \cdot \mathbf{B}_0$ of the proton has been neglected because the mass of the proton is roughly 2000 times larger than the mass of the electron. Determine the energies of this system. Examine your results in the limiting cases $A \gg \hbar\omega_0$ and $A \ll \hbar\omega_0$ by expanding the energy eigenvalues in a Taylor series or binomial expansion through first nonvanishing order.

**5.2.** Express the total-spin $s = 1$ states of two spin-$\frac{1}{2}$ particles given in (5.30a) and (5.30c) in terms of the states $|+\mathbf{x}, +\mathbf{x}\rangle$, $|+\mathbf{x}, -\mathbf{x}\rangle$, $|-\mathbf{x}, +\mathbf{x}\rangle$, and $|-\mathbf{x}, -\mathbf{x}\rangle$.

**5.3.** Express the total-spin $s = 0$ state of two spin-$\frac{1}{2}$ particles given in (5.31) in terms of the states $|+\mathbf{n}, -\mathbf{n}\rangle$ and $|-\mathbf{n}, +\mathbf{n}\rangle$, where for a single spin-$\frac{1}{2}$ particle

$$|+\mathbf{n}\rangle = \cos\frac{\theta}{2}|+\mathbf{z}\rangle + e^{i\phi}\sin\frac{\theta}{2}|-\mathbf{z}\rangle$$

$$|-\mathbf{n}\rangle = \sin\frac{\theta}{2}|+\mathbf{z}\rangle - e^{i\phi}\cos\frac{\theta}{2}|-\mathbf{z}\rangle$$

**5.4.** In the Einstein-Podolsky-Rosen experiment described in Section 5.4, the orientation of A's SG device is at angle $\theta$ in the $x$-$z$ plane, while the orientation of B's SG device is along the $z$ axis. Show that 50 percent of B's measurements yield $S_{2z} = \hbar/2$ and 50 percent yield $S_{2z} = -\hbar/2$.

**5.5.** At time $t = 0$, an electron and a positron are formed in a state with total spin angular momentum equal to zero, perhaps from the decay of a spinless particle. The particles are situated in a uniform magnetic field $B_0$ in the $z$ direction.

  (a) If interaction between the electron and the positron may be neglected, show that the spin Hamiltonian of the system may be written as

$$\hat{H} = \omega_0(\hat{S}_{1z} - \hat{S}_{2z})$$

  where $\hat{\mathbf{S}}_1$ is the spin operator of the electron, $\hat{\mathbf{S}}_2$ is the spin operator of the positron, and $\omega_0$ is a constant.

  (b) Show that the state of the system oscillates between a spin-0 and a spin-1 state. Determine the period of oscillation. What is the spin state of the system at time $t$?

  (c) At time $t$, measurements are made of $S_{1x}$ and $S_{2x}$. Calculate the probability that *both* of these measurements yield the value $\hbar/2$.

**5.6.** Take the spin Hamiltonian of the positronium atom (a bound state of an electron and a positron) in an external magnetic field in the $z$ direction to be

$$\hat{H} = \frac{2A}{\hbar^2}\hat{\mathbf{S}}_1 \cdot \hat{\mathbf{S}}_2 + \omega_0(\hat{S}_{1z} - \hat{S}_{2z})$$

Determine the energy eigenvalues.

**5.7.** The annihilation of positronium in its ground state produces two photons that travel back to back in the positronium rest frame along an axis taken to be the $z$ axis. The polarization state of the two-photon system is given by

$$|\psi\rangle = \frac{1}{\sqrt{2}}|R, R\rangle - \frac{1}{\sqrt{2}}|L, L\rangle$$

(a) What is the probability that a measurement of the circular polarization state of the two photons will find them both right-handed? Both left-handed?

(b) What is the probability that photon 1 will be found to be $x$ polarized *and* photon 2 will be found to be $y$ polarized, that is, that the system is in the state $|x, y\rangle$? What is the probability that the system is in the state $|y, x\rangle$?

(c) Compare the probability for the two photons to be in the state $|x, x\rangle$ or in the state $|y, y\rangle$ with what you would obtain if the two-photon state were *either* $|R, R\rangle$ *or* $|L, L\rangle$ rather than the superposition $|\psi\rangle$.

# CHAPTER
# 6

## WAVE MECHANICS IN ONE DIMENSION

Thus far, our discussion of quantum mechanics has concentrated on two-state systems, with most of the emphasis on the spin states of a spin-$\frac{1}{2}$ particle. But particles have more degrees of freedom than just intrinsic spin. We will now begin to discuss the results of measuring a particle's position or momentum. In this chapter we will concentrate on one dimension and neglect the spin degrees of freedom. This marks the beginning of our discussion of wave mechanics.

## 6.1 POSITION EIGENSTATES AND THE WAVE FUNCTION

When we wanted to analyze the results of measuring the intrinsic spin $S_z$ of a spin-$\frac{1}{2}$ particle, we expressed the state $|\psi\rangle$ of the particle as a superposition of the eigenstates $|+\mathbf{z}\rangle$ and $|-\mathbf{z}\rangle$ of the operator $\hat{S}_z$. If we are interested in measuring the position $x$ of the particle, it is natural to introduce position states $|x\rangle$ satisfying

$$\hat{x}|x\rangle = x|x\rangle \tag{6.1}$$

where $\hat{x}$ is the position operator and the value of $x$ runs over all possible values of the position of the particle, that is, from $-\infty$ to $+\infty$.

Strictly speaking, such position eigenstates are a mathematical abstraction. In contrast to the measurement of the intrinsic spin $S_z$ of a spin-$\frac{1}{2}$ particle, where we always obtain either $\hbar/2$ or $-\hbar/2$, we cannot obtain a single value for the position of a particle when we try to measure it. As an example, Fig. 6.1 shows a schematic of a microscope that might be used to determine the position of a particle. Light scattered by the particle is focused by the lens on a screen. The resolution of the microscope, that is, the precision with which the position of the particle can be determined, is given by

$$\Delta x \sim \frac{\lambda}{\sin \phi} \tag{6.2}$$

where $\lambda$ is the wavelength of the light and the angle $\phi$ is shown in the figure. The physical cause of this inherent uncertainty in the position is the diffraction pattern that is formed on the screen when light passes through the lens. We can make the resolution sharper and sharper by using light of shorter and shorter wavelength, but no matter how high the energy of the photons that are used in the microscope, we will never do better than localizing the position of the particle to some range $\Delta x$ in position. Thus, as this example suggests, we cannot prepare a particle in a state with a definite position by making a position measurement. In fact, as our discussion of the Einstein-Podolsky-Rosen paradox emphasized, we should not try to view the particle as having a definite position at all.

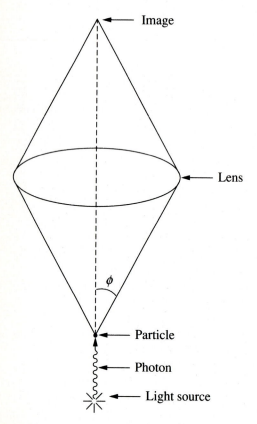

**FIGURE 6.1**
A microscope for determining the position of a particle.

Although it is not possible to obtain a single value for the measurement of the position of the particle, nonetheless kets such as $|x\rangle$ in which the particle has a single position are very useful. We may think of the physical states that occur in nature as a superposition of these position eigenstates. How should we express this superposition? If we try to mimic the formalism that we used for intrinsic spin with its discrete eigenstates and write

$$|\psi\rangle = \sum_i |x_i\rangle\langle x_i|\psi\rangle$$

where the sum runs over all the values of the position, we can quickly see that we have a serious problem. Writing the bra $\langle\psi|$ as

$$\langle\psi| = \sum_j \langle\psi|x_j\rangle\langle x_j|$$

allows us to calculate

$$\langle\psi|\psi\rangle = \sum_{ij} \langle\psi|x_j\rangle\langle x_j|x_i\rangle\langle x_i|\psi\rangle$$

Again, if we mimic our earlier formalism and use $\langle x_j|x_i\rangle = \delta_{ij}$, which states that the probability amplitude is equal to unity if the positions are the same and equal to zero if they are different, we obtain

$$\langle\psi|\psi\rangle = \sum_i \langle\psi|x_i\rangle\langle x_i|\psi\rangle = \sum_i |\langle x_i|\psi\rangle|^2$$

Let's consider a point $x_i = a$ where $\langle a|\psi\rangle$ is not zero. Then we may choose a sufficiently nearby point $x_i = a + \Delta x$ where $\langle a + \Delta x|\psi\rangle$ is also not zero. However, there are *infinitely* many points between $a$ and $a + \Delta x$, no matter how small we choose $\Delta x$. Thus

$$\sum_i |\langle x_i|\psi\rangle|^2 = \infty$$

and we are unable to satisfy the condition $\langle\psi|\psi\rangle = 1$.

Our way of expressing $|\psi\rangle$ as a discrete sum of position states cannot work when we are dealing with a variable like position that takes on a *continuum* of values. Instead of a sum, we need an *integral*:

$$|\psi\rangle = \int_{-\infty}^{\infty} dx\, |x\rangle\langle x|\psi\rangle \tag{6.3}$$

Now the coefficient of the position ket $|x\rangle$ is $dx\,\langle x|\psi\rangle$ so that if we integrate from $a$ to $a + \Delta x$ only, we obtain a contribution to $|\psi\rangle$ of

$$\int_a^{a+\Delta x} dx\, |x\rangle\langle x|\psi\rangle$$

which vanishes for vanishingly small $\Delta x$. Examination of (6.3) shows that the generalization of the completeness relation (2.48) to kets like those of position that take on a continuum rather than a discrete set of values is

$$\int_{-\infty}^{\infty} dx \, |x\rangle\langle x| = 1 \tag{6.4}$$

In order to see how these position kets should be normalized, let's consider the special case where the ket $|\psi\rangle$ in (6.3) is itself a position state $|x'\rangle$. Then

$$|x'\rangle = \int_{-\infty}^{\infty} dx \, |x\rangle\langle x|x'\rangle \tag{6.5}$$

which implies that $\langle x|x'\rangle = \delta(x - x')$, where $\delta(x - x')$ is a Dirac delta function.[1] It is comforting to note that when $x \neq x'$, the amplitude to find a position state $|x'\rangle$ at position $x$ vanishes. It is, however, somewhat disquieting to see that when $x = x'$, the amplitude is infinite. Let's recall that we are not able to make measurements of a single position out of the continuum of possible positions. Thus an amplitude like $\langle x|x'\rangle$ is not directly related to a physically observable quantity. Dirac delta functions always appear within an integral whenever we are calculating anything physical. For example, the bra corresponding to the ket (6.3) is

$$\langle\psi| = \int_{-\infty}^{\infty} dx \, \langle\psi|x\rangle\langle x| \tag{6.6}$$

We now calculate

$$\begin{aligned}
\langle\psi|\psi\rangle &= \iint dx' \, dx \, \langle\psi|x\rangle\langle x|x'\rangle\langle x'|\psi\rangle \\
&= \iint dx' \, dx \, \langle\psi|x\rangle\delta(x - x')\langle x'|\psi\rangle \\
&= \int dx \, \langle\psi|x\rangle\langle x|\psi\rangle = \int dx \, |\langle x|\psi\rangle|^2
\end{aligned} \tag{6.7}$$

where we have used a different dummy integration variable for the bra equation (6.6) than for the corresponding ket equation (6.3) because there are two *separate* integrals to be carried out when evaluating $\langle\psi|\psi\rangle$. Note that we could also have obtained this result by inserting the identity operator (6.4) between the bra and the ket in $\langle\psi|\psi\rangle$. Unless stated otherwise, the integrals are presumed to run over all space. The requirement that $\langle\psi|\psi\rangle = 1$ becomes

$$1 = \int dx \, \langle\psi|x\rangle\langle x|\psi\rangle = \int dx \, |\langle x|\psi\rangle|^2 \tag{6.8}$$

---

[1] Dirac delta functions are discussed in Appendix C.

It is natural to identify

$$dx \, |\langle x|\psi\rangle|^2 \tag{6.9}$$

with the probability of finding the particle between $x$ and $x + dx$ if a measurement of position is carried out, as first suggested by M. Born. The requirement that $\langle\psi|\psi\rangle = 1$ then ensures that the total probability of finding the particle in position space is unity. The complex number $\langle x|\psi\rangle$ is the amplitude to find a particle in the state $|\psi\rangle$ at position $x$. This amplitude will, in general, have different values for each different value of $x$; namely, it is a function, which we call the *wave function* $\psi(x)$ of wave mechanics:

$$\langle x|\psi\rangle \equiv \psi(x) \tag{6.10}$$

In terms of $\psi(x)$, (6.8) may be written:

$$1 = \int dx \, \psi^*(x)\psi(x) = \int dx \, |\psi(x)|^2 \tag{6.11}$$

We evaluate the amplitude $\langle\varphi|\psi\rangle$ by inserting the identity operator (6.4) between the bra and the ket vector:

$$\langle\varphi|\psi\rangle = \int dx \, \langle\varphi|x\rangle\langle x|\psi\rangle = \int dx \, \varphi^*(x)\psi(x) \tag{6.12}$$

Finally, as we saw in Chapter 2, for an observable $A$ the expectation value is given by

$$\langle A\rangle = \langle\psi|\hat{A}|\psi\rangle \tag{6.13}$$

Thus the average position of the particle is given by

$$\langle x\rangle = \langle\psi|\hat{x}|\psi\rangle = \int dx \, \langle\psi|\hat{x}|x\rangle\langle x|\psi\rangle$$

$$= \int dx \, \langle\psi|x\rangle x\langle x|\psi\rangle$$

$$= \int dx \, \psi^*(x)x\psi(x) = \int dx \, |\psi(x)|^2 x \tag{6.14}$$

which is the "sum" over all positions $x$ times the probability of finding the particle between $x$ and $x + dx$.

## 6.2 THE TRANSLATION OPERATOR

The natural *operation* to perform on our one-dimensional position basis states is to translate them:

$$\hat{T}(a)|x\rangle = |x + a\rangle \tag{6.15}$$

The operator $\hat{T}(a)$ changes a state of a particle in which the particle has position $x$ to one in which the particle has position $x + a$. In order to determine the action of the translation operator on an arbitrary state $|\psi\rangle$,

$$|\psi'\rangle = \hat{T}(a)|\psi\rangle \tag{6.16}$$

we express $|\psi\rangle$ as a superposition of position states. Then

$$\hat{T}(a)|\psi\rangle = \hat{T}(a)\int dx'\,|x'\rangle\langle x'|\psi\rangle = \int dx'\,|x' + a\rangle\langle x'|\psi\rangle \tag{6.17}$$

We have called the dummy variable in the integral $x'$ because we want to calculate the amplitude to find this translated state at the position $x$:

$$\psi'(x) = \langle x|\psi'\rangle = \langle x|\hat{T}(a)|\psi\rangle$$

$$= \int dx'\,\langle x|x' + a\rangle\langle x'|\psi\rangle = \int dx'\,\delta[x - (x' + a)]\langle x'|\psi\rangle$$

$$= \langle x - a|\psi\rangle = \psi(x - a) \tag{6.18}$$

At first it might seem strange that $\psi'(x) \neq \psi(x + a)$. But if, for example, $\psi(x)$ has its maximum at $x = b$, then $\psi'(x) = \psi(x - a)$ has its maximum at $x - a = b$, or $x = a + b$, as shown in Fig. 6.2. The state has indeed been translated in the positive $x$ direction.

Notice that the translation operator must be a unitary operator, since translating a state should not affect its normalization, that is,

$$\langle \psi'|\psi'\rangle = \langle \psi|\hat{T}^\dagger(a)\hat{T}(a)|\psi\rangle = \langle \psi|\psi\rangle \tag{6.19}$$

which requires

$$\hat{T}^\dagger(a)\hat{T}(a) = 1 \tag{6.20}$$

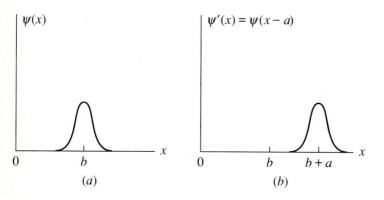

**FIGURE 6.2**
($a$) A real wave function $\psi(x) = \langle x|\psi\rangle$ and ($b$) the wave function of the translated state $\psi'(x) = \langle x|\psi'\rangle = \psi(x - a)$.

We can take advantage of this result to derive (6.18) in another way. From (6.15) we know that

$$\langle x|\hat{T}^\dagger(a) = \langle x + a| \tag{6.21}$$

But since the translation operator is unitary, $\hat{T}^\dagger$ is the inverse of $\hat{T}$. Thus if $\hat{T}^\dagger$ translates a bra vector by $a$, then $\hat{T}$ translates it by $-a$:

$$\langle x|\hat{T}(a) = \langle x - a| \tag{6.22}$$

Therefore

$$\langle x|\hat{T}(a)|\psi\rangle = \langle x - a|\psi\rangle = \psi(x - a) \tag{6.23}$$

as before.

This result is reminiscent of our discussion in Section 2.5 on active versus passive rotations. Here we have introduced the translation operator $\hat{T}(a)$ as an operator that translates the ket $|\psi\rangle$ by $a$ in the positive $x$ direction, creating a new ket $|\psi'\rangle$. When we examine the wave function of this translated state, we see that we can also consider this wave function as the amplitude for the original ket $|\psi\rangle$ to be in position states that have been translated by $a$ in the negative $x$ direction. Thus an active translation on the state itself is equivalent to a passive translation in the opposite direction on the basis states that are used to construct the wave function.[2]

## 6.3   THE GENERATOR OF TRANSLATIONS

We next consider the infinitesimal translation operator

$$\hat{T}(dx) = 1 - \frac{i}{\hbar}\hat{p}_x\,dx \tag{6.24}$$

whose action on a position ket is given by

$$\hat{T}(dx)|x\rangle = |x + dx\rangle \tag{6.25}$$

to first order in the infinitesimal $dx$. The operator $\hat{p}_x$ is called the *generator of translations*. We can generate a finite translation by $a$ through the application of an infinite number of infinitesimal translations:

$$\hat{T}(a) = \lim_{N\to\infty}\left[1 - \frac{i}{\hbar}\hat{p}_x\left(\frac{a}{N}\right)\right]^N = e^{-i\hat{p}_x a/\hbar} \tag{6.26}$$

Recall that the infinitesimal rotation operator is given by $\hat{R}(d\phi\,\mathbf{k}) = 1 - i\hat{J}_z\,d\phi/\hbar$,

---

[2] In some texts it is not uncommon to define the translation operator as one that shifts the *argument* of the wave function in the positive $x$ direction by $a$. In those texts, the translation operator will be the inverse of the one we have introduced here and will actually translate the physical state in the negative $x$ direction.

where the generator of rotations $\hat{J}_z$ has the dimensions of angular momentum and is the operator for the $z$ component of the angular momentum. Also, the infinitesimal time-translation operator is given by $\hat{U}(dt) = 1 - i\hat{H}\, dt/\hbar$, where the generator $\hat{H}$ has the dimensions of energy and is the energy operator. Since the dimensions of the generator of translations $\hat{p}_x$ are those of linear momentum, you will probably not be surprised to discover that it is indeed the operator for the $x$ component of the linear momentum.

In order to justify this assertion, we need to examine two important properties of the generator of translations. First, note that unitarity of the translation operator requires that the generator of translations must satisfy

$$\hat{p}_x = \hat{p}_x^{\dagger} \tag{6.27}$$

that is, it is a Hermitian operator. Second, we can establish that the generator of translations does not commute with the position operator. Consider an infinitesimal translation by $\delta x$.[3] Then

$$\hat{x}\hat{T}(\delta x) - \hat{T}(\delta x)\hat{x} = \hat{x}\left(1 - \frac{i}{\hbar}\hat{p}_x\,\delta x\right) - \left(1 - \frac{i}{\hbar}\hat{p}_x\,\delta x\right)\hat{x}$$

$$= \left(\frac{-i\,\delta x}{\hbar}\right)(\hat{x}\,\hat{p}_x - \hat{p}_x\hat{x})$$

$$= \left(\frac{-i\,\delta x}{\hbar}\right)[\hat{x},\,\hat{p}_x] \tag{6.28}$$

This is a relationship between operators and therefore means that for any state $|\psi\rangle$

$$\left(\hat{x}\hat{T}(\delta x) - \hat{T}(\delta x)\hat{x}\right)|\psi\rangle = \left(\frac{-i\,\delta x}{\hbar}\right)[\hat{x},\,\hat{p}_x]|\psi\rangle \tag{6.29}$$

If we use the expansion (6.3) for $|\psi\rangle$ in the position basis to evaluate the left-hand side of (6.29), we obtain

$$\left(\hat{x}\hat{T}(\delta x) - \hat{T}(\delta x)\hat{x}\right)\int dx\,|x\rangle\langle x|\psi\rangle$$

$$= \hat{x}\int dx\,|x + \delta x\rangle\langle x|\psi\rangle - \hat{T}(\delta x)\int dx\,x|x\rangle\langle x|\psi\rangle$$

$$= \int dx\,(x + \delta x)|x + \delta x\rangle\langle x|\psi\rangle - \int dx\,x|x + \delta x\rangle\langle x|\psi\rangle$$

$$= \delta x\int dx\,|x + \delta x\rangle\langle x|\psi\rangle = \delta x\int dx\,|x\rangle\langle x|\psi\rangle = \delta x|\psi\rangle \tag{6.30}$$

---

[3] We call the infinitesimal translation $\delta x$ instead of $dx$ to avoid confusion with the integration variable in (6.30).

where in the next to last step we have kept just the leading order in the infinitesimal $\delta x$.[4] Comparing (6.29) and (6.30), we see that the position operator and the generator of translations obey the commutation relation

$$[\hat{x}, \hat{p}_x] = i\hbar \tag{6.31}$$

Given the pivotal role that the commutation relations (3.14) played in our discussion of angular momentum, it is probably not surprising to find that the commutation relation (6.31) plays a very important role in our discussion of wave mechanics.

In order to ascertain the physical significance of the generator of translations, we next examine the time evolution of a particle of mass $m$ moving in one dimension. Continuing to neglect the spin degrees of freedom of the particle,[5] we can write the Hamiltonian as

$$\hat{H} = \frac{\hat{p}_x^2}{2m} + V(\hat{x}) \tag{6.32}$$

where we have expressed the kinetic energy of the particle in terms of the momentum and added a potential energy term $V$. Note that we are denoting the momentum operator by the same symbol as the generator of translations. We will now show that for quantum mechanics to yield predictions about the time evolution that are in accord with classical physics when appropriate, it is necessary that the momentum operator satisfy the commutation relation (6.31). Using (4.16), we can calculate the time rate of change of the expectation value of the position of the particle:

$$\frac{d\langle x \rangle}{dt} = \frac{i}{\hbar}\langle\psi|[\hat{H}, \hat{x}]|\psi\rangle = \frac{i}{\hbar}\langle\psi|\left[\frac{\hat{p}_x^2}{2m}, \hat{x}\right]|\psi\rangle$$

$$= \frac{i}{2m\hbar}\langle\psi|(\hat{p}_x[\hat{p}_x, \hat{x}] + [\hat{p}_x, \hat{x}]\hat{p}_x)|\psi\rangle$$

$$= \frac{\langle\psi|\hat{p}_x|\psi\rangle}{m} = \frac{\langle p_x \rangle}{m} \tag{6.33}$$

Moreover, you may also check (see Problem 6.1) that

$$\frac{d\langle p_x \rangle}{dt} = \frac{i}{\hbar}\langle\psi|[\hat{H}, \hat{p}_x]|\psi\rangle$$

$$= \left\langle -\frac{dV}{dx} \right\rangle \tag{6.34}$$

---

[4] If this step bothers you, see also the discussion in going from (6.38) to (6.39). Here too, we can shift the integration variable ($x' = x + \delta x$), expand the wave function in a Taylor series, and retain only the leading-order term.

[5] It's hard to worry much about angular momentum in a one-dimensional world.

In deriving these results, we have had to assume that the commutation relation (6.31) is satisfied. Thus, a necessary condition for the expectation values of position and momentum to obey the laws of classical physics is that the momentum operator and the generator of translations satisfy the same commutation relation with the position operator. Although it may seem somewhat abstract, the best way to define the momentum operator is as the generator of translations, just as we defined the angular momentum operators as the generators of rotations and the Hamiltonian, or energy operator, as the generator of time translations. Note that the momentum will be a constant of the motion when the Hamiltonian and the momentum operator commute. But since $\hat{p}_x$ is the generator of translations, this means the Hamiltonian is translationally invariant, which is the case when $V(x)$ is independent of $x$ [in accord with (6.34)].

A word of caution about (6.33) and (6.34), which are often referred to as Ehrenfest's theorem, is in order. These equations do not mean that the motion of all particles is essentially classical. If, as in classical physics, we call $-dV/dx = F(x)$, then by expanding the force $F(x)$ in a Taylor series about $x = \langle x \rangle$, we obtain

$$F(x) = F(\langle x \rangle) + (x - \langle x \rangle)\left(\frac{dF}{dx}\right)_{x=\langle x \rangle} + \frac{(x - \langle x \rangle)^2}{2}\left(\frac{d^2F}{dx^2}\right)_{x=\langle x \rangle} + \cdots \quad (6.35)$$

and therefore

$$\frac{d\langle p_x \rangle}{dt} = \langle F(x) \rangle = F(\langle x \rangle) + \frac{\Delta x^2}{2}\left(\frac{d^2F}{dx^2}\right)_{x=\langle x \rangle} + \cdots \quad (6.36)$$

The first term on the right-hand side of (6.36) shows the expectation values obeying Newton's second law, while the other terms constitute corrections. When are these corrections negligible? For example, we may certainly neglect the second term in comparison with the first if the uncertainty $\Delta x$ is microscopic in scale and the force varies appreciably only over macroscopic distances. In fact, this is true whether the particle itself is macroscopic or microscopic, and it accounts for our being able to use classical physics to analyze the motion of the particles in the Stern-Gerlach experiments of Chapter 1.

An immediate consequence of the commutation relation (6.31) that follows from the uncertainty relation (3.74) is

$$\Delta x \Delta p_x \geq \frac{\hbar}{2} \quad (6.37)$$

the famous Heisenberg uncertainty principle. We will return to discuss this important relation in Section 6.7, but first we need to venture briefly into momentum space.

## 6.4 THE MOMENTUM OPERATOR IN THE POSITION BASIS

We begin by using the action of the infinitesimal translation operator on an arbitrary state $|\psi\rangle$ to determine the representation of the momentum operator in position space. As before, we first express $|\psi\rangle$ as a superposition of position kets $|x\rangle$, since we know the action of the translation operator on each of these kets:

$$\hat{T}(\delta x)|\psi\rangle = \hat{T}(\delta x)\int dx\,|x\rangle\langle x|\psi\rangle$$

$$= \int dx\,|x + \delta x\rangle\langle x|\psi\rangle$$

$$= \int dx'\,|x'\rangle\langle x' - \delta x|\psi\rangle \tag{6.38}$$

where in the last step we have made the change of integration variable $x + \delta x = x'$. Expanding $\langle x' - \delta x|\psi\rangle = \psi(x' - \delta x)$ in a Taylor series about $x'$ to first order, we obtain

$$\psi(x' - \delta x) = \psi(x') - \delta x \frac{\partial}{\partial x'}\psi(x') = \langle x'|\psi\rangle - \delta x \frac{\partial}{\partial x'}\langle x'|\psi\rangle \tag{6.39}$$

Substituting this result into (6.38), we have

$$\hat{T}(\delta x)|\psi\rangle = \int dx'\,|x'\rangle\left(\langle x'|\psi\rangle - \delta x \frac{\partial}{\partial x'}\langle x'|\psi\rangle\right)$$

$$= |\psi\rangle - \delta x \int dx'\,|x'\rangle\frac{\partial}{\partial x'}\langle x'|\psi\rangle$$

$$= \left(1 - \frac{i}{\hbar}\hat{p}_x\,\delta x\right)|\psi\rangle \tag{6.40}$$

where the last step follows from the explicit form (6.24) of the infinitesimal translation operator. Therefore

$$\hat{p}_x|\psi\rangle = \frac{\hbar}{i}\int dx'\,|x'\rangle\frac{\partial}{\partial x'}\langle x'|\psi\rangle \tag{6.41}$$

If we take the inner product with the bra $\langle x|$, we obtain a very useful result:

$$\langle x|\hat{p}_x|\psi\rangle = \frac{\hbar}{i}\int dx'\,\langle x|x'\rangle\frac{\partial}{\partial x'}\langle x'|\psi\rangle$$

$$= \frac{\hbar}{i}\int dx'\,\delta(x - x')\frac{\partial}{\partial x'}\langle x'|\psi\rangle$$

$$= \frac{\hbar}{i}\frac{\partial}{\partial x}\langle x|\psi\rangle \tag{6.42}$$

If we choose the state $|\psi\rangle = |x'\rangle$, we obtain the matrix elements of the momentum operator in the position basis:[6]

$$\langle x|\hat{p}_x|x'\rangle = \frac{\hbar}{i}\frac{\partial}{\partial x}\langle x|x'\rangle = \frac{\hbar}{i}\frac{\partial}{\partial x}\delta(x - x') \tag{6.43}$$

We can also obtain a standard result from wave mechanics by taking the inner product of (6.41) with the bra $\langle\psi|$:

$$\langle p_x\rangle = \langle\psi|\hat{p}_x|\psi\rangle = \int dx'\,\langle\psi|x'\rangle\frac{\hbar}{i}\frac{\partial}{\partial x'}\langle x'|\psi\rangle$$

$$= \int dx'\,\psi^*(x')\frac{\hbar}{i}\frac{\partial}{\partial x'}\psi(x') = \int dx\,\psi^*(x)\frac{\hbar}{i}\frac{\partial}{\partial x}\psi(x) \tag{6.44}$$

The results (6.42), (6.43), and (6.44) all suggest that in position space the momentum operator takes the form

$$\hat{p}_x \xrightarrow[x\text{ basis}]{} \frac{\hbar}{i}\frac{\partial}{\partial x} \tag{6.45}$$

## 6.5 MOMENTUM SPACE

Having introduced the momentum operator as the generator of translations, we now consider a new set of states, momentum eigenstates, satisfying

$$\hat{p}_x|p\rangle = p|p\rangle \tag{6.46}$$

Momentum, like position, is a continuous variable. We can express an arbitrary state $|\psi\rangle$ as a superposition of momentum states:

$$|\psi\rangle = \int dp\,|p\rangle\langle p|\psi\rangle \tag{6.47}$$

where the integral again runs from $-\infty$ to $+\infty$, but here we are integrating over momentum instead of position. Since

$$\langle p'|p\rangle = \delta(p' - p) \tag{6.48}$$

[see the discussion surrounding (6.5)], then

$$1 = \langle\psi|\psi\rangle = \int dp\,|\langle p|\psi\rangle|^2 \tag{6.49}$$

and we can identify

$$dp\,|\langle p|\psi\rangle|^2 \tag{6.50}$$

---

[6] It is a somewhat unusual matrix since the row and column vectors have a continuous label.

as the probability that a particle in the state $|\psi\rangle$ has its momentum between $p$ and $p + dp$. Just as we refer to $\langle x|\psi\rangle$ as the wave function in position space, we call $\langle p|\psi\rangle$ the wave function in momentum space.

We can now determine $\langle x|p\rangle$, the momentum eigenstate's position-space wave function. Take the ket $|\psi\rangle$ in (6.42) to be a momentum eigenket $|p\rangle$. We obtain

$$\langle x|\hat{p}_x|p\rangle = p\langle x|p\rangle = \frac{\hbar}{i}\frac{\partial}{\partial x}\langle x|p\rangle \tag{6.51}$$

where we have taken advantage of the momentum eigenket relation (6.46) in the middle step. The differential equation (6.51) is easily solved to yield $\langle x|p\rangle = Ne^{ipx/\hbar}$. We can determine the constant $N$ up to an overall phase by requiring the momentum state to be properly normalized. First we express the state $|p\rangle$ in terms of the position basis as

$$|p\rangle = \int dx\, |x\rangle\langle x|p\rangle \tag{6.52}$$

Then

$$\langle p'|p\rangle = \int dx\, \langle p'|x\rangle\langle x|p\rangle$$

$$= N^*N \int_{-\infty}^{\infty} dx\, e^{i(p-p')x/\hbar}$$

$$= N^*N(2\pi\hbar)\delta(p - p') \tag{6.53}$$

where we have used the representation (C.17) of the Dirac delta function. Thus we choose $N = 1/\sqrt{2\pi\hbar}$ so that the normalized momentum eigenfunction is given by

$$\langle x|p\rangle = \frac{1}{\sqrt{2\pi\hbar}}e^{ipx/\hbar} \tag{6.54}$$

The Euler identity

$$e^{ipx/\hbar} = \cos(px/\hbar) + i\sin(px/\hbar) \tag{6.55}$$

emphasizes the complex character of this oscillatory function. Because when $x$ changes by a wavelength $\lambda$, the phase changes by $2\pi$ (namely, $p\lambda/\hbar = 2\pi$)

$$\lambda = \frac{h}{p} \tag{6.56}$$

the famous de Broglie relation.

Although we have called (6.54) the momentum-state wave function in position space (or the momentum eigenfunction), it is important to realize that the amplitudes $\langle x|p\rangle$ provide us with the necessary ingredients to transform back and forth between the position and the momentum bases, just as the amplitudes such as $\langle +\mathbf{z}|+\mathbf{x}\rangle$ permitted us to go back and forth between the $S_z$ and the $S_x$ bases in Chapter 2. Here, instead of a $2 \times 2$ matrix and matrix multiplication, we have integrals to evaluate:

$$\langle p|\psi\rangle = \int dx \,\langle p|x\rangle\langle x|\psi\rangle = \int dx \,\frac{1}{\sqrt{2\pi\hbar}} e^{-ipx/\hbar}\langle x|\psi\rangle \qquad (6.57a)$$

$$\langle x|\psi\rangle = \int dp \,\langle x|p\rangle\langle p|\psi\rangle = \int dp \,\frac{1}{\sqrt{2\pi\hbar}} e^{ipx/\hbar}\langle p|\psi\rangle \qquad (6.57b)$$

where both in position and momentum space the integrals run from $-\infty$ to $+\infty$. These equations show that $\langle p|\psi\rangle$ and $\langle x|\psi\rangle$ form a Fourier transform pair.

## 6.6   A GAUSSIAN WAVE PACKET

The form of the position-space momentum eigenfunction (6.54) gives us another way to see why a single momentum state is not a physically allowed state. Such a state clearly has a definite momentum $p$ and therefore $\Delta p = 0$, or zero momentum uncertainty. The probability of finding the particle between $x$ and $x + dx$,

$$\left|\langle x|p\rangle\right|^2 dx = \frac{dx}{2\pi\hbar} \qquad (6.58)$$

is independent of $x$. Thus the particle has a completely indefinite position: $\Delta x = \infty$. There is no physical measurement that we can carry out that can put a particle in such a state. How then do we generate physically acceptable states, even for a free particle? Since the particle is not permitted to have a definite momentum, (6.57b) suggests that we should superpose momentum states to obtain a physically allowed state $|\psi\rangle$ satisfying $\langle\psi|\psi\rangle = 1$. Such a superposition is called a wave packet because in position space the momentum eigenfunctions (6.54) are oscillating functions characteristic of waves. We are familiar with such superpositions for classical waves like sound waves; a clap of thunder is a localized disturbance that audibly contains many different frequencies or wavelengths.

We start with the Gaussian wave packet

$$\langle x|\psi\rangle = \psi(x) = N e^{-x^2/2a^2} \qquad (6.59)$$

which is a mathematically convenient lump in position space to play with. The normalization constant $N$ [a different $N$ than in (6.53)] is determined by the requirement that

$$\int dx \,\psi^*(x)\psi(x) = N^*N \int_{-\infty}^{\infty} dx \, e^{-x^2/a^2} = 1 \qquad (6.60)$$

Carrying out the Gaussian integral, we obtain[7]

---

[7] See Appendix D for techniques to evaluate all the Gaussian integrals in this section.

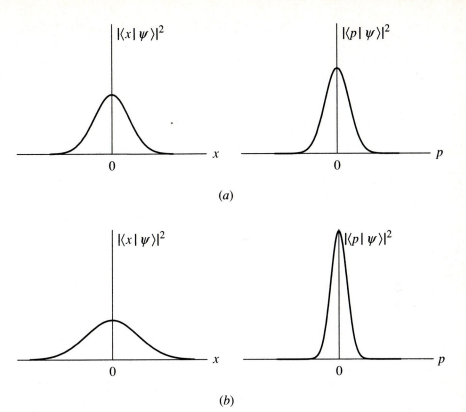

**FIGURE 6.3**
The probability densities (6.62) and (6.68) of a Gaussian wave packet in position space and momentum space, respectively. The value of $a$ in (6.62) is 50 percent larger for the position-space probability density in ($b$) than for that in ($a$).

$$N = \frac{1}{\sqrt{\sqrt{\pi}a}} \qquad (6.61)$$

The probability density

$$\psi^*(x)\psi(x) = \frac{1}{\sqrt{\pi}a}e^{-x^2/a^2} \qquad (6.62)$$

is plotted in Fig. 6.3$a$. By changing the parameter $a$, we can adjust the width of the Gaussian and therefore how much the particle is localized. Compare Fig. 6.3$b$ with 6.3$a$. The limiting case of (6.62) as $a \rightarrow 0$ even provides us with a representation of the Dirac delta function:

$$\delta(x) = \lim_{a \rightarrow 0} \frac{1}{\sqrt{\pi}a}e^{-x^2/a^2} \qquad (6.63)$$

In this limit the "function" is nonzero only at $x = 0$ and from (6.60) has unit area.

The Gaussian provides us with a probability distribution that is mathematically "nice" in that the integrals in both position and momentum space are easy to evaluate. We can relate the constant $a$ to the uncertainty $\Delta x$ in the position of the particle. Since

$$\langle x \rangle = \int_{-\infty}^{\infty} dx \, \frac{1}{\sqrt{\pi}a} e^{-x^2/a^2} x = 0 \tag{6.64}$$

because the integrand is an odd function of $x$, and

$$\langle x^2 \rangle = \int_{-\infty}^{\infty} dx \, \frac{1}{\sqrt{\pi}a} e^{-x^2/a^2} x^2 = \frac{a^2}{2} \tag{6.65}$$

the uncertainty

$$\Delta x = \sqrt{\langle x^2 \rangle - \langle x \rangle^2} = \frac{a}{\sqrt{2}} \tag{6.66}$$

We are now ready to determine the momentum-space wave function $\langle p|\psi \rangle$. Substituting the position-space wave function (6.59) into (6.57a), we obtain

$$\langle p|\psi \rangle = \int_{-\infty}^{\infty} dx \, \frac{1}{\sqrt{2\pi\hbar}} e^{-ipx/\hbar} \frac{1}{\sqrt{a\sqrt{\pi}}} e^{-x^2/2a^2} = \sqrt{\frac{a}{\hbar\sqrt{\pi}}} e^{-p^2a^2/2\hbar^2} \tag{6.67}$$

Note that *the Fourier transform of a Gaussian is another Gaussian*. This is a useful result that we will take full advantage of in Chapter 8. The probability of the particle having momentum between $p$ and $p + dp$ is given by

$$|\langle p|\psi \rangle|^2 \, dp = \frac{a}{\hbar\sqrt{\pi}} e^{-p^2a^2/\hbar^2} \, dp \tag{6.68}$$

The momentum probability density $|\langle p|\psi \rangle|^2$ is shown in Fig. 6.3. We can now easily calculate

$$\langle p_x \rangle = \langle \psi|\hat{p}_x|\psi \rangle = \int dp \, |\langle p|\psi \rangle|^2 p = \frac{a}{\hbar\sqrt{\pi}} \int_{-\infty}^{\infty} dp \, e^{-p^2a^2/\hbar^2} p = 0 \tag{6.69}$$

and

$$\langle p_x^2 \rangle = \langle \psi|\hat{p}_x^2|\psi \rangle = \int dp \, |\langle p|\psi \rangle|^2 p^2 = \frac{a}{\hbar\sqrt{\pi}} \int_{-\infty}^{\infty} dp \, e^{-p^2a^2/\hbar^2} p^2 = \frac{\hbar^2}{2a^2} \tag{6.70}$$

Thus

$$\Delta p_x = \sqrt{\langle p_x^2 \rangle - \langle p_x \rangle^2} = \frac{\hbar}{\sqrt{2}a} \tag{6.71}$$

Notice from (6.66) and (6.71) that $\Delta x \Delta p_x = \hbar/2$. Thus the Gaussian wave function is the minimum uncertainty state.

We could, of course, have calculated $\langle p_x \rangle$ and $\langle p_x^2 \rangle$ using the position-space wave function directly. For example,

$$\langle p_x \rangle = \langle \psi | \hat{p}_x | \psi \rangle = \int dx\, \psi^*(x) \frac{\hbar}{i} \frac{\partial}{\partial x} \psi(x)$$

$$= -\frac{\hbar}{ia^2} \int_{-\infty}^{\infty} dx\, \frac{1}{\sqrt{\pi}a} e^{-x^2/a^2} x = 0 \qquad (6.72)$$

## Time Evolution of a Free Particle

One of the advantages of knowing $\langle p | \psi \rangle$, in addition to being able to determine the probability that the momentum of the particle is in some range of momenta, is that we can use this amplitude to determine the time evolution of the state of a free particle. For a free particle

$$\hat{H} = \frac{\hat{p}_x^2}{2m} \qquad (6.73)$$

Thus the momentum states $|p\rangle$ are also energy eigenstates. Therefore if we express the state of the particle as a superposition of momentum eigenstates in (6.47), we can work out how the state evolves in time:

$$|\psi(t)\rangle = e^{-i\hat{H}t/\hbar} \int dp\, |p\rangle\langle p | \psi \rangle$$

$$= \int dp\, e^{-i\hat{p}_x^2 t/2m\hbar} |p\rangle\langle p | \psi \rangle$$

$$= \int dp\, e^{-i p^2 t/2m\hbar} |p\rangle\langle p | \psi \rangle \qquad (6.74)$$

For the Gaussian wave packet (6.59), $\langle p | \psi \rangle$ is given by (6.67) and $\psi(x, t)$ is given by

$$\psi(x, t) = \langle x | \psi(t) \rangle = \int dp\, e^{-i p^2 t/2m\hbar} \langle x | p \rangle\langle p | \psi \rangle \qquad (6.75)$$

But since this is also a Gaussian integral, it is straightforward to evaluate (see Problem 6.4):

$$\psi(x, t) = \frac{1}{\sqrt{\sqrt{\pi}[a + (i\hbar t/ma)]}} e^{-x^2/2a^2[1+(i\hbar t/ma^2)]} \qquad (6.76)$$

Comparing $\psi^*(x, t)\psi(x, t)$ with its form (6.62) at $t = 0$, we see that the position uncertainty is given by

$$\Delta x = \frac{a}{\sqrt{2}} \left( 1 + \frac{\hbar^2 t^2}{m^2 a^4} \right)^{1/2} \tag{6.77}$$

Notice that $\Delta x$ grows as time increases. If we call $T$ the time such that

$$\frac{\hbar^2 T^2}{m^2 a^4} = 1 \tag{6.78}$$

that is, the time necessary for significant wave packet spreading, we see that for a macroscopic particle with $m = 1$ g and $a = 0.1$ cm, then $T = 10^{25}$ s $= 3 \times 10^{17}$ years, which explains why we do not see macroscopic particles "spread." However, for an electron, choosing $a = 10^{-8}$ cm (the size of an atom), we find $T = 10^{-16}$ s, so that spreading is a very natural fact of life for a microscopic *free* particle.[8] Also notice for a particular particle that smaller $a$ (and hence smaller initial uncertainty $\Delta x$ in the position of the particle) means more rapid wave packet spreading.

## 6.7  THE HEISENBERG UNCERTAINTY PRINCIPLE

Our analysis of the Gaussian wave packet has illustrated a number of features of the position-momentum uncertainty relation $\Delta x \Delta p_x \geq \hbar/2$. As we noted earlier, this relation follows directly from the position-momentum commutation relation (6.31). By adjusting the width $a$ of the Gaussian wave packet, we directly control the uncertainty in the position of the particle, as (6.66) shows. However, as we make the position-space wave packet broader by increasing $a$, the momentum-space wave function $\langle p | \psi \rangle$ becomes narrower [see (6.71)], maintaining the uncertainty relation (see Fig. 6.3). Of course, in the macroscopic world we never seem to notice that we cannot specify *both* the position and the momentum (or the velocity) of the particle with arbitrary precision. It is the smallness of $\hbar$ that protects our classical illusions. If a particle of mass 1 g is moving with a velocity of 1 cm/s and we specify its momentum to one part in a million, that is, $\Delta p_x \sim 10^{-6}$ g·cm/s, then $\Delta x \sim 10^{-21}$ cm, which is $10^{-8}$ times smaller than the radius of a proton. We would be hard pressed experimentally not to say the particle has a definite position. On the other hand, for an electron in an atom, with a typical velocity of $10^8$ cm/s, the momentum $p_x \sim 10^{-19}$ g·cm/s, and even if we allow $\Delta p_x$ to be as large as $p_x$, we find $\Delta x \sim 10^{-8}$ cm, which is roughly the size of the atom itself. Thus in the microscopic world the uncertainty clearly matters.

We can see the importance of the Heisenberg uncertainty principle at a fundamental level by examining the role it plays in the famous double-slit experiment. In this experiment, a beam of particles—for example, electrons—with a well-defined momentum is projected at an opaque screen with two narrow slits

---

[8] In case you are suddenly worried about the long-term survivability of atoms, remember that if an electron is bound within an atom in an energy eigenstate, it is in a stationary state, which does not spread out or change as time progresses.

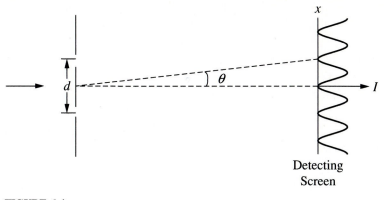

**FIGURE 6.4**
The double-slit experiment.

separated by a distance $d$, as shown in Fig. 6.4. Even if the intensity of the incident electron beam is so low that electrons arrive at a distant detecting screen one at a time, when a sufficiently large number of electrons has been counted, the intensity pattern on the screen is an interference pattern, with the location of the maxima satisfying $d \sin \theta = m\lambda$, where the wavelength $\lambda$ of the electrons is given by (6.56). The classical physicist is mystified by this result, thinking that surely a single electron passes through one slit *or* the other, and thus cannot understand how a *particle* like the electron can "interfere" with itself. The quantum physicist realizes that a single electron has an amplitude to reach any point on the detecting screen by taking two paths, one through the upper slit and one through the lower slit, and that these amplitudes can interfere with each other to produce the double-slit intensity pattern.[9]

If the classical physicist challenges this view by actually observing through which slit the electron passes by using a microscope, like the one in Fig. 6.1, and shining light on the two slits, the uncertainty relation (6.37) guarantees that the interference pattern disappears. If we call the direction along the screens the $x$ direction, determining through which slit the electron passes requires an uncertainty $\Delta x < d/2$ in the electron's position. This forces an uncertainty $\Delta p_x > 2h/d$ in the electron's momentum and hence an uncertainty in the angular deflection of the electron $\Delta \theta = \Delta p_x / p > (2h/d)/(h/\lambda) = 2\lambda/d$, which is of the same order as the angular spacing between interference maxima, wiping out the interference pattern.[10] From this analysis, we can see the pivotal role the uncertainty principle plays in maintaining the logical consistency of quantum mechanics: in this exper-

---

[9] We will justify this assertion more fully in Chapter 8.

[10] We are only making an order of magnitude estimate here, taking the right-hand side of the uncertainty relation to be of order $h$, Planck's constant. This has the advantage of freeing us from worrying about a detailed analysis of the position uncertainty associated with resolving which slit the electron went through and it keeps the algebra transparent.

iment it keeps us from knowing which slit the electron goes through and at the same time observing an interference pattern.[11]

## 6.8 GENERAL PROPERTIES OF SOLUTIONS TO THE SCHRÖDINGER EQUATION IN POSITION SPACE

So far, we have restricted our discussion of time evolution within one-dimensional wave mechanics to that of a free particle, for which the energy eigenstates are also momentum eigenstates. When the one-dimensional Hamiltonian, as given in (6.32), includes potential energy as well as kinetic energy, we start our analysis by projecting the equation of motion (4.8) into position space:

$$\langle x|\hat{H}|\psi(t)\rangle = i\hbar\langle x|\frac{d}{dt}|\psi(t)\rangle \tag{6.79}$$

Taking advantage of (6.42) and

$$\langle x|V(\hat{x}) = \langle x|V(x) \tag{6.80}$$

we can write

$$\langle x|\hat{H}|\psi(t)\rangle = \langle x|\left[\frac{\hat{p}_x^2}{2m} + V(\hat{x})\right]|\psi(t)\rangle$$

$$= \left[-\frac{\hbar^2}{2m}\frac{\partial^2}{\partial x^2} + V(x)\right]\langle x|\psi(t)\rangle \tag{6.81}$$

Thus (6.79) can be expressed as

$$\left[-\frac{\hbar^2}{2m}\frac{\partial^2}{\partial x^2} + V(x)\right]\psi(x, t) = i\hbar\frac{\partial}{\partial t}\psi(x, t) \tag{6.82}$$

where, as in Section 6.1, we have identified

$$\langle x|\psi(t)\rangle = \psi(x, t) \tag{6.83}$$

as the wave function. Equation (6.82) is the time-dependent Schrödinger equation in position space. Note that we have replaced the total time derivative of the ket $|\psi(t)\rangle$ in (6.79) with a partial time derivative of the wave function $\psi(x, t)$ because we are only calculating how the wave function evolves in time on the right-hand side of (6.82).

If we take the state $|\psi(t)\rangle$ in (6.79) to be an energy eigenstate, for which the time dependence is given by $|E\rangle e^{-iEt/\hbar}$, we can write the wave function as

$$\psi_E(x, t) = \langle x|E\rangle e^{-iEt/\hbar} \tag{6.84}$$

---

[11] However, a recent experiment with highly excited rubidium *atoms* as the projectiles and micro-maser cavities in front of the slits as detectors shows that it is possible to determine through which slit the atom passes without changing the momentum of the atom, thereby evading the limitations imposed by the Heisenberg uncertainty principle. Nonetheless, such measurements also destroy the interference pattern. See M. O. Scully, B.-G. Englert, and H. Walther, *Nature* **351**, 111 (1991).

Substituting this form for an energy eigenfunction into (6.82), we obtain

$$\left[-\frac{\hbar^2}{2m}\frac{\partial^2}{\partial x^2} + V(x)\right]\langle x|E\rangle = E\langle x|E\rangle \tag{6.85}$$

which is often referred to as the time-independent Schrödinger equation in position space. This equation also results from projecting the energy eigenvalue equation

$$\hat{H}|E\rangle = E|E\rangle \tag{6.86a}$$

into position space:

$$\langle x|\hat{H}|E\rangle = E\langle x|E\rangle \tag{6.86b}$$

It is common to write $\langle x|E\rangle = \psi_E(x)$. We will, however, drop the subscript $E$ and implicitly assume for the remainder of this chapter that the wave function $\psi(x)$ is an energy eigenfunction. Since we have factored out the time dependence, $\langle x|E\rangle$ is a function of $x$ only and we can replace the partial derivatives in (6.85) with ordinary derivatives:

$$\left[-\frac{\hbar^2}{2m}\frac{d^2}{dx^2} + V(x)\right]\psi(x) = E\psi(x) \tag{6.87}$$

Let's first take a specific example to illustrate some of the features of the solutions to this differential equation. Suppose that the potential energy $V(x)$ is the finite square well

$$V(x) = \begin{cases} 0 & |x| < a/2 \\ V_0 & |x| > a/2 \end{cases} \tag{6.88}$$

as shown in Fig. 6.5. For this particularly simple potential energy, which is piecewise constant, we can solve (6.87) analytically in the different regions: namely, inside the well ($|x| < a/2$) and outside the well (both $x < -a/2$ and $x > a/2$). We will restrict our attention in this section to solutions with energy $0 < E < V_0$. A classical particle would be bound strictly inside the well with

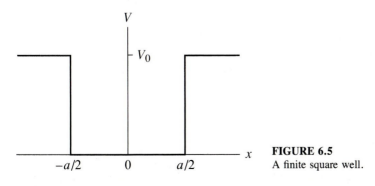

**FIGURE 6.5**
A finite square well.

this energy, since outside the well the potential energy would be greater than the energy, which classically would mean negative kinetic energy.[12]

The differential equation (6.87) can be expressed as

$$\frac{d^2\psi}{dx^2} = -\frac{2mE}{\hbar^2}\psi = -k^2\psi \qquad |x| < a/2 \tag{6.89}$$

$$\frac{d^2\psi}{dx^2} = -\frac{2m(E - V_0)}{\hbar^2}\psi = q^2\psi \qquad |x| > a/2 \tag{6.90}$$

Note that $k^2$ and $q^2$ are *positive* constants. We take

$$k = \sqrt{\frac{2mE}{\hbar^2}} \tag{6.91}$$

$$q = \sqrt{\frac{2m(V_0 - E)}{\hbar^2}} \tag{6.92}$$

According to (6.89) and (6.90), two derivatives of the wave function yield just a constant times the wave function; thus it is especially straightforward to solve these differential equations. In particular, within the well, where differentiating twice gives a *negative* constant times the wave function, the solutions can be written as

$$\psi(x) = A \sin kx + B \cos kx \qquad |x| < a/2 \tag{6.93}$$

while outside the well, where differentiating twice gives a *positive* constant times the wave function, the solutions are

$$\psi(x) = C e^{qx} + D e^{-qx} \qquad |x| > a/2 \tag{6.94}$$

Actually, since the solutions should satisfy the normalization condition (6.11), we must examine separately the regions $x < -a/2$ and $x > a/2$ and discard the exponential that blows up in each of these regions:

$$\psi(x) = C e^{qx} \qquad x < -a/2 \tag{6.95}$$

$$\psi(x) = D e^{-qx} \qquad x > a/2 \tag{6.96}$$

Thus we see that the solution oscillates inside the well, where $E > V$, and is exponentially damped outside the well, where $E < V$.

Since we are seeking a solution to a second-order differential equation, the different functions in the three regions must join up smoothly, that is, they must be continuous (so that the first derivative is well defined) and have a continuous first derivative everywhere. This condition on the continuity of the derivative follows directly from "integrating" the Schrödinger equation:

---

[12] We will discuss the unbound solutions to equations such as (6.87) in Section 6.10.

$$\left(\frac{d\psi}{dx}\right)_{x+\varepsilon} - \left(\frac{d\psi}{dx}\right)_{x-\varepsilon} = \int_{x-\varepsilon}^{x+\varepsilon} dx \, \frac{d}{dx}\frac{d\psi}{dx} = \int_{x-\varepsilon}^{x+\varepsilon} dx \, \frac{2m}{\hbar^2}(V - E)\psi \qquad (6.97)$$

since the right-hand side vanishes for well-behaved wave functions in the limit $\varepsilon \to 0$ unless the potential energy $V$ is infinite. We will see an example where the derivative is indeed discontinuous in the next section, when we consider the infinite potential well.[13]

If we start sketching from the left a bound-state wave function for the finite potential well, we see an exponential that rises as $x$ increases (Fig. 6.6). At the boundary of the potential well, this exponential (6.95) must match up with the oscillatory solution (6.93), with the wave function being continuous and having a continuous derivative across the boundary. This oscillating function must then join smoothly onto a damped exponential (6.96) at the $x = a/2$ boundary. This turns out to be a nontrivial accomplishment: only for special values of the energy will this matching be possible. Otherwise, the oscillatory function will join onto a combination of rising and damped exponentials, with the rising exponential blowing up as $x \to \infty$. This effect is readily seen if you integrate the Schrödinger equation (6.87) numerically.[14] Figure 6.7 shows the energies and corresponding eigenfunctions for a finite square well that admits four bound states. If you are interested in examining how to determine analytically the allowed values of the energy for the particular potential energy well (6.88), turn to Section 10.3, where this calculation is carried out for the three-dimensional spherically

---

[13] Also, see Problem 6.16.

[14] A numerical solution of the Schrödinger equation for a square well potential is discussed by R. Eisberg and R. Resnick. *Quantum Physics of Atoms, Molecules, Solids, Nuclei, and Particles*, 2nd ed., Wiley, New York, 1985, Appendix G.

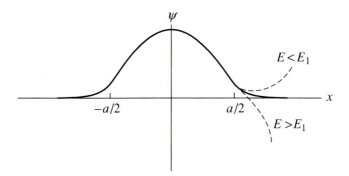

**FIGURE 6.6**
A schematic diagram of energy eigenfunctions of the finite square well for three different energies: an energy $E < E_1$, where $E_1$ is the ground-state energy, $E = E_1$, and $E > E_1$. Only for $E = E_1$ is the wave function normalizable.

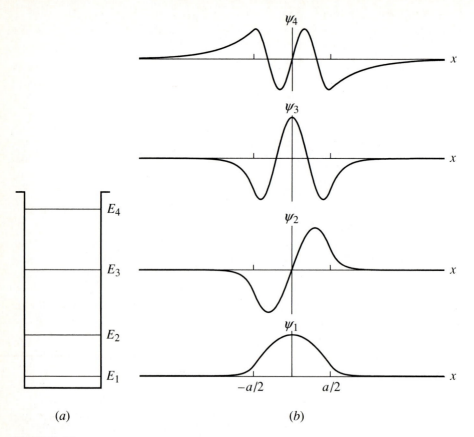

**FIGURE 6.7**
(a) The energies and (b) the corresponding energy eigenfunctions for a finite square well with four bound states.

symmetric square well; the mathematics is essentially the same as for solving the one-dimensional well.[15] In the next section we will examine how this quantization of the energy arises in a particularly simple example in which we let $V_0 \rightarrow \infty$.

Finally, we should note here that this combination of oscillatory and exponential-like behavior of the energy eigenfunction depending on whether the energy is greater than or less than the potential energy, respectively, is quite generally true, even when the potential energy is not a constant. For example, if $V = V(x)$, then when $E > V(x)$ we can write

$$\frac{d^2\psi}{dx^2} = -\frac{2m}{\hbar^2}[E - V(x)]\psi = -k^2(x)\psi \qquad (6.98)$$

---

[15] The major difference between the one- and three-dimensional problems is that in three dimensions the variable $r = \sqrt{x^2 + y^2 + z^2}$ replaces $x$. Clearly, $r$ cannot be less than zero, and in fact there is a boundary condition that eliminates the cosine piece of the one-dimensional solution (6.93).

Since $k$ is not a constant here, we cannot immediately write down the solution as in (6.93). However, note that if $\psi > 0$, then $d^2\psi/dx^2 < 0$. Thus if the wave function is positive, the second derivative is negative; that is, the function is concave down. It must therefore bend back toward the axis. Similarly, if $\psi < 0$, then $d^2\psi/dx^2 > 0$. Thus if the wave function is negative, the second derivative is positive and the function is concave up. In either case the function bends back toward the axis in an oscillatory manner. Also note for a particular value of $V(x)$ that the magnitude of the energy determines how rapidly the wave function oscillates. The larger the energy, the larger the value of $k^2(x)$, and the more rapidly the function bends back toward the axis. Thus the lower-energy eigenfunctions have the smaller curvature and, consequently, a fewer number of nodes. You can see this pattern in the energy eigenfunctions of the finite square well shown in Fig. 6.7.

In a region in which $V(x) > E$, on the other hand,

$$\frac{d^2\psi}{dx^2} = \frac{2m}{\hbar^2}[V(x) - E]\psi = q^2(x)\psi \tag{6.99}$$

Here, if the wave function $\psi$ is positive, the second derivative is positive as well, and the function is concave up; it bends *away* from the axis. We call such a solution an *exponential-like* solution. A similar bending away from the axis is seen if $\psi$ is negative. Thus there can be no physically meaningful solutions for which $V(x) > E$ everywhere, for then the wave function must eventually diverge. However, as long as there is some region for which $E > V(x)$, the exponential-like solution "turns over" into an oscillatory-type solution, and the wave function need not diverge. Notice that as we move from a region in which $E < V(x)$ to a region in which $E > V(x)$, we pass a value for $x$ such that $V(x) = E$, at which point the second derivative vanishes. Thus there is a point of inflection where the curvature changes as we move between the two regions.

These quite general characteristics of the energy eigenfunctions make it possible to sketch them in a rough way without actually solving the Schrödinger equation. In the general case in which $V$ depends on $x$ in other than a piecewise-constant manner, the eigenfunctions are *not* sines and cosines or exponentials. However, the eigenfunctions will still look roughly the same, exhibiting oscillatory behavior in regions in which $E > V(x)$ and exponential-like behavior in regions in which $E < V(x)$. A nice example to which we will turn in Chapter 7 is the harmonic oscillator, for which $V(x) = kx^2/2$. Some of the energy eigenfunctions in position space for the harmonic oscillator are shown in Fig. 7.7.

## 6.9 THE PARTICLE IN A BOX

A particularly easy but instructive energy eigenvalue equation to solve directly in position space is the one-dimensional infinite potential energy well[16]

---

[16] Mathematically, it is even easier if we choose our origin of coordinates to be at one edge of the box. See Problem 6.10.

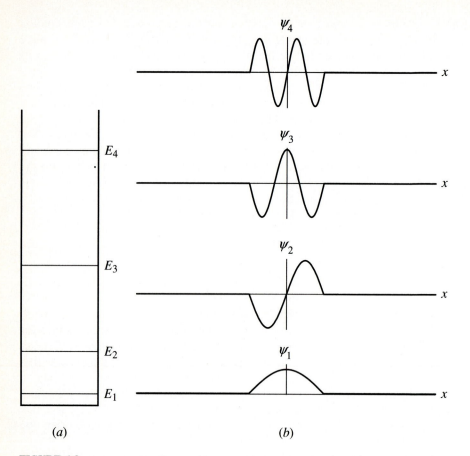

**FIGURE 6.8**
(*a*) The infinite potential energy well with the lowest four allowed energies. (*b*) The corresponding energy eigenfunctions. This potential well possesses an infinite number of bound states.

$$V(x) = \begin{cases} 0 & |x| < a/2 \\ \infty & |x| > a/2 \end{cases} \tag{6.100}$$

which is shown in Fig. 6.8*a*. Outside the well, the energy eigenfunction must vanish, as can be seen by examining the limit as $V_0 \to \infty$ for the wave functions (6.95) and (6.96) for the finite well. As for the finite well, the most general solution to the differential equation (6.89) inside the well is given by

$$\psi(x) = A \sin kx + B \cos kx \qquad |x| < a/2 \tag{6.101}$$

Since the wave function vanishes outside the well, the requirement that the wave function be continuous dictates that

$$\psi\left(\frac{a}{2}\right) = A \sin \frac{ka}{2} + B \cos \frac{ka}{2} = 0 \tag{6.102a}$$

and

$$\psi\left(-\frac{a}{2}\right) = A \sin \frac{-ka}{2} + B \cos \frac{-ka}{2}$$

$$= -A \sin \frac{ka}{2} + B \cos \frac{ka}{2} = 0 \qquad (6.102b)$$

These two equations can be expressed in matrix form as

$$\begin{pmatrix} \sin(ka/2) & \cos(ka/2) \\ -\sin(ka/2) & \cos(ka/2) \end{pmatrix}\begin{pmatrix} A \\ B \end{pmatrix} = 0 \qquad (6.103)$$

For a nontrivial solution to this set of homogeneous equations in the two unknowns $A$ and $B$, we must demand that the determinant of the coefficients vanishes:

$$\begin{vmatrix} \sin \dfrac{ka}{2} & \cos \dfrac{ka}{2} \\ -\sin \dfrac{ka}{2} & \cos \dfrac{ka}{2} \end{vmatrix} = 0 \qquad (6.104)$$

or simply

$$2 \sin \frac{ka}{2} \cos \frac{ka}{2} = \sin ka = 0 \qquad (6.105)$$

This equation is satisfied for

$$k_n a = n\pi \qquad n = \text{integer} \qquad (6.106)$$

where we have put a subscript $n$ on the $k$ that is specified by the particular integer $n$. For $n = 1, 3, 5, \ldots,$ $k_n a/2 = \pi/2, 3\pi/2, 5\pi/2, \ldots,$ and $\cos(k_n a/2) = 0$. Substituting this result into (6.103), we see that $A = 0$ and therefore

$$\psi_n(x) = B_n \cos \frac{n\pi x}{a} \qquad n = 1, 3, 5, \ldots \qquad (6.107a)$$

For $n = 2, 4, 6, \ldots,$ $k_n a/2 = \pi, 2\pi, 3\pi, \ldots,$ and $\sin(k_n a/2) = 0$. Substituting this result into (6.103), we find that $B = 0$ and therefore

$$\psi_n(x) = A_n \sin \frac{n\pi x}{a} \qquad n = 2, 4, 6, \ldots \qquad (6.107b)$$

We can determine the constants $A_n$ and $B_n$ by imposing the normalization condition (6.11), namely,

$$1 = \int dx\, \psi_n^*(x)\psi_n(x) = \begin{cases} \int_{-a/2}^{a/2} dx\, B_n^* B_n \cos^2 \dfrac{n\pi x}{a} & n = 1, 3, 5, \ldots \\ \int_{-a/2}^{a/2} dx\, A_n^* A_n \sin^2 \dfrac{n\pi x}{a} & n = 2, 4, 6, \ldots \end{cases}$$

$$(6.108)$$

Up to an overall phase, this tells us that $A_n = B_n = \sqrt{2/a}$ and therefore

$$
\psi_n(x) = \begin{cases} \sqrt{\dfrac{2}{a}} \cos \dfrac{n\pi x}{a} & n = 1, 3, 5, \ldots \\[2mm] \sqrt{\dfrac{2}{a}} \sin \dfrac{n\pi x}{a} & n = 2, 4, 6, \ldots \end{cases} \qquad |x| < a/2 \qquad (6.109)
$$

Note that we have not included the $n = 0$ solution because for $n = 0$, $\psi = 0$, corresponding to no particle in the well. Also note that the negative integers in (6.106) merely change the wave functions (6.107b) into the negative of themselves, corresponding to just an overall phase change for these states and not to different states themselves.

In addition to labeling the energy eigenfunctions (shown in Fig. 6.8), the quantum number $n$ specifies the corresponding energies. Since

$$
k_n a = \sqrt{\frac{2mE_n}{\hbar^2}}\, a = n\pi \qquad n = 1, 2, 3, \ldots \qquad (6.110)
$$

we have

$$
E_n = \frac{\hbar^2 \pi^2 n^2}{2ma^2} \qquad n = 1, 2, 3, \ldots \qquad (6.111)
$$

For the particle in the box, it is especially easy to see why only discrete energies are permitted. The requirement that the wave functions vanish at the boundaries of the box means that we can only fit in those waves with nodes at $x = \pm a/2$.

### An Example

Suppose that a measurement of the energy is carried out on a particle in the box and that the ground-state energy $E_1 = \hbar^2\pi^2/2ma^2$ is obtained. We then know that the state of the particle is the ground state $|E_1 = \hbar^2\pi^2/2ma^2\rangle$, with the corresponding energy eigenfunction $\langle x|E_1 = \hbar^2\pi^2/2ma^2\rangle = \psi_1(x)$. What if we now change the potential energy well that is confining the particle and pull the walls of the well out rapidly so they are positioned at $x = \pm a$ instead of $x = \pm a/2$? In fact, we imagine pulling the walls out so rapidly that instantaneously the state of the particle doesn't change. As can be readily seen by comparing the wave function of the particle in this state with the energy eigenfunctions of the new, larger potential well (see Fig. 6.9), the particle is no longer in an energy eigenstate. Thus we can ask, for example, what the probability is that a subsequent measurement of the energy of the particle will yield a particular energy eigenvalue of the new well.[17]

If we call the initial state $|i\rangle$ and the final state $|f\rangle$, the amplitude to find a particle in the state $|i\rangle$ in the state $|f\rangle$ is $\langle f|i\rangle$. Since we have already calculated the

---

[17] For a more physical example, see Problem 10.6.

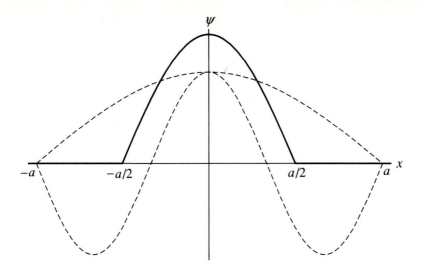

**FIGURE 6.9**
The ground-state energy eigenfunction for the small box of width $a$ (solid line) and the ground-state and second-excited-state energy eigenfunctions of the bigger box of width $2a$ (dashed lines).

position-space wave functions, it is convenient to calculate the amplitude $\langle f|i\rangle$ by inserting a complete set of position states (6.4) between the bra and the ket:

$$\langle f|i\rangle = \int dx \, \langle f|x\rangle\langle x|i\rangle \tag{6.112}$$

The amplitude $\langle x|i\rangle$, where the initial state is the ground state of the well of width $a$ ($|i\rangle = |E_1^{\text{width } a}\rangle$), is given by

$$\langle x|E_1^{\text{width } a}\rangle = \psi_1^{\text{width } a}(x) = \begin{cases} \sqrt{\dfrac{2}{a}}\cos\dfrac{\pi x}{a} & |x| < a/2 \\ 0 & |x| > a/2 \end{cases} \tag{6.113}$$

while the amplitude $\langle f|x\rangle$, where the final state is the ground state of the well with width $2a$ ($|f\rangle = |E_1^{\text{width } 2a}\rangle$), is given by

$$\langle E_1^{\text{width } 2a}|x\rangle = [\psi_1^{\text{width } 2a}(x)]^* = \begin{cases} \sqrt{\dfrac{1}{a}}\cos\dfrac{\pi x}{2a} & |x| < a \\ 0 & |x| > a \end{cases} \tag{6.114}$$

Thus

$$\langle E_1^{\text{width } 2a}|E_1^{\text{width } a}\rangle = \int dx \, \langle E_1^{\text{width } 2a}|x\rangle\langle x|E_1^{\text{width } a}\rangle$$

$$= \int_{-a/2}^{a/2} dx \sqrt{\frac{1}{a}}\cos\frac{\pi x}{2a}\sqrt{\frac{2}{a}}\cos\frac{\pi x}{a} = \frac{8}{3\pi} \tag{6.115}$$

where the integrand is nonzero only for $|x| < a/2$ because $\langle E_1^{\text{width } a}|x\rangle$ is nonzero only in this region. Thus the probability of finding the particle in the ground state of the bigger well is

$$|\langle E_1^{\text{width } 2a}|E_1^{\text{width } a}\rangle|^2 = \frac{64}{9\pi^2} = 0.72 \qquad (6.116)$$

In this way we could go on to calculate the probability of finding the particle in the other energy eigenstates of the bigger well. The form of the energy eigenfunction for the $n = 3$ state of the bigger well, as shown in Fig. 6.9, suggests that there is a significant overlap of the wave functions $\psi_1^{\text{width } a}$ and $\psi_3^{\text{width } 2a}$, and thus there should be a significant probability of finding the particle in this $n = 3$ state as well. On the other hand, we can also quickly see that there is zero amplitude of finding the particle in the even $n$ states. Since $\langle x|E_1^{\text{width } a}\rangle$ is an even function of $x$ [$\psi_n(-x) = \psi_n(x)$ for $n$ odd] while $\langle E_n^{\text{width } 2a}|x\rangle$ for $n$ even is an odd function of $x$ [$\psi_n(-x) = -\psi_n(x)$ for $n$ even], the product of an even and an odd function is of course an odd function, which vanishes when integrated from $-a/2$ to $a/2$. The evenness or oddness of the energy eigenfunctions, often referred to as their parity, turns out to be a general characteristic of the eigenfunctions of the Hamiltonian when the potential energy is even, that is, $V(-x) = V(x)$. We will discuss the reason for this more fully in Chapter 7.

Finally, we can ask how the system evolves in time after the walls of the potential energy well have been pulled out. Since the system is no longer in an energy eigenstate, it is no longer a stationary state, and thus can exhibit interesting time dependence. Since the initial state at $t = 0$ can be written as a superposition of the energy eigenstates:

$$|\psi(0)\rangle = |E_1^{\text{width } a}\rangle = \sum_n |E_n^{\text{width } 2a}\rangle\langle E_n^{\text{width } 2a}|E_1^{\text{width } a}\rangle \qquad (6.117)$$

then

$$|\psi(t)\rangle = e^{-i\hat{H}t/\hbar} \sum_n |E_n^{\text{width } 2a}\rangle\langle E_n^{\text{width } 2a}|E_1^{\text{width } a}\rangle$$

$$= \sum_n e^{-iE_n^{\text{width } 2a}t/\hbar}|E_n^{\text{width } 2a}\rangle\langle E_n^{\text{width } 2a}|E_1^{\text{width } a}\rangle \qquad (6.118)$$

and therefore

$$\langle x|\psi(t)\rangle = \sum_n e^{-i\hbar n^2\pi^2 t/8ma^2}\langle x|E_n^{\text{width } 2a}\rangle\langle E_n^{\text{width } 2a}|E_1^{\text{width } a}\rangle$$

$$= \sum_n e^{-i\hbar n^2\pi^2 t/8ma^2}\psi_n^{\text{width } 2a}(x)\langle E_n^{\text{width } 2a}|E_1^{\text{width } a}\rangle \qquad (6.119)$$

Once the amplitudes $\langle E_n^{\text{width } 2a}|E_1^{\text{width } a}\rangle$ have been calculated [as in (6.115)], it is probably best to carry out the sum in (6.119) numerically and let the computer show us how the wave function actually evolves in time.

## 6.10   SCATTERING IN ONE DIMENSION

Let's turn our attention to solutions of the Schrödinger equation for energies such that the particle is not bound, or confined, in a potential well. For example, for the potential energy well (6.88) we consider solutions with $E > V_0$, which are oscillatory everywhere, just like the momentum eigenfunctions (6.54). As for the momentum states, we will see in explicitly solving the Schrödinger equation that the energy eigenvalues take on a *continuum* of values in these cases. As for the free particle that we treated in Section 6.6, the way to generate physically acceptable states is to superpose these continuum energy solutions to form a wave packet. Such a wave packet will exhibit time dependence. We can form a wave packet that is, for example, initially localized far from the potential well and then propagates to the right, eventually interacting with the well and producing nonzero amplitudes for the wave packet to be both transmitted and reflected, as shown in Fig. 6.10. This is a typical scattering experiment in which particles are projected at a target, interact with the target, and are scattered. Scattering in one dimension is relatively straightforward because the only options for the particle are reflection or transmission. Although the right way to analyze scattering is in terms of wave packets,[18] often the wave packet is sharply peaked at a particular value of the energy and thus sufficiently broad in position space in comparison with the size of the region over which the potential energy varies that we can treat it as a plane wave when analyzing the scattering. This turns out to be a big simplification, but it raises the question: How do we calculate the probabilities of reflection and transmission when we are dealing with an energy eigenstate that is

---

[18] For a discussion of one-dimensional scattering in terms of wave packets, see R. Shankar, *Principles of Quantum Mechanics*, Plenum Press, New York, 1980, Section 5.4.

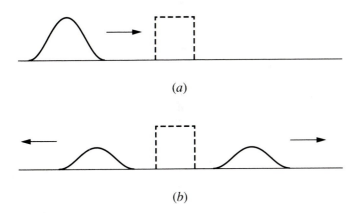

(a)

(b)

**FIGURE 6.10**
A schematic diagram showing (a) a wave packet incident on a potential energy barrier and (b) the reflected and transmitted waves.

a stationary state and thus doesn't show any time dependence? The answer is that we can think of scattering in terms of a steady-state situation in which particles are being continually projected at the target; some of this incident flux is reflected and some of the flux is transmitted. We can relate this flux of particles to a probability current that is needed to ensure local conservation of probability.

## The Probability Current

To see how this probability current arises, consider $\partial(\psi^*\psi)/\partial t$, the time rate of change of the probability density. Using the time-dependent Schrödinger equation (6.82), we see that the time derivative of the wave function is given by

$$\frac{\partial \psi(x, t)}{\partial t} = \frac{1}{i\hbar}\left[-\frac{\hbar^2}{2m}\frac{\partial^2 \psi(x, t)}{\partial x^2} + V(x)\psi(x, t)\right] \tag{6.120}$$

Therefore

$$\frac{\partial \psi^*(x, t)}{\partial t} = \frac{1}{-i\hbar}\left[-\frac{\hbar^2}{2m}\frac{\partial^2 \psi^*(x, t)}{\partial x^2} + V(x)\psi^*(x, t)\right] \tag{6.121}$$

and

$$
\begin{aligned}
\frac{\partial \psi^*\psi}{\partial t} &= \psi\frac{\partial \psi^*}{\partial t} + \psi^*\frac{\partial \psi}{\partial t} \\
&= \frac{\psi}{-i\hbar}\left[-\frac{\hbar^2}{2m}\frac{\partial^2 \psi^*}{\partial x^2} + V(x)\psi^*\right] + \frac{\psi^*}{i\hbar}\left[-\frac{\hbar^2}{2m}\frac{\partial^2 \psi}{\partial x^2} + V(x)\psi\right] \\
&= \frac{\psi}{-i\hbar}\left(-\frac{\hbar^2}{2m}\frac{\partial^2 \psi^*}{\partial x^2}\right) + \frac{\psi^*}{i\hbar}\left(-\frac{\hbar^2}{2m}\frac{\partial^2 \psi}{\partial x^2}\right)
\end{aligned}
\tag{6.122}
$$

This equation can be expressed in the form

$$\frac{\partial \psi^*\psi}{\partial t} = -\frac{\partial j_x}{\partial x} \tag{6.123}$$

where

$$j_x = \frac{\hbar}{2mi}\left(\psi^*\frac{\partial \psi}{\partial x} - \psi\frac{\partial \psi^*}{\partial x}\right) \tag{6.124}$$

This is just the form that we expect for a local conservation law.[19] For example, if we integrate (6.123) between $x = a$ and $x = b$, we obtain

---

[19] In three dimensions, local conservation of charge is contained in the relation

$$\frac{\partial \rho}{\partial t} + \nabla \cdot \mathbf{j} = 0$$

where $\rho$ is the charge density. A similar relation holds for the probability density in three dimensions. See Section 13.1.

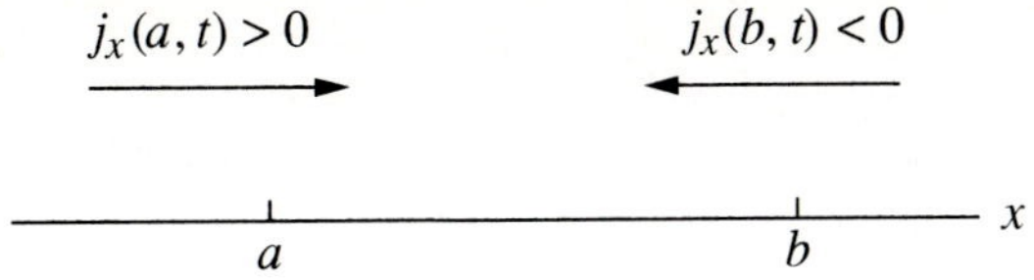

$$j_x(a, t) > 0 \qquad\qquad j_x(b, t) < 0$$

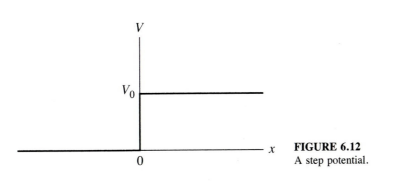

FIGURE 6.11

The probability of the particle being in the one-dimensional region between $a$ and $b$ increases with probability flowing into the region either at $a$ or at $b$.

$$\frac{d}{dt}\int_a^b dx\,\psi^*\psi = -j_x(b, t) + j_x(a, t) \tag{6.125}$$

If the probability of finding the particle between $a$ and $b$ increases, it does so because of a net probability current flowing into the region, either at $a$ [positive current flows in the positive $x$ direction and hence $j_x(a, t) > 0$ means inward flow at $a$] or at $b$ [negative current means current in the negative $x$ direction and hence $j_x(b, t) < 0$ means inward flow at $b$]. See Fig. 6.11. Thus the probability in a region of space increases or decreases because there is a net probability flow into or out of that region.

## A Potential Step

Let's take the particularly simple example of scattering from the potential energy step

$$V(x) = \begin{cases} 0 & x < 0 \\ V_0 & x > 0 \end{cases} \tag{6.126}$$

shown in Fig. 6.12 to illustrate how we relate the probability current to the probability of reflection and transmission. We wish to determine the energy eigenstates, for which

$$\psi_E(x, t) = \psi(x)e^{-iEt/\hbar} \tag{6.127}$$

where $\psi(x)$ satisfies (6.87). To the left of the barrier

$$-\frac{\hbar^2}{2m}\frac{d^2\psi(x)}{dx^2} = E\psi(x) \qquad x < 0 \tag{6.128}$$

FIGURE 6.12

A step potential.

which has solutions

$$\psi(x) = Ae^{ikx} + Be^{-ikx} \qquad x < 0 \qquad (6.129)$$

with

$$k = \sqrt{\frac{2mE}{\hbar^2}} \qquad (6.130)$$

Notice that we have chosen to write the oscillatory solutions of (6.128) in terms of complex exponentials instead of sines and cosines, as in (6.93). The reason becomes apparent when we evaluate the probability current (6.124) for the wave function (6.129):

$$j_x = \frac{\hbar k}{m} \left( |A|^2 - |B|^2 \right) \qquad x < 0 \qquad (6.131)$$

We thus can identify

$$j_{\text{inc}} = \frac{\hbar k}{m} |A|^2 \qquad (6.132)$$

as the probability current incident on the barrier from the left and

$$j_{\text{ref}} = \frac{\hbar k}{m} |B|^2 \qquad (6.133)$$

as the magnitude of the probability current reflected from the barrier, showing that the probability of reflection is given by

$$R = \frac{j_{\text{ref}}}{j_{\text{inc}}} = \frac{|B|^2}{|A|^2} \qquad (6.134)$$

The probability of transmission for this scattering experiment is given by

$$T = \frac{j_{\text{trans}}}{j_{\text{inc}}} \qquad (6.135)$$

where $j_{\text{trans}}$ is the probability current to the right of the step. In order to evaluate $R$ and $T$, we need to solve for the wave function for $x > 0$ and then satisfy the boundary conditions at $x = 0$. The wave equation to the right of the step is given by

$$\frac{d^2\psi(x)}{dx^2} = \frac{2m}{\hbar^2}(V_0 - E)\psi(x) \qquad x > 0 \qquad (6.136)$$

We consider two cases.

**Case 1:** $E > V_0$
Since the energy is greater than the potential energy for $x > 0$, the solutions to (6.136) are given by

$$\psi(x) = Ce^{iQx} + De^{-iQx} \qquad x > 0 \qquad (6.137)$$

where

$$Q = \sqrt{\frac{2m}{\hbar^2}(E - V_0)} \tag{6.138}$$

The $D$ term generates a probability current flowing to the left for $x > 0$. Such a term would be generated physically if the experiment in question involved projecting particles at the potential step from the right. Clearly, the solutions to the differential equation should permit this possibility. However, if we restrict our attention to an "experiment" in which particles are incident on the potential step only from the left, we are free to set $D = 0$ in (6.137). In this case

$$\psi(x) = C e^{iQx} \qquad x > 0 \tag{6.139}$$

and substituting the wave function (6.139) into (6.124), we find

$$j_{\text{trans}} = \frac{\hbar Q}{m}|C|^2 \tag{6.140}$$

Thus for the step potential the transmission coefficient is given by

$$T = \frac{Q}{k}\frac{|C|^2}{|A|^2} \tag{6.141}$$

Let's determine the reflection and transmission coefficients in terms of $V_0$ and $E$. In passing from the region to the left of the step to the region to the right of the step, we must require that the energy eigenfunction be continuous and have a continuous first derivative:

$$A + B = C$$
$$ik(A - B) = iQC \tag{6.142}$$

which yield

$$C = \frac{2k}{k + Q}A \qquad \text{and} \qquad B = \frac{k - Q}{k + Q}A \tag{6.143}$$

Note that we have satisfied the boundary conditions for any value of the energy. Therefore the allowed energies do take on a continuum of values. Using (6.134) and (6.141), we find

$$R = \frac{(k - Q)^2}{(k + Q)^2} \qquad T = \frac{4kQ}{(k + Q)^2} \tag{6.144}$$

Note that

$$R + T = 1 \tag{6.145}$$

as it must for probability to be conserved.

**Case 2: $E < V_0$**
Here the solution for $x < 0$ is the same as (6.129), but for $x > 0$ we have

$$\frac{d^2\psi}{dx^2} = \frac{2m}{\hbar^2}(V_0 - E)\psi = q^2\psi \qquad x > 0 \qquad (6.146)$$

with $q^2 > 0$. Now we *must* choose the solution

$$\psi = Ce^{-qx} \qquad (6.147)$$

since the increasing exponential would cause the wave function to blow up as $x \to \infty$. Rather than match the wave function at the $x = 0$ boundary again, a comparison of the wave function (6.147) with (6.139) shows that we can obtain the solution for $E < V_0$ from the solution for $E > V_0$ by the transcription $iQ \to -q$. Thus from (6.134) and (6.143), we obtain for $E < V_0$

$$R = \frac{|k - iq|^2}{|k + iq|^2} = \frac{k^2 + q^2}{k^2 + q^2} = 1 \qquad (6.148)$$

Conservation of probability requires that the transmission coefficient must vanish for $E < V_0$ even though

$$C = \frac{2k}{k + iq}A \neq 0 \qquad (6.149)$$

and the wave function penetrates into the potential energy barrier. Note that we cannot just make the transcription $iQ \to -q$ for the transmission coefficient in (6.144) because, unlike the reflection coefficient, which is given by (6.134) whether the energy is greater than or less than the height of the barrier, the transmission coefficient is determined by the probability current for $x > 0$, and when the argument of the exponential in the wave function is real, $j_{\text{trans}} = 0$. This emphasizes that the transmission coefficient is given by (6.135), and not by (6.141) in general.

## Tunneling

Suppose that we consider particles with energy $E < V_0$ incident on a potential energy barrier of height $V_0$, but this time we chop off the end of the barrier so that

$$V = \begin{cases} 0 & x < 0 \\ V_0 & 0 < x < a \\ 0 & x > a \end{cases} \qquad (6.150)$$

as shown in Fig. 6.13. Now the energy eigenfunction is given by

$$\psi = \begin{cases} Ae^{ikx} + Be^{-ikx} & x < 0 \\ Fe^{qx} + Ge^{-qx} & 0 < x < a \\ Ce^{ikx} & x > a \end{cases} \qquad (6.151)$$

with $k$ and $q$ given by (6.91) and (6.92), respectively. Note that both the rising and the falling exponential appear as part of the solution for $0 < x < a$ because the barrier is of finite width $a$ and therefore the rising exponential cannot diverge.

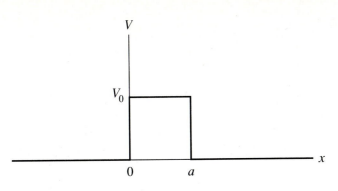

**FIGURE 6.13**
A square potential barrier.

The procedure for determining the transmission coefficient is straightforward, if somewhat laborious. Satisfying the boundary conditions on the continuity of the wave function and its first derivative leads to four equations:

$$A + B = F + G$$

$$ik(A - B) = q(F - G)$$

$$Fe^{qa} + Ge^{-qa} = Ce^{ika} \qquad (6.152)$$

$$q(Fe^{qa} - Ge^{-qa}) = ikCe^{ika}$$

The unknowns $B$, $F$, and $G$ can be eliminated from these equations, yielding

$$T = \frac{j_{x>a}}{j_{\text{inc}}} = \frac{\dfrac{\hbar k}{m}|C|^2}{\dfrac{\hbar k}{m}|A|^2} = \frac{1}{1 + \left(\dfrac{k^2 + q^2}{2kq}\right)^2 \sinh^2 qa} \qquad (6.153)$$

for the probability that the particle will tunnel through the potential barrier. For typical microscopic parameters such as an electron with 5 eV of kinetic energy tunneling through a barrier 10 eV high and 0.53 Å wide (the Bohr radius), the transmission probability is 0.68. Thus tunneling is a common occurrence on the microscopic level.

A useful limiting case of the transmission coefficient occurs for $qa \gg 1$. In this case

$$\sinh qa = \frac{e^{qa} - e^{-qa}}{2} \xrightarrow[qa \gg 1]{} \frac{e^{qa}}{2} \qquad (6.154)$$

and

$$T \xrightarrow[qa \gg 1]{} \left(\frac{4kq}{k^2 + q^2}\right)^2 e^{-2qa} \qquad (6.155)$$

We can quickly see why tunneling is not a common macroscopic occurrence if we plug in some typical macroscopic parameters such as $V_0 - E = 1$ erg, $a = 1$ cm, and $m = 1$ g. Then $qa \sim 10^{27}$, so that $T \sim e^{-10^{27}}$, an incredibly small number.

Even on the microscopic level, there are many situations where $qa$ is sufficiently large that we can take advantage of the approximation (6.155) for the transmission coefficient. Notice that if we evaluate the natural log of the transmission coefficient (6.155), we find

$$\ln T \xrightarrow[qa \gg 1]{} \ln \left( \frac{4kq}{k^2 + q^2} \right)^2 - 2qa \xrightarrow[qa \gg 1]{} - 2qa \tag{6.156}$$

where we have dropped the logarithm relative to $qa$ since ln(almost anything) is not very large. In the limit that (6.156) is a good approximation, we can use it to calculate the probability of transmission through a non-square barrier, such as that depicted in Fig. 6.14. When we only include the exponential term in (6.155), the probability of transmission through a barrier of width $2a$ is just the product of the individual transmission coefficients for two barriers of width $a$. Thus, if the barrier is sufficiently smooth so that we can approximate it by a series of square barriers (each of width $\Delta x$) that are not too thin for (6.156) to hold, then for the barrier as a whole

$$\ln T \simeq \ln \prod_i T_i = \sum_i \ln T_i \simeq -2 \sum_i q_i \Delta x \tag{6.157}$$

If we now assume that we can approximate this last term as an integral, we find

$$T \simeq \exp\left( -2 \sum_i q_i \Delta x \right) \simeq \exp\left( -2 \int dx \sqrt{\frac{2m}{\hbar^2}[V(x) - E]} \right) \tag{6.158}$$

where the integration is over the region for which the square root is real. You may have a somewhat queasy feeling about the derivation of (6.158). Clearly, the approximations we have made break down near the turning points, where $E = V(x)$. Nonetheless, a more detailed treatment using the WKB approximation

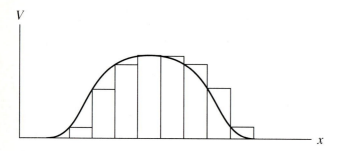

**FIGURE 6.14**
A non-square barrier can be approximated by a sequence of square barriers if the potential energy $V(x)$ does not vary too rapidly with position.

shows that (6.158) works reasonably well.[20] As an example, we can use it to estimate the currents generated by field emission for a metal (see Problem 6.21).

## 6.11   SUMMARY

In this chapter we have turned our attention to variables such as position and momentum that take on a continuum of values, instead of the discrete set of values characteristic of variables like angular momentum. Thus instead of expressing a ket $|\psi\rangle$ as a discrete sum of eigenstates as in (1.32), we write it as

$$|\psi\rangle = \int da \, |a\rangle\langle a|\psi\rangle \qquad (6.159)$$

where the ket $|a\rangle$ is an eigenket of the operator $\hat{A}$ corresponding to the observable $A$:

$$\hat{A}|a\rangle = a|a\rangle \qquad (6.160)$$

From (6.159) we see that the identity operator is given by

$$\int da \, |a\rangle\langle a| = 1 \qquad (6.161)$$

Substituting $|a'\rangle$ for $|\psi\rangle$ in (6.159), we find that the states of a continuous variable satisfy the normalization condition

$$\langle a|a'\rangle = \delta(a - a') \qquad (6.162)$$

where $\delta(a - a')$ is a Dirac delta function. On the other hand, a physical state $|\psi\rangle$ satisfies

$$1 = \langle \psi|\psi\rangle = \int da \, \langle \psi|a\rangle\langle a|\psi\rangle = \int da \, |\langle a|\psi\rangle|^2 \qquad (6.163)$$

indicating that we should identify

$$da \, |\langle a|\psi\rangle|^2 \qquad (6.164)$$

as the probability of finding the variable $A$ in the range between $a$ and $a + da$ if a measurement of $A$ is carried out.

We have restricted our attention in this chapter to one-dimensional position states, for which

$$\hat{x}|x\rangle = x|x\rangle \qquad (6.165)$$

and one-dimensional momentum states, for which

---

[20] The WKB approximation and its application to tunneling is discussed by L. Schiff, *Quantum Mechanics*, 3rd ed., McGraw-Hill, New York, 1968, Chapter 8, Section 34.

$$\hat{p}_x|p\rangle = p|p\rangle \tag{6.166}$$

Just as angular momentum made its appearance in Chapter 3 in the form of an operator that generated rotations and energy entered in Chapter 4 in the form of an operator that generated time translations, here linear momentum enters in the form of an operator that generates translations in space. The translation operator is given by

$$\hat{T}(a) = e^{-i\hat{p}_x a/\hbar} \tag{6.167}$$

where the action of the translation operator on a position ket $|x\rangle$ is given by

$$\hat{T}(a)|x\rangle = |x + a\rangle \tag{6.168}$$

In order for probability to be conserved under translations, the translation operator must be unitary:

$$\hat{T}^\dagger(a)\hat{T}(a) = e^{i\hat{p}_x^\dagger a/\hbar} e^{-i\hat{p}_x a/\hbar} = 1 \tag{6.169}$$

and therefore the linear momentum operator must be Hermitian:

$$\hat{p}_x^\dagger = \hat{p}_x \tag{6.170}$$

A consequence of the momentum operator being the generator of translations is that the position and momentum operators do not commute [compare $\hat{T}(a)\hat{x}|x\rangle = x|x + a\rangle$ with $\hat{x}\hat{T}(a)|x\rangle = (x + a)|x + a\rangle$] but rather satisfy the commutation relation

$$[\hat{x}, \hat{p}_x] = i\hbar \tag{6.171}$$

leading to the Heisenberg uncertainty relation

$$\Delta x \Delta p_x \geq \frac{\hbar}{2} \tag{6.172}$$

A further consequence is that the action of the momentum operator $\hat{p}_x$ in position space is given by

$$\langle x|\hat{p}_x|\psi\rangle = \frac{\hbar}{i}\frac{\partial}{\partial x}\langle x|\psi\rangle \tag{6.173}$$

and therefore

$$\langle p_x \rangle = \langle \psi|\hat{p}_x|\psi\rangle = \int_{-\infty}^{\infty} dx \, \langle \psi|x\rangle \frac{\hbar}{i}\frac{\partial}{\partial x}\langle x|\psi\rangle \tag{6.174}$$

Equation (6.173) indicates that in position space we can *represent* the momentum operator by a differential operator

$$\hat{p}_x \xrightarrow[x \text{ basis}]{} \frac{\hbar}{i}\frac{\partial}{\partial x} \tag{6.175}$$

Thus the equation of motion

$$\hat{H}|\psi(t)\rangle = i\hbar \frac{d}{dt}|\psi(t)\rangle \tag{6.176}$$

for the Hamiltonian

$$\hat{H} = \frac{\hat{p}_x^2}{2m} + V(\hat{x}) \tag{6.177}$$

becomes in position space

$$\langle x|\hat{H}|\psi(t)\rangle = \left[-\frac{\hbar^2}{2m}\frac{\partial^2}{\partial x^2} + V(x)\right]\langle x|\psi(t)\rangle = i\hbar\frac{\partial}{\partial t}\langle x|\psi(t)\rangle \tag{6.178}$$

the usual time-dependent Schrödinger equation. We make the identification

$$\langle x|\psi(t)\rangle = \psi(x, t) \tag{6.179}$$

since the amplitude to find a particle in the state $|\psi(t)\rangle$ at the position $x$ is just what we usually call the wave function in position space. Thus, according to (6.164), $dx\,|\psi(x, t)|^2$ is the probability of finding the particle between $x$ and $x + dx$, the usual Born interpretation of wave mechanics.

The energy eigenvalue equation

$$\hat{H}|\psi\rangle = E|\psi\rangle \tag{6.180}$$

in position space becomes

$$\left[-\frac{\hbar^2}{2m}\frac{\partial^2}{\partial x^2} + V(x)\right]\langle x|E\rangle = E\langle x|E\rangle \tag{6.181a}$$

or simply

$$\left[-\frac{\hbar^2}{2m}\frac{\partial^2}{\partial x^2} + V(x)\right]\psi_E(x) = E\psi_E(x) \tag{6.181b}$$

This differential equation can be solved to determine the energy eigenstates for one-dimensional potentials. See Sections 6.8 through 6.10.

The connection between the position-space wave function $\langle x|\psi\rangle$ and the momentum-space wave function $\langle p|\psi\rangle$ is through the set of amplitudes

$$\langle x|p\rangle = \frac{1}{\sqrt{2\pi\hbar}}e^{ipx/\hbar} \tag{6.182}$$

These amplitudes can be used to transform back and forth between position and momentum space:

$$\langle p|\psi\rangle = \int dx\,\langle p|x\rangle\langle x|\psi\rangle = \int dx\,\frac{1}{\sqrt{2\pi\hbar}}e^{-ipx/\hbar}\langle x|\psi\rangle \tag{6.183a}$$

$$\langle x|\psi\rangle = \int dp\,\langle x|p\rangle\langle p|\psi\rangle = \int dp\,\frac{1}{\sqrt{2\pi\hbar}}e^{ipx/\hbar}\langle p|\psi\rangle \tag{6.183b}$$

Thus the position-space and momentum-space wave functions form a Fourier transform pair.

## PROBLEMS

**6.1.** (a) Use induction to show that $\left[\hat{x}^n, \hat{p}_x\right] = i\hbar n \hat{x}^{n-1}$.

(b) By expanding $F(x)$ in a Taylor series, show that

$$\left[F(\hat{x}), \hat{p}_x\right] = i\hbar \frac{\partial F}{\partial x}(\hat{x})$$

(c) For the one-dimensional Hamiltonian

$$\hat{H} = \frac{\hat{p}_x^2}{2m} + V(\hat{x})$$

show that

$$\frac{d\langle p_x \rangle}{dt} = \left\langle -\frac{dV}{dx} \right\rangle$$

**6.2.** Show that

$$\langle p|\hat{x}|\psi \rangle = i\hbar \frac{\partial}{\partial p} \langle p|\psi \rangle$$

and

$$\langle \varphi|\hat{x}|\psi \rangle = \int dp \, \langle p|\varphi \rangle^* i\hbar \frac{\partial}{\partial p} \langle p|\psi \rangle$$

What do these results suggest for how you should represent the position operator in momentum space?

**6.3.** Show for infinitesimal translations $|\psi \rangle \rightarrow |\psi'\rangle = \hat{T}(\delta x)|\psi \rangle$ that $\langle x \rangle \rightarrow \langle x \rangle + \delta x$ and $\langle p_x \rangle \rightarrow \langle p_x \rangle$.

**6.4.** (a) Show for a free particle of mass $m$ initially in the state

$$\psi(x) = \langle x|\psi \rangle = \frac{1}{\sqrt{\sqrt{\pi} a}} e^{-x^2/2a^2}$$

that

$$\psi(x, t) = \langle x|\psi(t)\rangle = \frac{1}{\sqrt{\sqrt{\pi}[a + (i\hbar t/ma)]}} e^{-x^2/2a^2[1+(i\hbar t/ma^2)]}$$

and therefore

$$\Delta x = \frac{a}{\sqrt{2}} \sqrt{1 + \left(\frac{\hbar t}{ma^2}\right)^2}$$

(b) What is $\Delta p_x$ at time $t$? *Suggestion:* Use the momentum-space wave function to evaluate $\Delta p_x$.

**6.5.** Consider a wave packet defined by

$$\langle p|\psi\rangle = \begin{cases} 0 & p < -P/2 \\ N & -P/2 < p < P/2 \\ 0 & p > P/2 \end{cases}$$

(a) Determine a value for $N$ such that $\langle\psi|\psi\rangle = 1$ using the momentum-space wave function directly.

(b) Determine $\langle x|\psi\rangle = \psi(x)$.

(c) Sketch $\langle p|\psi\rangle$ and $\langle x|\psi\rangle$. Use reasonable estimates of $\Delta p_x$ from the form of $\langle p|\psi\rangle$ and $\Delta x$ from the form of $\langle x|\psi\rangle$ to estimate the product $\Delta x \Delta p_x$. Check that your result is independent of the value of $P$. *Note:* Simply estimate rather than actually calculate the uncertainties.

**6.6.** (a) Show that $\langle p_x\rangle = 0$ for a state with a *real* wave function $\langle x|\psi\rangle$.

(b) Show that if the wave function $\langle x|\psi\rangle$ is modified by a position-dependent phase

$$\langle x|\psi\rangle \rightarrow e^{ip_0 x/\hbar}\langle x|\psi\rangle$$

then

$$\langle x\rangle \rightarrow \langle x\rangle \qquad \text{and} \qquad \langle p_x\rangle \rightarrow \langle p_x\rangle + p_0$$

**6.7.** Establish that the position operator $\hat{x}$ is Hermitian (a) by showing that $\langle\varphi|\hat{x}|\psi\rangle = \langle\psi|\hat{x}|\varphi\rangle^*$ or (b) by taking the adjoint of the position-momentum commutation relation (6.31).

**6.8.** Without using exact mathematics—that is, using only arguments of curvature, symmetry, and semiquantitative estimates of wavelength—sketch the energy eigenfunctions for the ground state, first-excited state, and second-excited state of a particle in the potential energy well $V(x) = a|x|$ shown in Fig. 6.15. This potential energy has been suggested as arising from the force exerted by one quark on another quark.

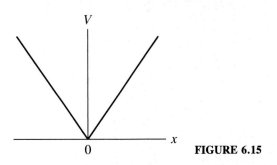

**FIGURE 6.15**

**6.9.** The one-dimensional time-independent Schrödinger equation for the potential energy discussed in Problem 6.8 is

$$\frac{d^2\psi}{dx^2} + \frac{2m}{\hbar^2}(E - a|x|)\psi = 0$$

Define $E = \varepsilon(\hbar^2 a^2/m)^{1/3}$ and $x = z(\hbar^2/ma)^{1/3}$.

(a) Show that $\varepsilon$ and $z$ are dimensionless.

(b) Show that the Schrödinger equation can be expressed in the form

$$\frac{d^2\psi}{dz^2} + 2(\varepsilon - |z|)\psi = 0$$

(c) Numerically integrate this equation for various values of $\varepsilon$, beginning with $d\psi/dz = 0$ at $z = 0$, to find the value of $\varepsilon$ corresponding to the ground-state eigenfunction.

**6.10.** Solve the energy eigenvalue equation in position space for a particle of mass $m$ in the potential energy well

$$V(x) = \begin{cases} 0 & 0 < x < L \\ \infty & \text{elsewhere} \end{cases}$$

Show that the energy eigenvalues are given by

$$E_n = \frac{\hbar^2\pi^2 n^2}{2mL^2} \qquad n = 1, 2, 3, \ldots$$

with the corresponding normalized energy eigenfunctions

$$\psi_n(x) = \begin{cases} \sqrt{\dfrac{2}{L}}\sin\dfrac{n\pi x}{L} & 0 < x < L \\ 0 & \text{elsewhere} \end{cases} \qquad n = 1, 2, 3, \ldots$$

**6.11.** Determine $\Delta x$, $\Delta p_x$, and $\Delta x \Delta p_x$ for a particle of mass $m$ in the ground state of the potential energy well

$$V(x) = \begin{cases} 0 & 0 < x < L \\ \infty & \text{elsewhere} \end{cases}$$

**6.12.** A particle of mass $m$ in the one-dimensional potential energy well

$$V(x) = \begin{cases} 0 & 0 < x < L \\ \infty & \text{elsewhere} \end{cases}$$

is at time $t = 0$ in the state

$$\psi(x) = \begin{cases} \left(\dfrac{1+i}{2}\right)\sqrt{\dfrac{2}{L}}\sin\dfrac{\pi x}{L} + \dfrac{1}{\sqrt{2}}\sqrt{\dfrac{2}{L}}\sin\dfrac{2\pi x}{L} & 0 < x < L \\ 0 & \text{elsewhere} \end{cases}$$

(a) What is $\psi(x, t)$?
(b) What is $\langle E \rangle$ for this state at time t?
(c) What is the probability that a measurement of the energy will yield the value $\hbar^2\pi^2/2mL^2$?
(d) Without detailed computation, give an argument that $\langle x \rangle$ is time dependent.

**6.13.** A particle of mass $m$ is in lowest-energy (ground) state of the infinite potential energy well

$$V(x) = \begin{cases} 0 & 0 < x < L \\ \infty & \text{elsewhere} \end{cases}$$

At time $t = 0$, the wall located at $x = L$ is suddenly pulled back to a position at $x = 2L$. This change occurs so rapidly that instantaneously the wave function does not change.

(a) Calculate the probability that a measurement of the energy will yield the ground-state energy of the *new* well. What is the probability that a measurement of the energy will yield the first-excited energy of the *new* well?

(b) Describe the procedure you would use to determine the time development of the system. Is the system in a stationary state?

**6.14.** A particle in the potential energy well

$$V(x) = \begin{cases} 0 & 0 < x < L \\ \infty & \text{elsewhere} \end{cases}$$

is in the state

$$\psi(x) = \begin{cases} Nx(x - L) & 0 < x < L \\ 0 & \text{elsewhere} \end{cases}$$

(a) Determine the value of $N$ so that the state is properly normalized.

(b) What is the probability that a measurement of the energy yields the ground-state energy of the well?

(c) What is $\langle E \rangle$ for this state?

**6.15.** (a) What is the magnitude of the ground-state energy for the infinite well if the confined particle is an electron and the width of the well is an angstrom, a typical size of an atom?

(b) If the particle is a neutron or a proton and the width of the well is a characteristic size of a nucleus, what is the magnitude of the ground-state energy?

**6.16.** An interesting limiting case of the finite square well discussed in Section 6.8 is the case where the well depth approaches infinity but the width of the well goes to zero such that $V_0 a$ remains constant. Such a well may be represented by the potential energy satisfying

$$\frac{2m}{\hbar^2} V(x) = -\frac{\lambda}{b} \delta(x)$$

where $\delta(x)$ is the Dirac delta function. Note that $\lambda/b$ is a constant having the units of inverse length and we have taken the top of the well to be at $V = 0$.

(a) Show by integrating the time-independent Schrödinger equation that the derivative of the energy eigenfunction is not continuous, but satisfies

$$\left(\frac{d\psi}{dx}\right)_{0+} - \left(\frac{d\psi}{dx}\right)_{0-} = -\frac{\lambda}{b} \psi(0)$$

(b) Determine the energy eigenvalue(s) for this well. Sketch the energy eigenfunction(s).

*Suggestion:* Solve the Schrödinger equation for $E < 0$ in the regions $x < 0$ and $x > 0$ and join the solutions together, making sure that the boundary conditions are satisfied.

**6.17.** Calculate the reflection and transmission coefficients for scattering from the potential energy barrier

$$\frac{2mV}{\hbar^2} V(x) = \frac{\lambda}{b} \delta(x)$$

Note the discussion in the preceding problem on the boundary conditions.

**6.18.** Show that the reflection and transmission coefficients for scattering from the step potential shown in Fig. 6.12 are given by (6.144) even when the particles are incident on the step from the right instead of from the left.

**6.19.** Derive the expression (6.153) for the transmission coefficient for tunneling through a square barrier.

**6.20.** (a) Show that the transmission coefficient for scattering from the potential energy well

$$V(x) = \begin{cases} 0 & x < 0 \\ -V_0 & 0 < x < a \\ 0 & x > a \end{cases}$$

is given by

$$T = \left[ 1 + \frac{\sin^2 \sqrt{\dfrac{2m}{\hbar^2}(E + V_0)a}}{4\dfrac{E}{V_0}\dfrac{(E + V_0)}{V_0}} \right]^{-1}$$

*Suggestion:* What is the transcription required to change the wave function (6.151) into the one appropriate for this problem? What happens to the transmission coefficient (6.153) under this transcription?

(b) Show that for certain incident energies there is 100 percent transmission. Suppose that we model an atom as a one-dimensional square well with a width of 1 Å and that an electron with 0.7 eV of kinetic energy encounters the well. What must the depth of the well be for 100 percent transmission? This absence of scattering is observed when the target atoms are composed of noble gases such as krypton.

**6.21.** Electrons in a metal are bound by a potential that may be approximated by a finite square well. Electrons fill up the energy levels of this well up to an energy called the Fermi energy, as indicated in Fig. 6.16a. The difference between the Fermi energy and the top of the well is the work function $W$ of the metal. Photons with energies exceeding the work function can eject electrons from the metal—the

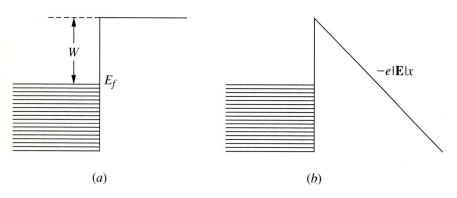

(a)                                             (b)

**FIGURE 6.16**
(a) A finite square well is an approximation to the potential well confining electrons within a metal. (b) Applying a negative voltage to the metal alters the potential well, permitting electrons to tunnel out.

photoelectric effect. Another way to pull out electrons is through application of an external uniform electric field **E**, which alters the potential energy as shown in Fig. 6.16b. Show that the transmission coefficient for electrons at the Fermi energy is given by

$$T \simeq \exp\left(\frac{-4\sqrt{2m}\,W^{3/2}}{3e|\mathbf{E}|\hbar}\right)$$

How would you expect the field-emission current to vary with the applied voltage?

# CHAPTER

# 7

# THE ONE-DIMENSIONAL HARMONIC OSCILLATOR

In this chapter we turn our attention to a system in which a particle experiences a potential energy $V(x)$ that varies with position in a nontrivial way—namely, the simple harmonic oscillator. Not only is this a system for which we can determine exactly the energy eigenvalues and eigenstates in a number of different ways, but it is also a system with an extremely broad physical significance.

## 7.1 THE IMPORTANCE OF THE HARMONIC OSCILLATOR

What gives the harmonic oscillator such a broad significance? First let's consider a specific example familiar from classical mechanics. A mass $m$ attached to a string of length $L$ is free to pivot under the influence of gravity about the point $O$, as shown in Fig. 7.1. The energy of the system can be expressed as

$$E = \frac{1}{2}mv^2 + mgh = \frac{1}{2}mv^2 + mgL(1 - \cos\theta) \tag{7.1}$$

If the angle $\theta$ is small, we can expand $\cos\theta$ in a Taylor series and retain the leading terms to obtain

$$E = \frac{1}{2}mv^2 + \frac{1}{2}mgL\theta^2 = \frac{1}{2}mv^2 + \frac{1}{2}\frac{mg}{L}x^2 \tag{7.2}$$

where we called the arc length $L\theta = x$. Thus, provided the oscillations are small, the system behaves like a harmonic oscillator with a spring constant $k = mg/L$

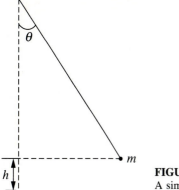

**FIGURE 7.1**
A simple pendulum.

and, therefore, a spring frequency $\omega = \sqrt{k/m} = \sqrt{g/L}$. Notice that there is no physical spring actually attached to the mass in this case.

Let's now examine *any* potential energy function $V(x)$ that has a minimum at a position that we call $x_0$, as shown in Fig. 7.2. Expanding $V(x)$ in a Taylor series about the minimum, we obtain

$$V(x) = V(x_0) + \left(\frac{dV}{dx}\right)_{x=x_0}(x - x_0) + \frac{1}{2!}\left(\frac{d^2V}{dx^2}\right)_{x=x_0}(x - x_0)^2 + \cdots \quad (7.3)$$

Since $x_0$ is the location of the minimum of the potential energy, the first derivative vanishes there and

$$V(x) = V(x_0) + \tfrac{1}{2}k(x - x_0)^2 + \cdots \quad (7.4)$$

where $k$ is a positive constant. Since it is only differences in potential energy that matter physically, we can choose the zero of potential energy such that $V(x_0) = 0$. If we now position the origin of our coordinates at $x_0$, then

$$V(x) = \tfrac{1}{2}kx^2 + \cdots \quad (7.5)$$

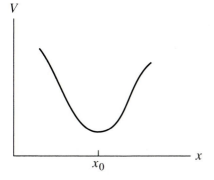

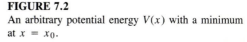

**FIGURE 7.2**
An arbitrary potential energy $V(x)$ with a minimum at $x = x_0$.

Thus, provided the system is undergoing sufficiently small oscillations about the equilibrium point, we can neglect the higher-order terms in the Taylor series expansion, and the effective potential energy is that of a harmonic oscillator. Good examples of systems that behave like harmonic oscillators on a microscopic scale are the vibrations of nuclei within diatomic molecules and the vibrations of atoms in a crystalline solid about their equilibrium positions.

In Section 7.2 we will solve the harmonic oscillator using operator methods reminiscent of those that we used in Chapter 3 to determine the eigenstates of angular momentum. In the example of angular momentum, we utilized only the commutation relations (3.14), without having to specify the type of angular momentum. This approach generated solutions that included intrinsic spin in addition to the more familiar orbital angular momentum, which we will analyze in Chapter 9. Here too, solving the harmonic oscillator with operator methods allows for more abstract solutions in which the variable $x$ may not be the usual position at all. In Chapter 14 we will see that the Hamiltonian of the electromagnetic field may be expressed as a collection of such abstract harmonic oscillators. Planck's resolution of the ultraviolet catastrophe in the analysis of the blackbody spectrum, which amounted to treating these oscillators as quantum oscillators, can be considered as the starting point of quantum field theory.[1]

## 7.2 OPERATOR METHODS

Our goal is to determine the eigenkets and eigenvalues of the Hamiltonian

$$\hat{H} = \frac{\hat{p}_x^2}{2m} + \frac{1}{2}m\omega^2\hat{x}^2 \tag{7.6}$$

where we have expressed the kinetic energy of the particle in terms of its momentum, as in Chapter 6. In addition to the expression for $\hat{H}$, the only other ingredient required for a solution using operator methods is the commutation relation

$$[\hat{x}, \hat{p}_x] = i\hbar \tag{7.7}$$

Notice that the Hamiltonian is quadratic in both the position $\hat{x}$ and the momentum $\hat{p}_x$, just as the operator $\hat{\mathbf{J}}^2$ is quadratic in the individual components of the angular momentum. For the harmonic oscillator we introduce two non-Hermitian operators

$$\hat{a} = \sqrt{\frac{m\omega}{2\hbar}}\left(\hat{x} + \frac{i}{m\omega}\hat{p}_x\right) \tag{7.8}$$

---

[1] Perhaps it is not so surprising that P. A. M. Dirac, who invented the elegant operator approach to the harmonic oscillator in 1925, went on to apply these same techniques to the quantization of the electromagnetic field as early as 1927, while the details of nonrelativistic quantum mechanics were still being worked out. At least it is not surprising if you happen to be as clever as Dirac, who in 1928 also developed the relativistic wave equation for spin-$\frac{1}{2}$ particles, the famous Dirac equation. Not bad for three years' work.

and

$$\hat{a}^\dagger = \sqrt{\frac{m\omega}{2\hbar}} \left( \hat{x} - \frac{i}{m\omega} \hat{p}_x \right) \tag{7.9}$$

in a fashion similar to the way we introduced $\hat{J}_+ = \hat{J}_x \pm i\hat{J}_y$. Since $\hat{x}$ and $\hat{p}_x$ have different dimensions, we cannot just add $\hat{x} \pm i\hat{p}_x$, as we did for angular momentum. The factor of $\sqrt{m\omega/2\hbar}$ in front is inserted so that the operators (7.8) and (7.9) are dimensionless. Using (7.7), we can verify that these operators satisfy the simple commutation relation

$$\left[ \hat{a}, \hat{a}^\dagger \right] = 1 \tag{7.10}$$

Inverting (7.8) and (7.9), we obtain

$$\hat{x} = \sqrt{\frac{\hbar}{2m\omega}} (\hat{a} + \hat{a}^\dagger) \tag{7.11}$$

and

$$\hat{p}_x = -i\sqrt{\frac{m\omega\hbar}{2}} (\hat{a} - \hat{a}^\dagger) \tag{7.12}$$

which can be used to express the Hamiltonian (7.6) as

$$\hat{H} = \frac{\hbar\omega}{2} (\hat{a}^\dagger \hat{a} + \hat{a}\hat{a}^\dagger) = \hbar\omega \left( \hat{a}^\dagger \hat{a} + \frac{1}{2} \right) \tag{7.13}$$

In the last step we took advantage of the commutation relation (7.10). Thus finding the eigenstates of $\hat{H}$ is equivalent to finding the eigenstates of

$$\hat{N} = \hat{a}^\dagger \hat{a} \tag{7.14}$$

often called the number operator, for reasons that will be apparent shortly.

Let's temporarily denote the eigenstates of $\hat{N}$ by $|\eta\rangle$:

$$\hat{N}|\eta\rangle = \eta|\eta\rangle \tag{7.15}$$

The expectation value of the number operator in an eigenstate is given by

$$\langle\eta|\hat{N}|\eta\rangle = \langle\eta|\hat{a}^\dagger \hat{a}|\eta\rangle = \eta\langle\eta|\eta\rangle \tag{7.16}$$

Calling

$$\hat{a}|\eta\rangle = |\psi\rangle \tag{7.17}$$

we can express equation (7.16) as

$$\langle\psi|\psi\rangle = \eta\langle\eta|\eta\rangle \tag{7.18}$$

Since $\langle\psi|\psi\rangle \geq 0$ and $\langle\eta|\eta\rangle \geq 0$, (7.18) shows that the eigenvalue $\eta \geq 0$, as might be expected from the semi-positive definite nature of the Hamiltonian (7.6).

It is the commutation relations

$$[\hat{N}, \hat{a}] = -\hat{a} \tag{7.19}$$

and

$$[\hat{N}, \hat{a}^\dagger] = \hat{a}^\dagger \tag{7.20}$$

which follow from (7.10) and (7.14), that make the operators $\hat{a}$ and $\hat{a}^\dagger$ so useful. Compare with the similar relations (3.39) for angular momentum. To see the action of $\hat{a}^\dagger$ on the ket $|\eta\rangle$, we evaluate $\hat{N}\hat{a}^\dagger|\eta\rangle$. In order to let the operator $\hat{N}$ act on its eigenstate, we use the commutation relation (7.20) to switch the order of the operators, picking up an extra term because the operators do not commute:

$$\begin{aligned}
\hat{N}\hat{a}^\dagger|\eta\rangle &= (\hat{a}^\dagger\hat{N} + \hat{a}^\dagger)|\eta\rangle \\
&= (\hat{a}^\dagger\eta + \hat{a}^\dagger)|\eta\rangle \\
&= (\eta + 1)\hat{a}^\dagger|\eta\rangle
\end{aligned} \tag{7.21a}$$

We can make the action of the operator $\hat{a}^\dagger$ more transparent with the addition of some parentheses to this equation:

$$\hat{N}(\hat{a}^\dagger|\eta\rangle) = (\eta + 1)(\hat{a}^\dagger|\eta\rangle) \tag{7.21b}$$

indicating that

$$\hat{a}^\dagger|\eta\rangle = c_+|\eta + 1\rangle \tag{7.22}$$

that is, $\hat{a}^\dagger|\eta\rangle$ is an eigenket of $\hat{N}$ with eigenvalue $\eta + 1$. Thus $\hat{a}^\dagger$ is a *raising operator*. Similarly,

$$\begin{aligned}
\hat{N}\hat{a}|\eta\rangle &= (\hat{a}\hat{N} - \hat{a})|\eta\rangle \\
&= (\eta - 1)\hat{a}|\eta\rangle
\end{aligned} \tag{7.23}$$

so $\hat{a}$ is a *lowering operator*:

$$\hat{a}|\eta\rangle = c_-|\eta - 1\rangle \tag{7.24}$$

Unlike the case of angular momentum, where there were limits on *both* how far we could raise and how far we could lower the eigenvalues of $\hat{J}_z$, the only limitation here comes from the requirement that $\eta \geq 0$. Thus there must exist a lowest eigenvalue, which we call $\eta_{min}$. The ket with this eigenvalue must satisfy

$$\hat{a}|\eta_{min}\rangle = 0 \tag{7.25}$$

for otherwise $\hat{a}|\eta_{min}\rangle = c|\eta_{min} - 1\rangle$, violating our assumption that $\eta_{min}$ is the lowest eigenvalue. However, if we apply the raising operator to (7.25), we obtain

$$\hat{a}^\dagger\hat{a}|\eta_{min}\rangle = \eta_{min}|\eta_{min}\rangle = 0 \tag{7.26}$$

where the middle step follows from the fact that $|\eta_{min}\rangle$ is an eigenstate of $\hat{a}^\dagger\hat{a}$. Since the ket $|\eta_{min}\rangle$ exists, (7.26) requires that $\eta_{min} = 0$. Thus we label the lowest state simply as $|0\rangle$. Applying the raising operator $n$ times, where $n$ must clearly

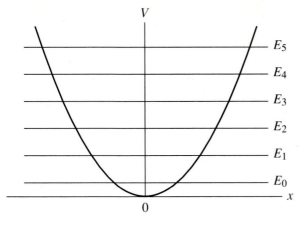

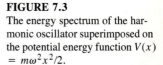

**FIGURE 7.3**
The energy spectrum of the harmonic oscillator superimposed on the potential energy function $V(x) = m\omega^2 x^2/2$.

be an integer, generates the state $|n\rangle$ satisfying

$$\hat{N}|n\rangle = n|n\rangle \qquad n = 0, 1, 2, \ldots \tag{7.27}$$

Thus the eigenvalues of the number operator are the integers—hence the name for this operator. The eigenvalues of the Hamiltonian are determined by

$$\hat{H}|n\rangle = \hbar\omega(\hat{N} + \tfrac{1}{2})|n\rangle = \hbar\omega\left(n + \tfrac{1}{2}\right)|n\rangle = E_n|n\rangle \qquad n = 0, 1, 2, \ldots \tag{7.28}$$

The energy of the harmonic oscillator is thus quantized, taking on only discrete values. This characteristic energy spectrum of the harmonic oscillator is shown in Fig. 7.3. Notice that there is a uniform spacing between levels.

## 7.3 AN EXAMPLE: TORSIONAL OSCILLATIONS OF THE ETHYLENE MOLECULE

An interesting system that exhibits the energy spectrum of the harmonic oscillator is the torsional oscillation of a molecule like the ethylene molecule, $C_2H_4$. In the ground state, the six atoms of this molecule all lie in a plane, as shown in Fig. 7.4. The angle between each of the adjacent C—H and C—C bonds is roughly 120°. It is possible to rotate one of the $CH_2$ groups with respect to the other by an angle $\phi$ about the C—C axis, as shown in Fig. 7.5a. If we rotate by an angle $\phi = \pi$, the molecule returns to a configuration that is indistinguishable from the

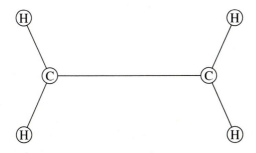

**FIGURE 7.4**
The planar structure of the ethylene molecule $C_2H_4$.

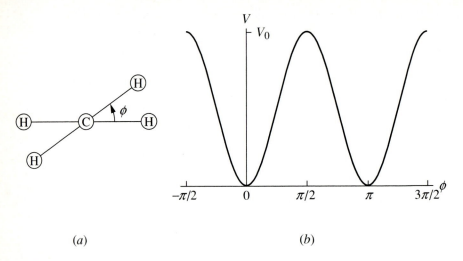

(a)                                                              (b)

**FIGURE 7.5**
(a) A view along the C—C axis of the $C_2H_4$ molecule showing one of the $CH_2$ groups rotated relative to the other by angle $\phi$. (b) The potential energy of the molecule as a function of $\phi$.

$\phi = 0$ configuration. Thus $\phi = \pi$, as well as $\phi = 0$, must be a minimum of the potential energy of orientation. These two minima are separated by a potential barrier, as shown in Fig. 7.5b. A simple approximation for this potential energy function is given by

$$V(\phi) = \frac{V_0}{2}(1 - \cos 2\phi) \tag{7.29}$$

In the vicinity of each of the minima, the system behaves like a harmonic os-

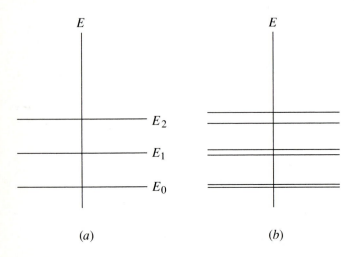

(a)                                                              (b)

**FIGURE 7.6**
(a) The three lowest energy levels of the $C_2H_4$ molecule, neglecting tunneling. (b) The energy spectrum with tunneling taken into account.

cillator. However, since the potential barrier between these configurations is not infinite, the $CH_2$ group can tunnel between them. As in the ammonia molecule, this tunneling causes each of the two-fold degenerate energy levels to split into two distinct energy levels, with a small spacing between them proportional to the tunneling amplitude. Since both the distance in energy below the top of the barrier and the width of the barrier decrease as the energy of the system increases, the magnitude of this splitting grows [see (6.158)] as the quantum number $n$ increases (see Fig. 7.6).

## 7.4 MATRIX ELEMENTS OF THE RAISING AND LOWERING OPERATORS

It is useful for us to determine the constants $c_+$ in (7.22) and $c_-$ in (7.24). For example, the bra equation corresponding to the ket equation

$$\hat{a}^\dagger|n\rangle = c_+|n + 1\rangle \tag{7.30a}$$

is given by

$$\langle n|\hat{a} = c_+^*\langle n + 1| \tag{7.30b}$$

Taking the inner product of these equations, we obtain

$$\langle n|\hat{a}\hat{a}^\dagger|n\rangle = \langle n|(\hat{a}^\dagger\hat{a} + 1)|n\rangle = (n + 1)\langle n|n\rangle$$

$$= c_+^* c_+\langle n + 1|n + 1\rangle \tag{7.31}$$

If the eigenstates are normalized, that is, they satisfy $\langle n|n\rangle = 1$ for all $n$, we can choose $c_+ = \sqrt{n + 1}$, or

$$\hat{a}^\dagger|n\rangle = \sqrt{n + 1}|n + 1\rangle \tag{7.32}$$

Similarly, we can establish that

$$\hat{a}|n\rangle = \sqrt{n}|n - 1\rangle \tag{7.33}$$

Thus the matrix elements of the raising and lowering operators are given by

$$\langle n'|\hat{a}^\dagger|n\rangle = \sqrt{n + 1}\delta_{n',n+1} \tag{7.34}$$

and

$$\langle n'|\hat{a}|n\rangle = \sqrt{n}\delta_{n',n-1} \tag{7.35}$$

The matrix representations of the raising and lowering operators using the energy eigenstates as a basis are then given by the infinite-dimensional matrices

$$\hat{a}^\dagger \rightarrow \begin{pmatrix} 0 & 0 & 0 & \cdots \\ \sqrt{1} & 0 & 0 & \\ 0 & \sqrt{2} & 0 & \\ 0 & 0 & \sqrt{3} & \\ & \vdots & & \end{pmatrix} \tag{7.36}$$

$$\hat{a} \rightarrow \begin{pmatrix} 0 & \sqrt{1} & 0 & 0 & \cdots \\ 0 & 0 & \sqrt{2} & 0 & \\ 0 & 0 & 0 & \sqrt{3} & \\ \vdots & & & & \end{pmatrix} \tag{7.37}$$

It is straightforward to construct the matrix representations of the position and the momentum operators using (7.36) and (7.37).

We can also establish (see Problem 7.3) that a normalized ket $|n\rangle$ can be expressed as

$$|n\rangle = \frac{(\hat{a}^{\dagger})^n}{\sqrt{n!}}|0\rangle \tag{7.38}$$

Finally, notice that increasing the energy of the harmonic oscillator from $E_n$ to $E_{n+1}$ requires the addition of energy $\hbar\omega$ to the oscillator. In Chapter 14 we will see that the electromagnetic field is composed of abstract harmonic oscillators. In that case the natural interpretation is that the state with energy $E_n$ is composed of $n$ photons and the state with energy $E_{n+1}$ is composed of $n + 1$ photons. The additional energy $\hbar\omega$ is exactly the quantum of energy that we expect for a photon with angular frequency $\omega$. For photons, instead of referring to $\hat{a}^{\dagger}$ as a raising operator, we will call it a *creation operator*. Similarly, the operator $\hat{a}$ will be referred to as an *annihilation operator*, since when $\hat{a}$ acts on a state $|n\rangle$, it decreases the number of quanta in the state from $n$ to $n - 1$.

## 7.5 POSITION-SPACE WAVE FUNCTIONS

It might appear that we are far removed from the wave functions of wave mechanics, but in fact we can obtain the position-space (and momentum-space) energy eigenfunctions for the harmonic oscillator easily from the results that we have obtained so far. We start with the ground state. The ground-state ket satisfies

$$\hat{a}|0\rangle = 0 \tag{7.39}$$

Projecting this equation into position space, we obtain

$$\langle x|\hat{a}|0\rangle = \sqrt{\frac{m\omega}{2\hbar}}\langle x|\left(\hat{x} + \frac{i}{m\omega}\hat{p}_x\right)|0\rangle = 0 \tag{7.40}$$

Recall from (6.42) that

$$\langle x|\hat{p}_x|0\rangle = \frac{\hbar}{i}\frac{\partial\langle x|0\rangle}{\partial x} \tag{7.41}$$

where $\langle x|0\rangle$ is the amplitude to find a particle in the ground state with position $x$. Also

$$\langle x|\hat{x}|0\rangle = x\langle x|0\rangle \tag{7.42}$$

where we have allowed the Hermitian operator $\hat{x}$ to act to the left on its eigenbra. Thus (7.40) reduces to the first-order differential equation[2]

$$\frac{d\langle x|0\rangle}{dx} = -\frac{m\omega x}{\hbar}\langle x|0\rangle \tag{7.43}$$

which is easily solved:

$$\langle x|0\rangle = Ne^{-m\omega x^2/2\hbar} \tag{7.44a}$$

Normalizing [see (6.61)], we obtain

$$\langle x|0\rangle = \left(\frac{m\omega}{\pi\hbar}\right)^{1/4} e^{-m\omega x^2/2\hbar} \tag{7.44b}$$

Once we have determined the ground-state wave function, we can take advantage of (7.38) to determine all of the position-space energy eigenfunctions:

$$\langle x|n\rangle = \frac{1}{\sqrt{n!}}\langle x|(a^\dagger)^n|0\rangle$$

$$= \frac{1}{\sqrt{n!}}\left(\sqrt{\frac{m\omega}{2\hbar}}\right)^n \left(x - \frac{\hbar}{m\omega}\frac{d}{dx}\right)^n \left(\frac{m\omega}{\pi\hbar}\right)^{1/4} e^{-m\omega x^2/2\hbar} \tag{7.45}$$

For example,

$$\langle x|1\rangle = \left[\frac{4}{\pi}\left(\frac{m\omega}{\hbar}\right)^3\right]^{1/4} x e^{-m\omega x^2/2\hbar} \tag{7.46}$$

$$\langle x|2\rangle = \left(\frac{m\omega}{4\pi\hbar}\right)^{1/4}\left(2\frac{m\omega}{\hbar}x^2 - 1\right) e^{-m\omega x^2/2\hbar} \tag{7.47}$$

The energy eigenfunctions and the corresponding position-space probability densities $|\langle x|n\rangle|^2$ for these states, as well as those with $n = 3$, 4, and 5, are plotted in Fig. 7.7.

These eigenfunctions exhibit a number of properties worthy of note. The number of nodes, or zeros, of $\langle x|n\rangle$ is $n$. The increasingly oscillatory character of the functions as $n$ increases reflects the increasing kinetic energy of each of these states. The expectation value of the kinetic energy is given by

$$\frac{1}{2m}\langle p_x^2\rangle = -\frac{\hbar^2}{2m}\int_{-\infty}^{\infty} dx\,\langle n|x\rangle\frac{d^2}{dx^2}\langle x|n\rangle \tag{7.48}$$

---

[2] Since $\langle x|0\rangle$ is a function of $x$ only, we can replace the partial derivative with a total derivative.

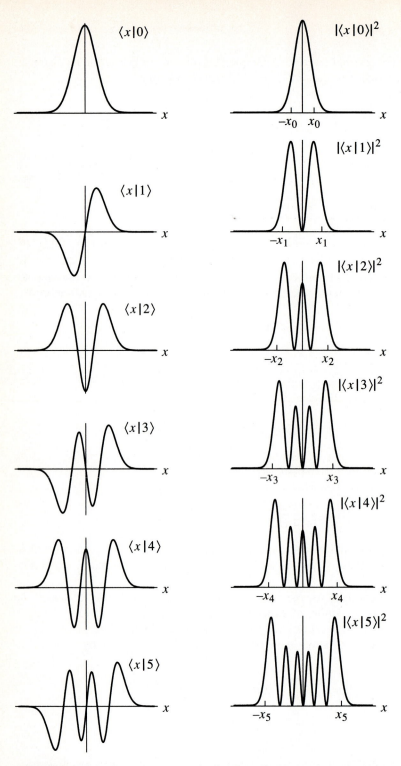

**FIGURE 7.7**
The wave functions $\langle x|n \rangle$ and the probability densities $|\langle x|n \rangle|^2$ plotted for the first six energy eigenstates of the harmonic oscillator. The classical turning points at $x_n = \sqrt{(2n + 1)(\hbar/m\omega)}$ are determined from (7.59).

As the number of nodes of $\langle x|n \rangle$ increases, the curvature of the eigenfunction increases and hence the second derivative in (7.48) takes on larger values. The expectation value of the potential energy

$$\langle V(x) \rangle = \tfrac{1}{2}m\omega^2 \int_{-\infty}^{\infty} dx \, \langle n|x \rangle x^2 \langle x|n \rangle = \tfrac{1}{2}m\omega^2 \int_{-\infty}^{\infty} dx \, x^2 |\langle x|n \rangle|^2 \qquad (7.49)$$

also increases with increasing $n$ as the region over which the eigenfunction is appreciable increases.

## 7.6 THE ZERO-POINT ENERGY

One of the most striking features of the harmonic oscillator is the existence of a nonzero ground-state energy $E_0 = \hbar\omega/2$, known as the zero-point energy. In classical mechanics the lowest-energy state occurs when the particle is at rest ($p_x = 0$ and hence zero kinetic energy) with the spring unstretched ($x = 0$ and hence zero potential energy). In the real world this configuration is forbidden by the Heisenberg uncertainty relation, which enters into the solution of the harmonic oscillator through the commutation relation (7.7). The particle in the ground state, and in fact in any of the eigenstates, has a nonzero position uncertainty $\Delta x$ ($\Delta x^2 = \langle x^2 \rangle - \langle x \rangle^2$) as well as a nonzero momentum uncertainty $\Delta p_x$ ($\Delta p_x^2 = \langle p_x^2 \rangle - \langle p_x \rangle^2$). It is straightforward to see how these uncertainties affect the value of the ground-state energy. For any state

$$\langle E \rangle = \frac{\langle p_x^2 \rangle}{2m} + \frac{1}{2}m\omega^2 \langle x^2 \rangle = \frac{\Delta p_x^2 + \langle p_x \rangle^2}{2m} + \frac{1}{2}m\omega^2(\Delta x^2 + \langle x \rangle^2) \qquad (7.50)$$

There are a number of ways to establish that $\langle x \rangle$ and $\langle p_x \rangle$ both vanish in an energy eigenstate of the harmonic oscillator. One way is through explicit evaluation:

$$\langle n|\hat{x}|n \rangle = \sqrt{\frac{\hbar}{2m\omega}} \langle n|(\hat{a} + \hat{a}^\dagger)|n \rangle$$

$$= \sqrt{\frac{\hbar}{2m\omega}} \left( \sqrt{n}\langle n|n-1 \rangle + \sqrt{n+1}\langle n|n+1 \rangle \right) = 0 \qquad (7.51)$$

$$\langle n|\hat{p}_x|n \rangle = -i\sqrt{\frac{m\omega\hbar}{2}} \langle n|(\hat{a} - \hat{a}^\dagger)|n \rangle$$

$$= -i\sqrt{\frac{m\omega\hbar}{2}} \left( \sqrt{n}\langle n|n-1 \rangle - \sqrt{n+1}\langle n|n+1 \rangle \right) = 0 \qquad (7.52)$$

Thus in an energy eigenstate

$$\langle E \rangle = \frac{\Delta p_x^2}{2m} + \frac{1}{2}m\omega^2 \Delta x^2 \qquad (7.53)$$

How does nature keep the ground-state energy as small as possible? Clearly, localizing the particle at the origin and minimizing the potential energy will not work because as $\Delta x \to 0$, $\Delta p_x \to \infty$ in order to satisfy $\Delta x \Delta p_x \geq \hbar/2$. Similarly, trying to put the particle in a state with zero momentum to minimize the kinetic energy implies $\Delta p_x \to 0$, which forces $\Delta x \to \infty$. Thus nature must choose a tradeoff in which the particle has both nonzero $\Delta x$ and $\Delta p_x$ and, therefore, nonzero energy. Explicitly, for the ground state

$$
\begin{aligned}
\Delta x^2 &= \frac{\hbar}{2m\omega}\langle 0|(\hat{a} + \hat{a}^\dagger)^2|0\rangle \\[2mm]
&= \frac{\hbar}{2m\omega}\langle 0|[\hat{a}^2 + (\hat{a}^\dagger)^2 + \hat{a}\hat{a}^\dagger + \hat{a}^\dagger\hat{a}]|0\rangle \\[2mm]
&= \frac{\hbar}{2m\omega}\langle 0|\hat{a}\hat{a}^\dagger|0\rangle = \frac{\hbar}{2m\omega}\langle 1|1\rangle = \frac{\hbar}{2m\omega}
\end{aligned}
\tag{7.54}
$$

and

$$
\begin{aligned}
\Delta p_x^2 &= -\frac{m\omega\hbar}{2}\langle 0|(\hat{a} - \hat{a}^\dagger)^2|0\rangle \\[2mm]
&= -\frac{m\omega\hbar}{2}\langle 0|[\hat{a}^2 + (\hat{a}^\dagger)^2 - \hat{a}\hat{a}^\dagger - \hat{a}^\dagger\hat{a}]|0\rangle \\[2mm]
&= \frac{m\omega\hbar}{2}\langle 0|\hat{a}\hat{a}^\dagger|0\rangle = \frac{m\omega\hbar}{2}\langle 1|1\rangle = \frac{m\omega\hbar}{2}
\end{aligned}
\tag{7.55}
$$

Notice that $\Delta x \Delta p_x = \hbar/2$ for the ground state. That the ground state is a minimum uncertainty state was already apparent from the Gaussian form of the ground-state wave function (7.44), given the discussion in Section 6.6. For the excited states, we can establish in a similar fashion that

$$
\Delta x = \sqrt{\left(n + \tfrac{1}{2}\right)\frac{\hbar}{m\omega}}
\tag{7.56}
$$

$$
\Delta p_x = \sqrt{\left(n + \tfrac{1}{2}\right)m\omega\hbar}
\tag{7.57}
$$

and

$$
\Delta x \Delta p_x = \left(n + \tfrac{1}{2}\right)\hbar
\tag{7.58}
$$

A good illustration of the effects of this zero-point energy is the unusual behavior of helium. Helium is the only substance that does not solidify at sufficiently low temperatures at atmospheric pressure. Rather, it is necessary to apply a pressure of at least 25 atmospheres. For substances other than helium, the uncertainty in the position of the nuclei in the ground state is in general quite small compared to the spacing between the nuclei, which is why these substances solidify at atmospheric pressure at sufficiently low temperature. In fact, increasing the temperature populates the higher vibrational states and increases the uncertainty, as (7.56) indicates. These substances melt when the uncertainty becomes comparable to the spacing between the nuclei in the solid. For helium, even in the ground state, the uncertainty is large because of two factors: the small mass of helium and the small value of $\omega$ (because of weak attraction between the helium atoms in the solid). Thus $\Delta x$ is too large for helium to solidify at low temperature at a pressure of one atmosphere. Increasing the pressure reduces the separation between the helium atoms, thereby increasing $\omega$ and reducing $\Delta x$, so that at high pressure helium solidifies.

## 7.7 THE CLASSICAL LIMIT

The existence of a zero-point energy and the *discrete* energy spectrum (7.28) of the harmonic oscillator are purely quantum phenomena. Why don't we notice this discreteness in a macroscopic oscillator such as the pendulum of Section 7.1? The answer resides in the smallness of Planck's constant on a macroscopic scale. For example, the angular frequency of the pendulum is $\omega = \sqrt{g/L}$. Thus if $L = 10$ cm, $\omega$ is about 10 radian/s and the spacing $\hbar\omega$ between energy levels is $10^{-26}$ ergs. If the energy $E$ of the pendulum is a typical macroscopic value, such as 1 erg, then $\hbar\omega/E = 10^{-26}$ and the system appears to have a continuous energy spectrum, which is what we would expect classically. Note in this case the quantum number $n = 10^{26}$. This suggests that the classical limit is indeed reached in the large $n$ limit.

The classical motion of a particle in a state of definite energy $E_n$ is restricted to lie within the classical turning points, which are determined by the condition that at these points all the energy is potential energy, with zero kinetic energy:

$$E_n = \left(n + \tfrac{1}{2}\right)\hbar\omega = \tfrac{1}{2}m\omega^2 x_n^2 \tag{7.59}$$

as shown in Fig. 7.8. Examination of the energy eigenfunctions in Fig. 7.7 shows that the eigenfunctions extend beyond these classical turning points, but that these

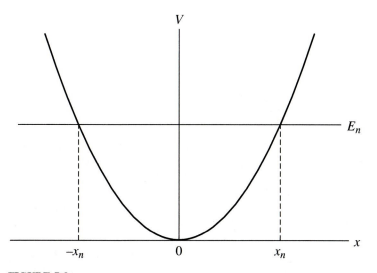

**FIGURE 7.8**
The classical turning points for an oscillator with energy $E_n$.

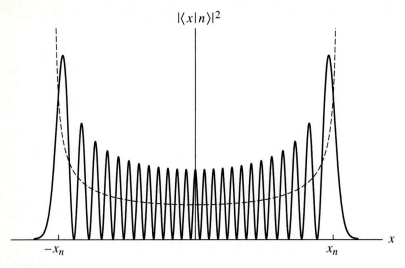

**FIGURE 7.9**

A plot of the probability density $|\langle x|n \rangle|^2$ for large $n$. The dashed line is a plot of the classical probability density from (7.61).

excursions become less pronounced as $n$ increases. Here again, we see the classical limit being reached in the limit of large $n$. The chance of finding a particle between $x$ and $x + dx$ for a classical oscillator with energy $E_n$ is proportional to the time $dx/v$ that it spends in the interval $dx$, where $v$ is the speed of the particle. Taking advantage of (7.59), we may express this classical probability as

$$P_{\text{cl}}\, dx \propto \frac{dx}{\sqrt{2E_n/m - \omega^2 x^2}} = \frac{dx}{\omega\sqrt{x_n^2 - x^2}} \tag{7.60}$$

Requiring that the total probability of finding the particle between $+x_n$ and $-x_n$ is unity determines the normalization:

$$P_{\text{cl}}\, dx = \frac{1}{\sqrt{\pi}} \frac{dx}{\omega\sqrt{x_n^2 - x^2}} \tag{7.61}$$

The probability density $|\langle x|n \rangle|^2$ for large $n$, as well as $P_{\text{cl}}$, is plotted in Fig. 7.9. As $n$ increases, the number of nodes of the wave function increases. For sufficiently large $n$, the quantum state is oscillating so rapidly on a macroscopic scale that only its mean value can be detected by any set of position measurements. In this case, the agreement between the predictions of quantum mechanics and classical mechanics is excellent. You can guess how good the agreement is for the pendulum example with $n = 10^{26}$.

## 7.8 TIME DEPENDENCE

A harmonic oscillator in an energy eigenstate is in a stationary state. Thus it will not exhibit the characteristic oscillatory behavior of a classical oscillator.

Time dependence for the harmonic oscillator results from the system being in a superposition of energy eigenstates with different energies. If we assume the initial state is a superposition of two *adjacent* energy states,

$$|\psi(0)\rangle = c_n|n\rangle + c_{n+1}|n + 1\rangle \tag{7.62}$$

then

$$|\psi(t)\rangle = e^{-i\hat{H}t/\hbar}|\psi(0)\rangle$$

$$= e^{-i(n+1/2)\omega t}\left(c_n|n\rangle + c_{n+1}e^{-i\omega t}|n + 1\rangle\right) \tag{7.63}$$

In particular, we can take advantage of the expression (7.11) for the position operator in terms of the raising and lowering operators to evaluate the expectation value of the position of the particle and show that it behaves as one would expect for a classical particle, namely,

$$\langle x \rangle = \langle\psi|\hat{x}|\psi\rangle = A\cos(\omega t + \delta) \tag{7.64}$$

See Problem 7.8.

## 7.9 SOLVING THE SCHRÖDINGER EQUATION IN POSITION SPACE

There is another technique for determining the energy eigenvalues and the position-space eigenfunctions of the harmonic oscillator that we will find particularly useful when we solve the three-dimensional Schrödinger equation in Chapter 10. Rather than take advantage of the operator techniques of Section 7.2, we solve the energy eigenvalue equation

$$\langle x|\hat{H}|E\rangle = \langle x|\left(\frac{\hat{p}_x^2}{2m} + \frac{1}{2}m\omega^2\hat{x}^2\right)|E\rangle = E\langle x|E\rangle \tag{7.65}$$

directly in position space, as in Chapter 6. Using the results of that chapter, we can express this equation as

$$-\frac{\hbar^2}{2m}\frac{d^2}{dx^2}\langle x|E\rangle + \frac{1}{2}m\omega^2 x^2\langle x|E\rangle = E\langle x|E\rangle \tag{7.66}$$

The position-space energy eigenvalue equation (7.66) is a nontrivial second-order differential equation. To make its structure a little more apparent, it is good to introduce the dimensionless variable

$$y = \sqrt{\frac{m\omega}{\hbar}}x \tag{7.67}$$

where the factor $\sqrt{m\omega/\hbar}$ is a factor with the dimensions of inverse length that occurs naturally in the problem. We call the wave function

$$\langle x|E\rangle = \psi(y) \tag{7.68}$$

where the energy eigenvalue $E$ is implicit on the right-hand side. Expressed in terms of these variables, the differential equation (7.66) becomes

$$\frac{d^2\psi}{dy^2} + (\varepsilon - y^2)\psi = 0 \tag{7.69}$$

where $\varepsilon = 2E/\hbar\omega$ is a dimensionless constant. An initial attempt at a power series solution to this differential equation leads to a three-term recursion relation, from which it is difficult to extract the physically acceptable solutions. A good procedure for resolving this difficulty is to explicitly factor out the behavior of the wave function as $|y| \to \infty$. In this limit we can neglect the term involving $\varepsilon$, and the differential equation becomes

$$\frac{d^2\psi}{dy^2} - y^2\psi = 0 \tag{7.70}$$

The solution to this equation is

$$\psi = Ae^{-y^2/2} + Be^{y^2/2} \tag{7.71}$$

We immediately discard the exponentially increasing solution as $|y| \to \infty$ because we are searching for a normalizable state satisfying $\langle\psi|\psi\rangle = 1$. In fact, in the limit of large $y$, we can take any power of $y$ times the decreasing exponential as an asymptotic solution of (7.70):

$$\frac{d^2}{dy^2}(Ay^m e^{-y^2/2}) = Ay^{m+2}\left[1 - \frac{2m+1}{y^2} + \frac{m(m-1)}{y^4}\right]e^{-y^2/2}$$

$$\xrightarrow[|y|\to\infty]{} Ay^{m+2}e^{-y^2/2} = y^2(Ay^m e^{-y^2/2}) \tag{7.72}$$

With this in mind, we express the wave function in the form

$$\psi(y) = h(y)e^{-y^2/2} \tag{7.73}$$

If we substitute (7.73) into (7.69), we find that $h(y)$ satisfies the differential equation

$$\frac{d^2h}{dy^2} - 2y\frac{dh}{dy} + (\varepsilon - 1)h = 0 \tag{7.74}$$

It should be stressed that we have not made any approximations in arriving at (7.74). We can think of (7.73) as just a definition of the function $h$. Although this equation for $h$ does not look any simpler than equation (7.69) for $\psi$, we can now obtain a power series solution of the form

$$h(y) = \sum_{k=0}^{\infty} a_k y^k \tag{7.75}$$

to this equation. To see this, we substitute (7.75) into (7.74) to obtain

$$\sum_{k=2}^{\infty} k(k-1)a_k y^{k-2} - 2\sum_{k=1}^{\infty} k a_k y^k + (\varepsilon-1)\sum_{k=0}^{\infty} a_k y^k = 0 \tag{7.76}$$

Letting $k - 2 = k'$ in the first term, we obtain

$$\sum_{k'=0}^{\infty} (k' + 2)(k' + 1)a_{k'+2}y^{k'} \tag{7.77}$$

Renaming the dummy summation variable $k$ instead of $k'$ in (7.77) allows us to write (7.76) as

$$\sum_{k=0}^{\infty} [(k + 2)(k + 1)a_{k+2} - 2ka_k + (\varepsilon - 1)a_k]y^k = 0 \tag{7.78}$$

Since the functions $y^k$ are linearly independent, the only way (7.78) can be satisfied is for the coefficient of each $y^k$ to vanish. Thus we obtain a *two-term* recursion relation:

$$\frac{a_{k+2}}{a_k} = \frac{2k + 1 - \varepsilon}{(k + 2)(k + 1)} \tag{7.79}$$

In general, this is an infinite power series, which for large $k$ behaves as

$$\frac{a_{k+2}}{a_k} \xrightarrow[k \to \infty]{} \frac{2}{k} \tag{7.80}$$

This is the same behavior that the function

$$e^{y^2} = \sum_{n=0}^{\infty} \frac{y^{2n}}{n!} = \sum_{k=0}^{\infty} b_k y^k \tag{7.81}$$

exhibits, since for this function $b_k = 1/(k/2)!$ and thus

$$\frac{b_{k+2}}{b_k} = \frac{(k/2)!}{[(k/2) + 1]!} = \frac{1}{(k/2) + 1} \xrightarrow[k \to \infty]{} \frac{2}{k} \qquad k = 0, 2, 4, \dots \tag{7.82}$$

In fact, this same asymptotic behavior is exhibited by any power of $y$ times $\exp(y^2)$. Since the large-$y$ behavior of these series is determined by the behavior for large $k$, the series solution for $h(y)$ generates the leading large-$y$ behavior

$$\psi \xrightarrow[|y| \to \infty]{} Ay^m e^{y^2} e^{-y^2/2} = Ay^m e^{y^2/2} \tag{7.83}$$

that we tried to discard when we attempted to find a solution of the form (7.73).

Are there, therefore, no solutions to the differential equation that are exponentially damped and hence satisfy the normalization requirement? The only way to evade (7.80) is for the series to terminate. Then $\psi$ is a *finite* polynomial in $y$ multiplied by the decreasing exponential. But from (7.79), we see that the series terminates when $k = n$, provided that

$$\varepsilon = 2n + 1 \qquad n = 0, 1, 2, \dots \tag{7.84}$$

and consequently

$$E_n = \left(n + \tfrac{1}{2}\right)\hbar\omega \qquad n = 0, 1, 2, \dots \tag{7.85}$$

This is the same result that we obtained earlier using operator methods.

The function $h(y)$ is thus a polynomial of order $n$, called a Hermite polynomial. We can determine the form of these polynomials either from the power series solution (7.79) or from our earlier result (7.45). The first three Hermite polynomials can be seen in the energy eigenfunctions (7.44), (7.46), and (7.47).

## 7.10 INVERSION SYMMETRY AND THE PARITY OPERATOR

One of the most obvious features of the energy eigenfunctions shown in Fig. 7.7 is that they are all either even functions satisfying $\psi(-x) = \psi(x)$ or odd functions satisfying $\psi(-x) = -\psi(x)$. The cause of this behavior is a symmetry in the Hamiltonian. We introduce the parity operator $\hat{\Pi}$, whose action on the position states is given by

$$\hat{\Pi}|x\rangle = |-x\rangle \tag{7.86}$$

The parity operator inverts states through the origin. An eigenstate of the parity operator satisfies

$$\hat{\Pi}|\psi_\lambda\rangle = \lambda|\psi_\lambda\rangle \tag{7.87}$$

Since inverting twice is the identity operator, we see that

$$\hat{\Pi}^2|\psi_\lambda\rangle = \lambda^2|\psi_\lambda\rangle = |\psi_\lambda\rangle \tag{7.88}$$

or $\lambda^2 = 1$. Thus the eigenvalues of the parity operator are $\lambda = \pm 1$.

We can evaluate the action of the parity operator on an arbitrary state $|\psi\rangle$ by projecting into position space:[3]

$$\langle x|\hat{\Pi}|\psi\rangle = \langle -x|\psi\rangle = \psi(-x) \tag{7.89}$$

Thus a parity eigenstate satisfies

$$\langle x|\hat{\Pi}|\psi_\lambda\rangle = \langle -x|\psi_\lambda\rangle = \psi_\lambda(-x) = \lambda\psi_\lambda(x) \tag{7.90}$$

where in the last step we have evaluated the action of the parity operator acting to the right on its eigenket. Thus if $\lambda = 1$, the eigenfunction is an even function of $x$, and if $\lambda = -1$, the eigenfunction is an odd function of $x$.

It is now easy to see for the harmonic oscillator that the parity operator and the Hamiltonian commute. Note that

$$\langle x|\hat{\Pi}\hat{H}|\psi\rangle = \langle -x|\hat{H}|\psi\rangle = \left(-\frac{\hbar^2}{2m}\frac{d^2}{dx^2} + V(-x)\right)\psi(-x)$$

$$= \left(-\frac{\hbar^2}{2m}\frac{d^2}{dx^2} + V(x)\right)\psi(-x) = \langle x|\hat{H}\hat{\Pi}|\psi\rangle \tag{7.91}$$

---

[3] The parity operator is Hermitian. See Problem 7.10.

provided $V(x) = V(-x)$. Thus for the harmonic oscillator, where $V(x) = m\omega^2 x^2/2$, we deduce that $\hat{\Pi}\hat{H} = \hat{H}\hat{\Pi}$, or

$$[\hat{\Pi}, \hat{H}] = 0 \tag{7.92}$$

This guarantees that the Hamiltonian and the parity operator have eigenstates in common, as we have seen.

The real advantage of this symmetry approach is that by observing the symmetry of the Hamiltonian under inversion, we can deduce some of the properties of the eigenstates—in this case, their evenness or oddness—*before* rather than *after* we have solved the eigenvalue equation. We will see the utility of this approach in Chapter 9, when we consider other symmetries of the Hamiltonian in three dimensions.

## 7.11  SUMMARY

The harmonic oscillator deserves a chapter all its own. In addition to the fact that an arbitrary potential energy function in the vicinity of its minimum resembles a harmonic oscillator (see Section 7.1), the harmonic oscillator is a nontrivial problem in one-dimensional wave mechanics with a nice exact solution (see Section 7.9). Moreover, the harmonic oscillator will also serve as the foundation of our approach to the quantum theory of the electromagnetic field in Chapter 14. One of the underlying reasons for such a broad significance of the harmonic oscillator is that we can determine the eigenstates and eigenvalues of the Hamiltonian

$$\hat{H} = \frac{\hat{p}_x^2}{2m} + \frac{1}{2}m\omega^2 \hat{x}^2 \tag{7.93}$$

in a completely representation-free way. We introduce the lowering and raising operators

$$\hat{a} = \sqrt{\frac{m\omega}{2\hbar}}\left(\hat{x} + \frac{i}{m\omega}\hat{p}_x\right) \tag{7.94}$$

and

$$\hat{a}^\dagger = \sqrt{\frac{m\omega}{2\hbar}}\left(\hat{x} - \frac{i}{m\omega}\hat{p}_x\right) \tag{7.95}$$

The position and momentum operators are then written in terms of the raising and lowering operators as

$$\hat{x} = \sqrt{\frac{\hbar}{2m\omega}}(\hat{a} + \hat{a}^\dagger) \tag{7.96}$$

and

$$\hat{p}_x = -i\sqrt{\frac{m\omega\hbar}{2}}(\hat{a} - \hat{a}^\dagger) \tag{7.97}$$

Using (7.96) and (7.97) as well as the commutation relation

$$[\hat{a}, \hat{a}^\dagger] = 1 \tag{7.98}$$

which follows from the commutation relation between the position and momentum operators, we can express the Hamiltonian in the form

$$\hat{H} = \hbar\omega\left(\hat{a}^\dagger \hat{a} + \tfrac{1}{2}\right) \tag{7.99}$$

The eigenstates of the Hamiltonian satisfy

$$\hat{H}|n\rangle = \left(n + \tfrac{1}{2}\right)\hbar\omega|n\rangle \qquad n = 0, 1, 2, \ldots \tag{7.100}$$

where the state $|n\rangle$ is obtained by letting the raising operator act $n$ times on the lowest-energy state $|0\rangle$:

$$|n\rangle = \frac{1}{\sqrt{n!}}(\hat{a}^\dagger)^n|0\rangle \tag{7.101}$$

The action of the raising and lowering operators on the energy eigenstates is given by

$$\hat{a}^\dagger|n\rangle = \sqrt{n+1}|n+1\rangle \tag{7.102}$$

and

$$\hat{a}|n\rangle = \sqrt{n}|n-1\rangle \tag{7.103}$$

which again follow from the commutation relation (7.98). These raising and lowering operators provide a powerful way to evaluate expectation values and matrix elements of the position and momentum operators (7.96) and (7.97), without having to work directly with wave functions either in position or momentum space. Since each increase or decrease in $n$ by a unit increases or decreases the energy of the oscillator by $\hbar\omega$, we can think of the oscillator as containing $n$ quanta of energy $\hbar\omega$, in addition to the zero-point energy $\hbar\omega/2$. The operator $\hat{a}^\dagger$ creates a quantum of energy and can hence be called a creation operator, while the operator $\hat{a}$ annihilates a quantum of energy and is called an annihilation operator.

## PROBLEMS

**7.1.** Show that the constant $c_- = \sqrt{n}$ in (7.24), that is,

$$\hat{a}\,|n\rangle = \sqrt{n}|n-1\rangle$$

using the procedure we used to establish that $c_+ = \sqrt{n+1}$, that is,

$$\hat{a}^\dagger|n\rangle = \sqrt{n+1}|n+1\rangle$$

**7.2.** Use the matrix representations (7.36) and (7.37) of the raising and lowering operators, respectively, to determine the matrix representations of the position and the momentum operators using the energy eigenstates as a basis. Verify using these matrix representations that the position-momentum commutation relation (7.7) is satisfied.

**7.3.** Show that properly normalized eigenstates of the harmonic oscillator are given by (7.38). *Suggestion:* Use induction.

**7.4.** Use $\hat{a}|0\rangle = 0$ and therefore $\langle p|\hat{a}|0\rangle = 0$ to solve directly for $\langle p|0\rangle$, the ground-state wave function of the harmonic oscillator in momentum space. Normalize the wave function. *Hint:* Recall the result of Problem 6.2,

$$\langle p|\hat{x}|\psi\rangle = i\hbar\frac{\partial}{\partial p}\langle p|\psi\rangle$$

**7.5.** Derive (7.56) and (7.57).

**7.6.** A particle of mass $m$ in the one-dimensional harmonic oscillator is in a state for which a measurement of the energy yields the values $\hbar\omega/2$ or $3\hbar\omega/2$, each with a probability of one-half. The average value of the momentum $\langle p_x\rangle$ at time $t = 0$ is $(m\omega\hbar/2)^{1/2}$. This information specifies the state of the particle completely. What is this state and what is $\langle p_x\rangle$ at time $t$?

**7.7.** (a) Determine the size of the classical turning point $x_0$ for a harmonic oscillator in its ground state with a mass of 1000 kg and a frequency of 1000 Hz. Compare your result with the size of a proton. A bar of aluminum of roughly this mass and tuned to roughly this frequency (called a Weber bar) is used in attempts to detect gravity waves.

(b) Suppose that the bar absorbs energy in the form of a graviton and makes a transition from a state with energy $E_n$ to a state with energy $E_{n+1}$. Show that the change in length of such a bar is given approximately by $x_0(2/n)^{1/2}$ for large $n$.

(c) To what $n$, on the average, is the oscillator excited by thermal energy if the bar is cooled to 1 K?

**7.8.** Show that in the superposition of adjacent energy states (7.63) the average value of the position of the particle is given by

$$\langle x\rangle = \langle\psi|\hat{x}|\psi\rangle = A\cos(\omega t + \delta)$$

and the average value of the momentum is given by

$$\langle p_x\rangle = \langle\psi|\hat{p}_x|\psi\rangle = -m\omega A\sin(\omega t + \delta)$$

in accord with Ehrenfest's theorem, (6.33) and (6.34).

**7.9.** A small cylindrical tube is drilled through the earth, passing through the center. A mass $m$ is released essentially at rest at the surface. Assuming the density of the earth is uniform, show that the mass executes simple harmonic motion and determine the frequency $\omega$. Determine the approximate quantum number $n$ for this state of the mass, using a typical macroscopic value for the magnitude of the mass $m$. Explain why a *single* quantum number $n$ is inadequate to specify the state.

**7.10.** Prove that the parity operator $\hat{\Pi}$ is Hermitian.

**7.11.** Substitute $\psi(x) = Ne^{-ax^2}$ into the position-space energy eigenvalue equation (7.66) and determine the value of the constant $a$ that makes this function an eigenfunction. What is the corresponding energy eigenvalue?

**7.12.** Calculate the probability that a particle in the ground state of the harmonic oscillator is located in a classically disallowed region, namely, where $V(x) > E$. Obtain a numerical value for the probability. *Suggestion:* Express your integral in terms of a dimensionless variable and compare with the tabulated values of the error function.

# CHAPTER
# 8

# PATH
# INTEGRALS

Our discussion of time evolution has emphasized the importance of the Hamiltonian as the generator of time translations. In the 1940s R. P. Feynman discovered a way to express quantum dynamics in terms of the Lagrangian instead of the Hamiltonian. His path-integral formulation of quantum mechanics provides us with a great deal of insight into quantum dynamics, which alone makes it worthy of study. The computational complexity of using this formulation for most problems in nonrelativistic quantum mechanics is sufficiently high, however, that the path-integral method remained something of a curiosity until more recently, when it was realized that it also provides an excellent approach to quantizing a relativistic system with an infinite number of degrees of freedom, a quantum field.

## 8.1 THE MULTISLIT, MULTISCREEN EXPERIMENT

We can get the spirit of the path-integral approach to quantum mechanics by considering a straightforward extension of the double-slit experiment. Recall that the interference pattern in the double-slit experiment, shown in Fig. 8.1, can be understood as a probability distribution with the probability density at a point on the detecting screen arising from the superposition of two amplitudes, one for the particle to reach the point going through one of the slits and the other for the particle to reach the point going through the other slit. Suppose we increase the number of slits from two to three. Then there will be three amplitudes (see Fig. 8.2a) that we must add together to determine the probability amplitude that the particle reaches a particular point on the detecting screen. Suppose we next insert another opaque screen with two slits behind the initial screen (Fig. 8.2b).

216

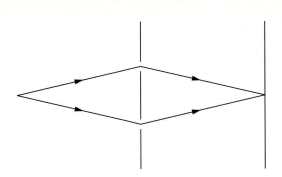

**FIGURE 8.1**
The two paths in the double-slit experiment. The amplitudes for these paths
add together to produce an interference
pattern on a distant detecting screen.

Now there are six possible paths that the particle can take to reach a point on
the detecting screen; thus we must add six amplitudes together to obtain the
total amplitude. One can imagine filling up the space between the source and the
detecting screen with an infinite series of opaque screens and then eliminating
these screens with an infinite number of slits in each screen. In this way, we see
that the probability amplitude for the particle to arrive at a point on the detecting
screen with no barriers in between the source and the detector must be the sum
of the amplitudes for the particle to take every path between the source and the
detection point.

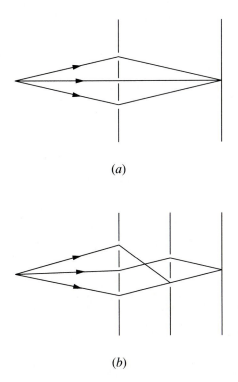

(a)

(b)

**FIGURE 8.2**
(a) The three paths for a triple-slit experiment.
(b) Three of the six paths that a particle may follow to reach a particular point on the detecting
screen when an additional screen with two slits is
inserted.

## 8.2   THE TRANSITION AMPLITUDE

We are now ready to see how we use quantum mechanics to evaluate the amplitude to take a particular path and how we add these amplitudes together to form a path integral.[1] In this chapter we will concentrate on a one-dimensional formulation of the path-integral formalism. The extension to three dimensions is straightforward.

   We start with the amplitude $\langle x', t'|x_0, t_0\rangle$ for a particle that is at position $x_0$ at time $t_0$ to be at the position $x'$ at time $t'$. In Chapter 4 when we introduced the subject of time evolution, we chose to set our clocks so that the initial state of the particle was specified at $t = 0$ and then considered the evolution for a time $t$. Here we are calling the initial time $t_0$ and considering the evolution for a time interval $t' - t_0$. Thus the transition amplitude is given by

$$\langle x', t'|x_0, t_0\rangle = \langle x'|\hat{U}(t' - t_0)|x_0\rangle = \langle x'|e^{-i\hat{H}(t' - t_0)/\hbar}|x_0\rangle \tag{8.1}$$

where $\hat{U}(t' - t_0)$ is the usual time-evolution operator and the Hamiltonian, which is assumed to be time-independent, is in general a function of the position and momentum operators: $\hat{H} = \hat{H}(\hat{p}_x, \hat{x})$. Of course, in the usual one-dimensional case

$$\hat{H} = \frac{\hat{p}_x^2}{2m} + V(\hat{x}) \tag{8.2}$$

   Once we know the amplitude (8.1), we can use it to determine how any state $|\psi\rangle$ evolves with time, since we can write the state $|\psi\rangle$ as a superposition of position eigenstates:

$$\langle x'|\psi(t')\rangle = \langle x'|e^{-i\hat{H}(t' - t_0)/\hbar}|\psi(t_0)\rangle$$

$$= \int_{-\infty}^{\infty} dx_0 \langle x'|e^{-i\hat{H}(t' - t_0)/\hbar}|x_0\rangle\langle x_0|\psi(t_0)\rangle$$

$$= \int_{-\infty}^{\infty} dx_0 \langle x', t'|x_0, t_0\rangle\langle x_0|\psi(t_0)\rangle \tag{8.3}$$

The amplitude $\langle x', t'|x_0, t_0\rangle$, which appears within the integral in (8.3), is often referred to in wave mechanics as the propagator; according to (8.3) we can use it to determine how an arbitrary state propagates in time.

---

[1] Our approach is not that initially followed by Feynman, who essentially postulated (8.28) in an *independent* formulation of quantum mechanics and then showed that it implied the Schrödinger equation. Here, we start with the known form for the time-development operator in terms of the Hamiltonian and from it derive (8.28), subject to certain conditions on the form of the Hamiltonian. For a discussion of Feynman's approach, see R. P. Feynman and A. R. Hibbs, *Path Integrals and Quantum Mechanics,* McGraw-Hill, New York, 1965. For Feynman's account of how he was influenced by Dirac's work on this subject, see *Nobel Lectures—Physics,* vol. III, Elsevier Publication, New York, 1972. For a very nice physical introduction to path integrals, see R. P. Feynman, *QED—The Strange Theory of Light and Matter,* Princeton University Press, Princeton, N. J., 1985.

As an example, let's evaluate the propagator for a free particle using our earlier formalism. The Hamiltonian for a free particle is given by

$$\hat{H} = \frac{\hat{p}_x^2}{2m} \tag{8.4}$$

Inserting a complete set of momentum states

$$\int_{-\infty}^{\infty} dp \, |p\rangle\langle p| = 1 \tag{8.5}$$

in (8.1), we obtain

$$\langle x', t'|x_0, t_0\rangle = \int_{-\infty}^{\infty} dp \, \langle x'|e^{-i\hat{p}_x^2(t'-t_0)/2m\hbar}|p\rangle\langle p|x_0\rangle$$

$$= \int_{-\infty}^{\infty} dp \, \langle x'|p\rangle\langle p|x_0\rangle e^{-i p^2(t'-t_0)/2m\hbar} \tag{8.6}$$

Using

$$\langle x|p\rangle = \frac{1}{\sqrt{2\pi\hbar}} e^{i px/\hbar} \tag{8.7}$$

we see that

$$\langle x', t'|x_0, t_0\rangle = \frac{1}{2\pi\hbar} \int_{-\infty}^{\infty} dp \, e^{i p(x'-x_0)/\hbar} e^{-i p^2(t'-t_0)/2m\hbar} \tag{8.8}$$

This is a Gaussian integral, which can be evaluated using (D.7):

$$\langle x', t'|x_0, t_0\rangle = \sqrt{\frac{m}{2\pi\hbar i(t'-t_0)}} e^{im(x'-x_0)^2/2\hbar(t'-t_0)} \tag{8.9}$$

Problem 8.1 illustrates how we can use this expression for the propagator to determine how a Gaussian wave packet for a free particle evolves in time.

## 8.3 EVALUATING THE TRANSITION AMPLITUDE FOR SHORT TIME INTERVALS

In order to evaluate the transition amplitude $\langle x', t'|x_0, t_0\rangle$ for the interacting case for a finite period of time using the path-integral formalism, we first break up the time interval $t' - t_0$ into $N$ intervals, each of size $\Delta t = (t' - t_0)/N$. We will eventually let $N \to \infty$ so that $\Delta t \to 0$. Thus we are interested first in evaluating the transition amplitude for very small time intervals. In this limit we can expand the exponential in the time-evolution operator in a Taylor series:

$$e^{-i\hat{H}\Delta t/\hbar} = 1 - \frac{i}{\hbar}\hat{H}(\hat{p}_x, \hat{x})\Delta t + O(\Delta t^2) \tag{8.10}$$

where the expression $O(\Delta t^2)$ includes the $\Delta t^2$ and higher powers of $\Delta t$ terms. If we now evaluate the amplitude for a particle at $x$ to be at $x'$ a time $\Delta t$ later, we obtain

$$\langle x'|e^{-i\hat{H}\Delta t/\hbar}|x\rangle = \langle x'|\left[1 - \frac{i}{\hbar}\hat{H}(\hat{p}_x, \hat{x})\Delta t\right]|x\rangle + O(\Delta t^2)$$

$$= \langle x'|\left[1 - \frac{i}{\hbar}\left(\frac{\hat{p}_x^2}{2m} + V(\hat{x})\right)\Delta t\right]|x\rangle + O(\Delta t^2) \quad (8.11)$$

It is easy to evaluate the action of $V(\hat{x})$ since the ket in (8.11) is an eigenstate of the position operator and therefore

$$V(\hat{x})|x\rangle = V(x)|x\rangle \quad (8.12)$$

In order to evaluate the action of the kinetic energy operator, it is convenient to insert the complete set of momentum states (8.5) between the bra vector and the operator in (8.11) and then take advantage of

$$\langle p|\hat{p}_x = \langle p|p \quad (8.13)$$

In this way we obtain

$$\langle x'|e^{-i\hat{H}\Delta t/\hbar}|x\rangle = \int_{-\infty}^{\infty} dp \, \langle x'|p\rangle\langle p|\left[1 - \frac{i}{\hbar}\left(\frac{p^2}{2m} + V(x)\right)\Delta t\right]|x\rangle + O(\Delta t^2)$$

$$= \int_{-\infty}^{\infty} dp \, \langle x'|p\rangle\langle p|\left[|1 - \frac{i}{\hbar}E(p, x)\Delta t\right]|x\rangle + O(\Delta t^2) \quad (8.14)$$

where

$$E(p, x) = \frac{p^2}{2m} + V(x) \quad (8.15)$$

We now take advantage of (8.10) in reverse to write

$$1 - \frac{i}{\hbar}E(p, x)\Delta t = e^{-iE(p,x)\Delta t/\hbar} + O(\Delta t^2) \quad (8.16)$$

Thus the transition amplitude (8.14) becomes

$$\langle x'|e^{-i\hat{H}\Delta t/\hbar}|x\rangle = \int_{-\infty}^{\infty} dp \, \langle x'|p\rangle\langle p|x\rangle e^{-iE(p,x)\Delta t/\hbar} + O(\Delta t^2)$$

$$= \frac{1}{2\pi\hbar}\int_{-\infty}^{\infty} dp \, e^{i\,p(x'-x)/\hbar}e^{-iE(p,x)\Delta t/\hbar} + O(\Delta t^2) \quad (8.17)$$

or simply

$$\langle x'|e^{-i\hat{H}\Delta t/\hbar}|x\rangle = \int_{-\infty}^{\infty} \frac{dp}{2\pi\hbar} \exp\left\{\frac{i}{\hbar}\left[p\frac{(x'-x)}{\Delta t} - E(p, x)\right]\Delta t\right\} + O(\Delta t^2)$$

$$(8.18)$$

Equation (8.18) is deceptively simple in appearance. Although we characterized (8.16) as (8.10) in reverse, the exponential (8.10) contains the Hamiltonian *operator*, while the exponential (8.16) involves no operators at all. Where have the operators gone? The answer is that we have avoided much of the complexity of having to deal with the exponential of an operator by retaining just the terms through first order in $\Delta t$ in (8.11). These complications are absorbed in the $O(\Delta t^2)$ term in (8.14). For example, if we were to try to calculate the $\Delta t^2$ term in (8.14), we would see that the fact that the position and momentum operators in the Hamiltonian do not commute prevents our replacing both these operators with ordinary numbers by inserting just a *single* complete set of momentum states. But if we consider the limit of the transition amplitude (8.18) as $\Delta t \rightarrow 0$, we can ignore these $O(\Delta t^2)$ complications. We will next see, however, that there is a penalty to pay for formulating quantum mechanics in a way that eliminates the operators that have been characteristic of our treatment of time development using the Hamiltonian formalism.

## 8.4 THE PATH INTEGRAL

We are now ready to evaluate the transition amplitude $\langle x', t'|x_0, t_0\rangle$ for a finite time interval. As we suggested earlier, we break up the interval $t' - t_0$ into $N$ equal-time intervals $\Delta t$ with intermediate times $t_1, t_2, \ldots, t_{N-1}$, as shown in Fig. 8.3. Therefore

$$\langle x', t'|x_0, t_0\rangle = \langle x'| \underbrace{e^{-i\hat{H}\Delta t/\hbar} \ldots e^{-i\hat{H}\Delta t/\hbar} e^{-i\hat{H}\Delta t/\hbar}}_{N \text{ times}} |x_0\rangle \tag{8.19}$$

We next insert complete sets of position states

$$\int_{-\infty}^{\infty} dx_i \, |x_i\rangle\langle x_i| = 1 \qquad i = 1, 2, \ldots, N - 1 \tag{8.20}$$

between each of these individual time-evolution operators:

$$\langle x', t'|x_0, t_0\rangle =$$

$$\int dx_1 \ldots \int dx_{N-1}\langle x'|e^{-i\hat{H}\Delta t/\hbar}|x_{N-1}\rangle\langle x_{N-1}|e^{-i\hat{H}\Delta t/\hbar}|x_{N-2}\rangle \ldots$$

$$\times \langle x_2|e^{-i\hat{H}\Delta t/\hbar}|x_1\rangle\langle x_1|e^{-i\hat{H}\Delta t/\hbar}|x_0\rangle \tag{8.21}$$

**FIGURE 8.3**
The interval $t' - t_0$ is broken into $N$ time intervals, each of length $\Delta t$.

where each of the integrals is understood to run from $-\infty$ to $\infty$, as indicated in (8.20). Reading this equation from right to left, we see the amplitude for the particle at position $x_0$ at time $t_0$ to be at position $x_1$ at time $t_1 = t_0 + \Delta t$, multiplied by the amplitude for a particle at position $x_1$ at time $t_0 + \Delta t$ to be at position $x_2$ at time $t_2 = t_0 + 2\Delta t$. This sequence concludes with the amplitude for the particle to be at $x'$ at time $t'$ when it is at position $x_{N-1}$ at a time $\Delta t$ earlier. Figure 8.4 shows a typical path in the $x$-$t$ plane for particular values of $x_1$, $x_2, \ldots, x_{N-1}$. Note that we are integrating over *all* values of $x_1, x_2, \ldots, x_{N-1}$ in (8.21). Thus, as we let $\Delta t \to 0$, we are effectively integrating over all paths that the particle can take in reaching the position $x'$ at time $t'$ when it starts at the position $x_0$ at time $t_0$.

We now use the expression (8.18) for the $N$ amplitudes $\langle x_{i+1}|e^{-i\hat{H}\Delta t/\hbar}|x_i\rangle$ in (8.21), provided we are careful to insert the appropriate values for the initial and final positions in each case. If we let $N \to \infty$, and correspondingly $\Delta t \to 0$, we can ignore the $O(\Delta t^2)$ piece in each of the individual amplitudes, and the expression for the full transition amplitude is exactly given by

$$\langle x', t'|x_0, t_0\rangle = \lim_{N\to\infty} \int dx_1 \ldots \int dx_{N-1} \int \frac{dp_1}{2\pi\hbar} \ldots$$

$$\times \int \frac{dp_N}{2\pi\hbar} \exp\left\{\frac{i}{\hbar}\sum_{i=1}^{N}\left[p_i\frac{(x_i - x_{i-1})}{\Delta t} - E(p_i, x_{i-1})\right]\Delta t\right\} \quad (8.22)$$

where we have called the final position $x' = x_N$ in the exponent.

We now face a task that appears rather daunting: evaluating an infinite number of integrals. In fact, (8.22) involves both an infinite number of momentum and an infinite number of position integrals. Fortunately, for a Hamiltonian of the form (8.2), each of the momentum integrals is a Gaussian integral, which can

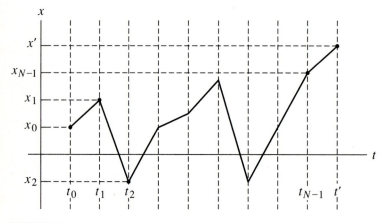

**FIGURE 8.4**

A possible path taken by the particle in going from position $x_0$ at time $t_0$ to a position $x'$ at time $t'$, with intermediate positions $x_1$ at time $t_1$, $x_2$ at time $t_2$, and so on.

be evaluated using (D.7) [with $a = i\Delta t/2m\hbar$ and $b = i(x_i - x_{i-1})/\hbar$]. A typical momentum integral is given by

$$\int \frac{dp_i}{2\pi\hbar} \exp\left[-i\frac{p_i^2\Delta t}{2m\hbar} + i\frac{p_i(x_i - x_{i-1})}{\hbar}\right]$$

$$= \sqrt{\frac{m}{2\pi\hbar i\Delta t}} \exp\left[\frac{i}{\hbar}\left(\frac{m\Delta t}{2}\right)\left(\frac{x_i - x_{i-1}}{\Delta t}\right)^2\right] \qquad (8.23)$$

After doing all of the $p$ integrals, we find

$$\langle x', t'|x_0, t_0\rangle = \lim_{N\to\infty} \int dx_1 \dots \int dx_{N-1} \left(\frac{m}{2\pi\hbar i\Delta t}\right)^{N/2}$$

$$\times \exp\left\{\frac{i}{\hbar}\Delta t \sum_{i=1}^{N}\left[\frac{m}{2}\left(\frac{x_i - x_{i-1}}{\Delta t}\right)^2 - V(x_{i-1})\right]\right\} \qquad (8.24)$$

Notice that as $N \to \infty$ and therefore $\Delta t \to 0$, the argument of the exponent becomes the standard definition of a Riemannian integral:

$$\lim_{N\to\infty, \Delta t\to 0} \frac{i}{\hbar}\Delta t \sum_{i=1}^{N}\left[\frac{m}{2}\left(\frac{x_i - x_{i-1}}{\Delta t}\right)^2 - V(x_{i-1})\right] = \frac{i}{\hbar}\int_{t_0}^{t'} dt\, L(x, \dot{x}) \qquad (8.25)$$

where

$$L(x, \dot{x}) = \frac{m}{2}\left(\frac{dx}{dt}\right)^2 - V(x) = \frac{1}{2}m\dot{x}^2 - V(x) \qquad (8.26)$$

is the usual Lagrangian familiar from classical mechanics.[2]

Finally, it is convenient to express the remaining infinite number of position integrals using the shorthand notation

$$\int_{x_0}^{x'} \mathcal{D}[x(t)] = \lim_{N\to\infty} \int dx_1 \dots \int dx_{N-1} \left(\frac{m}{2\pi\hbar i\Delta t}\right)^{N/2} \qquad (8.27)$$

which is a symbolic way of indicating that we are integrating over all paths connecting $x_0$ to $x'$. Then

$$\langle x', t'|x_0, t_0\rangle = \int_{x_0}^{x'} \mathcal{D}[x(t)]\, e^{iS[x(t)]/\hbar} \qquad (8.28)$$

where

$$S[x(t)] = \int_{t_0}^{t'} dt\, L(x, \dot{x}) \qquad (8.29)$$

---

[2] If you are not familiar with the Lagrangian and the principle of least action, a brief but fun introduction is given in *The Feynman Lectures on Physics*, Vol. II, Chap. 19.

is the value of the action evaluated for a particular path taken by the particle. An integral such as (8.28) is referred to as a functional integral. In summing over all possible paths, we are really integrating over all possible functions $x(t)$ that meet the boundary conditions $x(t_0) = x_0$ and $x(t') = x'$.

Summarizing, in order to determine the amplitude for a particle at position $x_0$ at time $t_0$ to be found at position $x'$ at time $t'$, we consider *all* paths in the $x$-$t$ plane connecting the two points. For each path $x(t)$, we evaluate the action $S[x(t)]$. Each path makes a contribution proportional to $e^{iS[x(t)]/\hbar}$, a factor that has unit modulus and depends on the path only through the phase factor $S[x(t)]/\hbar$. We then add up the contribution of each path. Note that in a formulation of quantum mechanics that starts with (8.28), operators need not be introduced at all. However, we must then face the issue of actually evaluating the path integral in order to determine the transition amplitude, or propagator. To give us some confidence that this is indeed feasible, at least in some cases, we first reconsider the evaluation of the transition amplitude (8.9) for a free particle, this time with the path-integral formalism. Then, in Section 8.6, we will use the path-integral formulation to examine the relationship between quantum and classical mechanics.

## 8.5   EVALUATION OF THE PATH INTEGRAL FOR A FREE PARTICLE

In order to evaluate the path integral (8.28) for a free particle, for which $V(x) = 0$, we retrace our derivation of (8.28) and break up the time interval $t' - t_0$ into $N$ discrete $\Delta t$ intervals:

$$\langle x', t'|x_0, t_0 \rangle = \lim_{N \to \infty} \int dx_1 \ldots$$

$$\times \int dx_{N-1} \left( \frac{m}{2\pi \hbar i \Delta t} \right)^{N/2} \exp\left[ \frac{i}{\hbar} \Delta t \sum_{i=1}^{N} \frac{m}{2} \left( \frac{x_i - x_{i-1}}{\Delta t} \right)^2 \right] \quad (8.30)$$

We introduce the dimensionless variables

$$y_i = x_i \sqrt{\frac{m}{2\hbar \Delta t}} \quad (8.31)$$

where again $x_N = x'$. Expressed in terms of these variables, the transition amplitude becomes

$$\langle x', t'|x_0, t_0 \rangle = \lim_{N \to \infty} \left( \frac{m}{2\pi \hbar i \Delta t} \right)^{N/2} \left( \frac{2\hbar \Delta t}{m} \right)^{(N-1)/2} \int_{-\infty}^{\infty} dy_1 \ldots$$

$$\times \int_{-\infty}^{\infty} dy_{N-1} \exp\left[ i \sum_{i=1}^{N} (y_i - y_{i-1})^2 \right] \quad (8.32)$$

Note that we have explicitly inserted the limits of integration.

Let's start with the $y_1$ integral, leaving aside for the moment the constants in front:

$$\int_{-\infty}^{\infty} dy_1 \, e^{i[(y_2-y_1)^2+(y_1-y_0)^2]} = \sqrt{\frac{i\pi}{2}} \, e^{i(y_2-y_0)^2/2} \tag{8.33}$$

where we have taken advantage of (D.7). Fortunately, evaluating this integral has left us with another Gaussian. We are thus able to tackle the $y_2$ integral, again with the aid of (D.7):

$$\sqrt{\frac{i\pi}{2}} \int_{-\infty}^{\infty} dy_2 \, e^{i(y_3-y_2)^2+(y_2-y_0)^2/2} = \sqrt{\frac{i\pi}{2}} \sqrt{\frac{i2\pi}{3}} \, e^{i(y_3-y_0)^2/3}$$

$$= \sqrt{\frac{(i\pi)^2}{3}} \, e^{i(y_3-y_0)^2/3} \tag{8.34}$$

A comparison of the result of the $y_1$ integral (8.33) with the result of having done both the $y_1$ and the $y_2$ integrals in (8.34) suggests that the result of $(N-1)$ $y$ integrals is just

$$\int_{-\infty}^{\infty} dy_1 \ldots \int_{-\infty}^{\infty} dy_{N-1} \exp\left[i \sum_{i=1}^{N}(y_i - y_{i-1})^2\right] = \sqrt{\frac{(i\pi)^{N-1}}{N}} \, e^{i(y_N-y_0)^2/N} \tag{8.35}$$

which can be established by induction. See Problem 8.2. Thus

$$\langle x', t'|x_0, t_0 \rangle = \lim_{N\to\infty} \left(\frac{m}{2\pi\hbar i \Delta t}\right)^{N/2} \left(\frac{2\hbar\Delta t}{m}\right)^{(N-1)/2} \left[\frac{(i\pi)^{N-1}}{N}\right]^{1/2} e^{i(y_N-y_0)^2/N}$$

$$= \lim_{N\to\infty} \sqrt{\frac{m}{2\pi\hbar i N\Delta t}} \, e^{im(x_N-x_0)^2/2\hbar N\Delta t}$$

$$= \sqrt{\frac{m}{2\pi\hbar i(t' - t_0)}} \, e^{im(x'-x_0)^2/2\hbar(t'-t_0)} \tag{8.36}$$

where we have used $t' - t_0 = N\Delta t$ in the last step. This result is the same as we obtained with considerably less effort in Section 8.2 using the Hamiltonian formalism.

There is a limited class of problems with Lagrangians of the form

$$L(x, \dot{x}) = a + bx + cx^2 + d\dot{x} + ex\dot{x} + f\dot{x}^2 \tag{8.37}$$

where the integrals in the path-integral formulation are all Gaussian and the procedure we have outlined for the free particle can also be applied to determine the transition amplitude. In general, this is a fairly cumbersome procedure, but there are some shortcuts that can be used to determine the amplitude in these cases. The interested reader is urged to consult Feynman and Hibbs, *Path Integrals and Quantum Mechanics*. The main utility of the path-integral approach in nonrelativistic quantum mechanics is not, as you can probably believe, in explicitly determining the transition amplitude but in the alternative way it gives us of viewing time evolution in quantum mechanics and in the insight it gives us into the classical limit of quantum mechanics.

## 8.6 WHY SOME PARTICLES FOLLOW THE PATH OF LEAST ACTION

Equation (8.28) is an amazing result. Not only does *every* path contribute to the amplitude, but each path makes a contribution of the same magnitude. The only thing that varies from one path to the next is the value of the phase factor $S[x(t)]/\hbar$. Since quantum mechanics applies to all particles, why then, for example, does a macroscopic particle seem to follow a particular path at all?

In order to see which paths "count," let's consider an example. Suppose that at $t = 0$ a *free* particle of mass $m$ is at the origin, $x = 0$, and that we are interested in the amplitude for the particle to be at $x = x'$ when $t = t'$. There are clearly an infinite number of possible paths between the initial and the final point. One such path, indicated in Fig. 8.5, is

$$x = \left(\frac{x'}{t'}\right)t \tag{8.38}$$

This is, of course, the path that a classical particle with no forces acting on it and moving at a constant speed $v = x'/t'$ would follow. For this path, $\dot{x} = x'/t'$, $L = m\dot{x}^2/2 = mx'^2/2t'^2$, and

$$S_{cl} = S[x_{cl}(t)] = \int_0^{t'} dt\, \frac{m}{2}\left(\frac{x'^2}{2t'^2}\right) = \frac{mx'^2}{2t'} \tag{8.39}$$

If we evaluate the phase $S_{cl}/\hbar$ for typical macroscopic parameters: $m = 1$ g, $x' = 1$ cm, and $t' = 1$ s, we find that the phase has the very large value of roughly $(1/2) \times 10^{27}$ radians.

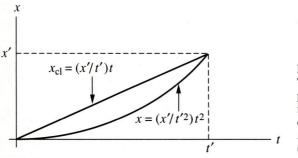

**FIGURE 8.5**
Two paths connecting the initial position $x = 0$ at $t = 0$ and the final position $x'$ at time $t'$: the classical path for a free particle, $x = (x'/t')t$, and the path $x = (x'/t'^2)t^2$.

We also choose another path, which is also depicted in Fig. 8.5, namely,

$$x = \left(\frac{x'}{t'^2}\right)t^2 \tag{8.40}$$

This path, which is characteristic of a particle undergoing uniform acceleration, is clearly *not* the classical path for a particle without any forces acting on it. For this path we find $L = m\dot{x}^2/2 = 2mx'^2t^2/t'^4$ and

$$S\left(x't^2/t'^2\right) = \int_0^{t'} dt \left(\frac{2mx'^2}{t'^4}\right)t^2 = \frac{2mx'^2}{3t'} \tag{8.41}$$

The value of the phase is roughly $(2/3) \times 10^{27}$ radians for the same macroscopic parameters.

Although the phases determined from (8.39) and (8.41) are different, what really distinguishes the classical path from any other is *not* the actual value of the action itself. Rather, the classical path is the path of least action, or, more precisely, the one for which the action is an extremum. To illustrate this explicitly, we consider a set of paths in the neighborhood of the two paths that we are using as examples. In the vicinity of the classical path, we take the set of paths

$$x = \frac{x'}{t'}\left[t + \varepsilon\frac{t(t - t')}{t'}\right] \tag{8.42}$$

where each value of the parameter $\varepsilon$ labels a different path that deviates slightly from the classical path if $\varepsilon$ is small. Notice that $x(t)$ still satisfies the initial and final conditions $x(0) = 0$ and $x(t') = x'$, respectively. It is straightforward to calculate the action:

$$S = \int_0^{t'} dt \frac{m\dot{x}^2}{2} = \int_0^{t'} dt \frac{m}{2}\left(\frac{x'}{t'}\right)^2\left[1 + \varepsilon\frac{(2t - t')}{t'}\right]^2$$

$$= \frac{m}{2}\frac{x'^2}{t'}\left(1 + \frac{\varepsilon^2}{3}\right) = S_{cl}\left(1 + \frac{\varepsilon^2}{3}\right) \tag{8.43}$$

The important thing to notice is that the change in the action depends on $\varepsilon^2$; there is no term linear in $\varepsilon$. The action is indeed a minimum; varying the path away from the classical path only increases the action from its value (8.39). Because the first-order contribution to the action vanishes:

$$\delta S = \left(\frac{\partial S}{\partial \varepsilon}\right)_{\varepsilon=0} \varepsilon = 0 \tag{8.44}$$

the contribution through first order of each of the paths to the path integral is proportional to

$$e^{i(S_{cl} + \delta S)/\hbar} = e^{iS_{cl}/\hbar} \tag{8.45}$$

Thus the amplitudes for the paths in the vicinity of the classical path will have roughly the same phase as does the classical path and will, therefore, add together constructively.

If, on the other hand, we consider the nonclassical path (8.40), we can also determine the action for a set of paths in its neighborhood,

$$x = \frac{x'}{t'^2}\left[t + \varepsilon \frac{t(t - t')}{t'}\right]^2 \tag{8.46}$$

which again satisfy $x(0) = 0$ and $x(t') = x'$. If we now calculate the action for (8.46), we obtain

$$S = \frac{2mx'^2}{3t'}\left(1 + \frac{\varepsilon}{2} + \cdots\right) = S\left(x't^2/t'^2\right)\left(1 + \frac{\varepsilon}{2} + \cdots\right) \tag{8.47}$$

Here, in agreement with the principle of least action, the first-order correction $\delta S = (\partial S/\partial \varepsilon)_{\varepsilon=0}\varepsilon \neq 0$. Some neighboring paths, in this case those with $\varepsilon < 0$, *reduce* the value of the action from its value for the path (8.40). The contribution through first order of the paths in the vicinity of the path (8.40) to the path integral is $e^{i(S+\delta S)/\hbar}$. Thus, in general, paths in the neighborhood of the nonclassical path may be out of phase with each other and may interfere destructively.

A useful pictorial way to show how this cancellation arises is in terms of phasors. For convenience, let's assume that we can label the paths discretely instead of continuously. When we add the complex numbers

$$e^{iS[x_1(t)]/\hbar} = \cos S[x_1(t)]/\hbar + i \sin S[x_1(t)]/\hbar \tag{8.48a}$$

and

$$e^{iS[x_2(t)]/\hbar} = \cos S[x_2(t)]/\hbar + i \sin S[x_2(t)]/\hbar \tag{8.48b}$$

together for two paths, we just add the real parts and the imaginary parts together separately. The magnitude of this complex number is of course given by

$$\left|e^{iS[x_1(t)]/\hbar} + e^{iS[x_2(t)]/\hbar}\right| \tag{8.49}$$

$$= \left(\{\cos S[x_1(t)]/\hbar + \cos S[x_2(t)]/\hbar\}^2 + \{\sin S[x_1(t)]\hbar + \sin S[x_2(t)]/\hbar\}^2\right)^{1/2}$$

We can recognize this as the same procedure we would use to find the length of an ordinary vector resulting from the addition of two vectors, $\mathbf{V} = \mathbf{V}_1 + \mathbf{V}_2$, namely,

$$|\mathbf{V}| = \left[(V_{1x} + V_{2x})^2 + (V_{1y} + V_{2y})^2\right]^{1/2} \tag{8.50}$$

Thus, if we indicate the complex amplitudes (8.48) by vectors in the complex plane, with the real part of the amplitude plotted along one axis and the imaginary part of the amplitude plotted along the other axis, the complex number resulting from the addition of the two amplitudes (8.48a) and (8.48b) is just the vector sum, as shown in Fig. 8.6.

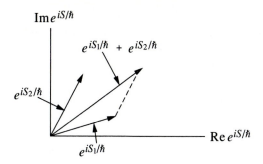

**FIGURE 8.6**
The addition of the amplitudes $e^{iS[x_1(t)]/\hbar}$ and $e^{iS[x_2(t)]/\hbar}$ is carried out using phasors. Each of the amplitudes is represented by an arrow of unit length in the complex plane, with an orientation angle, or phase angle, $S[x(t)]/\hbar$. The rule for adding the two amplitudes is the same as for ordinary vector addition.

What happens as we add up the contributions of the nonclassical path (8.40) and its neighbors? Notice that the first-order change in the action from (8.47) is proportional to the value $S$ of the action itself multiplied by $\varepsilon$. As $\varepsilon$ changes away from zero, the phase of the neighboring path changes. In the particular case (8.46), we see that when $\varepsilon S(x't^2/t'^2)/2\hbar = 2\pi$, the phase has returned to its initial value, modulo $2\pi$. Thus if $S/\hbar$ is $10^{27}$ for some typical macroscopic parameters, $\varepsilon$ need only reach the value $\varepsilon = 4\pi \times 10^{-27}$ to satisfy this condition. In Fig. 8.7a we add up the arrows for a discrete set of paths representing those between $\varepsilon = 0$ and $\varepsilon = 4\pi \times 10^{-27}$. These arrows form a closed "circle" and therefore sum to zero. Thus, the contribution from these paths cancel each other and hence do not contribute to the path integral (8.28). On the other hand, in the vicinity of the classical path (8.38), the first-order contribution to the action vanishes and thus the paths in the vicinity of the classical path have the same phase and add together constructively (Fig. 8.7b). This coherence will eventually break down, when the phase shift due to nearby paths reaches a value on the order of $\pi$. For our specific example (8.46), this means $S_{cl}\varepsilon^2/3\hbar \simeq \pi$, or $\varepsilon \simeq 10^{-13}$ for the macroscopic parameters. This is clearly a very tight constraint for a macroscopic particle, since the paths that count do not deviate far from the path of least action.

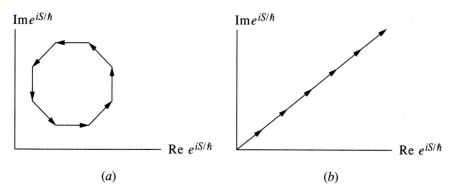

(a)   (b)

**FIGURE 8.7**
(a) The sum of a discrete set of amplitudes representing those in the vicinity of the nonclassical path. Since these arrows form a closed "circle," their sum vanishes. (b) In the vicinity of the classical path, the amplitudes, which all have the same phase to first order, sum to give a nonzero contribution to the path integral.

But the classical path is still important because *only* in its vicinity can many paths contribute to the path integral coherently. In the neighborhood of any other path, the contributions of neighboring paths cancel each other (see Fig. 8.8). Quantum mechanics thus allows us to understand how a particle knows to take the path of least action, at least in classical physics: the particle actually has an amplitude to take all paths.

Our numerical examples in this section so far have been entirely about a macroscopic particle. What happens if we replace the 1 g mass with an electron? Notice that the *phase difference* between the two paths in Fig. 8.5 is given by

$$\frac{\Delta S}{\hbar} = \frac{S(x't^2/t'^2) - S_{cl}}{\hbar} = \frac{mx'^2}{6t'\hbar} \tag{8.51}$$

While for $m = 1$ g with $x' = 1$ cm and $t' = 1$ s this phase difference is about $(1/6) \times 10^{27}$ radians, the phase difference between the two paths for the electron, for which $m \simeq 10^{-27}$ g, is only $\frac{1}{6}$ radian. Thus for an electron even the path $x = x't^2/t'^2$ is essentially coherent with the classical path $x = x't/t'$. Because there are many more paths that can contribute coherently to the path integral for the electron than there are for the macroscopic particle, the motion of the electron in this case should be extremely nonclassical in nature.

This last example with the electron is sufficient to cause us to wonder again about the double-slit experiment. Why do we see a clear interference pattern arising from the interference of the amplitudes to take just the two paths shown in Fig. 8.1? Why don't other paths, such as the one indicated in Fig. 8.9, contribute? The answer is that the action for the paths indicated in Fig. 8.1 is actually much larger than the previous example might lead you to think. For example, electrons with 50 eV of kinetic energy, a typical value for electron diffraction experiments, have a speed of $4 \times 10^8$ cm/s. Thus if we take $x' = 40$ cm as a typical size scale for the double-slit experiment and $t' = 10^{-7}$ s so that the speed has the proper

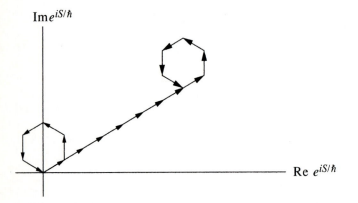

**FIGURE 8.8**
A schematic diagram using phasors showing for a macroscopic particle how the classical path *and* its neighbors dominate the path integral, while other paths give no net contribution as they and their neighbors cancel each other.

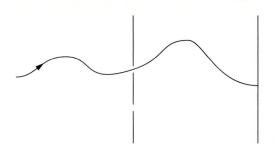

**FIGURE 8.9**
A path that does not contribute coherently
to the double-slit experiment illustrated in
Fig. 8.1.

magnitude, we find the phase $S/\hbar$ for the straight-line path in Fig. 8.5 to be
$7 \times 10^9$. When the phase is this large, only a small deviation away from the
classical path will cause coherence to be lost. The large size of the action is also
the reason that we can use classical physics to aim an electron gun in a television
set, where the electrons may have an energy of 5 KeV, or to describe the motion
of atoms through the magnets in the Stern-Gerlach experiments in Chapter 1.

## 8.7 QUANTUM INTERFERENCE DUE TO GRAVITY

We now show how we can use path integrals to analyze a striking experiment
illustrating the sensitivity of the neutron interferometer that we first introduced in
Section 4.3. An essentially monochromatic beam of thermal neutrons is split by
Bragg reflection by a perfect slab of silicon crystal at A. One of the beams follows
path ABD and the other follows path ACD, as shown in Fig. 8.10. In general,
there will be constructive or destructive interference at D depending on the path
difference between these two paths. Suppose that the interferometer initially lies
in a horizontal plane so that there are no gravitational effects. We then rotate the
plane formed by the two paths by angle $\delta$ about the segment AC. The segment BD
is now higher than the segment AC by $l_2 \sin \delta$. Thus there will be an additional
gravitational potential energy $mgl_2 \sin \delta$ along this path, which alters the action
and hence the amplitude to take the path BD by the factor

$$e^{-i(mgl_2 \sin \delta)T/\hbar} \tag{8.52}$$

where the action in the exponent is the negative of the potential energy multiplied
by the time $T$ it takes for the neutron to traverse the segment BD. Of course,
gravity also affects the action in traversing the segment AB, but this phase shift

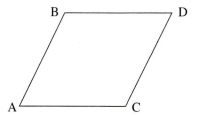

**FIGURE 8.10**
A schematic of the neutron interferometer. The interferometer, initially lying in a horizontal plane, can be
rotated vertically about the axis AC by an angle $\delta$.

is the same as for the segment CD, and thus the phase difference between the path ABD and the path ACD is given by

$$
\begin{aligned}
\frac{S[\text{ABD}] - S[\text{ACD}]}{\hbar} &= -\frac{m g l_2 T \sin \delta}{\hbar} \\
&= -\frac{m^2 g l_2 l_1 \sin \delta}{\hbar p} \\
&= -\frac{m^2 g l_2 l_1 \lambda \sin \delta}{2 \pi \hbar^2}
\end{aligned}
\tag{8.53}
$$

where we have used the de Broglie relation $p = h/\lambda$ to express this phase difference in terms of the wavelength of the neutrons. Figure 8.11 shows the interference fringes that are produced as $\delta$ varies from $-45°$ (BD below AC) to $+45°$ (BD above AC) for neutrons with $\lambda = 1.419$ Å. The contrast of the interference pattern dies out with increasing angle of rotation because the interferometer bends and warps slightly (on the scale of angstroms) under its own weight as it is rotated about the axis AC.

Notice in the classical limit that as $\hbar \to 0$, the spacing between the fringes in (8.53) becomes so small that the interference pattern effectively washes out. This interference is, in fact, the only gravitational effect that depends in a nontrivial way on quantum mechanics that has so far been observed.[3] Now, not surprisingly, neutrons are observed to "fall" in a gravitational field,[4] but from (6.33) and (6.34) we see for a gravitational field pointing in the negative $x$ direction that

$$
\frac{d^2 \langle x \rangle}{dt^2} = -g
\tag{8.54}
$$

---

[3] On a microscopic scale, where most quantum effects are observed, gravitation is an extremely weak force. For example, the ratio of the electromagnetic and the gravitational force between an electron and a proton is $G m_e m_p / e^2 = 4 \times 10^{-40}$.

[4] A. W. McReynolds, *Phys. Rev.* **83**, 172, 233 (1951); J. W. T. Dabbs, J. A. Harvey, D. Paya, and H. Horstmann, *Phys. Rev.* **139**, B756 (1965).

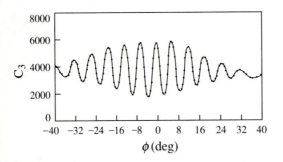

**FIGURE 8.11**
The interference pattern as a function of the angle $\delta$ [*From J.-L. Staudenmann, S. A. Werner, R. Colella, and A. W. Overhauser, Phys. Rev. A21, 1419 (1980)*.]

which does not depend on Planck's constant at all. Neither does (8.54) depend on value of the mass $m$. This lack of dependence on $m$ is a consequence of the equivalence of inertial mass $m_i$, which would appear on the left-hand side of (8.54) as the $m_i a$ of Newton's law, and the gravitational mass $m_g$, which appears in the right-hand side in the gravitational force.[5] All bodies fall at the same rate because of this equivalence. While this equivalence has been well tested in the classical regime, our result (8.53), which when expressed in terms of $m_i$ and $m_g$ becomes

$$\frac{S[ABD] - S[ACD]}{\hbar} = -\frac{m_i m_g g l_2 l_1 \lambda \sin \delta}{2\pi\hbar^2} \tag{8.55}$$

provides us with a test of the equivalence between inertial and gravitational mass at the quantum level. The determination of $m_i m_g$ from (8.55) is in complete agreement with the determination of $m_i^2$ from mass spectroscopy.

## 8.8   SUMMARY

The essence of Chapter 8 is contained in the expression

$$\langle x', t' | x_0, t_0 \rangle = \int_{x_0}^{x'} \mathcal{D}[x(t)] e^{iS[x(t)]/\hbar} \tag{8.56}$$

for the amplitude for a particle at position $x_0$ at time $t_0$ to be at position $x'$ at time $t'$. The right-hand side of (8.56) tells us that the amplitude is proportional to an integral of $e^{iS[x(t)]/\hbar}$ over *all paths* $x(t)$ connecting $x_0$ to $x'$, subject to the constraint that $x(t_0) = x_0$ and $x(t') = x'$, where

$$S[x(t)] = \int_{t_0}^{t'} dt\, L(x, \dot{x}) \tag{8.57}$$

is the value of the action evaluated for a particular path $x(t)$.

Although evaluating the path integral (8.56) is not especially practical in most problems, the path-integral approach does give us a useful way to think about quantum dynamics. For example, inserting an impenetrable screen with an aperture between a source of particles and a detector, as shown in Figure 8.12a, eliminates many of the paths that the particle could follow in moving between the two points, altering the amplitude for the particle to arrive at the detector from what it would have been in the absence of the screen. We call this phenomenon *diffraction*. If a second aperture is opened in the impenetrable screen, as shown in Fig. 8.12b, the paths for the particle to reach the detector by traveling through this second slit must be added to the paths to reach the detector by traveling through the first slit, generating an *interference* pattern. In fact, if you have doubts about

---

[5] Near the surface of the earth $m_g g = G m_g M / R^2$, where $G$ is the gravitational constant and $M$ is the mass and $R$ the radius of the earth.

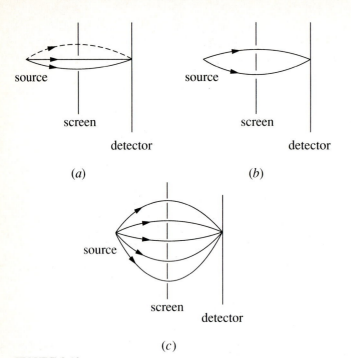

**FIGURE 8.12**
(*a*) A single-slit diffraction experiment. The path shown with the dashed line is an example of a path that is obstructed by the impenetrable screen and therefore does not contribute to the integral over all paths. (*b*) A double-slit interference experiment. (*c*) A diffraction-grating experiment.

the role played by paths such as the one blocked by the screen in Fig. 8.12*a*, consider opening a periodic array of apertures in the screen to allow the particle following a special set of these paths to reach the detector, as shown in Fig. 8.12*c*. The pattern will clearly differ from that obtained with a single or a double slit.

The path-integral approach also gives insight into the foundations of classical mechanics. Since the factor $e^{iS[x(t)]/\hbar}$ is a complex number of unit modulus, the only thing that differs from one path to another is the value of the phase $S[x(t)]/\hbar$. Figure 8.13 is a schematic diagram of the phase plotted as a function of the path $x(t)$. The particular path where the action is an extremum—$\delta S = 0$— is often called the "path of least action." This path of least action is the unique path $x_{cl}(t)$ that we expect a particle to follow in classical physics. In quantum mechanics, on the other hand, *all* paths contribute to the path integral (8.56). What makes $x_{cl}(t)$ special is that since it is the path for which the phase $S[x(t)]/\hbar$ is an extremum, the phase difference between the classical path and its neighbors changes less rapidly than for any other path and its neighbors. When we add up the contribution from all paths, only in the vicinity of the classical path do we find many paths that are in phase with each other and hence can add together coherently. In situations where $S[x(t)]/\hbar \gg 1$, such as for a macroscopic particle, this is a very tight constraint which indeed singles out the classical path and its

$S/\hbar$

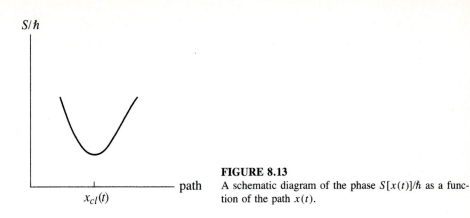

path

**FIGURE 8.13**
A schematic diagram of the phase $S[x(t)]/\hbar$ as a function of the path $x(t)$.

very nearby neighbors. However, when $S[x(t)]/\hbar \sim 1$, even paths that deviate significantly from the classical path can still be roughly in phase with it, and the behavior of the particle can no longer be adequately described by classical physics at all.

## PROBLEMS

**8.1.** Use the free-particle propagator (8.8) in (8.3) to determine how the Gaussian position-space wave packet (6.59) evolves in time. Check your result by comparing with (6.76).

**8.2.** Prove (8.35) by induction.

**8.3.** Determine, up to an overall multiplicative function of time, the transition amplitude, or propagator, for the harmonic oscillator using path integrals. See Feynman and Hibbs, *Path Integrals and Quantum Mechanics*, Sections 3-5 and 3-6.

**8.4.** Estimate the size of the action for free neutrons with $\lambda = 1.419$ Å traversing a distance of 10 cm.

**8.5.** For which of the following does classical mechanics give an adequate description of the motion? Explain.

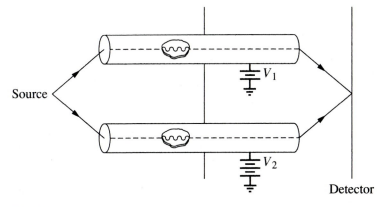

**FIGURE 8.14**

(a) An electron with a speed $v/c = 1/137$, which is typical in the ground state of the hydrogen atom, traversing a distance of 0.5 Å, which is a characteristic size of the atom.

(b) An electron with the same speed as in (a) traversing a distance of 1 cm.

**8.6.** A low-intensity beam of charged particles, each with charge $q$, is split into two parts. Each part then enters a very long metallic tube shown in Fig. 8.14. Suppose that the length of the wave packet for each of the particles is sufficiently smaller than the length of the tube so that for a certain time interval, say from $t_0$ to $t'$, the wave packet for the particle is definitely within the tubes. During this time interval, a constant potential $V_1$ is applied to the upper tube and a constant potential $V_2$ is applied to the lower tube. The rest of the time there is no voltage applied to the tubes. Determine how the interference pattern depends on the voltages $V_1$ and $V_2$ and explain physically why this dependence is completely incompatible with classical physics.

# CHAPTER
# 9

# TRANSLATIONAL AND ROTATIONAL SYMMETRY IN THE TWO-BODY PROBLEM

After spending Chapters 6, 7, and 8 in one dimension, we now return to the three-dimensional world and consider a system consisting of two bodies that interact through a potential energy that depends only on the magnitude of the distance between them. The Hamiltonian for this system is invariant under translations and rotations of *both* of the bodies, which leads to conservation of total linear momentum and relative orbital angular momentum, respectively. The relationship between an invariance, or a symmetry, in the system and a corresponding conservation law is one of the most fundamental and important in physics.

## 9.1 THE ELEMENTS OF WAVE MECHANICS IN THREE DIMENSIONS

Let's begin by extending our discussion of wave mechanics in Sections 6.1 through 6.5 to three dimensions.[1] The position eigenstate in three dimensions is given in Cartesian coordinates by

$$|\mathbf{r}\rangle = |x, y, z\rangle \tag{9.1}$$

where

---

[1] It would be good to review those sections of Chapter 6 before reading Section 9.1.

$$\hat{x}|\mathbf{r}\rangle = x|\mathbf{r}\rangle \qquad \hat{y}|\mathbf{r}\rangle = y|\mathbf{r}\rangle \qquad \hat{z}|\mathbf{r}\rangle = z|\mathbf{r}\rangle \qquad (9.2)$$

We express an arbitrary state $|\psi\rangle$ as a superposition of position states by

$$|\psi\rangle = \int\int\int dx\, dy\, dz\, |x, y, z\rangle\langle x, y, z|\psi\rangle = \int d^3r\, |\mathbf{r}\rangle\langle\mathbf{r}|\psi\rangle \qquad (9.3)$$

where the integrals run over all space. If we consider the special case where $|\psi\rangle = |x', y', z'\rangle$, we see that

$$\langle x, y, z|x', y', z'\rangle = \delta(x - x')\delta(y - y')\delta(z - z') \qquad (9.4)$$

or more compactly

$$\langle\mathbf{r}|\mathbf{r}'\rangle = \delta^3(\mathbf{r} - \mathbf{r}') \qquad (9.5)$$

The superscript on the Dirac delta function emphasizes that this is actually three delta functions.

Using the normalization condition, we see that

$$1 = \langle\psi|\psi\rangle = \int\int\int dx\, dy\, dz\, |\langle x, y, z|\psi\rangle|^2 = \int d^3r\, |\langle\mathbf{r}|\psi\rangle|^2 \qquad (9.6)$$

indicating that we should identify

$$dx\, dy\, dz\, |\langle x, y, z|\psi\rangle|^2 = d^3r\, |\langle\mathbf{r}|\psi\rangle|^2 \qquad (9.7)$$

with the probability of finding a particle in the state $|\psi\rangle$ in the volume $d^3r$ at $\mathbf{r}$ if a measurement of the position of the particle is carried out.

Just as we did in one dimension, we now introduce a three-dimensional translation operator that satisfies:

$$\hat{T}(a_x\mathbf{i})|x, y, z\rangle = |x + a_x, y, z\rangle \qquad (9.8a)$$

$$\hat{T}(a_y\mathbf{j})|x, y, z\rangle = |x, y + a_y, z\rangle \qquad (9.8b)$$

$$\hat{T}(a_z\mathbf{k})|x, y, z\rangle = |x, y, z + a_z\rangle \qquad (9.8c)$$

or, in short,

$$\hat{T}(\mathbf{a})|\mathbf{r}\rangle = |\mathbf{r} + \mathbf{a}\rangle \qquad (9.9)$$

As in (6.26), these translation operators can be expressed in terms of the three generators of translations $\hat{p}_x$, $\hat{p}_y$, and $\hat{p}_z$:

$$\hat{T}(a_x\mathbf{i}) = e^{-i\hat{p}_x a_x/\hbar} \qquad (9.10a)$$

$$\hat{T}(a_y\mathbf{j}) = e^{-i\hat{p}_y a_y/\hbar} \qquad (9.10b)$$

$$\hat{T}(a_z\mathbf{k}) = e^{-i\hat{p}_z a_z/\hbar} \qquad (9.10c)$$

In contrast to what we saw with rotations in Chapter 3, successive translations in different directions, such as in the $x$ and $y$ direction, clearly commute with each other (see Fig. 9.1). Thus

$$\hat{T}(a_y\mathbf{j})\hat{T}(a_x\mathbf{i}) = \hat{T}(a_x\mathbf{i})\hat{T}(a_y\mathbf{j}) \qquad (9.11)$$

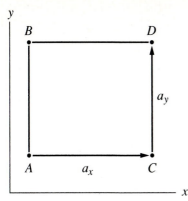

**FIGURE 9.1**

Translations along different directions commute: the translation operator $\hat{T}(a_x\mathbf{i})\hat{T}(a_y\mathbf{j})$, which is indicated by the path $ABD$, has the same effect as the translation operator $\hat{T}(a_y\mathbf{j})\hat{T}(a_x\mathbf{i})$, which is indicated by the path $ACD$.

If we substitute the series expansion

$$\hat{T}(a_x\mathbf{i}) = 1 - \frac{i\hat{p}_x a_x}{\hbar} - \frac{\hat{p}_x^2 a_x^2}{2\hbar^2} + \cdots \tag{9.12}$$

and the corresponding expression for $\hat{T}(a_y\mathbf{j})$ into (9.11) and retain terms through second order, we can show that the generators of translations along different directions commute:

$$\left[\hat{p}_x, \hat{p}_y\right] = 0 \tag{9.13}$$

See Problem 9.1. We can thus express the three-dimensional translation operator simply as[2]

$$\hat{T}(\mathbf{a}) = e^{-i\hat{p}_x a_x/\hbar}\, e^{-i\hat{p}_y a_y/\hbar}\, e^{-i\hat{p}_z a_z/\hbar} = e^{-i\hat{\mathbf{p}}\cdot\mathbf{a}/\hbar} \tag{9.14}$$

As we saw in Chapter 6, the generator of translations in a particular direction does not commute with the corresponding position operator. In three dimensions, this leads to the commutation relations

$$\left[\hat{x}, \hat{p}_x\right] = i\hbar \qquad \left[\hat{y}, \hat{p}_y\right] = i\hbar \qquad \left[\hat{z}, \hat{p}_z\right] = i\hbar \tag{9.15}$$

However, the generator of translations along an axis—for example, the $x$ axis—does commute with the position operator along an orthogonal direction, say the $y$ direction:

$$\hat{T}(a_x\mathbf{i})\hat{y}|\psi\rangle = \hat{T}(a_x\mathbf{i})\hat{y}\int\int\int dx\,dy\,dz\,|x,y,z\rangle\langle x,y,z|\psi\rangle$$

$$= \hat{T}(a_x\mathbf{i})\int\int\int dx\,dy\,dz\,y|x,y,z\rangle\langle x,y,z|\psi\rangle$$

---

[2] The product of two exponential operators can be replaced by the exponential of the sum of the two operators only when the two operators commute. See Problem 9.2.

$$= \int \int \int dx\, dy\, dz\, y |x + a_x, y, z\rangle\langle x, y, z|\psi\rangle$$

$$= \hat{y}\hat{T}(a_x\mathbf{i})|\psi\rangle \tag{9.16}$$

which indicates that

$$\left[\hat{y}, \hat{T}(a_x\mathbf{i})\right] = 0 \tag{9.17}$$

since $|\psi\rangle$ is arbitrary. Notice in this derivation that it is really adequate to verify that the operators commute when acting on an arbitrary position eigenstate $|x, y, z\rangle$ because, as (9.3) shows, we can express any state as a superposition of these position states. For (9.17) to be valid for arbitrary $a_x$,

$$\left[\hat{y}, \hat{p}_x\right] = 0 \tag{9.18}$$

In fact, the complete set of position-momentum commutation relations can be expressed in the shorthand form

$$\left[\hat{x}_i, \hat{p}_j\right] = i\hbar\delta_{ij} \tag{9.19}$$

where $i$ and $j$ each run over 1, 2, and 3, representing $x$, $y$, and $z$ components, respectively ($\hat{x}_1 = \hat{x}$, $\hat{x}_2 = \hat{y}$, and $\hat{x}_3 = \hat{z}$).

The generators of translations are of course the momentum operators. Since these operators commute with each other, we can form three-dimensional momentum states that are simultaneously eigenstates of $\hat{p}_x$, $\hat{p}_y$, and $\hat{p}_z$:

$$|p_x, p_y, p_z\rangle = |\mathbf{p}\rangle \tag{9.20}$$

where

$$\hat{p}_x|\mathbf{p}\rangle = p_x|\mathbf{p}\rangle \qquad \hat{p}_y|\mathbf{p}\rangle = p_y|\mathbf{p}\rangle \qquad \hat{p}_z|\mathbf{p}\rangle = p_z|\mathbf{p}\rangle \tag{9.21}$$

As with the position states, we normalize the momentum states by

$$\langle\mathbf{p}'|\mathbf{p}\rangle = \delta(p_x' - p_x)\delta(p_y' - p_y)\delta(p_z' - p_z) = \delta^3(\mathbf{p}' - \mathbf{p}) \tag{9.22}$$

and therefore

$$d^3p\, |\langle\mathbf{p}|\psi\rangle|^2 \tag{9.23}$$

is the probability of finding the momentum of a particle in the state $|\psi\rangle$ between $\mathbf{p}$ and $\mathbf{p} + d\mathbf{p}$.

Finally, we can establish that the generalization of (6.42) is given by

$$\langle\mathbf{r}|\hat{\mathbf{p}}|\psi\rangle = \frac{\hbar}{i}\nabla\langle\mathbf{r}|\psi\rangle \tag{9.24}$$

Taking $|\psi\rangle = |\mathbf{p}\rangle$, a momentum eigenket, we can solve this differential equation [as we did (6.51)] to obtain the three-dimensional momentum eigenfunction in position space:

$$\langle \mathbf{r}|\mathbf{p}\rangle = \left(\frac{1}{\sqrt{2\pi\hbar}}e^{i\,p_x x/\hbar}\right)\left(\frac{1}{\sqrt{2\pi\hbar}}e^{i\,p_y y/\hbar}\right)\left(\frac{1}{\sqrt{2\pi\hbar}}e^{i\,p_z z/\hbar}\right)$$

$$= \frac{1}{(2\pi\hbar)^{3/2}}e^{i\mathbf{p}\cdot\mathbf{r}/\hbar} \tag{9.25}$$

which is just the product of three momentum eigenfunctions like (6.54).

## 9.2  TRANSLATIONAL INVARIANCE AND CONSERVATION OF LINEAR MOMENTUM

The Hamiltonian for two bodies with a potential energy of interaction that depends on the magnitude of the distance separating the two bodies is given by

$$\hat{H} = \frac{\hat{\mathbf{p}}_1^2}{2m_1} + \frac{\hat{\mathbf{p}}_2^2}{2m_2} + V(|\hat{\mathbf{r}}_1 - \hat{\mathbf{r}}_2|) \tag{9.26}$$

where $\hat{\mathbf{p}}_1$ is the momentum operator for particle 1 and

$$\hat{\mathbf{p}}_1^2 = \hat{p}_{1x}^2 + \hat{p}_{1y}^2 + \hat{p}_{1z}^2 \tag{9.27}$$

Similarly, $\hat{\mathbf{p}}_2$ is the momentum operator for particle 2. It may seem strange to begin our discussion with a two-body problem instead of a one-body problem. However, any nontrivial Hamiltonian arises from the interaction of one body with at least one other body, so we might as well start with the two-body system. By far the most important example of a two-body system for which the Hamiltonian is in the form of (9.26) is the hydrogen atom, where the potential energy $V = -e^2/|\mathbf{r}_1 - \mathbf{r}_2|$. We will take advantage of what we learn in this chapter to solve the hydrogen atom, as well as some other two-body problems, in Chapter 10.

For the time being we are presuming it is safe to neglect any spin degrees of freedom, so we introduce just the two-body position basis states

$$|\mathbf{r}_1, \mathbf{r}_2\rangle = |\mathbf{r}_1\rangle_1 \otimes |\mathbf{r}_2\rangle_2 \tag{9.28}$$

The right-hand side expresses these two-particle states in terms of the direct product of single-particle position states, just as in Chapter 5 we expressed the two-particle spin states of two spin-$\frac{1}{2}$ particles as a direct product of single-particle spin states. Notice that we can translate the position of particle 1 leaving the position of particle 2 fixed:

$$\hat{T}_1(\mathbf{a})|\mathbf{r}_1, \mathbf{r}_2\rangle = e^{-i\hat{\mathbf{p}}_1\cdot\mathbf{a}/\hbar}|\mathbf{r}_1, \mathbf{r}_2\rangle = |\mathbf{r}_1 + \mathbf{a}, \mathbf{r}_2\rangle \tag{9.29a}$$

and similarly

$$\hat{T}_2(\mathbf{a})|\mathbf{r}_1, \mathbf{r}_2\rangle = e^{-i\hat{\mathbf{p}}_2\cdot\mathbf{a}/\hbar}|\mathbf{r}_1, \mathbf{r}_2\rangle = |\mathbf{r}_1, \mathbf{r}_2 + \mathbf{a}\rangle \tag{9.29b}$$

Thus we see that the generators commute:

$$\left[\hat{\mathbf{p}}_1, \hat{\mathbf{p}}_2\right] = 0 \tag{9.30}$$

and that the translation operator that translates *both* of the particles is given by

$$\hat{T}_1(\mathbf{a})\hat{T}_2(\mathbf{a}) = e^{-i\hat{\mathbf{p}}_1 \cdot \mathbf{a}/\hbar}\, e^{-i\hat{\mathbf{p}}_2 \cdot \mathbf{a}/\hbar} = e^{-i(\hat{\mathbf{p}}_1 + \hat{\mathbf{p}}_2)\mathbf{a}/\hbar} = e^{-i\hat{\mathbf{P}} \cdot \mathbf{a}/\hbar} \tag{9.31}$$

where

$$\hat{\mathbf{P}} = \hat{\mathbf{p}}_1 + \hat{\mathbf{p}}_2 \tag{9.32}$$

is the total momentum operator for the system.

Since translating both of the particles does not affect the distance between them, as indicated in Fig. 9.2, we expect that the two-particle translation operator should commute with the Hamiltonian (9.26). This is an important result, worth examining in detail. As noted in the previous section, it is sufficient to show that the operators commute when acting on an arbitrary two-particle position state, because we can express any two-particle state $|\psi\rangle$ as a superposition of the two-particle position states:

$$|\psi\rangle = \int\int d^3r_1\, d^3r_2\, |\mathbf{r}_1, \mathbf{r}_2\rangle\langle\mathbf{r}_1, \mathbf{r}_2|\psi\rangle \tag{9.33}$$

Thus

$$\begin{aligned}
\hat{T}_1(\mathbf{a})\hat{T}_2(\mathbf{a})V(|\hat{\mathbf{r}}_1 - \hat{\mathbf{r}}_2|)|\mathbf{r}_1, \mathbf{r}_2\rangle &= \hat{T}_1(\mathbf{a})\hat{T}_2(\mathbf{a})V(|\mathbf{r}_1 - \mathbf{r}_2|)|\mathbf{r}_1, \mathbf{r}_2\rangle \\
&= V(|\mathbf{r}_1 - \mathbf{r}_2|)|\mathbf{r}_1 + \mathbf{a}, \mathbf{r}_2 + \mathbf{a}\rangle \\
&= V(|\hat{\mathbf{r}}_1 - \hat{\mathbf{r}}_2|)|\mathbf{r}_1 + \mathbf{a}, \mathbf{r}_2 + \mathbf{a}\rangle \\
&= V(|\hat{\mathbf{r}}_1 - \hat{\mathbf{r}}_2|)\hat{T}_1(\mathbf{a})\hat{T}_2(\mathbf{a})|\mathbf{r}_1, \mathbf{r}_2\rangle \quad (9.34)
\end{aligned}$$

where in the next-to-last step we have taken advantage of the fact that

$$\hat{\mathbf{r}}_1|\mathbf{r}_1 + \mathbf{a}, \mathbf{r}_2 + \mathbf{a}\rangle = (\mathbf{r}_1 + \mathbf{a})|\mathbf{r}_1 + \mathbf{a}, \mathbf{r}_2 + \mathbf{a}\rangle \tag{9.35a}$$

$$\hat{\mathbf{r}}_2|\mathbf{r}_1 + \mathbf{a}, \mathbf{r}_2 + \mathbf{a}\rangle = (\mathbf{r}_2 + \mathbf{a})|\mathbf{r}_1 + \mathbf{a}, \mathbf{r}_2 + \mathbf{a}\rangle \tag{9.35b}$$

and thus

$$(\hat{\mathbf{r}}_1 - \hat{\mathbf{r}}_2)|\mathbf{r}_1 + \mathbf{a}, \mathbf{r}_2 + \mathbf{a}\rangle = (\mathbf{r}_1 - \mathbf{r}_2)|\mathbf{r}_1 + \mathbf{a}, \mathbf{r}_2 + \mathbf{a}\rangle \tag{9.36}$$

Equation (9.34) shows that

$$\left[V(|\hat{\mathbf{r}}_1 - \hat{\mathbf{r}}_2|), \hat{T}_1(\mathbf{a})\hat{T}_2(\mathbf{a})\right] = 0 \tag{9.37}$$

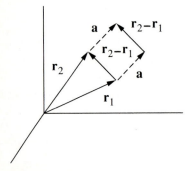

**FIGURE 9.2**
Translating both bodies in a two-body system by $\mathbf{a}$ leaves the distance between the bodies unchanged.

From the explicit form of $\hat{T}_1(\mathbf{a})\hat{T}_2(\mathbf{a})$ in terms of the momentum operators, it is also clear that

$$\left[ \frac{\hat{\mathbf{p}}_1^2}{2m_1} + \frac{\hat{\mathbf{p}}_2^2}{2m_2}, \hat{T}_1(\mathbf{a})\hat{T}_2(\mathbf{a}) \right] = 0 \qquad (9.38)$$

and therefore

$$\left[ \hat{H}, \hat{T}_1(\mathbf{a})\hat{T}_2(\mathbf{a}) \right] = 0 \qquad (9.39)$$

Thus from (9.31) we see that the Hamiltonian commutes with the operator that generates translations for both of the particles:

$$\left[ \hat{H}, \hat{\mathbf{P}} \right] = 0 \qquad (9.40)$$

Recall from (4.16) that

$$\frac{d\langle \mathbf{P} \rangle}{dt} = \frac{i}{\hbar} \langle \psi | \left[ \hat{H}, \hat{\mathbf{P}} \right] | \psi \rangle \qquad (9.41)$$

Thus the translational invariance of the Hamiltonian guarantees that the total momentum of the system is conserved. Translational invariance is another illustration of the deep connection between symmetries of the Hamiltonian and conservation laws. At the end of Chapter 7 we saw that the harmonic oscillator possesses inversion symmetry; the parity operator inverts the coordinates and leaves the Hamiltonian invariant. Thus the Hamiltonian and the parity operator commute and parity is conserved. Inversion symmetry is a discrete symmetry. Translation, on the other hand, is a continuous symmetry operation for the two-body Hamiltonian in that the Hamiltonian is invariant under translations by an arbitrary distance, leading to conservation of linear momentum.

Notice that if we look at how a particular state $|\psi(0)\rangle$ evolves with time,

$$|\psi(t)\rangle = e^{-i\hat{H}t/\hbar}|\psi(0)\rangle \qquad (9.42)$$

we see that the translated state $\hat{T}(\mathbf{a})|\psi(0)\rangle$ at time $t$ differs from the state $|\psi(t)\rangle$ by just a translation:

$$e^{-i\hat{H}t/\hbar}\hat{T}(\mathbf{a})|\psi(0)\rangle = \hat{T}(\mathbf{a})e^{-i\hat{H}t/\hbar}|\psi(0)\rangle = \hat{T}(\mathbf{a})|\psi(t)\rangle \qquad (9.43)$$

since the translation operator commutes with the Hamiltonian. Thus if you were to carry out experiments in a movable laboratory (without windows), you would not be able to determine whether the laboratory had been displaced based solely on experiments carried out within the laboratory.

In our analysis in this section, we have used translational invariance to argue that momentum is conserved. However, we can also turn the argument around: If momentum is conserved, the system is translationally invariant because the momentum operator is the generator of translations. What would break or destroy this translational symmetry? From classical physics we know the momentum of the system is not conserved if an external force acts on the system. Suppose that in our example of the hydrogen atom we insert a third charge $q$ at position $\mathbf{r}_3$, which interacts with both the proton at $\mathbf{r}_1$ and the electron at $\mathbf{r}_2$. The Hamiltonian of the three bodies including just their Coulomb interactions is then given by

$$\hat{H} = \frac{\hat{\mathbf{p}}_1^2}{2m_1} + \frac{\hat{\mathbf{p}}_2^2}{2m_2} + \frac{\hat{\mathbf{p}}_3^2}{2m_3} - \frac{e^2}{|\hat{\mathbf{r}}_1 - \hat{\mathbf{r}}_2|} + \frac{qe}{|\hat{\mathbf{r}}_1 - \hat{\mathbf{r}}_3|} - \frac{qe}{|\hat{\mathbf{r}}_2 - \hat{\mathbf{r}}_3|} \qquad (9.44)$$

We see that translating both the electron and the proton ($\mathbf{r}_1 \rightarrow \mathbf{r}_1 + \mathbf{a}$ and $\mathbf{r}_2 \rightarrow \mathbf{r}_2 + \mathbf{a}$) does not leave the Hamiltonian invariant. Therefore, total momentum of the electron-proton system is no longer conserved. However, if we enlarge our definition of the system to include all three particles, this three-particle system is invariant under translations of *all* the particles ($\mathbf{r}_1 \rightarrow \mathbf{r}_1 + \mathbf{a}$, $\mathbf{r}_2 \rightarrow \mathbf{r}_2 + \mathbf{a}$, *and* $\mathbf{r}_3 \rightarrow \mathbf{r}_3 + \mathbf{a}$), and thus the total momentum of the three-particle system is conserved. This translational invariance is not an accident but is built into the laws of electromagnetism, and not simply for static Coulomb interactions. In fact, all of the fundamental interactions—strong, weak, electromagnetic, and gravitational— seem to respect this translational symmetry. Thus, if we extend our definition of any system to include all of the bodies and fields that are interacting, we can be sure that the momentum of this system is conserved and that any experiment carried out on the system will give the same results as those carried out when the system is translated to a different position. This latter fact is often expressed by saying that space is homogeneous. Without this homogeneity we would have no confidence in our ability to apply the laws of physics as deduced, for example, from the behavior of hydrogen atoms here on earth to hydrogen atoms radiating in the distant interstellar medium.

## 9.3 RELATIVE AND CENTER-OF-MASS COORDINATES

The natural coordinates for the two-body problem when the Hamiltonian is of the form (9.26) are relative coordinates $\mathbf{r}$ and the center-of-mass coordinates $\mathbf{R}$, not the individual coordinates $\mathbf{r}_1$ and $\mathbf{r}_2$ of the bodies. The corresponding position operators are given by

$$\hat{\mathbf{r}} = \hat{\mathbf{r}}_1 - \hat{\mathbf{r}}_2 \qquad (9.45a)$$

$$\hat{\mathbf{R}} = \frac{m_1 \hat{\mathbf{r}}_1 + m_2 \hat{\mathbf{r}}_2}{m_1 + m_2} \qquad (9.45b)$$

Using the commutation relations (9.19) for each of the individual particles and (9.30), we see that the total momentum operator (9.32) satisfies the commutation relations

$$\left[\hat{x}_i, \hat{P}_j\right] = 0 \qquad (9.46)$$

which also follows from the invariance of the relative position under total translations. In addition,

$$\left[\hat{X}_i, \hat{P}_j\right] = i\hbar\delta_{ij} \qquad (9.47)$$

which shows that the total momentum and the position of the center of mass obey the usual canonical commutation relations of position and momentum. We also introduce the relative momentum operator

$$\hat{\mathbf{p}} = \frac{m_2\hat{\mathbf{p}}_1 - m_1\hat{\mathbf{p}}_2}{m_1 + m_2} \tag{9.48}$$

which satisfies the canonical commutation relations with $\hat{\mathbf{r}}$:

$$\left[\hat{x}_i, \hat{p}_j\right] = i\hbar\delta_{ij} \tag{9.49}$$

as well as the commutation relation

$$\left[\hat{X}_i, \hat{p}_j\right] = 0 \tag{9.50}$$

with $\hat{\mathbf{R}}$. Commutation relations (9.46) and (9.50) show that the relative and center-of-mass operators all commute with each other.

We will use the states $|\mathbf{r}, \mathbf{R}\rangle$ instead of $|\mathbf{r}_1, \mathbf{r}_2\rangle$ as a basis for our discussion of the two-body problem. The reason for this choice becomes apparent when we express the two-body Hamiltonian (9.26) in terms of the relative and center-of-mass operators. We find

$$\hat{H} = \frac{\hat{\mathbf{P}}^2}{2M} + \frac{\hat{\mathbf{p}}^2}{2\mu} + V(|\hat{\mathbf{r}}|) \tag{9.51}$$

where

$$M = m_1 + m_2 \tag{9.52a}$$

is the total mass of the system and

$$\mu = \frac{m_1 m_2}{m_1 + m_2} \tag{9.52b}$$

is the reduced mass. See Problem 9.5. The Hamiltonian (9.51) is the sum of the kinetic energy of the center of mass

$$\hat{H}_{\mathrm{cm}} = \frac{\hat{\mathbf{P}}^2}{2M} \tag{9.53}$$

and the energy of the relative motion of the two particles

$$\hat{H}_{\mathrm{rel}} = \frac{\hat{\mathbf{p}}^2}{2\mu} + V(|\hat{\mathbf{r}}|) \tag{9.54}$$

Since these two operators commute with each other, they have eigenstates $|E_{\mathrm{cm}}, E_{\mathrm{rel}}\rangle$ in common:

$$\hat{H}|E_{\mathrm{cm}}, E_{\mathrm{rel}}\rangle = (\hat{H}_{\mathrm{cm}} + \hat{H}_{\mathrm{rel}})|E_{\mathrm{cm}}, E_{\mathrm{rel}}\rangle = (E_{\mathrm{cm}} + E_{\mathrm{rel}})|E_{\mathrm{cm}}, E_{\mathrm{rel}}\rangle \tag{9.55}$$

and hence the energy eigenvalue of the two-body Hamiltonian is $E = E_{\mathrm{cm}} + E_{\mathrm{rel}}$.

The eigenstates of $\hat{H}_{\mathrm{cm}}$ are just those of the total momentum operator $\hat{\mathbf{P}}$. In position space, the total momentum eigenfunctions are given by

$$\langle \mathbf{R}|\mathbf{P}\rangle = \frac{1}{(2\pi\hbar)^{3/2}} e^{i\mathbf{P}\cdot\mathbf{R}/\hbar} \tag{9.56}$$

as in (9.25) except that here the momentum **P** is the momentum of the center of mass and the position variable **R** is the position of the center of mass. It is common to analyze the two-body problem in the center-of-mass frame, where **P** = 0 and therefore $E = E_{rel}$, since then the kinetic energy of the center of mass vanishes. Thus from now on we will concentrate our attention on just the Hamiltonian

$$\hat{H} = \frac{\hat{\mathbf{p}}^2}{2\mu} + V(|\hat{\mathbf{r}}|) \tag{9.57}$$

This Hamiltonian is the same as that for a single body in the central potential $V(r)$, provided the mass of the body is taken to be the reduced mass of the two-body system. This is the familiar result from classical mechanics, but here expressed in terms of operators. Thus in analyzing the Hamiltonian (9.57), we are analyzing a single body in a central potential as well as two bodies interacting through a potential energy that depends on the magnitude of the distance between them.

## 9.4   ESTIMATING GROUND-STATE ENERGIES USING THE UNCERTAINTY PRINCIPLE

Much of the remaining discussion in this chapter on orbital angular momentum will cover material that we will use in Chapter 10 in the determination of the energy eigenstates and eigenvalues of the Hamiltonian (9.57) for a number of specific central potentials. For now, it is useful to be able to estimate the energy scale for systems like the hydrogen atom without actually solving the energy eigenvalue equation. The Hamiltonian for the hydrogen atom, including only the predominant Coulomb interaction between the particles, is given by

$$\hat{H} = \frac{\hat{\mathbf{p}}^2}{2\mu} - \frac{e^2}{|\hat{\mathbf{r}}|} \tag{9.58}$$

with the reduced mass being that of the electron-proton system.[3]

The expectation value of the Hamiltonian (9.58) in the ground state is given by

$$E_1 = \langle \frac{\mathbf{p}^2}{2\mu} - \frac{e^2}{r} \rangle \tag{9.59}$$

We denote this energy by $E_1$ because, as we will find in Chapter 10, this state is labeled by the principal quantum number $n = 1$. Using dimensional analysis, we can express

$$\langle \frac{e^2}{r} \rangle = \frac{e^2}{a} \tag{9.60}$$

---

[3] In SI units, the potential energy is $-e^2/4\pi\varepsilon_0 r$. If you want to work in SI units, just consider $e^2$ a shorthand notation for $e^2/4\pi\varepsilon_0$.

where $a$ is a length, characteristic of the size of the atom, that we will now estimate. But if the atom has a finite size, the uncertainty in the relative position of the two particles is also at most on the order of $a$. A finite position uncertainty means there must be a finite momentum uncertainty as well. From the Heisenberg uncertainty relation, we expect that

$$|\Delta \mathbf{p}| \gtrsim \frac{\hbar}{a} \tag{9.61}$$

Note that we have not actually calculated the position uncertainty and thus the value we are taking for the momentum uncertainty is a rough estimate.

The expectation value of the kinetic energy is

$$\frac{\langle \mathbf{p}^2 \rangle}{2\mu} = \frac{\Delta \mathbf{p}^2 + \langle \mathbf{p} \rangle^2}{2\mu} = \frac{\Delta \mathbf{p}^2}{2\mu} \tag{9.62}$$

where in the last step we have taken $\langle \mathbf{p} \rangle = 0$, since $\langle \mathbf{p} \rangle$ is independent of time in a stationary state—the ground state is, of course, an energy eigenstate—and if $\langle \mathbf{p} \rangle \neq 0$, the system would not stay within a particular region of space. Our estimate of the total energy is thus given by

$$E_1 \gtrsim \frac{\hbar^2}{2\mu a^2} - \frac{e^2}{a} \tag{9.63}$$

Decreasing the value for $a$ decreases the potential energy. However, it also increases the kinetic energy. Clearly, there is an optimum value for $a$ that minimizes the energy.[4] Setting $dE_1/da = 0$, we find

$$a = \frac{\hbar^2}{m_e e^2} \tag{9.64}$$

and hence

$$E_1 \simeq -\frac{m_e e^4}{2\hbar^2} \tag{9.65}$$

where we have replaced the reduced mass by the electron mass since the two differ by only 1 part in 2000. If we put in numerical values for the mass and charge, we find that $a$ is on the order of angstroms and the energy is on the order of 10 eV. Although our estimates are strictly only order-of-magnitude estimates and we should be lucky to be within a factor of two of the exact ground-state energy, we have judiciously chosen (9.61) so that (9.65) turns out to be the exact value (13.6 eV) that we will find in Chapter 10.

The important thing to note at this point is how quantum mechanics has saved the atom from collapse. In classical physics, one could always lower the energy of the system by putting the proton and the electron closer together. In

---

[4] Also see the discussion in Section 12.2 on the variational method.

fact, before the discovery of quantum mechanics the stability of atoms was a puzzle. But we now see that making the atom smaller increases the kinetic energy of the system so that there is a natural resistance to compressing the atom into too small a region. Quantum mechanics with its own fundamental constant, Planck's constant, has set the natural length scale (9.64) for atomic physics, as well as the natural energy scale (9.65).

## 9.5 ROTATIONAL INVARIANCE AND CONSERVATION OF ANGULAR MOMENTUM

Let's continue our general analysis of the Hamiltonian (9.57). One of the first things we notice about this Hamiltonian is that it possesses rotational symmetry, since $\mathbf{p}^2 = \mathbf{p} \cdot \mathbf{p} = p_x^2 + p_y^2 + p_z^2$ and $r = |\mathbf{r}| = (\mathbf{r} \cdot \mathbf{r})^{1/2} = (x^2 + y^2 + z^2)^{1/2}$ are both invariant under rotations; they involve the length of a vector and the length of a vector doesn't change when it is rotated. The Hamiltonian, however, involves operators, not just ordinary numbers, and it is instructive to verify explicitly that it is rotationally invariant.

Let's consider the operator $\hat{R}(d\phi\,\mathbf{k})$, which rotates a position state counter-clockwise about the $z$ axis by an angle $d\phi$. Notice that there is nothing in the Hamiltonian that picks out a specific direction in space, so it is completely arbitrary which direction we choose to call the $z$ direction. Using (3.2), which shows how an arbitrary vector changes when rotated about the $z$ axis, we see that for an infinitesimal rotation

$$\hat{R}(d\phi\,\mathbf{k})|x, y, z\rangle = |x - y\,d\phi, y + x\,d\phi, z\rangle \tag{9.66}$$

We express the rotation operator $\hat{R}(d\phi\,\mathbf{k})$ in terms of the generator of rotations as

$$\hat{R}(d\phi\,\mathbf{k}) = 1 - \frac{i}{\hbar}\hat{L}_z\,d\phi \tag{9.67}$$

where we have called the generator $\hat{L}_z$ instead of $\hat{J}_z$ as in Chapter 3 because, as we will now see, this generator is the *orbital* angular momentum. Taking advantage of (9.8), (9.10), and the form of the translation operator (9.12) when the translation is infinitesimal, we can write through first order in the infinitesimal angle $d\phi$

$$|x - y\,d\phi, y + x\,d\phi, z\rangle = \left[1 - \frac{i}{\hbar}\hat{p}_x(-y\,d\phi)\right]\left[1 - \frac{i}{\hbar}\hat{p}_y(x\,d\phi)\right]|x, y, z\rangle$$

$$= \left[1 - \frac{i}{\hbar}(\hat{x}\,\hat{p}_y - \hat{y}\,\hat{p}_x)\,d\phi\right]|x, y, z\rangle \tag{9.68}$$

Thus the generator of rotations about the $z$ axis in position space is simply

$$\hat{L}_z = \hat{x}\,\hat{p}_y - \hat{y}\,\hat{p}_x \tag{9.69}$$

which is the $z$ component of the orbital angular momentum operator $\hat{\mathbf{L}} = \hat{\mathbf{r}} \times \hat{\mathbf{p}}$. We finally see *orbital* angular momentum entering as the generator of rotations because we have turned our attention to rotations that move *position* states around.

One way to confirm that the Hamiltonian is rotationally invariant is to check that it commutes with the generator of rotations. Using the position-momentum commutation relations (9.49) and the form (9.69) for $\hat{L}_z$, we find that

$$\left[\hat{L}_z, \hat{p}_x\right] = \left[\hat{x}\,\hat{p}_y - \hat{y}\,\hat{p}_x,\ \hat{p}_x\right] = \left[\hat{x}\,\hat{p}_y,\ \hat{p}_x\right]$$

$$= \left[\hat{x},\ \hat{p}_x\right]\hat{p}_y = i\hbar\,\hat{p}_y \qquad (9.70a)$$

$$\left[\hat{L}_z, \hat{p}_y\right] = \left[\hat{x}\,\hat{p}_y - \hat{y}\,\hat{p}_x,\ \hat{p}_y\right] = -\left[\hat{y}\,\hat{p}_x,\ \hat{p}_y\right]$$

$$= -\left[\hat{y},\ \hat{p}_y\right]\hat{p}_x = -i\hbar\,\hat{p}_x \qquad (9.70b)$$

$$\left[\hat{L}_z, \hat{p}_z\right] = \left[\hat{x}\,\hat{p}_y - \hat{y}\,\hat{p}_x,\ \hat{p}_z\right] = 0 \qquad (9.70c)$$

and therefore

$$\left[\hat{L}_z, \hat{\mathbf{p}}^2\right] = \left[\hat{L}_z,\ \hat{p}_x^2 + \hat{p}_y^2 + \hat{p}_z^2\right]$$

$$= \hat{p}_x\left[\hat{L}_z, \hat{p}_x\right] + \left[\hat{L}_z, \hat{p}_x\right]\hat{p}_x + \hat{p}_y\left[\hat{L}_z, \hat{p}_y\right] + \left[\hat{L}_z, \hat{p}_y\right]\hat{p}_y$$

$$= 2i\hbar\,\hat{p}_x\,\hat{p}_y - 2i\hbar\,\hat{p}_x\,\hat{p}_y = 0 \qquad (9.71)$$

Similarly, we can establish

$$\left[\hat{L}_z, \hat{x}\right] = i\hbar\hat{y} \qquad (9.72a)$$

$$\left[\hat{L}_z, \hat{y}\right] = -i\hbar\hat{x} \qquad (9.72b)$$

$$\left[\hat{L}_z, \hat{z}\right] = 0 \qquad (9.72c)$$

and therefore

$$\left[\hat{L}_z,\ \hat{x}^2 + \hat{y}^2 + \hat{z}^2\right] = 0 \qquad (9.73)$$

as well.[5] Thus $\hat{L}_z$ commutes with a potential energy that is a function of the magnitude of the radius vector.

There is another instructive way to establish that $V(|\hat{\mathbf{r}}|)$ is invariant under rotations. We express a particular point in position space in terms of the spherical coordinates shown in Fig. 9.3:

$$|x, y, z\rangle = |r, \theta, \phi\rangle \qquad (9.74)$$

The advantage of these coordinates is that the action of the operator $\hat{R}(d\phi\,\mathbf{k})$ on these states is transparent. Namely,

$$\hat{R}(d\phi\,\mathbf{k})|r, \theta, \phi\rangle = |r, \theta, \phi + d\phi\rangle \qquad (9.75)$$

is clearly a rotation of the state by angle $d\phi$ about the $z$ axis. Thus

---

[5] Notice the similarity in how $\hat{\mathbf{r}}$ and $\hat{\mathbf{p}}$ transform under rotations. In fact, the commutation relations (9.80) of the orbital angular momentum operators can be cast in a similar form as well:

$$[\hat{L}_z, \hat{L}_x] = i\hbar\hat{L}_y \qquad [\hat{L}_z, \hat{L}_y] = -i\hbar\hat{L}_x \qquad [\hat{L}_z, \hat{L}_z] = 0$$

*All* vector operators must behave the same way when they are rotated. See Problem 9.6.

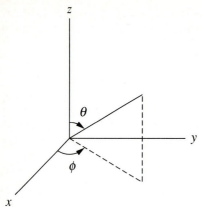

**FIGURE 9.3**
Spherical coordinates.

$$\hat{R}(d\phi\,\mathbf{k})V(|\hat{\mathbf{r}}|)|r,\theta,\phi\rangle = \hat{R}(d\phi\,\mathbf{k})V(r)|r,\theta,\phi\rangle$$
$$= V(r)|r,\theta,\phi+d\phi\rangle$$
$$= V(|\hat{\mathbf{r}}|)\hat{R}(d\phi\,\mathbf{k})|r,\theta,\phi\rangle \qquad (9.76)$$

Since the two operators commute when acting on an arbitrary position state, the two operators commute in general.

We have established that the Hamiltonian (9.57) commutes with the generator of rotations about the $z$ axis:

$$\left[\hat{H},\,\hat{L}_z\right] = 0 \qquad (9.77)$$

What we have chosen to call the $z$ axis could just as easily be called the $x$ or the $y$ axis by someone else. Therefore it must be true that

$$\left[\hat{H},\,\hat{L}_x\right] = 0 \qquad (9.78)$$

and

$$\left[\hat{H},\,\hat{L}_y\right] = 0 \qquad (9.79)$$

The system is invariant under rotations about any axis. Since the Hamiltonian commutes with the angular momentum operators, angular momentum is conserved. Rotational symmetry has led to a conservation law—conservation of angular momentum! Moreover, since the Hamiltonian commutes with the rotation operator, rotating a system does not affect the time evolution of the system, provided everything interacting with that system is rotated as well. We say this rotational invariance is a reflection of the isotropy of space, just as translational invariance is a reflection of its homogeneity.

## 9.6   A COMPLETE SET OF COMMUTING OBSERVABLES

Our goal in this section is to take advantage of what we have learned in the preceding section about the symmetry of the Hamiltonian to specify the energy eigenstates as well as write out the energy eigenvalue equation in position space.

First, let us note that since $\hat{L}_x$, $\hat{L}_y$, and $\hat{L}_z$ are the generators of rotations, they must satisfy the general commutation relations

$$\left[\hat{L}_x, \hat{L}_y\right] = i\hbar\hat{L}_z \qquad \left[\hat{L}_y, \hat{L}_z\right] = i\hbar\hat{L}_x \qquad \left[\hat{L}_z, \hat{L}_x\right] = i\hbar\hat{L}_y \qquad (9.80)$$

as discussed in Chapter 3. A useful way to express these commutation relations, as well as the commutation relations of any angular momentum operators, is in terms of the completely antisymmetric tensor $\varepsilon_{ijk}$. Complete antisymmetry means that the tensor changes sign if any two of the indices are interchanged: $\varepsilon_{ijk} = -\varepsilon_{jik}$ and so on. The other defining relation is that $\varepsilon_{123} = 1$. The complete antisymmetry determines all of the other components. For example, $\varepsilon_{132} = -\varepsilon_{123} = -1$ and if any two of the indices are the same, the tensor vanishes ($\varepsilon_{112} = -\varepsilon_{112} = 0$). The commutation relations (9.80) may now be put in the more compact form

$$\left[\hat{L}_i, \hat{L}_j\right] = i\hbar \sum_{k=1}^{3} \varepsilon_{ijk}\hat{L}_k \qquad (9.81)$$

where $i$, $j$, and $k$ take on the values from 1 to 3 and $\hat{L}_1$, $\hat{L}_2$, and $\hat{L}_3$ stand for $\hat{L}_x$, $\hat{L}_y$, and $\hat{L}_z$, respectively. In fact, using $\varepsilon_{ijk}$, we can express the ordinary cross product in component form as

$$\hat{L}_i = \sum_{j=1}^{3} \sum_{k=1}^{3} \varepsilon_{ijk}\hat{x}_j\hat{p}_k \qquad (9.82)$$

Since the generators of rotations about different axes do not commute with each other, we cannot choose more than one of them to simultaneously label the eigenstates of the Hamiltonian. However, since each of the generators as well as the Hamiltonian commutes with

$$\hat{\mathbf{L}}^2 = \hat{L}_x^2 + \hat{L}_y^2 + \hat{L}_z^2 \qquad (9.83)$$

we can form simultaneous eigenstates of $\hat{H}$, $\hat{\mathbf{L}}^2$, and one of the components of the angular momentum, which is generally taken to be $\hat{L}_z$. We then label these eigenstates by $|E, l, m\rangle$, where

$$\hat{H}|E, l, m\rangle = E|E, l, m\rangle \qquad (9.84a)$$

$$\hat{\mathbf{L}}^2|E, l, m\rangle = l(l + 1)\hbar^2|E, l, m\rangle \qquad (9.84b)$$

$$\hat{L}_z|E, l, m\rangle = m\hbar|E, l, m\rangle \qquad (9.84c)$$

The Hamiltonian (9.57) actually involves the operator $\hat{\mathbf{L}}^2$, although it is hidden away as part of the rotational kinetic energy. To see this, we first use the identity (see Problem 9.7)

$$\hat{\mathbf{L}}^2 = \hat{\mathbf{r}} \times \hat{\mathbf{p}} \cdot \hat{\mathbf{r}} \times \hat{\mathbf{p}} = \hat{r}^2\hat{p}^2 - (\hat{\mathbf{r}} \cdot \hat{\mathbf{p}})^2 + i\hbar\hat{\mathbf{r}} \cdot \hat{\mathbf{p}} \qquad (9.85)$$

Since we wish to express the energy eigenvalue equation in position space, we now evaluate

$$\langle\mathbf{r}|\hat{\mathbf{L}}^2|\psi\rangle = \langle\mathbf{r}|[\hat{r}^2\hat{p}^2 - (\hat{\mathbf{r}} \cdot \hat{\mathbf{p}})^2 + i\hbar\hat{\mathbf{r}} \cdot \hat{\mathbf{p}}]|\psi\rangle \qquad (9.86)$$

Note that

$$\langle \mathbf{r}|\hat{r}^2\hat{\mathbf{p}}^2|\psi\rangle = r^2\langle \mathbf{r}|\hat{\mathbf{p}}^2|\psi\rangle \tag{9.87}$$

and

$$\langle \mathbf{r}|\hat{\mathbf{r}} \cdot \hat{\mathbf{p}}|\psi\rangle = \mathbf{r} \cdot \frac{\hbar}{i}\nabla\langle \mathbf{r}|\psi\rangle = \frac{\hbar}{i}r\frac{\partial}{\partial r}\langle \mathbf{r}|\psi\rangle \tag{9.88}$$

Thus

$$\langle \mathbf{r}|(\hat{\mathbf{r}} \cdot \hat{\mathbf{p}})^2|\psi\rangle = -\hbar^2 r\frac{\partial}{\partial r}\left(r\frac{\partial}{\partial r}\right)\langle \mathbf{r}|\psi\rangle \tag{9.89}$$

Combining these results, we see that

$$\frac{1}{2\mu}\langle \mathbf{r}|\hat{\mathbf{p}}^2|\psi\rangle = -\frac{\hbar^2}{2\mu}\left(\frac{\partial^2}{\partial r^2} + \frac{2}{r}\frac{\partial}{\partial r}\right)\langle \mathbf{r}|\psi\rangle + \frac{\langle \mathbf{r}|\hat{\mathbf{L}}^2|\psi\rangle}{2\mu r^2} \tag{9.90}$$

and thus the position-space energy eigenvalue equation is given by

$$\langle \mathbf{r}|\frac{\hat{\mathbf{p}}^2}{2\mu}|\psi\rangle + \langle \mathbf{r}|V(|\hat{\mathbf{r}}|)|\psi\rangle$$

$$= -\frac{\hbar^2}{2\mu}\left(\frac{\partial^2}{\partial r^2} + \frac{2}{r}\frac{\partial}{\partial r}\right)\langle \mathbf{r}|\psi\rangle + \frac{\langle \mathbf{r}|\hat{\mathbf{L}}^2|\psi\rangle}{2\mu r^2} + V(r)\langle \mathbf{r}|\psi\rangle = E\langle \mathbf{r}|\psi\rangle \tag{9.91}$$

The kinetic energy (9.90) has two parts. One of the parts is easily recognizable as the rotational kinetic energy $\hat{L}^2/2I$, with a moment of inertia $I = \mu r^2$ — just the moment of inertia that you would expect for a mass $\mu$ rotating a distance $r$ from a center of force. The other part of the kinetic energy must be the radial part. We can express this part in a form familiar from classical mechanics if we define the radial component of the momentum operator

$$\langle \mathbf{r}|\hat{p}_r|\psi\rangle = \frac{\hbar}{i}\left(\frac{\partial}{\partial r} + \frac{1}{r}\right)\langle \mathbf{r}|\psi\rangle \tag{9.92a}$$

or

$$\hat{p}_r \rightarrow \frac{\hbar}{i}\left(\frac{\partial}{\partial r} + \frac{1}{r}\right) \tag{9.92b}$$

in position space.[6] Expressed in terms of this operator, the radial part of the kinetic energy becomes

$$-\frac{\hbar^2}{2\mu}\left(\frac{\partial^2}{\partial r^2} + \frac{2}{r}\frac{\partial}{\partial r}\right)\langle \mathbf{r}|\psi\rangle = \frac{\langle \mathbf{r}|\hat{p}_r^2|\psi\rangle}{2\mu} \tag{9.93}$$

If we choose the state $|\psi\rangle$ to be a simultaneous eigenstate $\hat{H}$, $\hat{L}^2$, and $\hat{L}_z$, that is, $|\psi\rangle = |E, l, m\rangle$, (9.91) becomes

---

[6] The form for $\hat{p}_r$ in position space may seem a little strange. See Problem 10.1.

$$\left[-\frac{\hbar^2}{2\mu}\left(\frac{\partial^2}{\partial r^2} + \frac{2}{r}\frac{\partial}{\partial r}\right) + \frac{l(l+1)\hbar^2}{2\mu r^2} + V(r)\right]\langle \mathbf{r}|E, l, m\rangle = E\langle \mathbf{r}|E, l, m\rangle \quad (9.94)$$

Note that the expression in brackets in the left-hand side of this equation depends only on $r$, not on the angles $\theta$ and $\phi$. If we express the wave function in the form

$$\langle \mathbf{r}|E, l, m\rangle = R(r)\Theta(\theta)\Phi(\phi) \quad (9.95)$$

we obtain the radial differential equation

$$\left[-\frac{\hbar^2}{2\mu}\left(\frac{d^2}{dr^2} + \frac{2}{r}\frac{d}{dr}\right) + \frac{l(l+1)\hbar^2}{2\mu r^2} + V(r)\right] R(r) = ER(r) \quad (9.96)$$

where we have divided out the angular part of the wave function.

We will devote Chapter 10 to solving this equation for a number of specific central potentials. For now, we note that if we introduce the function $u(r)$ through

$$R(r) = \frac{u(r)}{r} \quad (9.97)$$

the radial equation simplifies to

$$\left[-\frac{\hbar^2}{2\mu}\frac{d^2}{dr^2} + \frac{l(l+1)\hbar^2}{2\mu r^2} + V(r)\right] u(r) = Eu(r) \quad (9.98)$$

This radial equation has the same form as the one-dimensional Schrödinger equation

$$\left[-\frac{\hbar^2}{2m}\frac{d^2}{dx^2} + V(x)\right]\langle x|E\rangle = E\langle x|E\rangle \quad (9.99)$$

but with an effective potential energy

$$V_{\text{eff}}(r) = \frac{l(l+1)\hbar^2}{2\mu r^2} + V(r) \quad (9.100)$$

This means that you can carry over any techniques, numerical or otherwise, that you know from solving the one-dimensional Schrödinger equation to help solve the radial equation.

Note that the lack of dependence of the energy eigenvalue equation (9.96) [or (9.98)] on the eigenvalue $m$ of $\hat{L}_z$ is a direct manifestation of the rotational invariance of the Hamiltonian. In essence, there is no preferred axis picked out in space, such as would be the case if an external magnetic field were applied to the atom in, for example, the $z$ direction, in which case the energy would indeed depend on the projection of the angular momentum along this axis. Nonetheless, we still need the $m$ value in (9.84) in order to specify each state uniquely—there are, after all, $2l + 1$ different $m$ states for each value of $l$. The set of operators that commute with each other that are necessary to label each state uniquely is referred to as a *complete set of commuting observables*. For a given system, there may exist several complete sets of commuting observables. For example, for the

Hamiltonian (9.57) we could use $\hat{H}$, $\hat{\mathbf{L}}^2$, and $\hat{L}_x$ to label the states instead of $\hat{H}$, $\hat{\mathbf{L}}^2$, and $\hat{L}_z$. However, for a real hydrogen atom neither of these sets of operators is complete, since it is necessary to specify both the electron and proton's intrinsic spin state in order to label the states uniquely. Assuming that the Hamiltonian is one of our complete set of commuting observables, we need to know how the spin operators $\hat{\mathbf{S}}_1$ of the electron and $\hat{\mathbf{S}}_2$ of the proton enter into this Hamiltonian before we can determine the other members of the complete set of operators that commute with $\hat{H}$. For example, the form of the hyperfine spin-spin interaction of the electron and proton given in (5.9) shows that neither $\hat{S}_{1z}$ nor $\hat{S}_{2z}$ commutes with the Hamiltonian because they do not commute with $\hat{\mathbf{S}}_1 \cdot \hat{\mathbf{S}}_2$. In this case we would choose the total spin operators $\hat{\mathbf{S}}^2 = (\hat{\mathbf{S}}_1 + \hat{\mathbf{S}}_2)^2$ and $\hat{S}_z = \hat{S}_{1z} + \hat{S}_{2z}$ as well as $\hat{H}$, $\hat{\mathbf{L}}^2$, and $\hat{L}_z$. As we will see in Chapter 11, the Hamiltonian for the real hydrogen atom is even more complex, involving the coupling of the spins of the particles to the relative orbital angular momentum.

## 9.7 VIBRATIONS AND ROTATIONS OF A DIATOMIC MOLECULE

An interesting two-body system in which, to a first approximation, the radial motion of the particles decouples from the angular motion is formed by the nuclei of a diatomic molecule, such as HCl. A schematic diagram of the potential energy $V(r)$ of such a molecule is shown in Fig. 9.4. At large distances the atoms in the molecule attract each other through van der Waals forces, while at short distances, when the electrons in the atoms overlap, there is a strong repulsion.[7] In between there is a minimum in the potential energy. As we argued in Chapter 7, in the vicinity of the potential energy minimum at $r_0$, the system behaves like a harmonic

---

[7] Because of their small mass, the electrons in the molecule move rapidly in comparison with the nuclei and thus readjust their positions very quickly when the nuclear positions change.

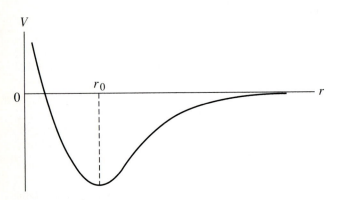

**FIGURE 9.4**
A schematic diagram of the potential energy $V(r)$ of a diatomic molecule.

oscillator and we can write

$$V(r) = V(r_0) + \frac{1}{2}\left(\frac{d^2V}{dr^2}\right)_{r=r_0}(r - r_0)^2 + \cdots$$

$$= V(r_0) + \frac{1}{2}\mu\omega^2(r - r_0)^2 + \cdots \qquad (9.101)$$

where $\mu$, the reduced mass of the two nuclei, is on the order of $M_N$, the nuclear mass.

In general, the potential energy of the molecule is on the order of $e^2/a$, where the size $a$ is roughly the same as for atomic systems, namely, $a = \hbar^2/m_e e^2$. Since the size scale is set by $a$, then by dimensional analysis

$$\frac{d^2V}{dr^2} \sim \frac{e^2}{a^3} \qquad (9.102)$$

The spacing between vibrational energy levels is thus given by

$$\hbar\omega = \hbar\left[\frac{1}{\mu}\left(\frac{d^2V}{dr^2}\right)_{r=r_0}\right]^{1/2} \sim \left(\frac{m_e}{M_N}\right)^{1/2}\left(\frac{m_e e^4}{\hbar^2}\right) \qquad (9.103)$$

The term in parentheses is the electronic energy scale given in (9.65). Since the factor $(m_e/M_N)^{1/2}$ is on the order of 1/40 for a diatomic molecule such as HCl, the wavelength of photons emitted or absorbed when the system changes from one vibrational energy level to an adjacent one is roughly 40 times longer than for a typical atomic transition and is thus in the infrared portion of the electromagnetic spectrum. The purely vibrational energies are given by[8]

$$E_{n_\nu} = (n_\nu + \tfrac{1}{2})\hbar\omega \qquad n_\nu = 0, 1, 2, \ldots \qquad (9.104)$$

Examination of the energy eigenfunctions of the harmonic oscillator shows that for states of low excitation the diatomic molecule vibrates over a distance scale given by

$$\sqrt{\frac{\hbar}{\mu\omega}} \sim \left(\frac{m_e}{M_N}\right)^{1/4}\frac{\hbar^2}{m_e e^2} = \left(\frac{m_e}{M_N}\right)^{1/4}a \qquad (9.105)$$

Thus the amplitude of vibration (in states of low vibrational excitation) is only a small fraction of the equilibrium separation $r_0$ of the nuclei in the molecule. For this reason, we can say that the molecule is fairly rigid, and we can treat the rotational motion separately from the vibrational motion.

In a particular state of vibration, the molecule is still free to rotate about its center of mass, forming a rigid rotator (see Fig. 9.5). The Hamiltonian for such

---

[8] We will call the vibrational quantum number $n_\nu$ instead of $n$, as was done in Chapter 7, because for atoms and molecules the principal electronic quantum number is generally referred to as $n$. See Section 10.2.

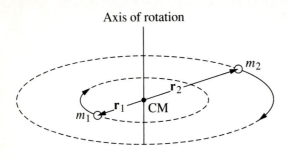

**FIGURE 9.5**
A classical model of a diatomic molecule rotating about its center of mass.

a rotator is given by

$$\hat{H} = \frac{\hat{\mathbf{L}}^2}{2I} \tag{9.106}$$

where the moment of inertia $I = \mu r_0^2$. This is exactly the form of the rotational part of the kinetic energy operator in (9.91), with the value of the radius replaced by the equilibrium separation. The eigenstates of this Hamiltonian are just the

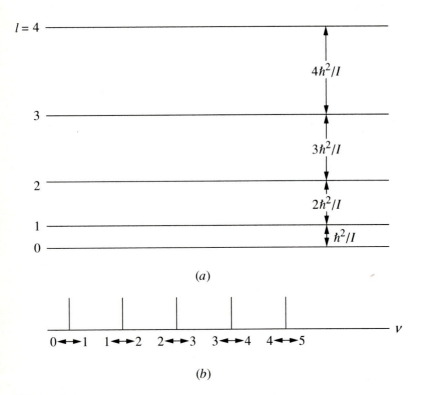

**FIGURE 9.6**
(*a*) An energy-level diagram of a three-dimensional rigid rotator. (*b*) Transitions between adjacent energy levels generate the rotational spectrum.

angular momentum eigenstates:

$$\frac{\hat{\mathbf{L}}^2}{2I}|l, m\rangle = \frac{l(l + 1)\hbar^2}{2I}|l, m\rangle = E_l|l, m\rangle \tag{9.107}$$

An energy-level diagram is shown in Fig. 9.6. The spacing between adjacent energy levels is given by

$$E_l - E_{l-1} = \frac{l(l + 1)\hbar^2}{2I} - \frac{(l - 1)l\hbar^2}{2I} = \frac{l\hbar^2}{I} \tag{9.108}$$

Notice how this energy spacing increases with increasing $l$, in contrast to the constant spacing between levels characteristic of the harmonic oscillator. The magnitude of this energy spacing is on the order of

$$\frac{l\hbar^2}{\mu r_0^2} \sim l\frac{\hbar^2}{M_N a^2} = l\frac{m_e}{M_N}\left(\frac{m_e e^4}{\hbar^2}\right) \tag{9.109}$$

The predominant electric dipole transitions obey the selection rule $\Delta l = \pm 1$, as we will see in Chapter 14. Thus, comparing (9.103) with (9.109), we see that the wavelength of photons emitted or absorbed in transitions between adjacent rotational energy levels of low $l$ is a factor of $(M_N/m_e)^{1/2}$ longer than that for the vibrational transitions. Purely rotational transitions reside in the very far infrared or short microwave portion of the electromagnetic spectrum. The energy spacing between levels is on the order of $10^{-2}$–$10^{-3}$ eV. Since $kT$ at room temperature is 1/40 eV, many of these levels will be excited at this temperature.

Figure 9.7 shows the purely rotational absorption spectrum of HCl. Notice that the values of $l$ are all integral. Setting $E_l - E_{l-1} = h\nu = hc/\lambda$, we see that

$$\lambda l = 2\pi I c/\hbar = 2\pi \mu r_0^2 c/\hbar \tag{9.110}$$

The observed values of $\lambda l$ are given in Table 9.1. Note that the constant value of this parameter is consistent with our treatment of the molecule as a rigid

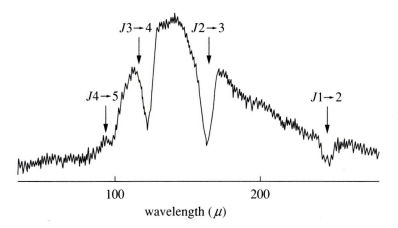

**FIGURE 9.7**
The absorption spectrum of HCl. [*From D. Bloor et al.*, Proc. Roy. Soc. **A260**, *510 (1961).*]

**TABLE 9.1**
**Rotational absorption transitions in HCl**

| Transition $l - 1 \rightarrow l$ | $\lambda$ (microns) | $\nu = c/\lambda$ $(10^9$ Hz) | $\nu/l$ $(10^9$ Hz) | $\lambda l$ (cm) | $h\nu$ (eV) |
|---|---|---|---|---|---|
| $(0 \rightarrow 1)^a$ | (479) | (626) | (626) | (0.0479) | (0.0026) |
| $1 \rightarrow 2$ | 243 | 1235 | 618 | 0.0486 | 0.0051 |
| $2 \rightarrow 3$ | 162 | 1852 | 617 | 0.0486 | 0.0077 |
| $3 \rightarrow 4$ | 121 | 2479 | 620 | 0.0484 | 0.0103 |
| $4 \rightarrow 5$ | 96 | 3125 | 625 | 0.0480 | 0.0129 |

$^a$ This transition not shown in Figure 9.7

*Source:* A. P. French and E. F. Taylor, *An Introduction to Quantum Physics*, W. W. Norton, New York, 1978, p. 492.

rotator.[9] From these values we can deduce that for HCl the internuclear distance $r_0 = 1.27$ Å. This is an example of how we can use the information contained in the rotational spectrum to learn about the structure of molecules.

In practice, it is difficult to produce in the far infrared or short microwave region the continuous spectrum of radiation that is required for observations in absorption of purely rotational transitions, like those shown in Fig. 9.7. However, the combined vibrational *and* rotational energies of a diatomic molecule are given by

$$E_{n_\nu, l} = \left(n_\nu + \frac{1}{2}\right)\hbar\omega + \frac{l(l+1)\hbar^2}{2I} \qquad (9.111)$$

Figure 9.8a shows an energy-level diagram. If the molecule, like HCl, possesses a permanent dipole moment, there is a vibrational selection rule $\Delta n_\nu = \pm 1$ for electric dipole transitions.[10] In addition, satisfying the rotational selection rule $\Delta l = \pm 1$ leads to the set of allowed vibration-rotation frequencies shown in Fig. 9.8b. These frequencies are in the easily accessible infrared part of the spectrum (see Fig. 9.9).

In concluding this discussion of diatomic molecules, we should note that the small energy-spacing between the rotational levels makes diatomic molecules interesting low-temperature thermometers. In 1941 A. McKellar observed absorption of light coming from the star $\zeta$ Ophiuchi by an interstellar cloud containing cyanogen radicals.[11] In the CN molecule there is a transition at 3874 Å from the ground electronic configuration to an excited electronic configuration. Just as with vibrational transitions, this change in electronic states may be accompanied by a

---

[9] For many molecules the separation distance is observed to increase slightly for increasing values of $l$, as you would expect in a centrifuge. It is often possible to observe 40 to 50 rotational energy levels between each vibrational level.

[10] We will see how such selection rules arise in Chapter 14. In particular, see Section 14.6 for an example involving electromagnetic transitions between states of the harmonic oscillator.

[11] A. McKellar, *Publs. Dominion Astrophys. Observatory* (Victoria, B.C.) **7**, 251 (1941).

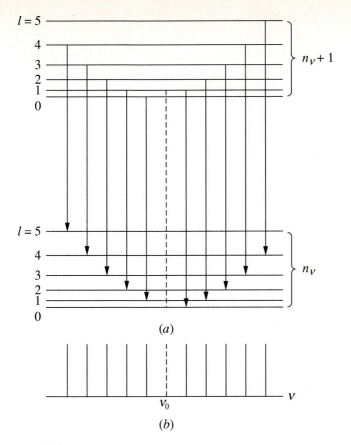

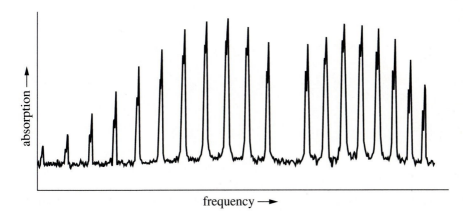

**FIGURE 9.8**
(a) Vibrational-rotational transitions for a diatomic molecule. (b) A schematic diagram of the resulting spectrum.

**FIGURE 9.9**
A vibrational-rotational absorption spectrum of HCl. Each peak is double because of the presence of two isotopes of chlorine in the gas—Cl$^{35}$ and the less abundant Cl$^{37}$. (Data is from M. Liu and W. Sly, Harvey Mudd College.)

change in the rotational level of the molecule as well. For CN the $l = 0$ ground state and $l = 1$ state are separated in energy by $E_{l=1} - E_{l=0} = hc/\lambda$ with the wavelength $\lambda = 2.64$ mm. McKellar's observations of the relative strengths of two absorption lines, one from the $l = 0$ state and the other from the $l = 1$ state, allowed him to deduce a population for the $l = 1$ rotational level that corresponded to the molecule being in a thermal bath at temperature $T = 2.3$ K, if no other special excitation mechanism was present. The significance of McKellar's observation was not appreciated until after 1965, when Penzias and Wilson, using a radio telescope, observed the cosmic background radiation resulting from the initial Big Bang at $\lambda = 7.35$ mm. The currently accepted temperature for this background radiation is 2.7 K. Subsequent reexamination of the CN absorption spectrum confirmed that no special mechanism for exciting the $l = 1$ state seemed to be in action and that the background temperature at 2.64 mm was consistent with that of the cosmic background radiation. Before high-altitude balloon flights took place in the 1970s, observations of the populations of diatomic molecules such as CN and CH provided the only information about the blackbody spectrum at wavelengths shorter than 3 mm, because the earth's atmosphere is strongly absorbing in this portion of the spectrum.

## 9.8   POSITION-SPACE REPRESENTATIONS OF $\hat{\mathbf{L}}$ IN SPHERICAL COORDINATES

Since the orbital angular momentum operators generate rotations in position space, we can determine position-space representations of these operators. Given the fact that we are dealing with orbital *angular* momentum, it is probably not surprising that the most convenient position-space states are expressed in spherical coordinates, where the angles appear explicitly. Note that the bra equation corresponding to the ket equation (9.75) is

$$\langle r, \theta, \phi | \hat{R}^\dagger (d\phi \mathbf{k}) = \langle r, \theta, \phi + d\phi | \qquad (9.112)$$

Since the rotation operators are unitary, $\hat{R}\hat{R}^\dagger = 1$, $\hat{R}^\dagger$ is the inverse of $\hat{R}$, and therefore

$$\langle r, \theta, \phi | \hat{R}(d\phi \mathbf{k}) = \langle r, \theta, \phi - d\phi | \qquad (9.113)$$

Thus

$$\langle r, \theta, \phi | \hat{R}(d\phi \mathbf{k}) | \psi \rangle = \langle r, \theta, \phi - d\phi | \psi \rangle$$

$$= \langle r, \theta, \phi | \psi \rangle - d\phi \frac{\partial \langle r, \theta, \phi | \psi \rangle}{\partial \phi} \qquad (9.114)$$

where the last step comes from expanding the wave function $\langle r, \theta, \phi - d\phi | \psi \rangle$ in a Taylor series. Since

$$\langle r, \theta, \phi | \hat{R}(d\phi \mathbf{k}) | \psi \rangle = \langle r, \theta, \phi | \left( 1 - \frac{i}{\hbar} \hat{L}_z \, d\phi \right) | \psi \rangle \qquad (9.115)$$

we see that

$$\langle r, \theta, \phi | \hat{L}_z | \psi \rangle = \frac{\hbar}{i} \frac{\partial}{\partial \phi} \langle r, \theta, \phi | \psi \rangle \tag{9.116}$$

Thus the $z$ component of the orbital *angular* momentum operator is represented in position space by

$$\hat{L}_z \rightarrow \frac{\hbar}{i} \frac{\partial}{\partial \phi} \tag{9.117}$$

which should be compared with the representation of the *linear* momentum operator:

$$\hat{p}_x \rightarrow \frac{\hbar}{i} \frac{\partial}{\partial x} \tag{9.118}$$

The important thing to note is that the orbital angular momentum is represented by a differential operator in position space. This has profound consequences. To help see why, let's return to (9.116) and consider the special case where $|\psi\rangle = |l, m\rangle$, namely, the state is an eigenstate of orbital angular momentum. Since $\hat{L}_z |l, m\rangle = m\hbar |l, m\rangle$, we have

$$\langle r, \theta, \phi | \hat{L}_z | l, m \rangle = \frac{\hbar}{i} \frac{\partial}{\partial \phi} \langle r, \theta, \phi | l, m \rangle$$

$$= m\hbar \langle r, \theta, \phi | l, m \rangle \tag{9.119}$$

Solving this differential equation, we find that the $\phi$ dependence of an eigenfunction is of the form $e^{im\phi}$. We must require that the eigenfunctions be single-valued:

$$e^{im\phi} = e^{im(\phi + 2\pi)} \tag{9.120}$$

Otherwise, how would we determine the derivative of a wave function at angle $\phi$ when the wave function approaches different values depending on the direction from which we approach $\phi$? Also, recall from (2.78) that for $\hat{L}_z$ to be Hermitian, it must satisfy

$$\langle \psi | \hat{L}_z | \chi \rangle = \langle \chi | \hat{L}_z | \psi \rangle^* \tag{9.121}$$

which in position space becomes[12]

$$\int_0^{2\pi} d\phi \, \psi^*(\phi) \frac{\hbar}{i} \frac{\partial}{\partial \phi} \chi(\phi) = \left[ \int_0^{2\pi} d\phi \, \chi^*(\phi) \frac{\hbar}{i} \frac{\partial}{\partial \phi} \psi(\phi) \right]^* \tag{9.122}$$

As integration by parts shows, for (9.122) to hold for all $\psi(\phi)$, the wave functions must satisfy $\psi(0) = \psi(2\pi)$. Note that if $\hat{L}_z$ were not Hermitian, the rotation operator $\hat{R}(\phi\mathbf{k})$ would not be unitary, and probability would not be conserved upon rotation of a state.

---

[12] We ignore all but the $\phi$ dependence for notational simplicity.

This single-valuedness requirement (9.120) is satisfied only if

$$m = 0, \pm 1, \pm 2, \ldots \tag{9.123}$$

But from our general analysis of angular momentum in Chapter 3, we know that the $m$ value for a state with a particular $l$ value runs from $l$ to $-l$ in integral steps. Thus the values of $l$ must also be integral:

$$l = 0, 1, 2, \ldots \tag{9.124}$$

as we also saw in the spectrum of the diatomic molecule. The fact that orbital angular momentum has representations in position space has restricted the possible values of the angular momentum quantum number $l$ to integer values. In particular, no half-integral values are permitted, unlike intrinsic spin angular momentum.

The position-space representations in spherical coordinates of the other components of the angular momentum are not as simple as that for $\hat{L}_z$; only for rotations about the $z$ axis is there a single angle that can be used to express the rotation as in (9.113). There are a number of techniques that can be used to determine the position-space representations for $\hat{L}_x$ and $\hat{L}_y$. One approach is to express the position-space representation

$$\hat{\mathbf{L}} = \hat{\mathbf{r}} \times \hat{\mathbf{p}} \rightarrow \mathbf{r} \times \frac{\hbar}{i}\boldsymbol{\nabla} \tag{9.125}$$

in Cartesian coordinates and then make a change of variables to spherical coordinates. However, it is a little easier to express the gradient in spherical coordinates directly:

$$\hat{\mathbf{L}} \rightarrow r\mathbf{u}_r \times \frac{\hbar}{i}\left(\mathbf{u}_r \frac{\partial}{\partial r} + \mathbf{u}_\theta \frac{1}{r}\frac{\partial}{\partial \theta} + \mathbf{u}_\phi \frac{1}{r\sin\theta}\frac{\partial}{\partial \phi}\right)$$

$$= \frac{\hbar}{i}\left(\mathbf{u}_\phi \frac{\partial}{\partial \theta} - \mathbf{u}_\theta \frac{1}{\sin\theta}\frac{\partial}{\partial \phi}\right) \tag{9.126}$$

Taking the $x$ and $y$ components of the unit vectors $\mathbf{u}_\theta$ and $\mathbf{u}_\phi$, we obtain

$$\hat{L}_x \rightarrow \frac{\hbar}{i}\left(-\sin\phi\frac{\partial}{\partial \theta} - \cot\theta\cos\phi\frac{\partial}{\partial \phi}\right) \tag{9.127}$$

$$\hat{L}_y \rightarrow \frac{\hbar}{i}\left(\cos\phi\frac{\partial}{\partial \theta} - \cot\theta\sin\phi\frac{\partial}{\partial \phi}\right) \tag{9.128}$$

We can then combine the results (9.117), (9.127), and (9.128) to obtain the representation of $\hat{\mathbf{L}}^2$:

$$\hat{\mathbf{L}}^2 \rightarrow -\hbar^2\left[\frac{1}{\sin\theta}\frac{\partial}{\partial \theta}\left(\sin\theta\frac{\partial}{\partial \theta}\right) + \frac{1}{\sin^2\theta}\frac{\partial^2}{\partial \phi^2}\right] \tag{9.129}$$

This form for $\hat{\mathbf{L}}^2$ may remind you of the angular part of the Laplacian in spherical coordinates. In fact, if we express the energy eigenvalue equation in position space in terms of the Laplacian:

$$\langle \mathbf{r}| \frac{\hat{\mathbf{p}}^2}{2\mu} + V(|\hat{\mathbf{r}}|)|\psi\rangle = \left[ -\frac{\hbar^2}{2\mu}\nabla^2 + V(r) \right]\langle \mathbf{r}|\psi\rangle = E\langle \mathbf{r}|\psi\rangle \tag{9.130}$$

and then express the Laplacian in spherical coordinates, we obtain

$$\left( -\frac{\hbar^2}{2\mu} \left\{ \frac{\partial^2}{\partial r^2} + \frac{2}{r}\frac{\partial}{\partial r} + \frac{1}{r^2}\left[ \frac{1}{\sin\theta}\frac{\partial}{\partial\theta}\left( \sin\theta\frac{\partial}{\partial\theta} \right) + \frac{1}{\sin^2\theta}\frac{\partial^2}{\partial\phi^2} \right] \right\} + V(r) \right)\langle \mathbf{r}|\psi\rangle$$

$$= E\langle \mathbf{r}|\psi\rangle \tag{9.131}$$

This equation agrees with (9.91), provided we make the identification (9.129).

## 9.9 ORBITAL ANGULAR MOMENTUM EIGENFUNCTIONS

One of the most evident features of the position-space representations (9.117), (9.127), and (9.128) of the angular momentum operators is that they depend only on the angles $\theta$ and $\phi$, not at all on the magnitude $r$ of the position vector. Rotating a position eigenstate changes its direction but not its length. Thus we can isolate the angular dependence and determine $\langle \theta, \phi|l, m\rangle$, the amplitude for a state of definite angular momentum to be at the angles $\theta$ and $\phi$. These amplitudes, which are functions of the angles, are called the *spherical harmonics* and denoted by

$$\langle \theta, \phi|l, m\rangle = Y_{l,m}(\theta, \phi) \tag{9.132}$$

Expressed in terms of these amplitudes, the energy eigenfunctions of the Hamiltonian (9.57) are given by

$$\langle r, \theta, \phi|E, l, m\rangle = R(r)Y_{l,m}(\theta, \phi) \tag{9.133}$$

We might have been led to an expression such as (9.133) for the eigenfunctions by solving the partial differential equation (9.131) by separation of variables. Here, however, we have been guided by the rotational symmetry of the problem to write the angular part of the eigenfunction $\Theta(\theta)\Phi(\phi) = Y_{l,m}(\theta, \phi)$ directly.

Let's address the question of how we should normalize these eigenfunctions. Clearly, we want

$$\int d^3r \,|\langle r, \theta, \phi|E, l, m\rangle|^2 = \int d^3r \,|R(r)|^2 |Y_{l,m}(\theta, \phi)|^2 = 1 \tag{9.134}$$

since the probability of finding the particle somewhere in position space should sum to one.[13] The differential volume element $d^3r$ in spherical coordinates is

---

[13] We are assuming that we are interested in states that yield a discrete energy spectrum, as opposed to a continuous eigenvalue spectrum, which would require Dirac delta function normalization similar to that which we used for the position and momentum eigenstates. We will consider the continuum solutions to the Schrödinger equation when we come to scattering in Chapter 13.

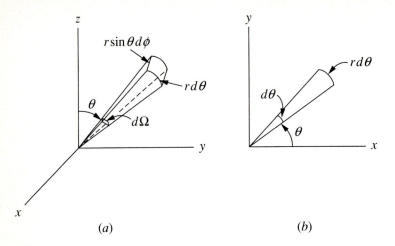

**FIGURE 9.10**
(a) The solid angle $d\Omega$ in three dimensions is defined as the surface area $dS$ subtended divided by the radius squared: $dS/r^2 = (r\,d\theta)(r\sin\theta\,d\phi)/r^2 = d\Omega$. (b) The ordinary angle $d\theta$ in two dimensions is defined as the arc length $ds$ subtended divided by the radius: $ds/r = d\theta$.

given by

$$d^3r = r^2\,dr\,\sin\theta\,d\theta\,d\phi = r^2\,dr\,d\Omega \qquad (9.135)$$

where the solid angle $d\Omega = \sin\theta\,d\theta\,d\phi$ is the angle subtended by the differential surface area $dS$ shown in Fig. 9.10$a$. Notice the definition of solid angle is in direct analogy with the definition of ordinary angles in radian measure as the angle subtended by the differential arc length shown in Fig. 9.10$b$. The solid angle subtended by a sphere (or any closed surface) is

$$\int_{\text{all directions}} d\Omega = \int_0^{2\pi} d\phi \int_0^\pi \sin\theta\,d\theta = 4\pi \qquad (9.136)$$

Since we want

$$\int_0^\infty r^2\,dr\,|R(r)|^2 \int_0^\pi \sin\theta\,d\theta \int_0^{2\pi} d\phi\,|Y_{l,m}(\theta,\phi)|^2 = 1 \qquad (9.137)$$

we can choose to normalize separately the radial and the angular parts of the eigenfunction:

$$\int_0^\infty r^2\,dr\,|R(r)|^2 = 1 \qquad (9.138)$$

so that the total probability of finding the particle between $r = 0$ and $r = \infty$ is one, and

$$\int_0^\pi \sin\theta\,d\theta \int_0^{2\pi} d\phi\,|Y_{l,m}(\theta,\phi)|^2 = 1 \qquad (9.139)$$

Thus we can interpret

$$|\langle \theta, \phi | l, m \rangle|^2 \, d\Omega = |Y_{l,m}(\theta, \phi)|^2 \, d\Omega \tag{9.140}$$

as the probability of finding a particle in the state $|l, m\rangle$ within the solid angle $d\Omega$ at the angles $\theta$ and $\phi$.

To obtain the orbital angular momentum eigenfunctions $Y_{l,m}(\theta, \phi)$ themselves, we start with the equation[14]

$$\hat{L}_+ | l, l \rangle = 0 \tag{9.141}$$

Since $\hat{L}_\pm = \hat{L}_x \pm i\hat{L}_y$, using (9.127) and (9.128), we can represent the raising and lowering operators in position space by the differential operators

$$\hat{L}_\pm \rightarrow \frac{\hbar}{i} e^{\pm i\phi} \left( \pm i \frac{\partial}{\partial \theta} - \cot\theta \frac{\partial}{\partial \phi} \right) \tag{9.142}$$

Thus in position space (9.141) becomes

$$\langle \theta, \phi | \hat{L}_+ | l, l \rangle = \frac{\hbar}{i} e^{i\phi} \left( i \frac{\partial}{\partial \theta} - \cot\theta \frac{\partial}{\partial \phi} \right) \langle \theta, \phi | l, l \rangle = 0 \tag{9.143}$$

Inserting the known $e^{il\phi}$ dependence, we can solve the differential equation

$$\left( \frac{\partial}{\partial \theta} - l \cot\theta \right) \langle \theta, \phi | l, l \rangle = 0 \tag{9.144}$$

to obtain

$$\langle \theta, \phi | l, l \rangle = c_l e^{il\phi} \sin^l \theta \tag{9.145}$$

Satisfying the normalization condition (9.139), we find

$$\langle \theta, \phi | l, l \rangle = Y_{l,l}(\theta, \phi) = \frac{(-1)^l}{2^l l!} \sqrt{\frac{(2l+1)!}{4\pi}} e^{il\phi} \sin^l \theta \tag{9.146}$$

We now apply the lowering operator to determine the remaining spherical harmonics. From Chapter 3 we know that

$$\hat{L}_- | l, m \rangle = \sqrt{l(l+1) - m(m-1)} \, \hbar | l, m-1 \rangle \tag{9.147}$$

Combining (9.146) and (9.147), we find (see Problem 9.18) for $m \geq 0$

$$Y_{l,m}(\theta, \phi) = \frac{(-1)^l}{2^l l!} \sqrt{\frac{(2l+1)(l+m)!}{4\pi(l-m)!}} e^{im\phi} \frac{1}{\sin^m \theta} \frac{d^{l-m}}{d(\cos\theta)^{l-m}} \sin^{2l} \theta \tag{9.148}$$

---

[14] This approach is similar to the one we use in Section 7.5 for determination of the position-space eigenfunctions of the harmonic oscillator. For an alternative technique in which the spherical harmonics are determined by solving a second-order partial differential equation by separation of variables, see Problem 9.17.

The choice of the phase factor $(-1)^l$ is taken to ensure that $Y_{l,0}(\theta, \phi)$, which is independent of $\phi$, has a real positive value for $\theta = 0$. In fact,

$$Y_{l,0}(\theta, \phi) = \sqrt{\frac{2l+1}{4\pi}} P_l(\cos\theta) \tag{9.149}$$

where $P_l(\cos\theta)$ is the standard Legendre polynomial. The spherical harmonics for $m < 0$ are given by

$$Y_{l,-m}(\theta, \phi) = (-1)^m \left[Y_{l,m}(\theta, \phi)\right]^* \tag{9.150}$$

It is useful to list the spherical harmonics with $l = 0$, 1, and 2:

$$Y_{0,0}(\theta, \phi) = \sqrt{\frac{1}{4\pi}} \tag{9.151}$$

$$Y_{1,\pm1}(\theta, \phi) = \mp\sqrt{\frac{3}{8\pi}} e^{\pm i\phi} \sin\theta \tag{9.152a}$$

$$Y_{1,0}(\theta, \phi) = \sqrt{\frac{3}{4\pi}} \cos\theta \tag{9.152b}$$

$$Y_{2,\pm2}(\theta, \phi) = \sqrt{\frac{15}{32\pi}} e^{\pm 2i\phi} \sin^2\theta \tag{9.153a}$$

$$Y_{2,\pm1}(\theta, \phi) = \mp\sqrt{\frac{15}{8\pi}} e^{\pm i\phi} \sin\theta \cos\theta \tag{9.153b}$$

$$Y_{2,0}(\theta, \phi) = \sqrt{\frac{5}{16\pi}} (3\cos^2\theta - 1) \tag{9.153c}$$

Figure 9.11 shows plots of $|Y_{l,m}(\theta, \phi)|^2$ as a function of $\theta$ and $\phi$. Since the spherical harmonics depend on $\phi$ through $e^{im\phi}$, these plots are all independent of $\phi$. The $l = 0$ state, often called an $s$ state, is spherically symmetric. Thus if a rotator, such as the diatomic molecule discussed in Section 9.6, is in an $s$ state, a measurement of the orientation of the rotator is equally likely to find it oriented in any direction. The $l = 1$ states are known as $p$ states. The states with $m = \pm1$ have a probability density that tends to reside in the $x$-$y$ plane, which is just the sort of behavior that you might expect for an object rotating around the $z$ axis with nonzero angular momentum. This effect becomes more pronounced for the maximum $m$ values with increasing values of $l$. The $l = 1$, $m = 0$ state is often referred to as a $p_z$ state, or orbital; the probability density is oriented along the $z$ axis. Since $z = r\cos\theta$, the function $Y_{1,0}$ may be expressed as

$$Y_{1,0}(\theta, \phi) = \sqrt{\frac{3}{4\pi}} \frac{z}{r} \tag{9.154}$$

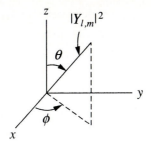

$l = 0, m = 0$

$l = 1, m = 0$      $l = 1, m = \pm 1$

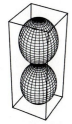

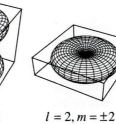

$l = 2, m = 0$      $l = 2, m = \pm 1$      $l = 2, m = \pm 2$

$l = 3, m = 0$      $l = 3, m = \pm 1$      $l = 3, m = \pm 2$      $l = 3, m = \pm 3$

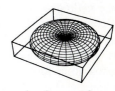

**FIGURE 9.11**
Plots of $|Y_{l,m}(\theta, \phi)|^2$ for $l = 0$, 1, 2, and 3.

Using $x = r \sin\theta \cos\phi$ and $y = r \sin\theta \sin\phi$, we see that

$$Y_{1,\pm1}(\theta,\phi) = \mp\sqrt{\frac{3}{8\pi}} \frac{(x \pm iy)}{r} \tag{9.155}$$

Thus we can find linear combinations of the $Y_{1,\pm1}$ states, namely,

$$\frac{(Y_{1,-1} - Y_{1,1})}{\sqrt{2}} = \sqrt{\frac{3}{4\pi}}\frac{x}{r} \quad \text{and} \quad \frac{i(Y_{1,1} + Y_{1,-1})}{\sqrt{2}} = \sqrt{\frac{3}{4\pi}}\frac{y}{r} \tag{9.156}$$

that can be naturally termed $p_x$ and $p_y$ orbitals, with probability densities oriented along the $x$ and $y$ axes, respectively.

In Chapter 10 we will solve the hydrogen atom and see how the principal quantum number $n$ enters the energy eigenvalue equation. In a later chapter we will examine how quantum mechanics allows us to understand the valence properties of multielectron atoms as well. It is worth getting a little ahead to point out that the directional properties of molecular bonds are to a large extent determined by the shape of the orbital angular momentum eigenfunctions. For example, oxygen, with eight electrons, has two electrons in $n = 1$ $s$ states, two in $n = 2$ $s$ states, and four in $n = 2$ $p$ states. As the electrons fill up these $p$ states, the first three go into the $p_x$, $p_y$, and $p_z$ states. This tends to keep the electrons apart, minimizing their Coulomb repulsion. The fourth electron is forced to go into one of these $p$ states, say the $p_z$ state, leaving the $p_x$ and $p_y$ states with one electron each. When oxygen binds with hydrogen to form $H_2O$, each of the hydrogen atoms shares its electron with oxygen, helping to fill the oxygen $n = 2$ $p_x$ and $n = 2$ $p_y$ states. Thus the two hydrogen atoms in the water molecule should make a right angle with respect to the oxygen. Actually, since the hydrogen atoms are sharing their electrons with the oxygen atom, each ends up with a net positive charge and these positive charges repel each other, pushing the hydrogen atoms somewhat further apart. The observed angle between the hydrogen atoms turns out to be 105°.

Lastly, we reexamine our old friend the ammonia molecule, $NH_3$. Nitrogen has seven electrons; as with oxygen, three of them are in $p_x$, $p_y$, and $p_z$ states. Thus nitrogen has room for three additional electrons, which it acquires from the three hydrogen atoms in forming the $NH_3$ molecule. The three hydrogens should all come out at right angles to each other, but here again the repulsion between the hydrogen atoms forces the angles to be somewhat larger than 90°. We now see why the ammonia molecule is not a flat, or planar, structure, a necessary precondition for the tunneling action of the nitrogen atom that we discussed in Chapter 4.

## 9.10   SUMMARY

Physicists have learned a lot about nature by analyzing the two-body problem. In this chapter we have focused on two bodies that interact through a central potential that depends only on the magnitude $|\mathbf{r}| = |\mathbf{r}_1 - \mathbf{r}_2|$ of the distance separating the two bodies. In this case there are a number of symmetries of the system that we can take advantage of to determine the eigenstates of the Hamiltonian. First, the Hamiltonian is invariant under translations of both of the bodies ($\mathbf{r}_1 \rightarrow \mathbf{r}_1 + \mathbf{a}$, $\mathbf{r}_2 \rightarrow \mathbf{r}_2 + \mathbf{a}$), and therefore the generator of total translations, the total momentum operator $\hat{\mathbf{P}} = \hat{\mathbf{p}}_1 + \hat{\mathbf{p}}_2$, commutes with the Hamiltonian. Thus the Hamiltonian and the total momentum operator have eigenstates in common.

It is common to work in the center-of-mass frame where $\mathbf{P} = 0$, in which case we can restrict our analysis to the relative Hamiltonian

$$\hat{H} = \frac{\hat{\mathbf{p}}^2}{2\mu} + V(|\hat{\mathbf{r}}|) \tag{9.157}$$

which is invariant under rotations. The operator

$$\hat{R}(d\phi\,\mathbf{k}) = 1 - \frac{i}{\hbar}\hat{L}_z\,d\phi \tag{9.158}$$

that rotates position states about the $z$ axis:

$$\hat{R}(d\phi\,\mathbf{k})|x, y, z\rangle = |x - y\,d\phi, y + x\,d\phi, z\rangle \tag{9.159}$$

has a generator

$$\hat{L}_z = \hat{x}\,\hat{p}_y - \hat{y}\,\hat{p}_x \tag{9.160}$$

which is the $z$ component of the orbital angular momentum. Thus the invariance of the Hamiltonian under rotations means the relative *orbital* angular momentum operators $\hat{\mathbf{L}} = \hat{\mathbf{r}} \times \hat{\mathbf{p}}$ that generate these rotations commute with the Hamiltonian. We therefore deduce that $\hat{H}$, $\hat{\mathbf{L}}^2$, and one of the components, say $\hat{L}_z$, have eigenstates $|E, l, m\rangle$ in common. In position space, the energy eigenvalue equation is given by

$$\langle\mathbf{r}|\hat{H}|E, l, m\rangle = \left[-\frac{\hbar^2}{2\mu}\left(\frac{\partial^2}{\partial r^2} + \frac{2}{r}\frac{\partial}{\partial r}\right) + \frac{l(l+1)\hbar^2}{2\mu r^2} + V(r)\right]\langle\mathbf{r}|E, l, m\rangle$$

$$= E\langle\mathbf{r}|E, l, m\rangle \tag{9.161}$$

where $l(l+1)\hbar^2$ is the eigenvalue of $\hat{\mathbf{L}}^2$.

The operator $\hat{L}_z$ is represented in position space by

$$\hat{L}_z \rightarrow \frac{\hbar}{i}\frac{\partial}{\partial\phi} \tag{9.162}$$

and the $\phi$ dependence of the orbital angular momentum eigenfunctions $\langle\theta, \phi|l, m\rangle = Y_{l,m}(\theta, \phi)$ is given by $e^{im\phi}$. In order that the action of differential operators such as (9.162) be well defined, the position-space wave functions must be single-valued, and therefore the $m$ values must be integral. Since $m$ runs from $-l$ to $l$ in integral steps, $l$ must be integral as well. Thus, we see here clearly that the intrinsic spin of a spin-$\frac{1}{2}$ particle, for example, does not arise from the body literally spinning about an axis, for if it did, the spin angular momentum would necessarily be of the $\mathbf{r} \times \mathbf{p}$ type, and the angular momentum operators that generate rotations of this body could be represented by differential operators in position space, which cannot lead to half-integral values for $l$.

In our analysis of the two-body problem, we have started with a Hamiltonian that exhibits translational and rotational symmetries and used these symmetries to determine its eigenstates. Since $[\hat{H}, \hat{\mathbf{P}}] = 0$ and $[\hat{H}, \hat{\mathbf{L}}] = 0$, these symmetries also tell us that the total momentum and the relative orbital angular momentum of the system are conserved. The importance of this connection between symmetries and conservation laws becomes really apparent if you continue your study of quantum mechanics through quantum field theory. There you will see, for example, how conservation of charge and conservation of color can actually be used to determine, through symmetry principles, the form of the laws governing electromagnetic (quantum electrodynamics) and strong (quantum chromodynamics) interactions, respectively.

## PROBLEMS

**9.1.** Follow the suggestion after (9.11) to show that $[\hat{T}(a_x\mathbf{i}), \hat{T}(a_y\mathbf{j})] = 0$ implies $[\hat{p}_x, \hat{p}_y] = 0$.

**9.2.** Show that *if* $[\hat{A}, \hat{B}] = 0$, $e^{\hat{A}+\hat{B}} = e^{\hat{A}}e^{\hat{B}}$. *Suggestion:* Expand the exponentials. If you are feeling inspired, try showing that if $[\hat{A}, \hat{B}] \neq 0$, but both $\hat{A}$ and $\hat{B}$ commute with their commutator, then

$$e^{\hat{A}+\hat{B}} = e^{\hat{A}}e^{\hat{B}}e^{-[\hat{A}, \hat{B}]/2}$$

**9.3.** Explain the nature of the symmetry that is responsible for conservation of energy.

**9.4.** Use the commutation relations of $\hat{\mathbf{r}}_1, \hat{\mathbf{r}}_2, \hat{\mathbf{p}}_1$, and $\hat{\mathbf{p}}_2$ to establish that the center-of-mass and relative position and momentum operators satisfy

$$[\hat{x}_i, \hat{P}_j] = 0 \qquad [\hat{X}_i, \hat{P}_j] = i\hbar\delta_{ij}$$
$$[\hat{x}_i, \hat{p}_j] = i\hbar\delta_{ij} \qquad [\hat{X}_i, \hat{p}_j] = 0$$

**9.5.** Show explicitly that

$$\frac{\hat{\mathbf{p}}_1^2}{2m_1} + \frac{\hat{\mathbf{p}}_2^2}{2m_2} = \frac{\hat{\mathbf{P}}^2}{2M} + \frac{\hat{\mathbf{p}}^2}{2\mu}$$

where

$$\hat{\mathbf{p}} = \frac{m_2\hat{\mathbf{p}}_1 - m_1\hat{\mathbf{p}}_2}{m_1 + m_2} \qquad \hat{\mathbf{P}} = \hat{\mathbf{p}}_1 + \hat{\mathbf{p}}_2 \qquad M = m_1 + m_2 \qquad \text{and} \qquad \mu = \frac{m_1 m_2}{m_1 + m_2}$$

**9.6.** Use the fact that all vectors must transform the same way under rotations to establish that any vector operator $\hat{\mathbf{V}}$ must satisfy the commutation relations $[\hat{L}_z, \hat{V}_x] = i\hbar\hat{V}_y$, $[\hat{L}_z, \hat{V}_y] = -i\hbar\hat{V}_x$, and $[\hat{L}_z, \hat{V}_z] = 0$.

**9.7.** Use the identity

$$\sum_{i=1}^{3} \varepsilon_{ijk}\varepsilon_{ilm} = \delta_{jl}\delta_{km} - \delta_{jm}\delta_{kl}$$

together with the commutation relations (9.19) of the position and momentum operators and the expression (9.82) for the orbital angular momentum operators to verify that

$$\hat{\mathbf{L}}^2 = \hat{\mathbf{r}} \times \hat{\mathbf{p}} \cdot \hat{\mathbf{r}} \times \hat{\mathbf{p}} = \hat{\mathbf{r}}^2\hat{\mathbf{p}}^2 - (\hat{\mathbf{r}} \cdot \hat{\mathbf{p}})^2 + i\hbar\hat{\mathbf{r}} \cdot \hat{\mathbf{p}}$$

**9.8.** Use the commutation relations $\left[\hat{x}_i, \hat{p}_j\right] = i\hbar\delta_{ij}$ to verify that the angular momentum operators $\hat{\mathbf{L}} = \hat{\mathbf{r}} \times \hat{\mathbf{p}}$, or, in component form,

$$\hat{L}_i = \sum_{j=1}^{3} \sum_{k=1}^{3} \varepsilon_{ijk}\hat{x}_j\hat{p}_k$$

satisfy the commutation relations

$$\left[\hat{L}_i, \hat{L}_j\right] = i\hbar \sum_{k=1}^{3} \varepsilon_{ijk}\hat{L}_k$$

**9.9.** The carbon monoxide molecule, CO, absorbs a photon with a frequency of $1.15 \times 10^{11}$ Hz, making a purely rotational transition from an $l = 0$ to $l = 1$ energy level. What is the internuclear distance for this molecule?

**9.10.** The energy spacing between the vibrational energy levels of HCl is 0.37 eV.
(a) What is the wavelength of a photon emitted in a vibrational transition?
(b) What is the effective spring constant $k$ for this molecule?
(c) What resolution is required for a spectrometer to resolve the presence of $HCl^{35}$ and $HCl^{37}$ molecules in the vibrational spectrum?

**9.11.** The ratio of the number of molecules in the rotational level $l$, with energy $E_l$, to the number in the $l = 0$ ground state, with energy $E_0$, in a sample of molecules in equilibrium at temperature $T$ is given by

$$(2l + 1)e^{-(E_l - E_0)/kT}$$

where the factor of $2l + 1$ reflects the number of rotational states with energy $E_l$, that is, the degeneracy of this energy level.
(a) Show that the population of rotational energy levels first increases and then decreases with increasing $l$.
(b) Which energy level will be occupied by the largest number of molecules for HCl at room temperature? Compare your result with the intensities of the absorption spectrum in Fig. 9.9. What do you deduce about the temperature of the gas?

**9.12.** The wave function for a particle is of the form $\psi(\mathbf{r}) = (x + y + z)f(r)$. What are the values that a measurement of $\mathbf{L}^2$ can yield? What values can be obtained by measuring $L_z$? What are the probabilities of obtaining these results? *Suggestion:* Express the wave function in spherical coordinates and then in terms of the $Y_{l,m}$'s.

**9.13.** A particle is in the orbital angular momentum state $|l, m\rangle$. Evaluate $\Delta L_x$ and $\Delta L_y$ for this state. Which states satisfy the equality in the uncertainty relation

$$\Delta L_x \Delta L_y \geq \frac{\hbar}{2}|\langle L_z\rangle|$$

*Suggestion:* One approach is to use $\hat{L}_x = (\hat{L}_+ + \hat{L}_-)/2$, and so on. Another is to take advantage of the symmetry of the expectation values of $L_x^2$ and $L_y^2$ in an eigenstate of $\hat{L}_z$.

**9.14.** Use the position-space representations of the orbital angular momentum operators $\hat{L}_x$, $\hat{L}_y$, and $\hat{L}_z$ given in (9.117), (9.127), and (9.128), respectively, to derive the position-space representation of the operator $\hat{L}^2$ given in (9.129).

**9.15.** Show that the spherical harmonics $Y_{l,m}$ are eigenfunctions of the parity operator with eigenvalue $(-1)^l$. *Note:* An inversion of coordinates in spherical coordinates is accomplished by $r \to r$, $\theta \to \pi - \theta$, and $\phi \to \phi + \pi$. Use (9.148). It may be wise to check the specific examples in (9.151) through (9.153) before attempting the general case.

**9.16.** (a) Obtain $Y_{1,0}$ by application of the lowering operator in (9.142) to $Y_{1,1}$.

(b) By direct application of the operator $\hat{\mathbf{L}}^2$ in position space [see (9.129)], verify that $Y_{1,1}$ is an eigenfunction with eigenvalue $2\hbar^2$.

**9.17.** Determine the spherical harmonics and the eigenvalues of $\hat{\mathbf{L}}^2$ by solving the eigenvalue equation $\hat{\mathbf{L}}^2|\lambda, m\rangle = \lambda\hbar^2|\lambda, m\rangle$ in position space,

$$-\hbar^2\left[\frac{1}{\sin\theta}\frac{\partial}{\partial\theta}\left(\sin\theta\frac{\partial}{\partial\theta}\right) + \frac{1}{\sin^2\theta}\frac{\partial^2}{\partial\phi^2}\right]\Theta_{\lambda,m}(\theta)e^{im\phi} = \lambda\hbar^2\Theta_{\lambda,m}(\theta)e^{im\phi}$$

Note that we have inserted the known $\phi$ dependence. To illustrate the procedure, restrict your attention to the $m = 0$ case. Rewrite the equation in terms of $u = \cos\theta$ and show that it becomes

$$(1 - u^2)\frac{d^2\Theta_{\lambda,0}}{du^2} - 2u\frac{d\Theta_{\lambda,0}}{du} + \lambda\Theta_{\lambda,0} = 0$$

which you may recognize as Legendre's equation. Try a power series solution

$$\Theta_{\lambda,0} = \sum_{k=0}^{\infty} a_k u^k$$

Show that the series diverges for $|u| \to 1$ ($\theta \to 0$ or $\theta \to \pi$) unless $\lambda = l(l + 1)$ with $l = 0, 1, 2, \ldots$, in which case the series terminates. The solutions $\Theta_{l,0}(u) = P_l(u)$ are just the Legendre polynomials. Compare the first few solutions with the spherical harmonics listed in equations (9.151), (9.152b), and (9.153c).

**9.18.** Apply the lowering operator to $Y_{l,l}$ as given in (9.146) to determine $Y_{l,l-1}$. Check your result against the general expression (9.148) for the $Y_{l,m}$'s.

**9.19.** The wave function of a rigid rotator with a Hamiltonian $\hat{H} = \hat{\mathbf{L}}^2/2I$ is given by

$$\langle\theta, \phi|\psi(0)\rangle = \sqrt{\frac{3}{4\pi}}\sin\theta\sin\phi$$

(a) What is $\langle\theta, \phi|\psi(t)\rangle$? *Suggestion:* Express the wave function in terms of the $Y_{l,m}$'s.

(b) What values of $L_z$ will be obtained if a measurement is carried out and with what probability will these values occur?

(c) What is $\langle L_x\rangle$ for this state?

(d) If a measurement of $L_x$ is carried out, what result(s) will be obtained? With what probability?

**9.20.** Suppose that the rigid rotator of Problem 9.19 is immersed in a uniform magnetic field $\mathbf{B} = B_0\mathbf{k}$, and that the Hamiltonian is given by

$$\hat{H} = \frac{\hat{\mathbf{L}}^2}{2I} + \omega_0\hat{L}_z$$

where $\omega_0$ is a constant. If

$$\langle\theta, \phi|\psi(0)\rangle = \sqrt{\frac{3}{4\pi}}\sin\theta\sin\phi$$

what is $\langle\theta, \phi|\psi(t)\rangle$? What is $\langle L_x\rangle$ at time $t$?

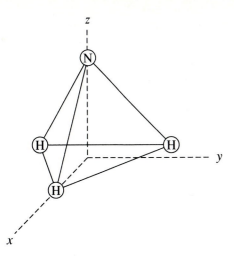

**FIGURE 9.12**

**9.21.** Treat the ammonia molecule, $NH_3$, shown in Fig. 9.12 as a symmetric rigid rotator. Call the moment of inertia about the $z$ axis $I_3$ and the moments about the pair of axes perpendicular to the $z$ axis $I_1$.

(a) Express the Hamiltonian of this rigid rotor in terms of $\hat{\mathbf{L}}$, $I_1$, and $I_3$.

(b) Show that $\left[\hat{H}, \hat{L}_z\right] = 0$.

(c) What are the eigenstates and eigenvalues of the Hamiltonian?

(d) Suppose that at time $t = 0$ the molecule is in the state

$$|\psi\rangle = \frac{1}{\sqrt{2}}|0, 0\rangle + \frac{1}{\sqrt{2}}|1, 1\rangle$$

What is $|\psi(t)\rangle$?

**9.22.** The Hamiltonian for a three-dimensional system with *cylindrical* symmetry is given by

$$\hat{H} = \frac{\hat{\mathbf{p}}^2}{2\mu} + V(\hat{\rho})$$

where $\rho = \sqrt{x^2 + y^2}$.

(a) Use symmetry arguments to establish that both $\hat{p}_z$ and $\hat{L}_z$, the $z$ component of the linear and angular momentum operators, respectively, commute with $\hat{H}$.

(b) Use the fact that $\hat{H}$, $\hat{p}_z$, and $\hat{L}_z$ have eigenstates in common to express the position-space eigenfunctions of the Hamiltonian in terms of those of $\hat{p}_z$ and $\hat{L}_z$.

(c) What is the radial equation? *Note:* The Laplacian in cylindrical coordinates is given by

$$\nabla^2 \psi = \frac{1}{\rho} \frac{\partial}{\partial \rho} \left( \rho \frac{\partial \psi}{\partial \rho} \right) + \frac{1}{\rho^2} \frac{\partial^2 \psi}{\partial \phi^2} + \frac{\partial^2 \psi}{\partial z^2}$$

# CHAPTER

# 10

# BOUND
# STATES
# OF CENTRAL
# POTENTIALS

In this chapter we solve the Schrödinger equation for the bound states of three systems—the Coulomb potential, the spherical well, and the three-dimensional harmonic oscillator—for which the potential energy $V = V(r)$. Figuratively, the Coulomb potential forms the centerpiece of our discussion. The exact solution of the two-body problem with a pure Coulomb interaction serves as the starting point for a detailed comparison in both this chapter and the next between theory and experiment for the hydrogen atom, a comparison that gives us much of our confidence in quantum mechanics.

## 10.1 THE BEHAVIOR OF THE RADIAL WAVE FUNCTION NEAR THE ORIGIN

In Chapter 9 we saw that for a potential energy with spherical symmetry, we can write the energy eigenfunctions as simultaneous eigenfunctions of $\hat{\mathbf{L}}^2$ and $\hat{L}_z$:

$$\langle \mathbf{r}|E, l, m \rangle = R_{E,l}(r)Y_{l,m}(\theta, \phi) \tag{10.1}$$

The Schrödinger equation for the radial wave function is given in (9.96),

$$\left[ -\frac{\hbar^2}{2\mu} \left( \frac{d^2}{dr^2} + \frac{2}{r}\frac{d}{dr} \right) + \frac{l(l+1)\hbar^2}{2\mu r^2} + V(r) \right] R_{E,l}(r) = E R_{E,l}(r) \tag{10.2}$$

where we have inserted subscripts indicating explicitly that the radial wave function depends on the values for $E$ and $l$. The lack of dependence of (10.2) on $m$

means that each energy state with fixed $E$ and fixed $l$ will have at least a $2l + 1$ degeneracy. As we did in Chapter 9, it is convenient to make the substitution

$$R_{E,l}(r) = \frac{u_{E,l}(r)}{r} \tag{10.3}$$

in which case (10.2) becomes

$$\left[ -\frac{\hbar^2}{2\mu} \frac{d^2}{dr^2} + \frac{l(l+1)\hbar^2}{2\mu r^2} + V(r) \right] u_{E,l}(r) = E u_{E,l}(r) \tag{10.4}$$

Expressed in terms of $u_{E,l}(r)$, the normalization condition (9.138) for bound states is given by[1]

$$\int_0^\infty r^2 \, dr \, R_{E,l}^* R_{E,l} = \int_0^\infty dr \, u_{E,l}^* u_{E,l} = 1 \tag{10.5}$$

Thus, as we remarked earlier, (10.4) has the same form as the one-dimensional Schrödinger equation (9.98) except that the variable $r$ runs from 0 to $\infty$, not from $-\infty$ to $\infty$. This naturally raises the question of what happens to the wave function when $r$ reaches the origin.

Provided the potential energy $V(r)$ is not more singular at the origin than $1/r^2$, the differential equation (10.4) has what is known as a regular singularity at $r = 0$, and we are guaranteed that solutions in the form of a power series about the origin exist. To determine the leading behavior of $u_{E,l}(r)$ for small $r$, we substitute $r^s$ into (10.4):

$$-\frac{\hbar^2}{2\mu} s(s-1)r^{s-2} + \frac{l(l+1)\hbar^2}{2\mu} r^{s-2} + V(r)r^s = E r^s \tag{10.6}$$

Notice that the $r^{s-2}$ terms dominate for small $r$ if $V(r)$ is less singular than $1/r^2$, that is,

$$r^2 V(r) \xrightarrow[r \to 0]{} 0 \tag{10.7}$$

Satisfying (10.6) for small $r$ requires the coefficients of $r^{s-2}$ to obey

$$-s(s-1) + l(l+1) = 0 \tag{10.8}$$

which shows that $s = l + 1$ or $s = -l$. However, we must discard those solutions that behave as $r^{-l}$ for small $r$. For $l \geq 1$, these solutions cannot satisfy the normalization condition (10.5) because the integral diverges at the lower limit. For $l = 0$, the leading behavior of $u$ for small $r$ is a constant and the integral (10.5) is finite. But if the leading behavior is a constant, the wave function $R$

---

[1] We restrict our analysis in this chapter to bound states, which have a discrete energy spectrum and can therefore obey this normalization condition. In Chapter 13 we will consider the continuum portion of the energy spectrum for central potentials.

behaves as $1/r$ near the origin, which is also unacceptable. To see this, return to the full three-dimensional Schrödinger equation in position space (9.130) and note that[2]

$$\nabla^2 \frac{1}{r} = -4\pi\delta^3(\mathbf{r}) \tag{10.9}$$

Therefore, a $1/r$ behavior for $R$ cannot satisfy the Schrödinger equation, since we are presuming that the potential energy does not have a delta function singularity at the origin. Thus we must discard the $r^{-l}$ solutions for all $l$; consequently, we deduce that the allowed behavior for small $r$ is given by

$$u_{E,l} \xrightarrow[r\to 0]{} r^{l+1} \tag{10.10}$$

and hence $u_{E,l}(r)$ must satisfy[3]

$$u_{E,l}(0) = 0 \tag{10.11}$$

Notice that as $l$ increases, the particle is less and less likely to be found in the vicinity of the origin. Recall that the "one-dimensional" Schrödinger equation (10.4) has an effective potential energy

$$V_{\text{eff}}(r) = \frac{l(l+1)\hbar^2}{2\mu r^2} + V(r) \tag{10.12}$$

The $l(l+1)/2\mu r^2$ term, known as the centrifugal barrier, increases with increasing $l$ and tends to keep the particle away from the origin, producing the small-$r$ behavior of the wave functions (see Fig. 10.1).

The behavior (10.10) implies that the radial wave functions

$$R_{E,l} \xrightarrow[r\to 0]{} r^l \tag{10.13}$$

and thus these wave functions vanish at the origin for all except $l = 0$, or $s$, states. A dramatic illustration is the annihilation of positronium, a hydrogen-like atom where the nucleus is a positron instead of a proton. For the electron and the positron to annihilate, they must overlap spatially, which can occur only

---

[2] One way to verify this result is to integrate both sides of the equation over a spherical volume including the origin and use Gauss's theorem:

$$\int d^3r\, \boldsymbol{\nabla}\cdot\boldsymbol{\nabla}\frac{1}{r} = \oint dS\, \mathbf{n}\cdot\boldsymbol{\nabla}\frac{1}{r} = 4\pi r^2\left(-\frac{1}{r^2}\right) = -4\pi = \int d^3r\,[-4\pi\delta^3(\mathbf{r})]$$

Another way is to note the solution of Poisson's equation in electrostatics, $\nabla^2\varphi = -4\pi\rho$, for a point charge $q$:

$$\nabla^2\frac{q}{r} = -4\pi q\delta^3(\mathbf{r})$$

[3] See also Problem 10.1.

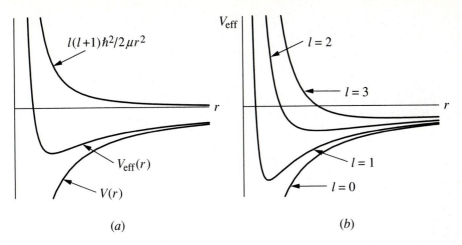

**FIGURE 10.1**

(a) The Coulomb potential $V(r) = -e^2/r$ and the centrifugal barrier $l(l + 1)\hbar^2/2\mu r^2$ add together to produce the effective potential energy

$$V_{\text{eff}} = \frac{l(l + 1)\hbar^2}{2\mu r^2} + V(r)$$

(b) The effective potential for several values of $l$.

in $s$ states. Positronium is often formed by the capture of a slow positron by an electron, generally in a highly excited state of the "atom." The atom then undergoes a sequence of radiative transitions, most often ending in the ground state, which we will see in the next section is an $l = 0$ state, where annihilation of the particle-antiparticle pair finally takes place.

## 10.2 THE COULOMB POTENTIAL AND THE HYDROGEN ATOM

The Hamiltonian for a hydrogenic atom is

$$\hat{H} = \frac{\hat{\mathbf{p}}^2}{2\mu} - \frac{Ze^2}{|\hat{\mathbf{r}}|} \tag{10.14}$$

Of course, the usual hydrogen atom has $Z = 1$, but by introducing the factor of $Z$, we can also consider atoms with a charge $Ze$ on the nucleus that have been ionized so that only a single electron remains. Examples include $\text{He}^+$, $\text{Li}^{++}$, and so on. The radial equation for the function $u_{E,l}$ is given by

$$\left[ -\frac{\hbar^2}{2\mu} \frac{d^2}{dr^2} + \frac{l(l + 1)\hbar^2}{2\mu r^2} - \frac{Ze^2}{r} \right] u_{E,l}(r) = E u_{E,l}(r) \tag{10.15}$$

Note that the potential energy of the system is negative because, as is customary, we have chosen the zero of potential energy when the two particles in the system are very far apart, that is, when $r \to \infty$. The potential energy $V(r)$ is then the work that *you* perform to bring the particles from infinity to a distance $r$ apart with no kinetic energy. Since the particles attract each other, you do negative work and therefore the potential energy is negative, as shown in Fig. 10.1. Classically, the two particles are bound together whenever the energy $E$ of the system is negative, because in this case there exists a radius beyond which the potential energy exceeds the total energy $E$, which would require negative kinetic energy (see Fig. 10.2). In this chapter we will restrict our attention to determining the bound states; in Chapter 13 we will examine the significance of the positive-energy solutions when we discuss scattering.

Since we are seeking solutions with negative energy to the differential equation (10.15), it is convenient to write $E = -|E|$ and to introduce the dimensionless variable

$$\rho = \sqrt{\frac{8\mu|E|}{\hbar^2}}\, r \tag{10.16}$$

Expressed in terms of this variable, (10.15) becomes[4]

$$\frac{d^2u}{d\rho^2} - \frac{l(l+1)}{\rho^2}u + \left(\frac{\lambda}{\rho} - \frac{1}{4}\right)u = 0 \tag{10.17}$$

where

$$\lambda = \frac{Ze^2}{\hbar}\sqrt{\frac{\mu}{2|E|}} \tag{10.18}$$

---

[4] We suppress the subscripts in this and succeeding equations for notational convenience. The factor of 8 in (10.16) is chosen to make (10.17) work out nicely.

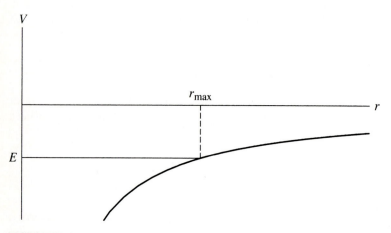

**FIGURE 10.2**
For a particular value of the total energy $E$, there is a maximum radius beyond which the particles cannot separate classically.

If we try to solve (10.17) through a power-series solution, we get a three-term recursion relation. However, in the limit $\rho \to \infty$, the equation simplifies to

$$\frac{d^2 u}{d\rho^2} - \frac{1}{4} u = 0 \tag{10.19}$$

which has solutions

$$u = A e^{-\rho/2} + B e^{\rho/2} \tag{10.20}$$

We discard the exponentially increasing solution because such a solution cannot satisfy the normalization condition (10.5). If we factor out the small-$\rho$ behavior (10.10) as well as the large-$\rho$ behavior, we can attempt to find a solution to (10.17) of the form

$$u(\rho) = \rho^{l+1} e^{-\rho/2} F(\rho) \tag{10.21}$$

in a manner similar to that which we used to solve the harmonic oscillator in Section 7.9.[5]

With this substitution, the differential equation (10.17) becomes

$$\frac{d^2 F}{d\rho^2} + \left( \frac{2l+2}{\rho} - 1 \right) \frac{dF}{d\rho} + \left( \frac{\lambda}{\rho} - \frac{l+1}{\rho} \right) F = 0 \tag{10.22}$$

Although this differential equation may seem more complicated than our starting point (10.17), it is now straightforward to obtain a power-series solution of the form

$$F(\rho) = \sum_{k=0}^{\infty} c_k \rho^k \tag{10.23}$$

with the restriction that $c_0 \neq 0$ so as not to violate (10.10). It should again be emphasized that we have made no approximations in deriving (10.22). It may appear that we have discarded one of the two possible behaviors of the function $u$ for large $\rho$, but, as we will now see, the exponentially increasing solution will resurrect itself unless we make a judicious choice of $\lambda$.

Substituting (10.23) into (10.22), we obtain

$$\sum_{k=2}^{\infty} k(k-1) c_k \rho^{k-2} + \sum_{k=1}^{\infty} (2l+2) k c_k \rho^{k-2} + \sum_{k=0}^{\infty} \left[ -k + \lambda - (l+1) \right] c_k \rho^{k-1} = 0 \tag{10.24}$$

Making the change of variables $k - 1 = k'$ in the first two terms and then renaming $k' = k$, we can express (10.24) as

$$\sum_{k=0}^{\infty} \left\{ \left[ k(k+1) + (2l+2)(k+1) \right] c_{k+1} + \left[ -k + \lambda - (l+1) \right] c_k \right\} \rho^{k-1} = 0 \tag{10.25}$$

---

[5] Factoring out the small-$\rho$ behavior is really just using the method of Frobenius to solve the differential equation by a power series.

leading to

$$\frac{c_{k+1}}{c_k} = \frac{k + l + 1 - \lambda}{(k + 1)(k + 2l + 2)} \tag{10.26}$$

Note that

$$\frac{c_{k+1}}{c_k} \xrightarrow[k \to \infty]{} \frac{1}{k} \tag{10.27}$$

which is the same asymptotic behavior as $e^\rho$. Thus, unless the series (10.23) terminates, the function $u(\rho)$ in (10.21) will grow exponentially like $e^{\rho/2}$. To avoid this fate, we must have

$$\lambda = 1 + l + n_r \tag{10.28}$$

where

$$n_r = 0, 1, 2, \ldots \tag{10.29}$$

determines the value of $k$ at which the series terminates. The function $F$ will thus be a polynomial of degree $n_r$, known as an associated Laguerre polynomial.

Quantizing $\lambda$ in (10.28) leads to a quantized energy from (10.18):

$$E = -\frac{\mu Z^2 e^4}{2\hbar^2(1 + l + n_r)^2} \tag{10.30}$$

Since $l$ and $n_r$ are both integers that are greater than or equal to zero, we define the principal quantum number $n$ by

$$l + 1 + n_r = n \tag{10.31}$$

Thus in terms of the principal quantum number

$$E_n = -\frac{\mu Z^2 e^4}{2\hbar^2 n^2} \qquad n = 1, 2, 3, \ldots \tag{10.32}$$

A useful way to express the result (10.32) is to introduce the speed of light $c$ to form

$$\frac{e^2}{\hbar c} = \alpha \tag{10.33}$$

a dimensionless quantity whose value is approximately $1/137$ and is known as the fine-structure constant, for reasons that will become apparent in Chapter 11.[6] In terms of $\alpha$, the allowed energies are given by

$$E_n = -\frac{\mu c^2 Z^2 \alpha^2}{2n^2} \tag{10.34}$$

---

[6] In SI units, $\alpha = e^2/4\pi\varepsilon_0\hbar c$.

Equation (10.34) is easy to remember. The quantity $\mu c^2$ carries the units of energy, and for hydrogen equals 0.511 MeV. The reason that the atomic energy scale is eV rather than MeV is the small value of $\alpha$ ($\alpha^2 = 5.33 \times 10^{-5}$). Numerically, the energy levels for hydrogen ($Z = 1$) are given by[7]

$$E_n = -\frac{13.6 \text{ eV}}{n^2} \tag{10.35}$$

and are indicated in Fig. 10.3.

When a hydrogen atom makes a transition from a state with principal quantum number $n_i$ to one with $n_f$ ($n_f < n_i$), the atom emits a photon with energy

---

[7] To reach a deep understanding of why the energy scale of atomic physics has the value it does, we need to understand both why $m_e c^2 = 0.511$ MeV and why the value of $\alpha$ has the value 1/137.0360. Although we do not know why elementary particles such as the electron have the masses that they do, the actual numerical value for the mass-energy of the electron cannot have any deep significance, since it depends on our choice of units, including, for example, the magnitude of the standard kilogram that is kept in Paris. The numerical value of $\alpha$, on the other hand, is completely independent of the choice of units. It is, therefore, a fair and important question to ask why $\alpha$ has this particular value. If you can provide the answer, you can skip past Chapter 14 and on to Stockholm to collect your Nobel prize.

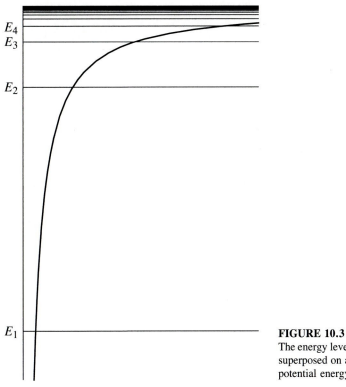

**FIGURE 10.3**
The energy levels of the hydrogen atom superposed on a graph of the Coulomb potential energy.

$$h\nu = E_{n_i} - E_{n_f} = \frac{\mu c^2 \alpha^2}{2}\left(\frac{1}{n_f^2} - \frac{1}{n_i^2}\right) \tag{10.36}$$

and therefore with inverse wavelength

$$\frac{1}{\lambda} = \frac{\mu c \alpha^2}{2h}\left(\frac{1}{n_f^2} - \frac{1}{n_i^2}\right) = R_{\mathrm{H}}\left(\frac{1}{n_f^2} - \frac{1}{n_i^2}\right) \tag{10.37}$$

The value of the Rydberg $R_{\mathrm{H}}$ for hydrogen, as determined by (10.37), is in complete agreement with experiment. Figure 10.4 shows the spectrum of hydrogen produced when transitions take place from a state with principal quantum number $n_i$ directly to a state with $n_f = 2$. These transitions, which are in the visible part of the spectrum, form the Balmer series. Transitions directly from excited states to the ground state ($n_f = 1$) emit more energetic photons in the ultraviolet portion of the spectrum, known as the Lyman series, while transitions from excited states to states with $n_f = 3$ emit less energetic photons in the infrared portion of the spectrum, known as the Paschen series.

Note from (10.34) that $E_n \ll m_e c^2$, which is consistent with our use of the nonrelativistic Schrödinger equation to describe the hydrogen atom. Of course, relativistic effects do exist. In Chapter 11 we will see that these effects produce a fine structure within the energy levels. There is also a different type of fine structure, whose origin is already apparent in (10.34), that is discernible in a typical hydrogen spectrum. This structure is due to the existence of an isotope of hydrogen in which the nucleus consists of a deuteron, a bound state of a proton and a neutron, instead of a single proton. Expressing the reduced mass $\mu$ in terms of the mass $M_N$ of the nucleus, we see that

$$E_n = -\frac{m_e c^2 Z^2 \alpha^2}{2n^2}\left(\frac{1}{1 + m_e/M_N}\right) \cong -\frac{m_e c^2 Z^2 \alpha^2}{2n^2}(1 - m_e/M_N) \tag{10.38}$$

where the last step follows since $m_e/m_N \ll 1$. Comparing this expression for hydrogen, where $M_N = m_p$, with that for deuterium, where the mass of the nu-

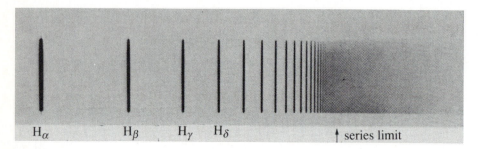

$H_\alpha$   $H_\beta$   $H_\gamma$   $H_\delta$   $\uparrow$ series limit

**FIGURE 10.4**
The visible spectrum of hydrogen, showing the Balmer series. (*From W. Finkelnburg,* Structure of Matter, *Springer-Verlag, Heidelberg, 1964.*)

cleus is roughly twice as large as for hydrogen, we see that the spectral lines of deuterium are shifted to slightly shorter wavelengths in comparison with those of hydrogen. For example, the $H_\alpha$ line at 6562.8 Å in the Balmer series, corresponding to the transition from $n_i = 3$ to $n_f = 2$, is shifted by about 1.8 Å, while the $H_\beta$ line at 4861.3 Å, corresponding to the transition from $n_i = 4$ to $n_f = 2$, is shifted by 1.3 Å. In naturally occurring hydrogen, this effect can be difficult to see because the natural abundance of deuterium is roughly 1 part in 7000. However, by increasing the concentration of the heavy isotope using thermal diffusion techniques, H. C. Urey discovered deuterium spectroscopically in 1932, the same year that the neutron was discovered.[8]

## The Hydrogenic Wave Functions

Since we can specify the energy eigenvalue by specifying the principal quantum number $n$, we can label the energy eigenfunctions by the quantum numbers $n$, $l$, and $m$:

$$\langle \mathbf{r}|n, l, m \rangle = R_{n,l}(r)Y_{l,m}(\theta, \phi) = \frac{u_{n,l}(r)}{r} Y_{l,m}(\theta, \phi) \tag{10.39}$$

Note that the dimensionless variable $\rho$ in (10.16) is given by

$$\rho = \sqrt{\frac{8\mu|E|}{\hbar^2}} r = \frac{2Z\mu c\alpha}{\hbar n} r = \frac{2Z}{n}\frac{r}{a_0} \tag{10.40}$$

where the length

$$a_0 = \frac{\hbar}{\mu c \alpha} \tag{10.41}$$

known as the Bohr radius, is a convenient length scale to use in expressing the wave functions. For hydrogen, $a_0$ has the magnitude 0.529 Å.

The ground state has $n = 1$ and, consequently from (10.31), $l = 0$ and $n_r = 0$. The power series (10.23) terminates after the first term, $F$ is just the constant $c_0$, and from (10.21)

$$u_{1,0}(\rho) = \rho e^{-\rho/2} c_0 \tag{10.42}$$

The normalized radial wave function is then given by

$$R_{1,0} = 2\left(\frac{Z}{a_0}\right)^{3/2} e^{-Zr/a_0} \tag{10.43}$$

We should emphasize that the ground state has zero orbital angular momentum.

---

[8] H. C. Urey, F. G. Brickwedde, and G. M. Murphy, *Phys. Rev.* **40**, 1 (1932). Urey received the Nobel prize in 1934 for this discovery.

This is to be contrasted with the early Bohr model, which preceded the development of quantum mechanics, in which the electron was believed to follow a definite orbit in each allowed stationary state. Each state in the Bohr model had a nonzero value of the orbital angular momentum, in an attempt to account for the stability of the atom. The only way to describe a bound state of zero orbital angular momentum with classical trajectories would be to have the electron traveling through the proton in the hydrogen atom in straight lines. Of course, as we saw in Section 9.4, quantum mechanics accounts naturally for the stability of the ground state through the uncertainty principle, and as we saw in Chapter 8, the concept of a classical trajectory is inappropriate for describing the motion of an electron within the atom. In particular, see Problem 8.5.

Let's examine the higher-energy states. The first-excited states have $n = 2$. Here we can have $l = 0$, $n_r = 1$, which means the power series (10.23) has two terms and is therefore a first-degree polynomial; or we can have $l = 1$, $n_r = 0$, which means the series (10.23) has only the first term, but in contrast with (10.40), we pick up an extra factor of $r$ from the $r^{l+1}$ in (10.21). The normalized radial wave functions for these states are given by

$$R_{2,0} = 2 \left( \frac{Z}{2a_0} \right)^{3/2} \left( 1 - \frac{Zr}{2a_0} \right) e^{-Zr/2a_0} \qquad (10.44a)$$

$$R_{2,1} = \frac{1}{\sqrt{3}} \left( \frac{Z}{2a_0} \right)^{3/2} \frac{Zr}{a_0} e^{-Zr/2a_0} \qquad (10.44b)$$

The second-excited states have $n = 3$. There are three possibilities: $l = 0$, $n_r = 2$, a second-degree polynomial for $F$; $l = 1$, $n_r = 1$, a first-degree polynomial for $F$; and $l = 2$, $n_r = 0$. The corresponding normalized radial wave functions are given by

$$R_{3,0} = 2 \left( \frac{Z}{3a_0} \right)^{3/2} \left[ 1 - \frac{2Zr}{3a_0} + \frac{2(Zr)^2}{27a_0^2} \right] e^{-Zr/3a_0} \qquad (10.45a)$$

$$R_{3,1} = \frac{4\sqrt{2}}{9} \left( \frac{Z}{3a_0} \right)^{3/2} \frac{Zr}{a_0} \left( 1 - \frac{Zr}{6a_0} \right) e^{-Zr/3a_0} \qquad (10.45b)$$

$$R_{3,2} = \frac{2\sqrt{2}}{27\sqrt{5}} \left( \frac{Z}{3a_0} \right)^{3/2} \left( \frac{Zr}{a_0} \right)^2 e^{-Zr/3a_0} \qquad (10.45c)$$

These are, of course, just the radial wave functions. The complete energy eigenfunctions also involve the spherical harmonics, as indicated in (10.39). As we saw in Chapter 9, these spherical harmonics can give rise to rather involved probability distributions as functions of the angles $\theta$ and $\phi$. However, if we only ask for the probability of finding the particle between $r$ and $r + dr$, we must integrate the three-dimensional probability density $|\langle \mathbf{r} | n, l, m \rangle|^2$ over all angles. Since the $Y_{l,m}$'s are themselves normalized according to (9.139), we are left with

$$\int\int \sin\theta\, d\theta\, d\phi\, r^2 dr\, |\langle \mathbf{r}|E, l, m\rangle|^2 = r^2 |R_{n,l}(r)|^2\, dr \tag{10.46}$$

as the probability of finding the particle between $r$ and $r + dr$. Note that the factor of $r^2$ in the radial probability density $r^2 |R_{n,l}(r)|^2$ comes from the volume element $d^3 r$. The radial wave functions, as well as the radial probability density, are plotted in Fig. 10.5 for the wave functions (10.43), (10.44), and (10.45).

As we have seen, $F(\rho)$ is a polynomial of degree $n_r = n - l - 1$. Thus it has $n_r$ radial nodes. The probability density $r^2 |R_{n,l}(r)|^2$ has $n - l$ "bumps." When, for a particular value of $n$, $l$ has its maximum value of $n - 1$, there is only one bump. Since $n_r = 0$ in this case, the wave function

$$R_{n,n-1} \propto r^{n-1} e^{-Zr/a_0 n} \tag{10.47}$$

and thus the probability density

$$r^2 |R_{n,n-1}|^2 \propto r^{2n} e^{-2Zr/a_0 n} \tag{10.48}$$

The location of the peak in the probability distribution can be found from

$$\frac{d}{dr} r^2 |R_{n,n-1}|^2 \propto \left(2n - \frac{2Z}{a_0 n} r\right) r^{2n-1} e^{-2Zr/a_0 n} = 0 \tag{10.49}$$

which yields

$$r = \frac{n^2 a_0}{Z} \tag{10.50}$$

As Fig. 10.5 shows, the states with different values of $l$ for a given energy do have differing radial probability densities. Even though the states with smaller values of $l$ for a particular $n$ have additional bumps in their probability distributions, the average position for each of the states with a given $n$ tends to reside in *shells* of increasing radius as $n$ increases. These extra bumps, which occur within the radius (10.50) for each shell, play a big role in determining the order that these states fill up in a multielectron atom and consequently in determining the structure of the periodic table. We will return to this issue in Chapter 12.

## Degeneracy

One of the most striking features of the hydrogen atom is the surprising degree of degeneracy, that is, the number of linearly independent states with the same energy. For each $n$, the allowed $l$ values are

$$l = 0, 1, \ldots, n - 1 \tag{10.51}$$

and for each $l$, there are $2l + 1$ states specified by the $m$ values. Thus the total degeneracy for a particular $n$ is

$$\sum_{l=0}^{n-1} (2l + 1) = 2\frac{(n - 1)n}{2} + n = n^2 \tag{10.52}$$

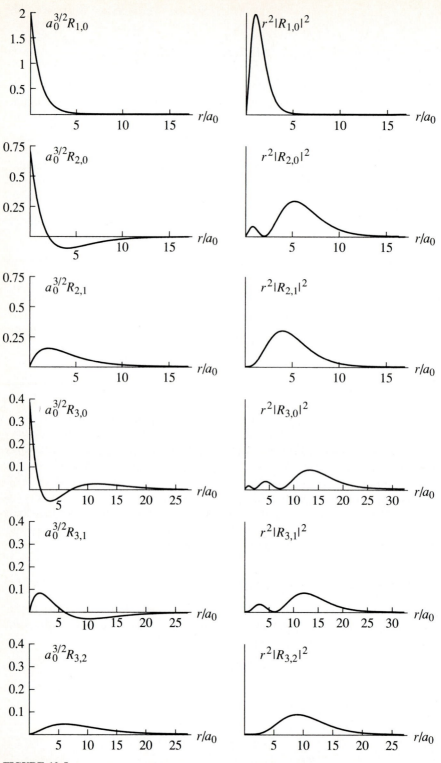

**FIGURE 10.5**
Plots of the radial wave function $R_{n,l}(r)$ and the radial probability density $r^2|R_{n,l}(r)|^2$ for the wave functions in (10.43), (10.44), and (10.45).

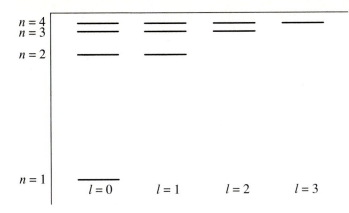

**FIGURE 10.6**
The $n = 1$ through $n = 4$ energy levels of the hydrogen atom, showing the degeneracy. States with $l = 0$ are called $s$ states, $l = 1$ $p$ states, $l = 2$ $d$ states, $l = 3$ $f$ states, and from then on the labeling is alphabetical. Historically, this nomenclature for the low values of $l$ arose from characteristics in the spectrum: *s*harp, *p*rincipal, and *d*iffuse.

The lack of dependence of the energy on $l$ is shown in Fig. 10.6. This degeneracy is unexpected, unlike the independence of the energy on the $m$ value, which we expected on the grounds of rotational symmetry. This rotational symmetry would disappear if, for example, we were to apply an external magnetic field that picks out a particular direction, such as the $z$ direction. In that case, the energy would depend on the projection of the angular momentum on the $z$ axis. Unlike the rotational symmetry that is responsible for the degeneracy of the different $m$ states, there is no "obvious" symmetry that indicates that states, such as (10.44a) and (10.44b), with different $l$'s should have exactly the same energy. In fact, if we examine the different effective potentials shown in Fig. 10.1 for these states, we see how unusual it is, for example, that the state with one node and no centrifugal barrier should have exactly the same energy as the state with no nodes and a $\hbar^2/\mu r^2$ centrifugal barrier. Historically, because the reason for the degeneracy of the $l$ values wasn't obvious, it was often termed an "accidental" degeneracy.[9]

Our discussion of degeneracy has ignored the spin of the electron and the spin of the proton, both spin-$\frac{1}{2}$ particles. Since there are two electron spin states and two proton spin states for each $|n, l, m\rangle$, we should multiply (10.52) by 4, yielding a total degeneracy of $4n^2$. Thus the ground state, for which $n = 1$, is really four-fold degenerate. It is this degeneracy that is partially split by the hyperfine interaction, which we discussed in Section 5.2.

---

[9] For a discussion of the dynamical symmetry associated with the hydrogen atom, see L. I. Schiff, *Quantum Mechanics,* 3rd ed., McGraw-Hill, New York, 1968, Section 30.

## 10.3 THE FINITE SPHERICAL WELL AND THE DEUTERON

Let's shift our attention from atomic physics to nuclear physics. For the hydrogen atom, the excellent agreement between the energy levels (actually the energy differences) and the spectrum of the photons emitted as the atom shifts from one energy level to another provides a detailed confirmation that the potential energy between an electron and a proton is indeed $-e^2/r$, even on the distance scale of angstroms. This is our first serious indication that Maxwell's equations describe physics on the microscopic as well as the macroscopic scale. When we examine the simplest two-body problem in nuclear physics, the neutron-proton bound state known as the deuteron, we find things are not so straightforward. The nuclear force between the proton and neutron is a short-range force: essentially, the proton and neutron interact strongly only if the particles touch each other. Thus we don't have macroscopic equations, like those in electromagnetism, that we can apply on the microscopic scale for nuclear physics. Instead we try to deduce the nuclear-force law by guessing or modeling the nature of the nuclear interaction and then comparing the results of our quantum mechanical calculations with experiment. As we will see in our analysis of the deuteron, this approach faces severe limitations.

The simplest model of the nuclear force between a proton and a neutron is a spherical well of finite range $a$ and finite depth $V_0$, shown in Fig. 10.7. The potential well, which looks square in Fig. 10.7a, is often referred to as a "square"

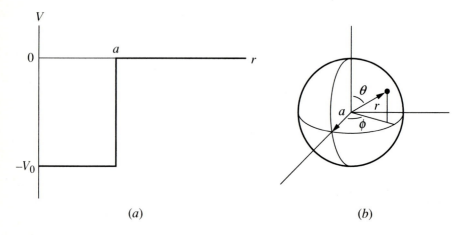

(a)                                                                  (b)

**FIGURE 10.7**
(a) A graph of the potential energy of the finite spherical well:

$$V = \begin{cases} -V_0 & r < a \\ 0 & r > a \end{cases}$$

(b) The region of the well shown in three dimensions.

well but is really a spherical well in three dimensions, as indicated in Fig. 10.7$b$. Unlike the hydrogen atom with its infinite set of bound states, experiment reveals that there is only a single bound state for the $n$-$p$ system. All the excited states of the two-nucleon system are unbound.

The ground state of the potential well

$$V = \begin{cases} -V_0 & r < a \\ 0 & r > a \end{cases} \tag{10.53}$$

is an $l = 0$ state, for which there is no centrifugal barrier. We thus solve the radial equation for an $l = 0$ bound state, one with energy $-V_0 < E < 0$:

$$-\frac{\hbar^2}{2\mu}\frac{d^2u}{dr^2} - V_0 u = Eu \qquad r < a \tag{10.54a}$$

$$-\frac{\hbar^2}{2\mu}\frac{d^2u}{dr^2} = Eu \qquad r > a \tag{10.54b}$$

We can express this equation in the form

$$\frac{d^2u}{dr^2} = -\frac{2\mu}{\hbar^2}(V_0 + E)u = -Q^2u \qquad r < a \tag{10.55a}$$

$$\frac{d^2u}{dr^2} = -\frac{2\mu E}{\hbar^2}u = q^2u \qquad r > a \tag{10.55b}$$

where

$$Q = \sqrt{\frac{2\mu}{\hbar^2}(V_0 + E)} \tag{10.56a}$$

and

$$q = \sqrt{-\frac{2\mu E}{\hbar^2}} \tag{10.56b}$$

so that $q$ and $Q$ are both real for the energy in the range $-V_0 < E < 0$.
The solutions to (10.55) are just

$$u = A\sin Qr + B\cos Qr \qquad r < a \tag{10.57a}$$

$$u = Ce^{-qr} + De^{qr} \qquad r > a \tag{10.57b}$$

The boundary condition $u(0) = 0$ tells us that $B = 0$, while the requirement that our solution satisfy the normalization condition (10.5) demands that we set $D = 0$. Thus the form for the radial wave function $u$ must be

$$u = A\sin Qr \qquad r < a \tag{10.58a}$$

$$u = Ce^{-qr} \qquad r > a \tag{10.58b}$$

Since the differential equation (10.54) is a second-order differential equation with a finite potential energy, the first derivative of $u$ must be continuous so that the second derivative exists and is finite everywhere. Correspondingly, in order for the first derivative to be well defined, the function $u$ must be continuous everywhere. Satisfying the continuity condition on $u$ at $r = a$ yields

$$A \sin Qa = Ce^{-qa} \tag{10.59a}$$

while making the derivative of $u$ continuous at $r = a$ yields

$$AQ \cos Qa = -qCe^{-qa} \tag{10.59b}$$

Dividing these equations, we find

$$\tan Qa = -\frac{Q}{q} \tag{10.60}$$

Equation (10.60) is a transcendental equation that determines the allowed values of the energy. A convenient aid to determining graphically the energy eigenvalues is to introduce the variables $Qa = \zeta$ and $qa = \eta$. Expressed in terms of these variables, (10.60) becomes

$$\zeta \cot \zeta = -\eta \tag{10.61}$$

Note that

$$\zeta^2 + \eta^2 = \frac{2\mu V_0 a^2}{\hbar^2} \tag{10.62}$$

which is independent of the energy $E$. Figure 10.8 shows a plot of (10.61) and (10.62) in the $\zeta - \eta$ plane. From the figure we see that there are no bound states unless

$$\frac{2\mu V_0 a^2}{\hbar^2} > \left(\frac{\pi}{2}\right)^2 \tag{10.63a}$$

For

$$\left(\frac{\pi}{2}\right)^2 < \frac{2\mu V_0 a^2}{\hbar^2} < \left(\frac{3\pi}{2}\right)^2 \tag{10.63b}$$

there is a single bound state, and so on. These results are to be contrasted with the results from the one-dimensional finite square well, for which there is at least one bound state no matter how shallow or narrow the well. Although (10.54) has the same form as the energy eigenvalue equation for a one-dimensional well, the boundary condition $u(0) = 0$ restricts the eigenvalue condition to (10.61), eliminating the $\zeta \tan \zeta = \eta$ curves that would fill in the "missing space" in Fig. 10.8 in the purely one-dimensional system (see Problem 10.8).

From analysis of the threshold for $\gamma + d \rightarrow n + p$, the photodisintegration of the deuteron, we know that the deuteron bound state has an energy $E = -2.2$ MeV. However, with a single bound state we cannot do more than determine the product of $V_0 a^2$. Let's take the value for $a$ to be roughly $1.7 \times 10^{-13}$ cm, as determined from scattering experiments, and determine the value for the depth $V_0$ of the nuclear potential well. To get started, let's assume that $|E| \ll V_0$, that is, the deuteron is just barely bound. If this is so, the curves in Fig. 10.8 intersect

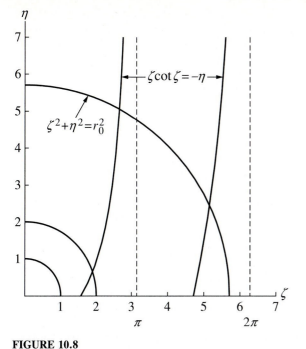

**FIGURE 10.8**
A plot of $\zeta \cot \zeta = -\eta$ and $\zeta^2 + \eta^2 = 2\mu V_0 a^2/\hbar^2 = r_0^2$ for three values of the radius $r_0$, which yield zero, one, and two bound states, respectively.

when $2\mu V_0 a^2/\hbar^2 \cong (\pi/2)^2$. Using the experimental value for $a$, we find $V_0 \cong 35$ MeV, in agreement with our assumption that the deuteron binding energy is much smaller than the depth of the potential well. In Fig. 10.9a we redraw the potential well and the energy of the bound state more to scale. We now have enough information to sketch the function $u$ in Fig. 10.9b. The two nucleons have a substantial probability of being separated by a distance that is greater than the range of the potential well and the shape of the wave function is, correspondingly, not very sensitive to the detailed nature of the potential.

Even though the square well is not an especially realistic model for the inter-nuclear potential, it does give us some useful information about the nuclear force. However, we now see the problem in trying to understand nuclear physics by studying the two-body bound-state problem. With just a single bound state we do not have enough information to gain a detailed picture of the nuclear interaction from a study of the bound-state spectrum. It is worth noting here, however, that this single bound state does give us some additional information. The deuteron has an intrinsic spin of one; the proton and the neutron do not bind together in a spin-0 state, indicating that the nuclear force is spin dependent. The deuteron has a magnetic moment and an electric quadrupole moment. The existence of an electric quadrupole moment tells us that the probability distribution in the ground state is not spherically symmetric. Thus this system is not strictly described by a central potential. However, the departure from spherical symmetry turns out not

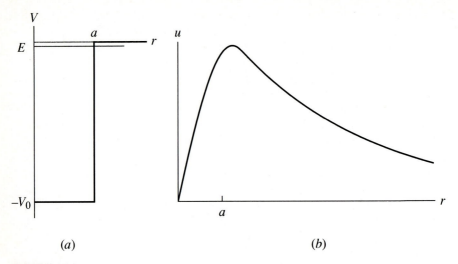

(a)                                                                 (b)

**FIGURE 10.9**
(a) The potential energy of a finite spherical well showing the energy of the bound state corresponding to the deuteron relative to the depth of the well. (b) A sketch of the radial wave function $u(r)$ for the deuteron. Note that the wave function extends significantly beyond the range $a$ of the well.

to be large. More detailed calculations show that the ground state of the deuteron is a mixture of 96 percent $l = 0$ and 4 percent $l = 2$ states. This mixing is due to a spin-orbit coupling in the nucleon-nucleon interaction that we have neglected in our simple model. We will see more evidence for this spin-orbit coupling in the next section when we look at a system of many nucleons, and we will see how spin-orbit coupling can arise in atomic systems in Chapter 11.

## 10.4   THE INFINITE SPHERICAL WELL

One way to learn more about nuclear interactions is to study systems containing a larger number of nucleons, that is, heavier nuclei. One convenient way to describe such a multibody system is in terms of a model in which each nucleon moves independently in a potential well due to its interactions with the other nucleons. A simple but useful model—known as the Fermi gas model—is to take the potential energy of each nucleon to be that of a spherical potential well like that shown in Fig. 10.7. As we add more and more nucleons to the well, the size of the well increases. Note that the effect of increasing $a$ for the $l = 0$ solutions to the finite potential well is the same as increasing the depth of the well $V_0$. In either case, the radius of the circle formed by $\zeta^2 + \eta^2$ grows, and the intersection points of the circle and $\zeta \cot \zeta = -\eta$ approach $\zeta = n\pi$ as the radius approaches infinity. In order to determine the energy levels of the $l \neq 0$ as well as the $l = 0$ states, we therefore examine the infinite potential well

$$V = \begin{cases} 0 & r < a \\ \infty & r > a \end{cases} \tag{10.64}$$

This well will give us an idea of the ordering of the energy levels for a spherical well, and it is much easier to solve than the finite potential well.

What is the behavior of the wave function in a system where the potential energy jumps discontinuously to infinity? Notice in the explicit form (10.58$b$) for the $l = 0$ wave function in the region outside the finite potential well that as the energy $E$ takes on values that are more negative and therefore further below the top of the potential well, the exponential tail of the wave function falls off more rapidly. In the limiting case that the energy is infinitely far below the top of the well, the wave function vanishes in the region $r > a$. As we saw when we discussed the one-dimensional particle in the box, in this limit the derivative of the wave function is discontinuous at the boundary, as shown in Fig. 6.8. In fact, in order to obtain a solution to the differential equation (10.2) in this limit, we want the derivative to be discontinuous so that when we evaluate the second derivative we do get infinity. Otherwise we cannot satisfy the differential equation at the point where $V$ jumps discontinuously to infinity. See (6.97).

Because the potential energy inside the well is zero, we are searching for the solutions inside the well to the Schrödinger equation for a free particle. However, since we must satisfy the boundary condition $\psi(r = a, \theta, \phi) = 0$, we need to find solutions in spherical coordinates of the form $R(r)Y_{l,m}(\theta, \phi)$; we cannot just use the plane wave solutions (9.25). Rather than start with the radial equation for the function $u(r)$, we return to the radial equation (10.2) for $R(r)$ for the case $V(r) = 0$. This equation takes the form

$$\frac{d^2R}{dr^2} + \frac{2}{r}\frac{dR}{dr} - \frac{l(l+1)}{r^2}R + k^2R = 0 \tag{10.65}$$

where

$$k = \sqrt{\frac{2\mu E}{\hbar^2}} \tag{10.66}$$

With no potential energy, we can obtain a power-series solution to (10.65) in the form of a two-term recursion relation. However, we don't need to follow this procedure in this case, since (10.65) can be solved in terms of simple functions. If we introduce the dimensionless variable $\rho = kr$, (10.65) becomes

$$\frac{d^2R}{d\rho^2} + \frac{2}{\rho}\frac{dR}{d\rho} + \left[1 - \frac{l(l+1)}{\rho^2}\right]R = 0 \tag{10.67}$$

known as the *spherical Bessel equation*. Solutions to this equation that are regular at the origin are called *spherical Bessel functions*,

$$j_l(\rho) = (-\rho)^l \left(\frac{1}{\rho}\frac{d}{d\rho}\right)^l \left(\frac{\sin\rho}{\rho}\right) \tag{10.68a}$$

while irregular solutions at the origin are called *spherical Neumann functions*,

$$\eta_l(\rho) = -(-\rho)^l \left(\frac{1}{\rho}\frac{d}{d\rho}\right)^l \left(\frac{\cos\rho}{\rho}\right) \tag{10.68b}$$

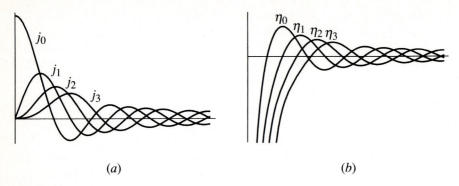

(a)                                                                        (b)

**FIGURE 10.10**
(a) Spherical Bessel functions and (b) spherical Neumann functions.

The first few functions, shown in Fig. 10.10, are

$$j_0(\rho) = \frac{\sin \rho}{\rho} \qquad\qquad \eta_0(\rho) = -\frac{\cos \rho}{\rho}$$

$$j_1(\rho) = \frac{\sin \rho}{\rho^2} - \frac{\cos \rho}{\rho} \qquad\qquad \eta_1(\rho) = -\frac{\cos \rho}{\rho^2} - \frac{\sin \rho}{\rho}$$

$$j_2(\rho) = \left(\frac{3}{\rho^3} - \frac{1}{\rho}\right)\sin \rho - \frac{3\cos \rho}{\rho^2} \qquad \eta_2(\rho) = -\left(\frac{3}{\rho^3} - \frac{1}{\rho}\right)\cos \rho - \frac{3\sin \rho}{\rho^2}$$

$$(10.69)$$

For the spherical well we must choose solutions to (10.67) that satisfy the boundary condition that $rR(r) = u(r)$ vanishes when $r = 0$. Thus we must discard the spherical Neumann functions. The energy eigenvalues are then determined by requiring that

$$j_l(ka) = 0 \tag{10.70}$$

Let's first examine the $l = 0$ condition,

$$j_0(ka) = \frac{\sin ka}{ka} = 0 \tag{10.71}$$

which is satisfied when $ka = n_r\pi$, where $n_r = 1, 2, 3, \ldots$. The $l = 0$ energies are given by

$$E_{n_r, l=0} = \frac{\hbar^2 k^2}{2\mu} = \frac{\hbar^2}{2\mu}\left(\frac{n_r \pi}{a}\right)^2 \qquad n_r = 1, 2, 3, \ldots \tag{10.72}$$

which agrees with our analysis of the finite well in the limit that the depth of the well approaches infinity.[10] The value of $n_r$ specifies the number of nodes

---

[10] To compare the results, redefine the bottom of the finite potential to be at $V = 0$ and the top at $V = V_0$ and then let $V_0 \to \infty$.

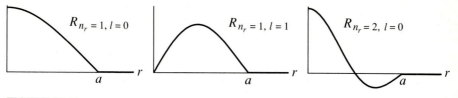

**FIGURE 10.11**
The ground-state, first-excited, and second-excited radial wave functions $R(r)$ for the infinite spherical potential well.

in the radial wave function, as indicated in Fig. 10.11. The ground state has the only node occurring at $r = a$, the first-excited $l = 0$ state has the second node occurring at $r = a$, and so on.

For the higher-order spherical Bessel functions we cannot determine the zeros by inspection, as we have done for the $l = 0$ state. However, these zeros are tabulated:[11]

|          | $l = 0$ | $l = 1$ | $l = 2$ | $l = 3$ |
|----------|---------|---------|---------|---------|
| $n_r = 1$ | 3.14    | 4.49    | 5.76    | 6.99    |
| $n_r = 2$ | 6.28    | 7.73    | 9.09    | 10.42   |
| $n_r = 3$ | 9.42    | 10.90   | 12.32   | 13.70   |

Notice that the lowest zero of an $l = 1$ energy state occurs when $ka = 4.49$, and thus the energy is given by

$$E_{n_r = 1, l = 1} = \frac{\hbar^2 k^2}{2\mu} = \frac{\hbar^2}{2\mu}\left(\frac{4.49}{a}\right)^2 \tag{10.73}$$

which is therefore intermediate between the $l = 0$ ground state [$n_r = 1$ in (10.72)] and the first-excited $l = 0$ state [$n_r = 2$ in (10.72)]. Figure 10.12 shows the energy spectrum for the infinite spherical well. Note the absence of any accidental degeneracy.

Let's now try making a nucleus by filling the energy levels with protons and neutrons. Since protons and neutrons are both spin-$\frac{1}{2}$ particles, we will find in Chapter 12 that we can put no more than two protons and two neutrons in each of these energy states. If we neglect the Coulomb repulsion between the protons as a first approximation, the energy levels for protons and neutrons are the same. If we fill the levels with protons, for example, we can put two protons in the $n_r = 1$, $l = 0$ ground state. The next level is an $n_r = 1$, $l = 1$ state into which we can put six protons, since there are three different $m$ values for $l = 1$. The next energy level is an $n_r = 1$, $l = 2$ state, which can fit $2 \times 5 = 10$ protons,

---

[11] P. M. Morse and H. Feshback, *Methods of Theoretical Physics,* McGraw-Hill, New York, 1953, p.1576.

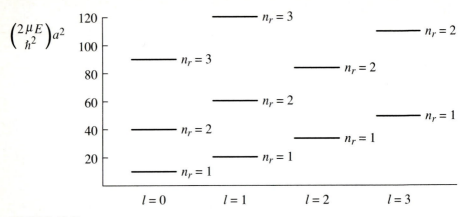

**FIGURE 10.12**
The energy levels of the infinite spherical well.

since there are five different $m$ values for $l = 2$. After we have filled this energy level, the next energy level is $n_r = 2$, $l = 0$, which can again accommodate two protons. In this way, we see that the energy levels will be completely filled when the number of protons is 2, 8 ($= 2 + 6$), 18 ($= 8 + 10$), 20 ($=18 + 2$), 34 ($= 20 + 14$), 40, 58, $\ldots$, with a similar sequence for neutrons. Real nuclei exhibit special properties that are associated with filled energy levels, or closed shells, with the "magic" numbers 2, 8, 20, 28, 50, 82, and 126. The differences between the observed magic numbers and those in our very simple model arise because, as for the deuteron, there is a strong "inverted" spin-orbit coupling that shifts the energy levels (see Fig. 10.13).

## 10.5  THE THREE-DIMENSIONAL ISOTROPIC HARMONIC OSCILLATOR

As our last example of analytically determining the energy eigenvalues of a central potential, we consider the three-dimensional simple harmonic oscillator, for which the potential energy

$$V(r) = \tfrac{1}{2}\mu\omega^2 r^2 = \tfrac{1}{2}\mu\omega^2(x^2 + y^2 + z^2) \tag{10.74}$$

Such an oscillator is often referred to as an isotropic oscillator, since the "spring constant" has the same magnitude in all directions.[12] One of the things that makes this isotropic oscillator especially noteworthy is that we can easily determine the eigenstates and eigenvalues in both spherical and Cartesian coordinates and then see the connection between two different sets of basis states.

---

[12] An anisotropic oscillator would have a potential energy of the form

$$V = \tfrac{1}{2}\mu(\omega_x^2 x^2 + \omega_y^2 y^2 + \omega_z^2 z^2)$$

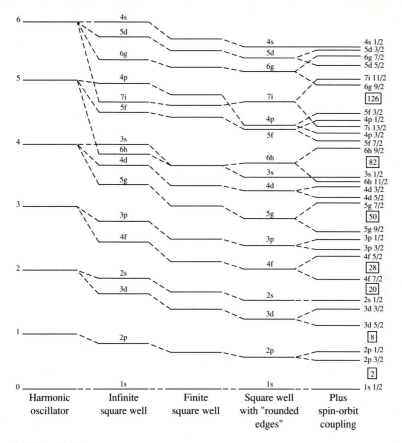

**FIGURE 10.13**
The ordering of the energy levels in a variety of potential energy wells. *(From B. T. Feld*, Ann. Rev. Nuclear Sci. **2,** *239 (1953), as reproduced by R. B. Leighton*, Principles of Modern Physics, *McGraw-Hill, New York, 1959.)*

## Cartesian Coordinates

The Hamiltonian for the three-dimensional isotropic oscillator is given by

$$\hat{H} = \frac{\hat{\mathbf{p}}^2}{2\mu} + \frac{1}{2}\mu\omega^2|\hat{\mathbf{r}}|^2 = \frac{\hat{p}_x^2 + \hat{p}_y^2 + \hat{p}_z^2}{2\mu} + \frac{1}{2}\mu\omega^2(\hat{x}^2 + \hat{y}^2 + \hat{z}^2) \quad (10.75)$$

which in the last step has been expressed in Cartesian coordinates. This Hamiltonian is a sum of three independent one-dimensional harmonic oscillators:

$$\hat{H} = \hat{H}_x + \hat{H}_y + \hat{H}_z \quad (10.76)$$

where

$$\hat{H}_x = \frac{\hat{p}_x^2}{2\mu} + \frac{1}{2}\mu\omega^2\hat{x}^2 \quad (10.77a)$$

$$\hat{H}_y = \frac{\hat{p}_y^2}{2\mu} + \frac{1}{2}\mu\omega^2\hat{y}^2 \tag{10.77b}$$

$$\hat{H}_z = \frac{\hat{p}_z^2}{2\mu} + \frac{1}{2}\mu\omega^2\hat{z}^2 \tag{10.77c}$$

Since these Hamiltonians commute with each other—$\left[\hat{H}_x, \hat{H}_y\right] = 0$, and so on—they have simultaneous eigenstates in common,

$$|E\rangle = |E_x, E_y, E_z\rangle \tag{10.78}$$

where

$$\hat{H}|E\rangle = (\hat{H}_x + \hat{H}_y + \hat{H}_z)|E_x, E_y, E_z\rangle = (E_x + E_y + E_z)|E_x, E_y, E_z\rangle \tag{10.79}$$

and hence $E = E_x + E_y + E_z$. Here we can take full advantage of the eigenstates of the one-dimensional harmonic oscillator that we found in Chapter 7. Namely, we can specify the eigenstates with the three integers

$$|E\rangle = |n_x, n_y, n_z\rangle \qquad n_x, n_y, n_z = 0, 1, 2, \ldots \tag{10.80}$$

where

$$
\begin{aligned}
E &= \left(n_x + \tfrac{1}{2}\right)\hbar\omega + \left(n_y + \tfrac{1}{2}\right)\hbar\omega + \left(n_z + \tfrac{1}{2}\right)\hbar\omega \\
&= \left(n_x + n_y + n_z + \tfrac{3}{2}\right)\hbar\omega \qquad n_x, n_y, n_z = 0, 1, 2, \ldots
\end{aligned} \tag{10.81}
$$

Setting $n_x + n_y + n_z = n$, we can express the total energy as

$$E_n = \left(n + \tfrac{3}{2}\right)\hbar\omega \qquad n = 0, 1, 2, \ldots \tag{10.82}$$

We can write the energy eigenfunctions $\langle x, y, z \,|\, n_x, n_y, n_z\rangle$ as a product of three one-dimensional eigenfunctions that we determined in Chapter 7. It is instructive, however, to see how these eigenfunctions arise by solving the three-dimensional Schrödinger equation directly in position space, because this provides a good illustration of the technique of separation of variables that we have alluded to several times. We write the energy eigenfunction as

$$\langle x, y, z \,|\, E\rangle = X(x)Y(y)Z(z) \tag{10.83}$$

and substitute it into the position-space energy eigenvalue equation:

$$\left[-\frac{\hbar^2}{2\mu}\left(\frac{\partial^2}{\partial x^2} + \frac{\partial^2}{\partial y^2} + \frac{\partial^2}{\partial z^2}\right) + \frac{1}{2}\mu\omega^2(x^2 + y^2 + z^2)\right]X(x)Y(y)Z(z)$$

$$= EX(x)Y(y)Z(z) \tag{10.84}$$

If we then divide this equation by the wave function $X(x)Y(y)Z(z)$, we obtain

$$\left[-\frac{1}{X}\frac{\hbar^2}{2\mu}\frac{d^2X}{dx^2} + \frac{1}{2}\mu\omega^2x^2\right] + \left[-\frac{1}{Y}\frac{\hbar^2}{2\mu}\frac{d^2Y}{dy^2} + \frac{1}{2}\mu\omega^2y^2\right]$$

$$+ \left[-\frac{1}{Z}\frac{\hbar^2}{2\mu}\frac{d^2Z}{dz^2} + \frac{1}{2}\mu\omega^2z^2\right] = E \tag{10.85}$$

This separation-of-variables approach (10.83) "works," since the partial differential equation (10.84) can now be expressed as the sum of three independent pieces: the term in the first bracket in (10.85) is solely a function of $x$, the second bracket is solely a function of $y$, and the third bracket is solely a function of $z$. Now $x$, $y$, and $z$ are independent variables, and hence each of the functions in the brackets can be varied independently. Thus the only way for this equation to hold for all $x$, $y$, and $z$ is for each of the terms in the brackets to be equal to a constant. With some foresight we call the constants $E_x$, $E_y$, and $E_z$. Then (10.85) breaks into three separate equations:

$$-\frac{\hbar^2}{2\mu}\frac{d^2X}{dx^2} + \frac{1}{2}\mu\omega^2x^2X = E_xX \tag{10.86a}$$

$$-\frac{\hbar^2}{2\mu}\frac{d^2Y}{dy^2} + \frac{1}{2}\mu\omega^2y^2Y = E_yY \tag{10.86b}$$

$$-\frac{\hbar^2}{2\mu}\frac{d^2Z}{dz^2} + \frac{1}{2}\mu\omega^2z^2Z = E_zZ \tag{10.86c}$$

where $E_x + E_y + E_z = E$. Each of the equations (10.86) is an energy eigenvalue equation for a one-dimensional harmonic oscillator with the eigenfunctions

$$X_{n_x}(x) = \langle x|n_x\rangle \qquad Y_{n_y}(y) = \langle y|n_y\rangle \qquad \text{and} \qquad Z_{n_z}(z) = \langle z|n_z\rangle \tag{10.87}$$

where we have used the same bra-ket notation for each independent oscillator that we used in Chapter 7. The energy eigenvalues (10.81) and the energy eigenstates then follow directly from those results.

## Spherical Coordinates

We next take advantage of the spherical symmetry of the Hamiltonian (10.75) to write the energy eigenfunction as

$$\langle \mathbf{r}|E\rangle = \langle r,\theta,\phi|E\rangle = R(r)Y_{l,m}(\theta,\phi) = \frac{u(r)}{r}Y_{l,m}(\theta,\phi) \tag{10.88}$$

The radial equation is then given by

$$-\frac{\hbar^2}{2\mu}\frac{d^2u}{dr^2} + \frac{l(l+1)\hbar^2}{2\mu r^2}u + \frac{1}{2}\mu\omega^2r^2u = Eu \tag{10.89}$$

Expressed in terms of the dimensionless variables

$$\rho = \sqrt{\frac{\mu\omega}{\hbar}}\,r \qquad \text{and} \qquad \lambda = \frac{2E}{\hbar\omega} \tag{10.90}$$

the differential equation (10.89) becomes

$$\frac{d^2u}{d\rho^2} - \frac{l(l+1)}{\rho^2}u - \rho^2u = -\lambda u \tag{10.91}$$

We can see that attempting a power-series solution to (10.91) will meet with a three-term recursion relation. However, for $\rho \to \infty$, the differential equation becomes

$$\frac{d^2u}{d\rho^2} = \rho^2 u \tag{10.92}$$

This suggests we search for a solution of the form

$$u = \rho^{l+1} e^{-\rho^2/2} f(\rho) \tag{10.93}$$

where the first factor indicates the known behavior for small $\rho$ and the exponential indicates the asymptotic behavior for large $\rho$. We can then find a two-term recursion relation for the power series

$$f(\rho) = \sum_{k=0}^{\infty} c_k \rho^k \tag{10.94}$$

It is straightforward to show that unless this power series terminates, it has the behavior of $e^{\rho^2}$ for large $\rho$ (see Problem 10.11). The energy quantization condition resulting from requiring termination of the power series is

$$E = \left(2n_r + l + \tfrac{3}{2}\right)\hbar\omega \qquad n_r = 0, 1, 2, \ldots \tag{10.95}$$

where $n_r$ is the number of the nodes of the function $f(\rho)$. Defining the principal quantum number $n = 2n_r + l$, we obtain

$$E_n = \left(n + \tfrac{3}{2}\right)\hbar\omega \qquad n = 2n_r + l \qquad n = 0, 1, 2, \ldots \tag{10.96}$$

in agreement with our earlier result. These energy levels are indicated in Fig. 10.14.

## Degeneracy

As with the hydrogen atom, one of the surprising features of the energy eigenvalues of the harmonic oscillator is the high degree of degeneracy. We can see

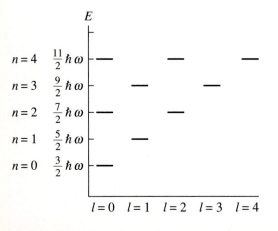

$n = 4 \quad \tfrac{11}{2}\hbar\omega$

$n = 3 \quad \tfrac{9}{2}\hbar\omega$

$n = 2 \quad \tfrac{7}{2}\hbar\omega$

$n = 1 \quad \tfrac{5}{2}\hbar\omega$

$n = 0 \quad \tfrac{3}{2}\hbar\omega$

$l = 0 \quad l = 1 \quad l = 2 \quad l = 3 \quad l = 4$

**FIGURE 10.14**
The energy levels $E_n = (n + \tfrac{3}{2})\hbar\omega$ of the isotropic harmonic oscillator, showing the degeneracy.

this in both approaches to the oscillator. In Cartesian coordinates there are different combinations of $n_x$, $n_y$, and $n_z$ in (10.81) that all yield the same energy, while in spherical coordinates, for a particular value of $n$, the states with $l = n, n - 2, \ldots, 1$ or $0$ all have the same energy [see (10.95)]. This degeneracy is illustrated for the first three energy states below:

| Cartesian coordinates | | | | | Spherical coordinates | | |
|---|---|---|---|---|---|---|---|

$$
\begin{array}{llll}
\text{Cartesian coordinates} & & & \\
n = 0 \quad n_x = 0 \quad n_y = 0 \quad n_z = 0 & \text{1 state} & n = 0 \quad l = 0 \quad m = 0
\end{array}
$$

$$
n = 1 \left\{ \begin{array}{lll} n_x = 1 & n_y = 0 & n_z = 0 \\ n_x = 0 & n_y = 1 & n_z = 0 \\ n_x = 0 & n_y = 0 & n_z = 1 \end{array} \right. \quad \text{3 states} \quad n = 1 \quad l = 1 \left\{ \begin{array}{l} m = 1 \\ m = 0 \\ m = -1 \end{array} \right.
$$

$$
n = 2 \left\{ \begin{array}{lll} n_x = 2 & n_y = 0 & n_z = 0 \\ n_x = 0 & n_y = 2 & n_z = 0 \\ n_x = 0 & n_y = 0 & n_z = 2 \\ n_x = 1 & n_y = 1 & n_z = 0 \\ n_x = 1 & n_y = 0 & n_z = 1 \\ n_x = 0 & n_y = 1 & n_z = 1 \end{array} \right. \quad \text{6 states} \quad n = 2 \left\{ \begin{array}{l} l = 0 \quad m = 0 \\ \\ l = 2 \left\{ \begin{array}{l} m = 2 \\ m = 1 \\ m = 0 \\ m = -1 \\ m = -2 \end{array} \right. \end{array} \right.
$$

If we look at the position-space wave function for the ground state, we see that

$$
\langle x, y, z | n_x = 0, \, n_y = 0, \, n_z = 0 \rangle = X_0(x) Y_0(y) Z_0(z)
$$

$$
= \left( \frac{m\omega}{\pi\hbar} \right)^{3/4} e^{-m\omega x^2/2\hbar} e^{-m\omega y^2/2\hbar} e^{-m\omega z^2/2\hbar}
$$

$$
= \left( \frac{m\omega}{\pi\hbar} \right)^{3/4} e^{-m\omega r^2/2\hbar} \tag{10.97}
$$

where we have used the form for the wave function (7.44b) for $X_0(x)$ and the corresponding expressions for $Y_0(y)$ and $Z_0(z)$. Notice that in the last step we have gone from an energy eigenfunction expressed in Cartesian coordinates to one expressed in spherical coordinates. The lack of angular dependence tells us that this is indeed a state with $l = 0$. However, if we take one of the three $n = 1$ eigenfunctions in Cartesian coordinates,

$$
\langle x, y, z | n_x = 1, \, n_y = 0, \, n_z = 0 \rangle = X_1(x) Y_0(y) Z_0(z)
$$

$$
= \left[ \frac{4}{\pi} \left( \frac{m\omega}{\hbar} \right)^3 \right]^{1/4} x \, e^{-m\omega x^2/2\hbar} \left( \frac{m\omega}{\pi\hbar} \right)^{1/2} e^{-m\omega y^2/2\hbar} e^{-m\omega z^2/2\hbar}
$$

$$
= \left( \frac{m\omega}{\pi\hbar} \right)^{1/2} \left[ \frac{4}{\pi} \left( \frac{m\omega}{\hbar} \right)^3 \right]^{1/4} r \sin\theta \cos\phi \, e^{-m\omega r^2/2\hbar} \tag{10.98}
$$

we can recognize the angular dependence as a linear combination of $Y_{1,1}$ and $Y_{1,-1}$, showing that the $n = 1$ states do have $l = 1$.

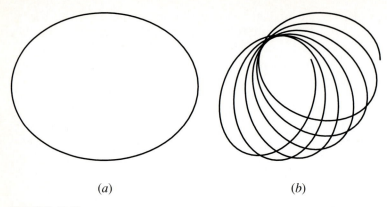

$(a)$                                    $(b)$

**FIGURE 10.15**
($a$) The classical orbits of a particle moving in a pure Coulomb or isotropic oscillator central potential close on themselves. ($b$) The classical orbits for other central potentials do not close and the orbit precesses.

The high degree of degeneracy for the isotropic harmonic oscillator is reminiscent of that for the hydrogen atom. Here too there is a "hidden" symmetry that is responsible.[13] In Chapter 9 we saw that symmetries lead to conservation laws, and so it is natural to ask what is conserved in these two central-force systems in addition to orbital angular momentum. Classically, conservation of orbital angular momentum means the orbital angular momentum points in a fixed direction. Consequently, the classical orbit must reside in a plane. In addition, the $1/r$ and $r^2$ central potentials share an unusual feature in classical mechanics: they are the only ones for which the orbits close upon themselves and do not precess (see Fig. 10.15). Thus within the plane of the orbit there is an additional constant of the motion for these two potentials—a vector pointing from the apogee to the perigee of the orbit maintains its orientation in space.

## 10.6  CONCLUSION

In this chapter we have examined almost all of the energy eigenvalue problems for a central potential that have exact solutions. In the case of the isotropic oscillator, we can solve the eigenvalue equation in two different coordinate systems. Surprisingly, the $1/r$ potential can also be solved in two different coordinate systems, parabolic as well as spherical. There is a certain irony in this because there are so few problems we can solve exactly, and we can solve each of these two in two different ways. Nonetheless, we should be grateful that we can solve these particular problems at all. After all, the solutions to the Coulomb potential form the foundation for our analysis of the hydrogen atom, which continues in Chapter 11.

---

[13] See the discussion of "accidental" degeneracies in R. Shankar, *Principles of Quantum Mechanics*, Plenum, New York, 1980.

# PROBLEMS

**10.1.** The position-space representation of the radial component of the momentum operator is given by

$$\hat{p}_r \rightarrow \frac{\hbar}{i}\left(\frac{\partial}{\partial r} + \frac{1}{r}\right)$$

Show that for its expectation value to be real: $\langle\psi|\hat{p}_r|\psi\rangle = \langle\psi|\hat{p}_r|\psi\rangle^*$, the radial wave function must satisfy the condition $u(0) = 0$. *Suggestion:* Express the expectation value in position space in spherical coordinates and integrate by parts.

**10.2.** An electron in the Coulomb field of the proton is in the state

$$|\psi\rangle = \frac{4}{5}|1, 0, 0\rangle + \frac{3i}{5}|2, 1, 1\rangle$$

where $|n, l, m\rangle$ are the usual energy eigenstates of hydrogen.
(a) What is $\langle E\rangle$ for this state? What are $\langle\hat{L}^2\rangle$ and $\langle\hat{L}_z\rangle$?
(b) What is $|\psi(t)\rangle$? Which of the expectation values in (a) vary with time?

**10.3.** Calculate the probability that an electron in the ground state of hydrogen is outside the classically allowed region.

**10.4.** What is the ground-state energy and Bohr radius for each of the following two-particle systems?
(a) $H^2$, a bound state of a deuteron and an electron
(b) $He^+$, singly ionized helium
(c) Positronium
(d) A bound state of a proton and a negative muon
(e) A gravitational bound state of two neutrons
What is the wavelength of the radiation emitted in the transition from the $n = 2$ state to the $n = 1$ state in each case? In what portion of the electromagnetic spectrum does this radiation reside?

**10.5.** Use the power-series solution of the hydrogen atom to determine $u_{3,0}(\rho)$. Ignore normalization. Compare your answer with (10.45a).

**10.6.** An electron is in the ground state of tritium, for which the nucleus is the isotope of hydrogen with one proton and two neutrons. A nuclear reaction instantaneously changes the nucleus into $He^3$, which consists of two protons and one neutron. Calculate the probability that the electron remains in the ground state of the new atom. Obtain a numerical answer.

**10.7.** Show that there are no allowed energies $E < -V_0$ for the potential well

$$V = \begin{cases} -V_0 & r < a \\ 0 & r > a \end{cases}$$

by explicitly solving the Schrödinger equation and attempting to satisfy all the appropriate boundary conditions.

**10.8.** Use the techniques illustrated in Section 10.3 to solve the one-dimensional potential well

$$V(x) = \begin{cases} -V_0 & |x| < a \\ 0 & |x| > a \end{cases}$$

Show that there always exists at least one bound state for this well.

**10.9.** Determine the ground-state energy of a particle of mass $\mu$ in the *cubic* potential well

$$V(x_i) = \begin{cases} 0 & 0 < x_i < a \\ \infty & \text{elsewhere} \end{cases} \qquad x_i = x, y, z$$

Compare the volume of this infinite well with the spherical one (10.64) and discuss in general terms whether the relative values of the ground-state energies for the two wells are consistent with the position-momentum uncertainty relation.

**10.10.** A particle of mass $\mu$ is in the *cylindrical* potential well

$$V(\rho) = \begin{cases} 0 & \rho < a \\ \infty & \rho > a \end{cases} \qquad \rho = \sqrt{x^2 + y^2}$$

(a) Determine the three lowest energy eigenvalues for states that also have $p_z$ and $L_z$ equal to zero.

(b) Determine the three lowest energy eigenvalues for states with $p_z$ equal to zero. The states may have nonzero $L_z$.

*Suggestion:* Work out Problem 9.22 before attempting this problem. Watch out for the appearance of Bessel's equation and ordinary Bessel functions when solving the radial equation.

**10.11.** (a) Substitute the expression (10.93) for the radial wave function of the three-dimensional isotropic oscillator into (10.91) to determine the differential equation that $f(\rho)$ obeys.

(b) Obtain a two-term recursion relation for the power series (10.94); show that this power series must terminate and that the energy eigenvalues of the oscillator are given by (10.95).

**10.12.** Expectation values are constant in time in an energy eigenstate. Hence

$$\frac{d\langle \mathbf{r} \cdot \mathbf{p} \rangle}{dt} = \frac{i}{\hbar} \langle E | [\hat{H}, \hat{\mathbf{r}} \cdot \hat{\mathbf{p}}] | E \rangle = 0$$

Use this result to show for the Hamiltonian

$$\hat{H} = \frac{\hat{\mathbf{p}}^2}{2\mu} + V(|\hat{\mathbf{r}}|)$$

that

$$\langle K \rangle = \langle \frac{\mathbf{p}^2}{2\mu} \rangle = \frac{1}{2} \langle \mathbf{r} \cdot \nabla V(r) \rangle$$

which can be considered a quantum statement of the virial theorem.

**10.13.** (a) Calculate $\langle V \rangle$ for the ground state of hydrogen. Show that $E = \langle V \rangle / 2$. What is $\langle K \rangle$, the expectation value of the kinetic energy, for the ground state? Show that these expectation values obey the virial theorem from classical mechanics.

(b) Calculate $\langle V \rangle$ for the ground state of the isotropic three-dimensional harmonic oscillator. How are $\langle K \rangle$ and $\langle V \rangle$ related for the oscillator? What do you expect based on the virial theorem? Explain.

**10.14.** Suppose that nucleons within the nucleus are presumed to move independently in a potential energy well in the form of an isotropic harmonic oscillator. What are the first five nuclear "magic numbers" within such a model?

**10.15.** The potential energy in a particular anisotropic harmonic oscillator with cylindrical symmetry is given by

$$V = \tfrac{1}{2} \mu \left[ \omega_1^2 (x^2 + y^2) + \omega_3^2 z^2 \right]$$

with $\omega_1 < \omega_3 < 2\omega_1$.

(a) Determine the energy eigenvalues and the degeneracies of the three lowest energy levels by using Cartesian coordinates.

(b) Solve the energy eigenvalue equation in cylindrical coordinates and check your results in comparison with those of (a).

**10.16.** Consider the Hamiltonian for the two-dimensional motion of a particle of mass $\mu$ in a harmonic oscillator potential:

$$\hat{H} = \frac{\hat{p}_x^2}{2\mu} + \frac{1}{2}\mu\omega^2\hat{x}^2 + \frac{\hat{p}_y^2}{2\mu} + \frac{1}{2}\mu\omega^2\hat{y}^2$$

(a) Show that the energy eigenvalues are given by $E_n = (n + 1)\hbar\omega$, where $n = n_1 + n_2$, with $n_1, n_2 = 0, 1, 2, \ldots$

(b) Express the operator $\hat{L}_z = \hat{x}\,\hat{p}_y - \hat{y}\,\hat{p}_x$ in terms of the lowering operators

$$\hat{a}_1 = \sqrt{\frac{\mu\omega}{2\hbar}}\left(\hat{x} + \frac{i}{\mu\omega}\,\hat{p}_x\right) \quad \text{and} \quad \hat{a}_2 = \sqrt{\frac{\mu\omega}{2\hbar}}\left(\hat{y} + \frac{i}{\mu\omega}\,\hat{p}_y\right)$$

and the corresponding raising operators $\hat{a}_1^\dagger$ and $\hat{a}_2^\dagger$. Give a symmetry argument showing that $[\hat{H}, \hat{L}_z] = 0$. Evaluate this commutator directly and confirm that it indeed vanishes.

(c) Determine the correct linear combination of the energy eigenstates with energy $E_1 = 2\hbar\omega$ that are eigenstates of $\hat{L}_z$ by diagonalizing the matrix representation of $\hat{L}_z$ restricted to this subspace of states.

**10.17.** The spherically symmetric potential energy of a particle of mass $\mu$ is given by

$$V(r) = \begin{cases} 0 & a < r < b \\ \infty & \text{elsewhere} \end{cases}$$

where $r = \sqrt{x^2 + y^2 + z^2}$.

(a) Determine the ground-state energy.

(b) What is the ground-state position-space eigenfunction up to an overall normalization constant? What condition would you impose to determine this constant?

(c) What is the energy of the first-excited $l = 0$ state? Explain why it would not be so straightforward to determine the energy of the $l = 1$ states.

**10.18.** The Hamiltonian for two spin-$\frac{1}{2}$ particles, one with mass $m_1$ and the other with mass $m_2$, is given by

$$\hat{H} = \frac{\hat{\mathbf{p}}_1^2}{2m_1} + \frac{\hat{\mathbf{p}}_2^2}{2m_1} + V_a(|\hat{\mathbf{r}}|) + \left(\frac{1}{4} - \frac{\hat{\mathbf{S}}_1 \cdot \hat{\mathbf{S}}_2}{\hbar^2}\right)V_b(|\hat{\mathbf{r}}|)$$

where $\hat{\mathbf{r}} = \hat{\mathbf{r}}_1 - \hat{\mathbf{r}}_2$ and

$$V_a(r) = \begin{cases} 0 & r < a \\ V_0 & r > a \end{cases} \qquad V_b(r) = \begin{cases} 0 & r < b \\ V_0 & r > b \end{cases}$$

with $b < a$ and $V_0$ very large and positive.

(a) Determine the normalized position-space energy eigenfunction for the ground state. What is the spin state of the ground state? What is the degeneracy? *Note:* Take $V_0$ to be infinite where appropriate to make the calculation as straightforward as possible.

(b) What can you say about the energy and spin state of the first-excited state? Does your result depend on how much larger $a$ is than $b$? Explain.

# CHAPTER
# 11

# TIME-INDEPENDENT PERTURBATIONS

Obtaining quantitative agreement between theory and experiment in the real world has its ups and downs. The bad news is that there aren't any interacting systems that have Hamiltonians for which we can determine the energy eigenvalues and eigenstates exactly. The good news is that because a number of extremely important physical systems are sufficiently close to ones that we *can* solve, such as the harmonic oscillator and the hydrogen atom, we can treat the differences as perturbations and deal with them in a systematic way. In the beginning of this chapter we will focus on the effect of an external electric field on a number of familiar systems—the ammonia molecule treated as a two-state system, the one-dimensional harmonic oscillator, and the hydrogen atom. We will then consider the effect of internal relativistic perturbations in the hydrogen atom, leading to the fine structure. We will also investigate the effect on the hydrogen atom of an external magnetic field, the Zeeman effect.

## 11.1 NONDEGENERATE PERTURBATION THEORY

We begin by expressing the Hamiltonian for some system in the form

$$\hat{H} = \hat{H}_0 + \hat{H}_1 \tag{11.1}$$

where the part of the Hamiltonian that is presumed to be "big" is $\hat{H}_0$, often called the *unperturbed* Hamiltonian, and $\hat{H}_1$ is the "small" part, often referred to as the *perturbing* Hamiltonian. For a perturbative approach to work, we must be able to determine the eigenstates and eigenvalues of $\hat{H}_0$:

$$\hat{H}_0|\varphi_n^{(0)}\rangle = E_n^{(0)}|\varphi_n^{(0)}\rangle \tag{11.2}$$

where $|\varphi_n^{(0)}\rangle = |E_n^{(0)}\rangle$ is the eigenstate with energy $E_n^{(0)}$. Of course, we are presuming that we are not able to determine the energy eigenstates and eigenvalues of the full Hamiltonian:

$$\hat{H}|\psi_n\rangle = E_n|\psi_n\rangle \tag{11.3}$$

so we will attempt a solution of (11.3) in the form of a perturbative expansion.

In order to keep track of the order of smallness in our perturbative expansion, it is convenient to introduce a parameter $\lambda$ into the Hamiltonian:

$$\hat{H} = \hat{H}_0 + \lambda\hat{H}_1 \tag{11.4}$$

Thus by adjusting the value of $\lambda$, we can adjust the Hamiltonian. In particular, as $\lambda \to 0$ and we turn off the perturbation, $\hat{H} \to \hat{H}_0$, while as $\lambda \to 1$, $\hat{H} \to \hat{H}_0 + \hat{H}_1$, the full Hamiltonian for the system.[1] We assume that we can express the exact eigenstates and eigenvalues as a power-series expansion in $\lambda$:

$$|\psi_n\rangle = |\varphi_n^{(0)}\rangle + \lambda|\varphi_n^{(1)}\rangle + \lambda^2|\varphi_n^{(2)}\rangle + \cdots \tag{11.5}$$

$$E_n = E_n^{(0)} + \lambda E_n^{(1)} + \lambda^2 E_n^{(2)} + \cdots \tag{11.6}$$

If this perturbative expansion is to be useful, successive terms in the series must grow progressively smaller, and we can then obtain a reasonable approximation to the full energy eigenvalue equation by retaining just the first few terms. In particular, note that we are presuming that as $\lambda \to 0$, $E_n \to E_n^{(0)}$ and $|\psi_n\rangle \to |\varphi_n^{(0)}\rangle$ smoothly, as indicated in Fig. 11.1.

As an example illustrating how a series expansion such as (11.6) might arise, let's first reexamine the two-state system of the ammonia molecule in an

---

[1] Some authors prefer to consider $\lambda$ as part of the real Hamiltonian, rather than just a parameter that is introduced to help keep track of smallness. The problem with this alternative approach is that it is sometimes difficult to see at the start a natural small dimensionless parameter in the system that can play the role of $\lambda$.

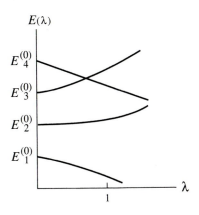

**FIGURE 11.1**
A schematic diagram showing how the energy levels of the Hamiltonian (11.4) might change as $\lambda$ varies between 0 and 1.

external electric field, which we analyzed in Section 4.5. There we noted that the matrix representation of the Hamiltonian can be expressed as

$$\hat{H} \rightarrow \begin{pmatrix} \langle 1|\hat{H}|1\rangle & \langle 1|\hat{H}|2\rangle \\ \langle 2|\hat{H}|1\rangle & \langle 2|\hat{H}|2\rangle \end{pmatrix} = \begin{pmatrix} E_0 + \mu_e|\mathbf{E}| & -A \\ -A & E_0 - \mu_e|\mathbf{E}| \end{pmatrix} \tag{11.7}$$

which has the exact eigenvalues

$$E = E_0 \pm \sqrt{(\mu_e|\mathbf{E}|)^2 + A^2} \tag{11.8}$$

For external electric fields that satisfy $\mu_e|\mathbf{E}| \ll A$, we can expand the square root to obtain the following power series for the energy:

$$E = E_0 \pm A \pm A \left[ \frac{1}{2}\left(\frac{\mu_e|\mathbf{E}|}{A}\right)^2 - \frac{1}{8}\left(\frac{\mu_e|\mathbf{E}|}{A}\right)^4 + \cdots \right] \tag{11.9}$$

Notice that as $\mu_e|\mathbf{E}| \rightarrow 0$, the energies go smoothly into the energies of the molecule in the absence of the electric field, namely, $E_I^{(0)} = E_0 - A$, which has the corresponding eigenstate $|I\rangle = (1/\sqrt{2})(|1\rangle + |2\rangle)$ that we found in Section 4.5, and $E_{II}^{(0)} = E_0 + A$, which has the corresponding eigenstate $|II\rangle = (1/\sqrt{2})(|1\rangle - |2\rangle)$. The exact eigenstates of the Hamiltonian (11.7) can also be expressed as a power series in the small quantity $\mu_e|\mathbf{E}|/A$, with the zeroth-order terms given by $|I\rangle$ and $|II\rangle$ (see Problem 11.5).[2]

Let's return to the general problem of determining the expansions (11.5) and (11.6) when we are not able to determine the eigenstates and eigenvalues exactly. Substituting (11.5) and (11.6) into the energy eigenvalue equation (11.3), we obtain

$$\left(\hat{H}_0 + \lambda\hat{H}_1\right)\left(|\varphi_n^{(0)}\rangle + \lambda|\varphi_n^{(1)}\rangle + \lambda^2|\varphi_n^{(2)}\rangle + \cdots\right)$$

$$= \left(E_n^{(0)} + \lambda E_n^{(1)} + \lambda^2 E_n^{(2)} + \cdots\right)\left(|\varphi_n^{(0)}\rangle + \lambda|\varphi_n^{(1)}\rangle + \lambda^2|\varphi_n^{(2)}\rangle + \cdots\right) \tag{11.10}$$

Since $\lambda$ is an arbitrary parameter, for (11.10) to hold, the coefficients of each power of $\lambda$ must separately satisfy the equation. The terms that are independent of $\lambda$, the $\lambda^0$ terms, are just (11.2), or

$$\hat{H}_0|\varphi_n^{(0)}\rangle = E_n^{(0)}|\varphi_n^{(0)}\rangle \tag{11.11a}$$

The $\lambda$ terms yield

$$\hat{H}_0|\varphi_n^{(1)}\rangle + \hat{H}_1|\varphi_n^{(0)}\rangle = E_n^{(0)}|\varphi_n^{(1)}\rangle + E_n^{(1)}|\varphi_n^{(0)}\rangle \tag{11.11b}$$

---

[2] We will continue our discussion of the ammonia molecule in an external electric field in Section 11.5.

while the $\lambda^2$ terms yield

$$\hat{H}_0|\varphi_n^{(2)}\rangle + \hat{H}_1|\varphi_n^{(1)}\rangle = E_n^{(0)}|\varphi_n^{(2)}\rangle + E_n^{(1)}|\varphi_n^{(1)}\rangle + E_n^{(2)}|\varphi_n^{(0)}\rangle \qquad (11.11c)$$

and so on. You can see the pattern that arises if we were to go on to consider higher-order terms.

## The First-Order Energy Shift

A useful procedure for extracting the information contained in these equations is to take the inner product with the complete set of basis bras $\langle\varphi_k^{(0)}|$. We start with (11.11$b$) and take the inner product with $\langle\varphi_n^{(0)}|$ to obtain

$$\langle\varphi_n^{(0)}|\hat{H}_0|\varphi_n^{(1)}\rangle + \langle\varphi_n^{(0)}|\hat{H}_1|\varphi_n^{(0)}\rangle = E_n^{(0)}\langle\varphi_n^{(0)}|\varphi_n^{(1)}\rangle + E_n^{(1)}\langle\varphi_n^{(0)}|\varphi_n^{(0)}\rangle \qquad (11.12)$$

Since

$$\langle\varphi_n^{(0)}|\hat{H}_0 = \langle\varphi_n^{(0)}|E_n^{(0)} \qquad (11.13)$$

and we are presuming that

$$\langle\varphi_k^{(0)}|\varphi_n^{(0)}\rangle = \delta_{kn} \qquad (11.14)$$

(11.12) becomes

$$E_n^{(1)} = \langle\varphi_n^{(0)}|\hat{H}_1|\varphi_n^{(0)}\rangle \qquad (11.15)$$

The first-order shift in the energy is simply the expectation value of the perturbing Hamiltonian in the unperturbed state corresponding to that energy.

## The First-Order Correction to the Energy Eigenstate

Taking the inner product of (11.11$b$) with $\langle\varphi_k^{(0)}|$ for $k \neq n$, we find

$$\langle\varphi_k^{(0)}|\hat{H}_0|\varphi_n^{(1)}\rangle + \langle\varphi_k^{(0)}|\hat{H}_1|\varphi_n^{(0)}\rangle = E_n^{(0)}\langle\varphi_k^{(0)}|\varphi_n^{(1)}\rangle \qquad (11.16)$$

or

$$\langle\varphi_k^{(0)}|\varphi_n^{(1)}\rangle = \frac{\langle\varphi_k^{(0)}|\hat{H}_1|\varphi_n^{(0)}\rangle}{E_n^{(0)} - E_k^{(0)}} \qquad k \neq n \qquad (11.17)$$

If we use the basis states $|\varphi_k^{(0)}\rangle$ to express $|\varphi_n^{(1)}\rangle$ as

$$|\varphi_n^{(1)}\rangle = \sum_k |\varphi_k^{(0)}\rangle\langle\varphi_k^{(0)}|\varphi_n^{(1)}\rangle = |\varphi_n^{(0)}\rangle\langle\varphi_n^{(0)}|\varphi_n^{(1)}\rangle + \sum_{k \neq n} |\varphi_k^{(0)}\rangle\langle\varphi_k^{(0)}|\varphi_n^{(1)}\rangle \qquad (11.18)$$

then (11.17) tells us how much of $|\varphi_n^{(1)}\rangle$ lies along each of the $|\varphi_k^{(0)}\rangle$ for $k \neq n$. What about $\langle\varphi_n^{(0)}|\varphi_n^{(1)}\rangle$? We return to (11.5) and require that

$$1 = \langle\psi_n|\psi_n\rangle = \langle\varphi_n^{(0)}|\varphi_n^{(0)}\rangle + \lambda\langle\varphi_n^{(0)}|\varphi_n^{(1)}\rangle + \lambda\langle\varphi_n^{(1)}|\varphi_n^{(0)}\rangle + \cdots \qquad (11.19)$$

Since $\langle \varphi_n^{(0)} | \varphi_n^{(0)} \rangle = 1$, through first order in $\lambda$ we must have

$$\langle \varphi_n^{(0)} | \varphi_n^{(1)} \rangle = ia \qquad a \text{ real} \tag{11.20}$$

and therefore

$$|\psi_n\rangle = |\varphi_n^{(0)}\rangle + ia\lambda|\varphi_n^{(0)}\rangle + \lambda \sum_{k \neq n} |\varphi_k^{(0)}\rangle\langle\varphi_k^{(0)}|\varphi_n^{(1)}\rangle + O(\lambda^2)$$

$$= e^{ia\lambda}|\varphi_n^{(0)}\rangle + \lambda \sum_{k \neq n} |\varphi_k^{(0)}\rangle\langle\varphi_k^{(0)}|\varphi_n^{(1)}\rangle + O(\lambda^2), \tag{11.21}$$

where in the last step we have taken advantage of the fact that $e^{ia\lambda} = 1 + ia\lambda + O(\lambda^2)$. Even after $|\psi_n\rangle$ is normalized, its phase can be chosen arbitrarily; thus it is convenient to require that $a = 0$, or

$$\langle \varphi_n^{(0)} | \varphi_n^{(1)} \rangle = 0 \tag{11.22}$$

so that, to this order in $\lambda$, $|\psi_n\rangle$ and $|\varphi_n^{(0)}\rangle$ have the same phase. This is a natural choice, since then the first-order correction $|\varphi_n^{(1)}\rangle$ is orthogonal to the state $|\varphi_n^{(0)}\rangle$ and the perturbative correction generates the state $|\psi_n\rangle$ which, to this order, "points" in a slightly different direction, as depicted in Fig. 11.2. Thus

$$|\psi_n\rangle = |\varphi_n^{(0)}\rangle + \lambda \sum_{k \neq n} |\varphi_k^{(0)}\rangle \frac{\langle\varphi_k^{(0)}|\hat{H}_1|\varphi_n^{(0)}\rangle}{E_n^{(0)} - E_k^{(0)}} + O(\lambda^2) \tag{11.23}$$

## The Second-Order Energy Shift

Let's go on to determine $E_n^{(2)}$. We take the inner product of (11.11c) with the bra $\langle \varphi_n^{(0)}|$:

$$\langle \varphi_n^{(0)} | \hat{H}_0 | \varphi_n^{(2)} \rangle + \langle \varphi_n^{(0)} | \hat{H}_1 | \varphi_n^{(1)} \rangle = E_n^{(0)}\langle \varphi_n^{(0)} | \varphi_n^{(2)} \rangle + E_n^{(1)}\langle \varphi_n^{(0)} | \varphi_n^{(1)} \rangle + E_n^{(2)}\langle \varphi_n^{(0)} | \varphi_n^{(0)} \rangle \tag{11.24}$$

Taking advantage of (11.13), (11.17), and (11.22), we obtain

$$E_n^{(2)} = \langle \varphi_n^{(0)} | \hat{H}_1 | \varphi_n^{(1)} \rangle$$

$$= \sum_{k \neq n} \frac{\langle\varphi_n^{(0)}|\hat{H}_1|\varphi_k^{(0)}\rangle\langle\varphi_k^{(0)}|\hat{H}_1|\varphi_n^{(0)}\rangle}{E_n^{(0)} - E_k^{(0)}} = \sum_{k \neq n} \frac{|\langle\varphi_k^{(0)}|\hat{H}_1|\varphi_n^{(0)}\rangle|^2}{E_n^{(0)} - E_k^{(0)}} \tag{11.25}$$

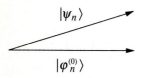

**FIGURE 11.2**
A pictorial representation of first-order nondegenerate perturbation theory using ordinary vectors. For perturbation theory to be effective, the "angle" between $|\varphi_n^{(0)}\rangle$ and $|\psi_n\rangle$ must be small. Remember that ket vectors are vectors in a complex vector space, so this picture with real angles should not be taken literally.

where in the last step we have used

$$\langle \varphi_n^{(0)} | \hat{H}_1 | \varphi_k^{(0)} \rangle = \langle \varphi_k^{(0)} | \hat{H}_1 | \varphi_n^{(0)} \rangle^* \tag{11.26}$$

since $\hat{H}_1$ is Hermitian.

Note that to calculate the first-order shift in the energy in (11.15), all we need is the zeroth-order state. Similarly, in order to calculate the second-order shift in the energy in (11.25), all we need to know is the first-order correction to the state. In general, calculating the energy to order $s$ requires knowledge of the state to order $s - 1$. Although we could go on to determine higher-order corrections, we will find (11.15) and (11.25) adequate for our purposes.

## 11.2  AN EXAMPLE INVOLVING THE ONE-DIMENSIONAL HARMONIC OSCILLATOR

Before turning our attention to fully three-dimensional systems, let's apply the results of Section 11.1 to our favorite one-dimensional system, the simple harmonic oscillator. We suppose that a particle with charge $q$ is in a harmonic oscillator potential and that we perturb the system by applying a constant electric field $\mathbf{E}$ that points in the positive $x$ direction. Since there is a constant force $q\mathbf{E}$ exerted on the particle, the additional contribution to the potential energy, which is the extra work that *we* must do to displace the particle by a distance $x$ from the origin, is $-q|\mathbf{E}|x$. Therefore the Hamiltonian of the system is given by

$$\hat{H} = \frac{\hat{p}_x^2}{2m} + \frac{1}{2} m\omega^2 \hat{x}^2 - q|\mathbf{E}|\hat{x} \tag{11.27}$$

We break up the Hamiltonian into two parts:

$$\hat{H}_0 = \frac{\hat{p}_x^2}{2m} + \frac{1}{2} m\omega^2 \hat{x}^2 \tag{11.28a}$$

$$\hat{H}_1 = -q|\mathbf{E}|\hat{x} \tag{11.28b}$$

Note that we can express this new addition (11.28b) as the usual electric dipole interaction Hamiltonian

$$\hat{H}_1 = -\hat{\boldsymbol{\mu}}_e \cdot \mathbf{E} \tag{11.29}$$

where the electric dipole moment operator $\hat{\boldsymbol{\mu}}_e = q\hat{x}$.[3]

---

[3] There is nothing wrong with introducing the electric dipole moment

$$\boldsymbol{\mu}_e = \sum_i q_i \mathbf{r}_i$$

of a system of charges that has a net charge $q$. However, in this case the dipole moment depends on where you locate the origin of your coordinates. If you prefer to deal with a neutral system for the one-dimensional harmonic oscillator, you can add a charge $-q$ on a heavy mass that effectively resides at the origin.

The energy eigenvalues of the unperturbed Hamiltonian are given by

$$E_n^{(0)} = (n + \tfrac{1}{2})\hbar\omega \tag{11.30}$$

There are a number of easy ways to evaluate the first-order shift in the energy. Using

$$\hat{x} = \sqrt{\frac{\hbar}{2m\omega}}\,(\hat{a} + \hat{a}^\dagger) \tag{11.31}$$

from Chapter 7, we find that

$$E_n^{(1)} = \langle n|\hat{H}_1|n\rangle$$

$$= -q|\mathbf{E}|\sqrt{\frac{\hbar}{2m\omega}}\,\langle n|(\hat{a} + \hat{a}^\dagger)|n\rangle = 0 \tag{11.32}$$

It is also instructive to evaluate the expectation value in position space:

$$E_n^{(1)} = -q|\mathbf{E}|\langle n|\hat{x}|n\rangle = -q|\mathbf{E}|\int_{-\infty}^{\infty} dx\,|\langle x|n\rangle|^2 x = 0 \tag{11.33}$$

Here the integral vanishes because the energy eigenfunction $\langle x|n\rangle$ is an even or odd function (see Section 7.10), and hence $|\langle x|n\rangle|^2$ is always even and $x|\langle x|n\rangle|^2$ always odd. Since the first-order shift in the energy is just proportional to the expectation value of the electric dipole moment operator $q\hat{x}$, the vanishing of (11.33) can be ascribed to the absence of a permanent electric dipole moment that can interact with the applied electric field.

The second-order shift in the energy is given by

$$E_n^{(2)} = \sum_{k \neq n} \frac{|\langle k|\hat{H}_1|n\rangle|^2}{(n + \tfrac{1}{2})\hbar\omega - (k + \tfrac{1}{2})\hbar\omega} \tag{11.34}$$

Since

$$\langle k|\hat{H}_1|n\rangle = -q|\mathbf{E}|\sqrt{\frac{\hbar}{2m\omega}}\,(\sqrt{n+1}\langle k|n+1\rangle + \sqrt{n}\langle k|n-1\rangle) \tag{11.35}$$

there are contributions to the sum in (11.34) when $k = n+1$ or $k = n-1$. Thus

$$E_n^{(2)} = \frac{q^2|\mathbf{E}|^2\hbar}{2m\omega}\left(\frac{n+1}{-\hbar\omega} + \frac{n}{\hbar\omega}\right) = -\frac{q^2|\mathbf{E}|^2}{2m\omega^2} \tag{11.36}$$

What is the physical source of this nonvanishing higher-order contribution? On average, the electric field causes the particle to be displaced from the origin, inducing a dipole moment proportional to the magnitude of the electric field. This induced dipole moment itself interacts with the applied field, giving a contribution to the energy that is proportional to the magnitude of the field squared.

In this particular problem we have a simple way to confirm the results (11.32) and (11.36). We really didn't need to use perturbation theory for the Hamiltonian

(11.27), because we can determine the eigenvalues and eigenstates exactly by "completing the square":

$$\hat{H} = \frac{\hat{p}_x^2}{2m} + \frac{1}{2} m\omega^2 \hat{x}^2 - q|\mathbf{E}|\hat{x}$$

$$= \frac{\hat{p}_x^2}{2m} + \frac{1}{2} m\omega^2 \left( \hat{x} - \frac{q|\mathbf{E}|}{m\omega^2} \right)^2 - \frac{q^2|\mathbf{E}|^2}{2m\omega^2} \tag{11.37}$$

Figure 11.3 shows a graph of the potential energy for this Hamiltonian. It is a pure harmonic oscillator potential, just shifted along the $x$ axis by $q|\mathbf{E}|/m\omega^2$ and shifted down in energy by $q^2|\mathbf{E}|^2/2m\omega^2$. In order to solve formally the quantum mechanical energy eigenvalue equation, we define the shifted position operator

$$\hat{x}_s = \hat{x} - \frac{q|\mathbf{E}|}{m\omega^2} \tag{11.38}$$

which satisfies the usual commutation relation $[\hat{x}_s, \hat{p}_x] = i\hbar$ with the momentum

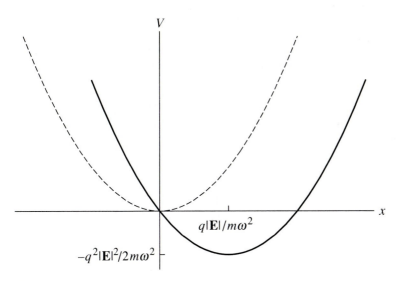

**FIGURE 11.3**
Graphs of the potential energy

$$V(x) = \tfrac{1}{2} m\omega^2 x^2$$

of the harmonic oscillator (dashed line) and the potential energy

$$V(x) = \frac{1}{2} m\omega^2 \left( x - \frac{q|\mathbf{E}|}{m\omega^2} \right)^2 - \frac{q^2|\mathbf{E}|^2}{2m\omega^2}$$

of the oscillator in an external electric field (solid line).

operator $\hat{p}_x$. Thus the exact eigenvalues of the Hamiltonian

$$\hat{H} = \frac{\hat{p}_x^2}{2m} + \frac{1}{2}m\omega^2\hat{x}_s^2 - \frac{q^2|\mathbf{E}|^2}{2m\omega^2} \tag{11.39}$$

are given by

$$E_n = \left(n + \frac{1}{2}\right)\hbar\omega - \frac{q^2|\mathbf{E}|^2}{2m\omega^2} \tag{11.40}$$

in agreement with our earlier perturbative results. The exact eigenstates are those of the usual harmonic oscillator, only shifted in position by $q|\mathbf{E}|/m\omega^2$. These eigenstates can thus be expressed in terms of the translation operator by

$$|\psi_n\rangle = \hat{T}(q|\mathbf{E}|/m\omega^2)|n\rangle = e^{-iq|\mathbf{E}|\hat{p}_x/m\omega^2\hbar}|n\rangle \tag{11.41}$$

You can verify that (11.41) agrees with the perturbative expansion (11.23) (see Problem 11.2).

## 11.3 DEGENERATE PERTURBATION THEORY

If we try to apply the formalism of perturbation theory when there is degeneracy, we face a crisis. In particular, the first-order correction to the eigenstate and, consequently, the second-order shift in the energy involve the quantity

$$\frac{\langle\varphi_k^{(0)}|\hat{H}_1|\varphi_n^{(0)}\rangle}{E_n^{(0)} - E_k^{(0)}} \tag{11.42}$$

which diverges if there exist states other than $|\varphi_n^{(0)}\rangle$ with energy $E_n^{(0)}$, that is, if there is degeneracy. In our earlier derivation we assumed that each unperturbed eigenstate $|\varphi_n^{(0)}\rangle$ turns smoothly into the exact eigenstate $|\psi_n\rangle$ as we turn on the perturbing Hamiltonian. However, if there are $N$ states

$$|\varphi_{n,i}^{(0)}\rangle \qquad i = 1, 2, \ldots, N \tag{11.43}$$

all with the same energy, it isn't clear which are the right linear combinations of the unperturbed states that become the exact eigenstates. For example, in the case of two-fold degeneracy, is it

$$|\varphi_{n,1}^{(0)}\rangle \qquad \text{and} \qquad |\varphi_{n,2}^{(0)}\rangle$$

or

$$\frac{1}{\sqrt{2}}\left(|\varphi_{n,1}^{(0)}\rangle + |\varphi_{n,2}^{(0)}\rangle\right) \qquad \text{and} \qquad \frac{1}{\sqrt{2}}\left(|\varphi_{n,1}^{(0)}\rangle - |\varphi_{n,2}^{(0)}\rangle\right)$$

or some other of the infinite number of linear combinations that we can construct from these two states? If we choose the wrong linear combination of unperturbed states as a starting point, even the small change in the Hamiltonian generated by turning on the perturbation with an infinitesimal $\lambda$ must produce a large change in the state. See Fig. 11.4.

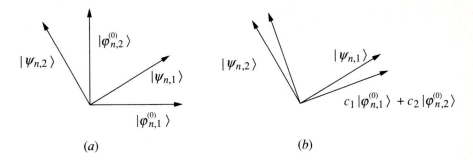

**FIGURE 11.4**

(a) The problem with degenerate perturbation theory for two-fold degeneracy. Neither the vector $|\varphi_{n,1}^{(0)}\rangle$ nor $|\varphi_{n,2}^{(0)}\rangle$ "points" sufficiently close to the exact vectors $|\psi_{n,1}\rangle$ or $|\psi_{n,2}\rangle$. (b) Degenerate perturbation theory selects the "right" linear combinations of states so that the perturbative correction is small. Remember that ket vectors are vectors in a complex vector space, so these pictures with real angles should not be taken literally.

In order to determine appropriate linear combinations of unperturbed states, we return to our expansion (11.5). Allowing for degeneracy, we write[4]

$$|\psi_n\rangle = \sum_{i=1}^{N} c_i |\varphi_{n,i}^{(0)}\rangle + \lambda |\varphi_n^{(1)}\rangle + \cdots \tag{11.44}$$

If we substitute this expression for the state into the eigenvalue equation (11.3), instead of (11.11b) we obtain

$$\hat{H}_0 |\varphi_n^{(1)}\rangle + \hat{H}_1 \sum_{i=1}^{N} c_i |\varphi_{n,i}^{(0)}\rangle = E_n^{(0)} |\varphi_n^{(1)}\rangle + E_n^{(1)} \sum_{i=1}^{N} c_i |\varphi_{n,i}^{(0)}\rangle \tag{11.45}$$

We then take the inner product of this equation with each of the $N$ bra vectors $\langle \varphi_{n,j}^{(0)} |$, leading to

$$\sum_{i=1}^{N} \langle \varphi_{n,j}^{(0)} | \hat{H}_1 | \varphi_{n,i}^{(0)} \rangle c_i = E_n^{(1)} \sum_{i=1}^{N} \langle \varphi_{n,j}^{(0)} | \varphi_{n,i}^{(0)} \rangle c_i = E_n^{(1)} \sum_{i=1}^{N} \delta_{ij} c_i \tag{11.46}$$

where the last step follows from the assumption that the degenerate states are orthonormal, that is, they satisfy

$$\langle \varphi_{n,j}^{(0)} | \varphi_{n,i}^{(0)} \rangle = \delta_{ij} \tag{11.47}$$

On the left-hand side of (11.46) we see the matrix elements

$$\langle \varphi_{n,j}^{(0)} | \hat{H}_1 | \varphi_{n,i}^{(0)} \rangle = (H_1)_{ji} \tag{11.48}$$

---

[4] Strictly, there are $N$ different first-order corrections for the $N$ different $|\psi_n\rangle$. We have suppressed an extra subscript in labeling these states for notational simplicity.

of the perturbing Hamiltonian in the subspace of degenerate states. In fact, in this subspace, (11.46) is just the standard eigenvalue equation. For example, in the case that $N = 2$, (11.46) can be written in matrix form as

$$\begin{pmatrix} (H_1)_{11} & (H_1)_{12} \\ (H_1)_{21} & (H_1)_{22} \end{pmatrix} \begin{pmatrix} c_1 \\ c_2 \end{pmatrix} = E_n^{(1)} \begin{pmatrix} c_1 \\ c_2 \end{pmatrix} \tag{11.49}$$

The first-order energy shifts will be the eigenvalues of this equation and the corresponding eigenstates will be the proper linear combinations of the degenerate states. Of course, if, by chance, we had initially chosen the proper linear combination of states, we would have found that the matrix representation is diagonal, with the first-order shifts in the energies as the diagonal matrix elements. Thus we can say that in determining these first-order shifts we are diagonalizing the perturbing Hamiltonian *in the subspace of degenerate states*.[5]

## 11.4   THE STARK EFFECT IN HYDROGEN

As an interesting illustration of degenerate perturbation theory, let's consider what happens when we apply an external electric field **E** to a hydrogen atom, producing the Stark effect. We expect a perturbing Hamiltonian of the form

$$\hat{H}_1 = -\hat{\boldsymbol{\mu}}_e \cdot \mathbf{E} = e\hat{\mathbf{r}} \cdot \mathbf{E} \tag{11.50}$$

where the electric dipole moment $\boldsymbol{\mu}_e$ of the hydrogen atom is $-e\mathbf{r}$, since the radius vector **r** points from the proton to the electron, while the dipole moment points from the negative to the positive charge. Of course the unperturbed Hamiltonian is just

$$\hat{H}_0 = \frac{\hat{\mathbf{p}}^2}{2\mu} - \frac{e^2}{|\hat{\mathbf{r}}|} \tag{11.51}$$

with eigenstates $|n,l,m\rangle$.

We choose to orient our coordinate axes so that the electric field points in the $z$ direction. The electric dipole Hamiltonian becomes

$$\hat{H}_1 = e|\mathbf{E}|\hat{z} \tag{11.52}$$

We first consider the ground state, $|1, 0, 0\rangle$, for which we can utilize nondegenerate perturbation theory to calculate the first-order shift in the energy,

$$E_1^{(1)} = e|\mathbf{E}|\langle 1, 0, 0|\hat{z}|1, 0, 0\rangle = 0 \tag{11.53}$$

The expectation value vanishes since eigenstates of the hydrogen atom with def-

---

[5] It should be emphasized that we are not diagonalizing the perturbing Hamiltonian in space formed by the (often infinite) *complete* set of eigenstates of $\hat{H}_0$. If we were able to carry out this diagonalization, we would be able to find the exact eigenstates of $\hat{H}_0 + \hat{H}_1$, and we would not need to resort to perturbation theory.

inite orbital angular momentum $l$ have definite parity $(-1)^l$ (see Problem 9.15). Thus, as for the harmonic oscillator in Section 11.2 [see(11.33)], the expectation value (11.53) in position space involves an odd function, which integrates to zero.

The second-order shift in the ground-state energy is given by

$$E_1^{(2)} = \sum_{n \neq 1, l, m} \frac{e^2 |\mathbf{E}|^2 |\langle n, l, m | \hat{z} | 1, 0, 0 \rangle|^2}{E_1^{(0)} - E_n^{(0)}} \tag{11.54}$$

Notice that the sum is over all states except the ground state. Although this sum is not as easy to evaluate as the one in Section 11.2 for the harmonic oscillator, the physics in the two cases is essentially the same.[6] Here again, the atom in the ground state does not have a dipole moment, as indicated by (11.53), but one is induced by the applied electric field, generating a shift in the energy proportional to $\mathbf{E}^2$.

Let's now turn our attention to the first-excited states, where the principal quantum number $n = 2$ and there is a four-fold degeneracy, ignoring spin. We first construct the $4 \times 4$ matrix representation of $\hat{H}_1$ using the four degenerate states $|2, 0, 0\rangle$, $|2, 1, 0\rangle$, $|2, 1, 1\rangle$, and $|2, 1, -1\rangle$ as a basis:

$$\begin{pmatrix} \langle 2, 0, 0 | \hat{H}_1 | 2, 0, 0 \rangle & \langle 2, 0, 0 | \hat{H}_1 | 2, 1, 0 \rangle & \langle 2, 0, 0 | \hat{H}_1 | 2, 1, 1 \rangle & \langle 2, 0, 0 | \hat{H}_1 | 2, 1, -1 \rangle \\ \langle 2, 1, 0 | \hat{H}_1 | 2, 0, 0 \rangle & \langle 2, 1, 0 | \hat{H}_1 | 2, 1, 0 \rangle & \langle 2, 1, 0 | \hat{H}_1 | 2, 1, 1 \rangle & \langle 2, 1, 0 | \hat{H}_1 | 2, 1, -1 \rangle \\ \langle 2, 1, 1 | \hat{H}_1 | 2, 0, 0 \rangle & \langle 2, 1, 1 | \hat{H}_1 | 2, 1, 0 \rangle & \langle 2, 1, 1 | \hat{H}_1 | 2, 1, 1 \rangle & \langle 2, 1, 1 | \hat{H}_1 | 2, 1, -1 \rangle \\ \langle 2, 1, -1 | \hat{H}_1 | 2, 0, 0 \rangle & \langle 2, 1, -1 | \hat{H}_1 | 2, 1, 0 \rangle & \langle 2, 1, -1 | \hat{H}_1 | 2, 1, 1 \rangle & \langle 2, 1, -1 | \hat{H}_1 | 2, 1, -1 \rangle \end{pmatrix} \tag{11.55}$$

We have chosen a particular order for the states in this matrix for reasons that will become apparent shortly.

Evaluating 16 matrix elements and then diagonalizing a $4 \times 4$ matrix is straightforward, but it does not seem like a particularly fun task. However, as is frequently the case in applications of degenerate perturbation theory, there are symmetry arguments that allow us to deduce without explicit computation that many of these matrix elements vanish. For example, as in (11.53), we can use the parity argument to deduce that all the diagonal matrix elements must vanish. In fact, since the evenness or oddness of the wave functions depends on the value of $l$ alone and not the value of $m$, we see that all the matrix elements where the ket and the bra have the same $l$ vanish. Thus the only nonzero matrix elements can be the off-diagonal matrix elements in the first row and first column of the matrix. Moreover, with the electric field pointing in the $z$ direction, the perturbing Hamiltonian is invariant under rotations about the $z$ axis, and thus the Hamiltonian commutes with the corresponding generator of rotations, $\hat{L}_z$. Explicitly, the perturbing Hamiltonian just involves the position operator $\hat{z}$, and from (9.72c) we see that

$$[\hat{H}_1, \hat{L}_z] = 0 \tag{11.56}$$

---

[6] For an exact calculation of the second-order Stark effect, see S. Borowitz, *Fundamentals of Quantum Mechanics*, W. A. Benjamin, 1967.

Consequently,

$$m'\hbar\langle n, l', m'|\hat{z}|n, l, m\rangle = \langle n, l', m'|\hat{L}_z\hat{z}|n, l, m\rangle$$
$$= \langle n, l', m'|\hat{z}\hat{L}_z|n, l, m\rangle$$
$$= m\hbar\langle n, l', m'|\hat{z}|n, l, m\rangle \tag{11.57}$$

and therefore

$$\langle n, l', m'|\hat{z}|n, l, m\rangle = 0 \qquad m \neq m' \tag{11.58}$$

The vanishing of the commutator (11.56) dictates that matrix elements of the perturbing Hamiltonian with different $m$'s vanish.

Thus the only matrix element in (11.55) that we need to evaluate explicitly is

$$\langle 2, 0, 0|\hat{H}_1|2, 1, 0\rangle = e|\mathbf{E}|\langle 2, 0, 0|\hat{z}|2, 1, 0\rangle \tag{11.59}$$

Using the position-space radial wave functions (10.44), the spherical harmonics (9.151) and (9.152b), and $z = r\cos\theta$, we find

$$\langle 2, 0, 0|\hat{H}_1|2, 1, 0\rangle = e|\mathbf{E}|\int_0^\infty r^2\,dr\int_0^\pi \sin\theta\,d\theta\int_0^{2\pi} d\varphi\, R_{2,0}^* Y_{0,0}^* r\cos\theta R_{2,1}Y_{1,0}$$

$$= -3e|\mathbf{E}|a_0 \tag{11.60}$$

where the length $a_0$ is just the Bohr radius of hydrogen. Therefore, the $4 \times 4$ matrix (11.55) is given by

$$\begin{pmatrix} 0 & -3e|\mathbf{E}|a_0 & 0 & 0 \\ -3e|\mathbf{E}|a_0 & 0 & 0 & 0 \\ 0 & 0 & 0 & 0 \\ 0 & 0 & 0 & 0 \end{pmatrix} \tag{11.61}$$

We have taken advantage of the Hermiticity of the Hamiltonian to relate the value of the matrix element in the second row, first column to that in the first row, second column.

Thus for the Stark effect in hydrogen, (11.46) can now be written as

$$\begin{pmatrix} 0 & -3e|\mathbf{E}|a_0 & 0 & 0 \\ -3e|\mathbf{E}|a_0 & 0 & 0 & 0 \\ 0 & 0 & 0 & 0 \\ 0 & 0 & 0 & 0 \end{pmatrix}\begin{pmatrix} c_1 \\ c_2 \\ c_3 \\ c_4 \end{pmatrix} = E_2^{(1)}\begin{pmatrix} c_1 \\ c_2 \\ c_3 \\ c_4 \end{pmatrix} \tag{11.62}$$

Recall that for this equation to possess a nontrivial solution, the following determinant must vanish:

$$\begin{vmatrix} -E_2^{(1)} & -3e|\mathbf{E}|a_0 & 0 & 0 \\ -3e|\mathbf{E}|a_0 & -E_2^{(1)} & 0 & 0 \\ 0 & 0 & -E_2^{(1)} & 0 \\ 0 & 0 & 0 & -E_2^{(1)} \end{vmatrix} = 0 \tag{11.63}$$

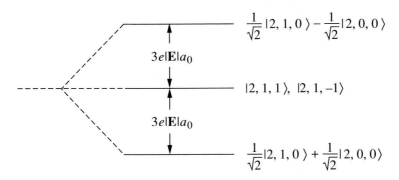

**FIGURE 11.5**
The first-order shifts in the energy levels of the $n = 2$ states of hydrogen in an external electric field.

The four values for the first-order shifts in the energy are

$$E_2^{(1)} = 0, \quad 0, \quad 3e|\mathbf{E}|a_0, \quad -3e|\mathbf{E}|a_0 \tag{11.64}$$

If we substitute these values into (11.62), we find that the corresponding linear combinations of the degenerate eigenstates are given by

$$|2, 1, 1\rangle, \quad |2, 1, -1\rangle, \quad \frac{1}{\sqrt{2}}(|2, 0, 0\rangle - |2, 1, 0\rangle), \quad \frac{1}{\sqrt{2}}(|2, 0, 0\rangle + |2, 1, 0\rangle) \tag{11.65}$$

respectively, as indicated in Fig. 11.5. Again, as a consequence of (11.56), we see that the two states with the same $m$ values are the only ones mixed together by the perturbation. Therefore, we could have chosen initially to concentrate our efforts in degenerate perturbation theory on these two states alone and formed at most the $2 \times 2$ matrix in the upper left-hand corner of (11.63). Finally, notice that when there is degeneracy, there is an energy shift linear in the applied field, as compared with the quadratic effect for the ground state. Although each of the states $|2, 0, 0\rangle$ and $|2, 1, 0\rangle$ has a definite parity, the linear combinations of these states in (11.65) do not. Consequently, these linear combinations can have a nonvanishing expectation value of the dipole moment, which can then interact directly with the applied electric field.

## 11.5 THE AMMONIA MOLECULE IN AN EXTERNAL ELECTRIC FIELD REVISITED

With these results in mind, let's return to the example of the $NH_3$ molecule in an external electric field with which we started our discussion of perturbation theory. First, using perturbation theory, we consider the case of a weak field. The eigenstates of

$$\hat{H}_0 \rightarrow \begin{pmatrix} \langle 1|\hat{H}_0|1\rangle & \langle 1|\hat{H}_0|2\rangle \\ \langle 2|\hat{H}_0|1\rangle & \langle 2|\hat{H}_0|2\rangle \end{pmatrix} = \begin{pmatrix} E_0 & -A \\ -A & E_0 \end{pmatrix} \tag{11.66}$$

are the states $|I\rangle$ and $|II\rangle$ given after equation (11.9). If we use these states as a basis, the matrix representation of $\hat{H}_0$ is diagonal,

$$\hat{H}_0 \rightarrow \begin{pmatrix} \langle I|\hat{H}_0|I\rangle & \langle I|\hat{H}_0|II\rangle \\ \langle II|\hat{H}_0|I\rangle & \langle II|\hat{H}_0|II\rangle \end{pmatrix} = \begin{pmatrix} E_0 - A & 0 \\ 0 & E_0 + A \end{pmatrix} \tag{11.67}$$

as we saw in Section 4.5, while the matrix representation of the perturbing Hamiltonian is given by

$$\hat{H}_1 \rightarrow \begin{pmatrix} \langle I|\hat{H}_1|I\rangle & \langle I|\hat{H}_1|II\rangle \\ \langle II|\hat{H}_1|I\rangle & \langle II|\hat{H}_1|II\rangle \end{pmatrix} = \begin{pmatrix} 0 & \mu_e|\mathbf{E}| \\ \mu_e|\mathbf{E}| & 0 \end{pmatrix} \tag{11.68}$$

Since the parity operator $\hat{\Pi}$ inverts states through the origin, the effect of applying the parity operator, indicated in Fig. 11.6, is to take the state $|1\rangle$ of the molecule, in which the N atom is above the plane formed by the H atoms, and change it into $|2\rangle$, in which the N atom is below the plane:

$$\hat{\Pi}|1\rangle = |2\rangle \tag{11.69a}$$

Similarly,

$$\hat{\Pi}|2\rangle = |1\rangle \tag{11.69b}$$

Thus both the ground state, $|I\rangle$, and the first-excited state, $|II\rangle$, are eigenstates of parity:

$$\hat{\Pi}|I\rangle = \hat{\Pi}\left(\frac{1}{\sqrt{2}}|1\rangle + \frac{1}{\sqrt{2}}|2\rangle\right) = \left(\frac{1}{\sqrt{2}}|2\rangle + \frac{1}{\sqrt{2}}|1\rangle\right) = |I\rangle \quad (11.70a)$$

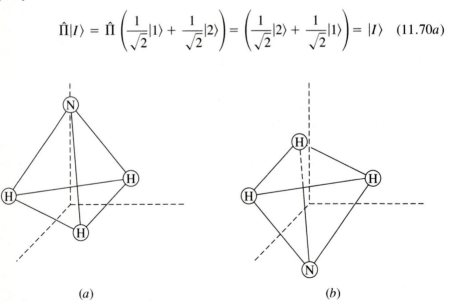

(a)                                              (b)

**FIGURE 11.6**
The action of the parity operator on state $|1\rangle$ of the $NH_3$ molecule, with the N atom above the plane formed by the three H atoms, as shown in (a), produces state $|2\rangle$, with the N atom below the plane, as shown in (b).

$$\hat{\Pi}|II\rangle = \hat{\Pi}\left(\frac{1}{\sqrt{2}}|1\rangle - \frac{1}{\sqrt{2}}|2\rangle\right) = \left(\frac{1}{\sqrt{2}}|2\rangle - \frac{1}{\sqrt{2}}|1\rangle\right) = -|II\rangle \quad (11.70b)$$

Therefore, as shown by the vanishing of the diagonal matrix elements of the perturbing Hamiltonian (11.68), the first-order shift in the energy due to an external electric field is zero, since the dipole moment operator has a vanishing expectation value in a state of definite parity. Our first-order results are in agreement with the exact result (11.9), showing that the molecule exhibits an energy shift that is quadratic rather than linear in the applied field.

What happens if the electric field is a strong field satisfying $\mu_e|\mathbf{E}| \gg A$? If we were still permitted to use nondegenerate perturbation theory with (11.68) as the perturbing Hamiltonian, the first-order shifts in the energies would vanish. However, from the exact eigenvalues (11.8) we see that

$$E = E_0 \pm \mu_e|\mathbf{E}| \pm \frac{A^2}{2\mu_e|\mathbf{E}|} \mp \cdots \qquad \mu_e|\mathbf{E}| \gg A \qquad (11.71)$$

which has a leading term that is linear in the field. The reason for this discrepancy is that for $\mu_e|\mathbf{E}| \gg A$ we really need to use degenerate perturbation theory. Although the states $|I\rangle$ and $|II\rangle$ are not strictly degenerate, they are close together in energy. The energy difference between them is $2A$, which is much less than the energy $\mu_e|\mathbf{E}|$ for strong fields. Thus the magnitude of the factor (11.42) is

$$\frac{\langle II|\hat{H}_1|I\rangle}{E_{II}^{(0)} - E_I^{(0)}} = \frac{\mu_e|\mathbf{E}|}{2A} \gg 1 \qquad (11.72)$$

and we cannot expect nondegenerate perturbation theory to work.

Let's see how we combine perturbation theory with matrix mechanics to work out the terms of the series (11.71). In the strong-field limit we can include the dipole moment interaction as part of $\hat{H}_0$ and break up the Hamiltonian matrix (11.7) in the $|1\rangle$-$|2\rangle$ basis as follows:

$$\hat{H}_0 \rightarrow \begin{pmatrix} \langle 1|\hat{H}_0|1\rangle & \langle 1|\hat{H}_0|2\rangle \\ \langle 2|\hat{H}_0|1\rangle & \langle 2|\hat{H}_0|2\rangle \end{pmatrix} = \begin{pmatrix} E_0 + \mu_e|\mathbf{E}| & 0 \\ 0 & E_0 - \mu_e|\mathbf{E}| \end{pmatrix} \quad (11.73a)$$

$$\hat{H}_1 \rightarrow \begin{pmatrix} \langle 1|\hat{H}_1|1\rangle & \langle 1|\hat{H}_1|2\rangle \\ \langle 2|\hat{H}_1|1\rangle & \langle 2|\hat{H}_1|2\rangle \end{pmatrix} = \begin{pmatrix} 0 & -A \\ -A & 0 \end{pmatrix} \quad (11.73b)$$

Clearly, the eigenstates of this $\hat{H}_0$ are just the states $|1\rangle$ and $|2\rangle$ with eigenvalues $E_1^{(0)} = E_0 + \mu_e|\mathbf{E}|$ and $E_2^{(0)} = E_0 - \mu_e|\mathbf{E}|$, respectively. The first-order shift in these energies in the strong-field limit vanishes:

$$E_1^{(1)} = \langle 1|\hat{H}_1|1\rangle = (1, 0)\begin{pmatrix} 0 & -A \\ -A & 0 \end{pmatrix}\begin{pmatrix} 1 \\ 0 \end{pmatrix} = 0 \qquad (11.74a)$$

$$E_2^{(1)} = \langle 2|\hat{H}_1|2\rangle = (0, 1)\begin{pmatrix} 0 & -A \\ -A & 0 \end{pmatrix}\begin{pmatrix} 0 \\ 1 \end{pmatrix} = 0 \qquad (11.74b)$$

while the second-order shift is given by

$$E_1^{(2)} = \frac{|\langle 2|\hat{H}_1|1\rangle|^2}{E_1^{(0)} - E_2^{(0)}} = \frac{\left|(0, 1)\begin{pmatrix} 0 & -A \\ -A & 0 \end{pmatrix}\begin{pmatrix} 1 \\ 0 \end{pmatrix}\right|^2}{E_0 + \mu_e|\mathbf{E}| - (E_0 - \mu_e|\mathbf{E}|)} = \frac{A^2}{2\mu_e|\mathbf{E}|} \qquad (11.75a)$$

$$E_2^{(2)} = \frac{|\langle 1|\hat{H}_1|2\rangle|^2}{E_2^{(0)} - E_1^{(0)}} = \frac{\left|(1, 0)\begin{pmatrix} 0 & -A \\ -A & 0 \end{pmatrix}\begin{pmatrix} 0 \\ 1 \end{pmatrix}\right|^2}{E_0 - \mu_e|\mathbf{E}| - (E_0 + \mu_e|\mathbf{E}|)} = -\frac{A^2}{2\mu_e|\mathbf{E}|} \qquad (11.75b)$$

These results agree with the expansion (11.71).

In the next section we will examine perturbations to the hydrogen atom due to internal relativistic effects. These perturbations partially break the degeneracy of the four $n = 2$ states, for which we used degenerate perturbation theory in the previous section to work out the Stark effect. But the message of this section is that these relativistic effects don't obviate the need for degenerate perturbation theory as long as the magnitude of the matrix element (11.59) of the perturbing Hamiltonian is large compared with the energy scale of these relativistic effects. In general, whenever the unperturbed states are "close" together in energy, we should include them in the subspace of states that we use to form the matrix representation of the perturbing Hamiltonian.

## 11.6  RELATIVISTIC PERTURBATIONS TO THE HYDROGEN ATOM

Although the agreement between the observed spectrum of hydrogen and our theoretical predictions of Section 10.2 is excellent, there is a fine structure to these energy levels that we haven't accounted for at all. Overall, there are three relativistic perturbations to the Hamiltonian (11.51) of the hydrogen atom that contribute to the fine structure: a relativistic correction to the electron's kinetic energy, a spin-orbit interaction, and the Darwin term. The spin-orbit interaction couples together the intrinsic spin and orbital angular momentum of the electron.

### The Relativistic Correction to the Kinetic Energy

One obvious relativistic perturbation is that the kinetic energy in (11.51) arises from a completely nonrelativistic approximation. Instead of expressing the kinetic energy operator of this two-body system as

$$\hat{K} = \frac{\hat{\mathbf{p}}_e^2}{2m_e} + \frac{\hat{\mathbf{p}}_p^2}{2m_p} \qquad (11.76)$$

we use the relativistically correct expression for the electron's kinetic energy, in which case

$$\hat{K} = \sqrt{\hat{\mathbf{p}}_e^2 c^2 + (m_e c^2)^2} - m_e c^2 + \frac{\hat{\mathbf{p}}_p^2}{2m_p}$$

$$= m_e c^2 \left( \sqrt{1 + (\hat{\mathbf{p}}_e^2/m_e^2 c^2)} - 1 \right) + \frac{\hat{\mathbf{p}}_p^2}{2m_p} \qquad (11.77)$$

Expanding the square root in a Taylor series, we find

$$\hat{K} = \frac{\hat{\mathbf{p}}_e^2}{2m_e} - \frac{(\hat{\mathbf{p}}_e^2)^2}{8m_e^3c^2} + \cdots + \frac{\hat{\mathbf{p}}_p^2}{2m_p} \tag{11.78}$$

In the center-of-mass frame (see Section 9.3), the kinetic energy operator can then be written as

$$\hat{K} = \frac{\hat{\mathbf{p}}^2}{2\mu} - \frac{(\hat{\mathbf{p}}^2)^2}{8m_e^3c^2} + \cdots \tag{11.79}$$

In deriving (11.79), we have ignored the relativistic correction to the proton's kinetic energy because $m_p \gg m_e$.

The unperturbed Hamiltonian for a *hydrogenic* atom is the usual

$$\hat{H}_0 = \frac{\hat{\mathbf{p}}^2}{2\mu} - \frac{Ze^2}{|\hat{\mathbf{r}}|} \tag{11.80}$$

In an energy eigenstate

$$\langle n, l, m| \frac{\hat{\mathbf{p}}^2}{2\mu} |n, l, m\rangle = -E_n^{(0)} = \frac{\mu c^2 Z^2 \alpha^2}{2n^2} \tag{11.81}$$

(see Problem 11.16). Because of the small value of $\alpha$, for modest values of $Z$ the average kinetic energy is much less than the rest-mass energy, and therefore the atom is quite nonrelativistic. We thus can treat

$$\hat{H}_K = -\frac{(\hat{\mathbf{p}}^2)^2}{8m_e^3c^2} \tag{11.82}$$

as a perturbation on the Hamiltonian (11.80). Notice that (11.82) is rotationally invariant and therefore

$$[\hat{H}_K, \hat{\mathbf{L}}] = 0 \tag{11.83}$$

Thus, although the eigenstates $|n,l,m\rangle$ of $\hat{H}_0$ are highly degenerate, the matrix representation of the perturbing Hamiltonian (11.82) in each degenerate subspace is already diagonal, and we can calculate the first-order energy shift as

$$E_{n,l}^{(1)} = -\langle n, l, m| \frac{(\hat{\mathbf{p}}^2)^2}{8m_e^3c^2} |n, l, m\rangle \tag{11.84}$$

We could evaluate (11.84) directly in position space, letting the operator $(\hat{\mathbf{p}}^2)^2 \rightarrow (-\hbar^2\nabla^2)^2$ differentiate the wave function $\langle \mathbf{r}|n, l, m\rangle = R_{n,l}(r)Y_{l,m}(\theta, \varphi)$, and so on. Fortunately, there is a better way. We can simplify the evaluation by rewriting the operator (11.82) in the form

$$-\frac{(\hat{\mathbf{p}}^2)^2}{8m_e^3c^2} = -\frac{1}{2m_ec^2}\left(\frac{\hat{\mathbf{p}}^2}{2m_e}\right)^2 = -\frac{1}{2m_ec^2}\left(\hat{H}_0 + \frac{Ze^2}{|\hat{\mathbf{r}}|}\right)\left(\hat{H}_0 + \frac{Ze^2}{|\hat{\mathbf{r}}|}\right) \tag{11.85}$$

where we have ignored the difference between the reduced mass of the hydrogen atom and the mass of the electron in the perturbation. Thus

$$E_{n,l}^{(1)} = -\frac{1}{2m_ec^2}\left[\left(E_n^{(0)}\right)^2 + 2E_n^{(0)}\langle n, l, m|\frac{Ze^2}{|\hat{\mathbf{r}}|}|n, l, m\rangle + \langle n, l, m|\frac{Z^2e^4}{|\hat{\mathbf{r}}|^2}|n, l, m\rangle\right] \tag{11.86}$$

From Problem 11.15

$$-\langle n, l, m | \frac{Ze^2}{|\hat{\mathbf{r}}|} | n, l, m \rangle = 2E_n^{(0)} \tag{11.87}$$

and from Problem 11.17

$$\langle n, l, m | \frac{Z^2 e^4}{|\hat{\mathbf{r}}|^2} | n, l, m \rangle = \frac{Z^4 e^4}{a_0^2 n^3 (l + \frac{1}{2})} = \frac{4(E_n^{(0)})^2 n}{l + \frac{1}{2}} \tag{11.88}$$

Thus the first-order shift in the energy due to the relativistic correction to the electron's kinetic energy is given by

$$E_K^{(1)} = -\frac{1}{2} m_e c^2 Z^4 \alpha^4 \left[ -\frac{3}{4n^4} + \frac{1}{n^3(l + \frac{1}{2})} \right] \tag{11.89}$$

## Spin-Orbit Coupling

In order to determine the form of the spin-orbit interaction, we start with a classical argument. In the rest frame of the electron, the motion of the proton generates a current, which, from the Biot-Savart law, produces a magnetic field

$$\mathbf{B} = \frac{-Ze\mathbf{v} \times \mathbf{r}}{cr^3} \tag{11.90}$$

where $-\mathbf{v}$ is the proton's velocity, which is equal and opposite to the velocity $\mathbf{v}$ of the electron in the proton's rest frame. The energy of interaction of the electron's intrinsic spin magnetic moment with this magnetic field is given by

$$-\boldsymbol{\mu} \cdot \mathbf{B} = -\left( -\frac{ge}{2m_e c} \mathbf{S} \cdot \frac{-Ze\mathbf{v} \times \mathbf{r}}{cr^3} \right) = \frac{Ze^2}{m_e^2 c^2 r^3} \mathbf{S} \cdot \mathbf{L} \tag{11.91}$$

where $\mathbf{L} = \mathbf{r} \times \mathbf{p}$ is the electron's orbital angular momentum. We have also taken $g = 2$ for the electron.[7]

Equation (11.91) might not seem like a truly *relativistic* effect. However, we can express the magnetic field (11.90) as

$$\mathbf{B} = -(\mathbf{v}/c) \times \mathbf{E} \tag{11.92}$$

---

[7] It is interesting to see how the factor of $1/c^2$ in (11.91) arises in SI units. Since $\boldsymbol{\mu} = -(e/m_e)\mathbf{S}$ and $\mathbf{B} = (\mu_0/4\pi)[(-Ze\mathbf{v}) \times \mathbf{r}/r^3]$ in SI units,

$$-\boldsymbol{\mu} \cdot \mathbf{B} = \frac{Ze^2(\mu_0 \varepsilon_0)}{4\pi\varepsilon_0} \frac{\mathbf{S} \cdot \mathbf{L}}{m_e^2 r^3} = \frac{Ze^2}{4\pi\varepsilon_0} \frac{\mathbf{S} \cdot \mathbf{L}}{m_e^2 c^2 r^3}$$

where we have used $\mu_0 \varepsilon_0 = 1/c^2$ in the last step. Thus, as is the case for expressions such as the potential energy $-Ze^2/r$ in Gaussian units, we can go from (11.91) to the corresponding expression in SI units with the replacement $e^2 \rightarrow e^2/4\pi\varepsilon_0$.

where $\mathbf{E}$ is the electric field in the electron's rest frame. Magnetic effects, which depend on the motion of charges, are all inherently relativistic, as the factor of $\mathbf{v}/c$ in (11.92) suggests. In fact, in "deriving" (11.91), we have made a relativistic error, which was first discovered by Thomas and is called the Thomas precession effect. This effect simply reduces the energy of interaction (11.91) by a factor of two. We will not derive the Thomas factor here.[8] The best way to obtain the full spin-orbit interaction Hamiltonian

$$\hat{H}_{\text{S-O}} = \frac{Ze^2}{2m_e^2 c^2 |\hat{\mathbf{r}}|^3} \hat{\mathbf{L}} \cdot \hat{\mathbf{S}} \qquad (11.93)$$

is from the nonrelativistic limit of the famous Dirac equation with a Coulomb potential energy. The Dirac equation is the fully relativistic wave equation of a spin-$\frac{1}{2}$ particle, such as the electron. This equation, for example, predicts that $g = 2$ for the electron, so relativistically we don't have to insert this factor by hand based on experimental results, as we have done so far.

We are now ready to treat the Hamiltonian (11.93) as a perturbation. Let's concentrate first on the $\hat{\mathbf{L}} \cdot \hat{\mathbf{S}}$ part of the interaction, which is reminiscent of the spin-spin interaction $\hat{\mathbf{S}}_1 \cdot \hat{\mathbf{S}}_2$ (see Chapter 5) that couples together the spin angular momentum states of two spin-$\frac{1}{2}$ particles. Here the story is essentially the same, except that one of the angular momentum operators is orbital and the other is intrinsic spin. We can form a basis as a direct product of the orbital angular momentum and intrinsic spin states:

$$|l, m, +\mathbf{z}\rangle = |l, m\rangle \otimes |+\mathbf{z}\rangle = |l, m\rangle \otimes |\tfrac{1}{2}, \tfrac{1}{2}\rangle \qquad (11.94a)$$

$$|l, m, -\mathbf{z}\rangle = |l, m\rangle \otimes |-\mathbf{z}\rangle = |l, m\rangle \otimes |\tfrac{1}{2}, -\tfrac{1}{2}\rangle \qquad (11.94b)$$

We can form simultaneous eigenstates of $\hat{\mathbf{L}}^2$ and $\hat{L}_z$ as well as $\hat{\mathbf{S}}^2$ and $\hat{S}_z$, since the orbital and spin angular momentum operators commute with each other. After all, $\hat{\mathbf{L}}$ generates rotations in position space, while $\hat{\mathbf{S}}$ generates rotations independently on spin states. Thus the operator that generates rotations of both the spatial and spin degrees of freedom is the total angular momentum operator

$$\hat{\mathbf{J}} = \hat{\mathbf{L}} + \hat{\mathbf{S}} \qquad (11.95)$$

Diagonalizing the interaction Hamiltonian (11.93) means finding the eigenstates of $\hat{\mathbf{L}} \cdot \hat{\mathbf{S}}$. Just as the eigenstates of $\hat{\mathbf{S}}_1 \cdot \hat{\mathbf{S}}_2$ are eigenstates of total spin, the eigenstates of $\hat{\mathbf{L}} \cdot \hat{\mathbf{S}}$ are eigenstates of total angular momentum $\hat{\mathbf{J}}^2$ and $\hat{J}_z$, where

$$\hat{\mathbf{J}}^2 = \hat{\mathbf{L}}^2 + \hat{\mathbf{S}}^2 + 2\hat{\mathbf{L}} \cdot \hat{\mathbf{S}} \qquad (11.96a)$$

---

[8] See, for example, R. Eisberg and R. Resnick, *Quantum Physics of Atoms, Molecules, Solids, Nuclei, and Particles,* 2nd ed., Wiley, New York, 1985, Appendix O. Thomas's discovery provided the mysterious factor of two necessary to make Goudsmit and Uhlenbeck's intrinsic spin hypothesis fit the spectrum of hydrogen. Uhlenbeck has noted that "it seemed unbelievable that a relativistic effect could give a factor of two instead of something of order $v/c$" and "even the cognoscenti of the relativity theory (Einstein included!) were quite surprised." *Physics Today,* June 1976, p.48.

$$\hat{J}_z = \hat{L}_z + \hat{S}_z \qquad (11.96b)$$

From the expression

$$2\hat{\mathbf{L}} \cdot \hat{\mathbf{S}} = \hat{\mathbf{J}}^2 - \hat{\mathbf{L}}^2 - \hat{\mathbf{S}}^2 \qquad (11.97)$$

it is easy to see that $\hat{J}_z$ commutes with $\hat{\mathbf{L}} \cdot \hat{\mathbf{S}}$, since $\hat{J}_z$ commutes with $\hat{\mathbf{J}}^2$, $\hat{L}_z$ commutes with $\hat{\mathbf{L}}^2$, and $\hat{S}_z$ commutes with $\hat{\mathbf{S}}^2$. Or we can evaluate the commutator explicitly, which shows that although *neither* $\hat{L}_z$ nor $\hat{S}_z$ commutes with $\hat{\mathbf{L}} \cdot \hat{\mathbf{S}}$, the operator $\hat{J}_z$ does:[9]

$$
\begin{aligned}
\left[ \hat{J}_z, \hat{\mathbf{L}} \cdot \hat{\mathbf{S}} \right] &= \left[ \hat{L}_z + \hat{S}_z, \hat{\mathbf{L}} \cdot \hat{\mathbf{S}} \right] = \left[ \hat{L}_z + \hat{S}_z, \hat{L}_x \hat{S}_x + \hat{L}_y \hat{S}_y + \hat{L}_z \hat{S}_z \right] \\
&= \left[ \hat{L}_z, \hat{L}_x \right] \hat{S}_x + \left[ \hat{L}_z, \hat{L}_y \right] \hat{S}_y + \left[ \hat{S}_z, \hat{S}_x \right] \hat{L}_x + \left[ \hat{S}_z, \hat{S}_y \right] \hat{L}_y \\
&= i\hbar \hat{L}_y \hat{S}_x - i\hbar \hat{L}_x \hat{S}_y + i\hbar \hat{S}_y \hat{L}_x - i\hbar \hat{S}_x \hat{L}_y = 0 \qquad (11.98)
\end{aligned}
$$

Since these operators commute, we can find eigenstates of $\hat{\mathbf{L}} \cdot \hat{\mathbf{S}}$ that are simultaneous eigenstates of $\hat{J}_z$.

For the hydrogen atom, this substantially simplifies the job of determining the linear combinations of degenerate states that diagonalize the perturbing Hamiltonian. Counting the intrinsic spin states of the electron, there are $2n^2$ degenerate states for any given $n$. However, (11.98) shows that only states with the same eigenvalue for $\hat{J}_z$ can be mixed together by the perturbation. Also, since $\hat{H}_{S-O}$ commutes with $\hat{\mathbf{L}}^2$, we can focus on states with the same value for $l$. For a fixed $l$, there are just two states with the eigenvalue of $\hat{J}_z$ equal to $(m + \frac{1}{2})\hbar$:

$$|l, m, +\mathbf{z}\rangle \qquad |l, m + 1, -\mathbf{z}\rangle \qquad (11.99)$$

assuming that $m \neq l$; otherwise, there is only a single state. In order to determine the two linear combinations of these states that are eigenstates of $2\hat{\mathbf{L}} \cdot \hat{\mathbf{S}}$, we use these two states as a basis to form the matrix representation of the operator. Using the identity

$$2\hat{\mathbf{L}} \cdot \hat{\mathbf{S}} = \hat{L}_+ \hat{S}_- + \hat{L}_- \hat{S}_+ + 2\hat{L}_z \hat{S}_z \qquad (11.100)$$

and the general results for the action of the angular momentum raising and lowering operators, we find that the matrix representation is given by

$$
2\hat{\mathbf{L}} \cdot \hat{\mathbf{S}} \rightarrow
\begin{array}{c}
\phantom{x} \\
\langle l, m, +\mathbf{z}| \\
\langle l, m + 1, -\mathbf{z}|
\end{array}
\hbar^2
\begin{pmatrix}
m & \sqrt{l(l + 1) - m(m + 1)} \\
\sqrt{l(l + 1) - m(m + 1)} & -(m + 1)
\end{pmatrix}
$$

with column labels $|l, m, +\mathbf{z}\rangle$ and $|l, m + 1, -\mathbf{z}\rangle$.

$$(11.101)$$

where we have indicated above and to the left of the matrix the kets and bras that are used to determine each of the matrix elements.

---

[9] One can also argue that $\hat{\mathbf{L}} \cdot \hat{\mathbf{S}}$ as a dot product of two vector operators is invariant under total rotations and that $\hat{J}_z$ is a generator of total rotations about the $z$ axis and must therefore commute with $\hat{\mathbf{L}} \cdot \hat{\mathbf{S}}$.

The eigenvalue equation

$$2\hat{\mathbf{L}} \cdot \hat{\mathbf{S}}|\lambda\rangle = \lambda\hbar^2|\lambda\rangle \tag{11.102}$$

has nontrivial solutions provided that

$$\begin{vmatrix} m - \lambda & \sqrt{l(l+1) - m(m+1)} \\ \sqrt{l(l+1) - m(m+1)} & -(m+1) - \lambda \end{vmatrix} = 0 \tag{11.103}$$

or

$$\lambda^2 + \lambda - l(l+1) = 0 \tag{11.104}$$

The two solutions are $\lambda = l$ and $\lambda = -(l+1)$. By substituting these eigenvalues into (11.102) in matrix form, we can determine the linear combinations of the states (11.99) that are eigenstates of $\hat{\mathbf{L}} \cdot \hat{\mathbf{S}}$. Since each of the states (11.99) is an eigenstate of $\hat{\mathbf{L}}^2$ with eigenvalue $l(l+1)\hbar^2$ and $\hat{\mathbf{S}}^2$ with eigenvalue $\frac{1}{2}(\frac{1}{2}+1)\hbar^2$, these linear combinations are also eigenstates of $\hat{\mathbf{J}}^2$, as given in (11.96a). The value of the total angular momentum quantum number $j$ is then determined by

$$j(j+1) = l(l+1) + \tfrac{1}{2}(\tfrac{1}{2}+1) + \begin{cases} l \\ -(l+1) \end{cases} \tag{11.105}$$

which yields the two solutions

$$j = \begin{cases} l + \tfrac{1}{2} \\ l - \tfrac{1}{2} \end{cases} \tag{11.106}$$

Thus the $\hat{\mathbf{L}} \cdot \hat{\mathbf{S}}$ interaction term has coupled the orbital angular momentum $l$ together with the spin angular momentum $\frac{1}{2}$ to produce a total angular momentum $j$ that takes on the values $l + \frac{1}{2}$ and $l - \frac{1}{2}$. The eigenstates are given by

$$|j = l + \tfrac{1}{2}, m_j\rangle = \sqrt{\frac{l+m+1}{2l+1}}|l, m, +\mathbf{z}\rangle + \sqrt{\frac{l-m}{2l+1}}|l, m+1, -\mathbf{z}\rangle \tag{11.107a}$$

$$|j = l - \tfrac{1}{2}, m_j\rangle = \sqrt{\frac{l-m}{2l+1}}|l, m, +\mathbf{z}\rangle - \sqrt{\frac{l+m+1}{2l+1}}|l, m+1, -\mathbf{z}\rangle \tag{11.107b}$$

with $m_j = m + \frac{1}{2}$. The right-hand side of these equations can be expressed directly in terms of $m_j$ as

$$|j = l \pm \tfrac{1}{2}, m_j\rangle = \sqrt{\frac{l \pm m_j + \tfrac{1}{2}}{2l+1}}|l, m_j - \tfrac{1}{2}, +\mathbf{z}\rangle$$

$$\pm \sqrt{\frac{l \mp m_j + \tfrac{1}{2}}{2l+1}}|l, m_j + \tfrac{1}{2}, -\mathbf{z}\rangle \tag{11.108}$$

We now know the linear combinations of the basis states (11.99) that diagonalize the perturbing Hamiltonian (11.93) in the subspace of degenerate states. The energy shift due to the spin-orbit interaction is given by the expectation value of this Hamiltonian for these states:

$$
\begin{aligned}
E_{\text{S-O}}^{(1)} &= \langle n, j, m_j | \frac{Z e^2}{2 m_e^2 c^2 |\hat{\mathbf{r}}|^3} \hat{\mathbf{L}} \cdot \hat{\mathbf{S}} | n, j, m_j \rangle \\[2mm]
&= \frac{Z e^2 \hbar^2}{2 m_e^2 c^2} \left\langle \frac{1}{r^3} \right\rangle_{n,l}
\begin{cases}
l & j = l + \frac{1}{2} \\
-(l+1) & j = l - \frac{1}{2}
\end{cases}
\end{aligned}
\tag{11.109}
$$

Since from Problem 11.18

$$
\left\langle \frac{1}{r^3} \right\rangle_{n,l} = \frac{Z^3}{a_0^3 n^3 l(l + \frac{1}{2})(l + 1)}
\tag{11.110}
$$

we can express the first-order shift in the energy due to the spin-orbit interaction as

$$
E_{\text{S-O}}^{(1)} = \frac{m_e c^2 Z^4 \alpha^4}{4 n^3 l(l + \frac{1}{2})(l + 1)}
\begin{cases}
l & j = l + \frac{1}{2} \\
-(l+1) & j = l - \frac{1}{2}
\end{cases}
\tag{11.111}
$$

The spectroscopic notation that is used to label these states, $1s$, $2s$, $2p$, and so on, where the number in front is the principal quantum number and the letter indicates the orbital angular momentum ($l = 0$ is $s$, $l = 1$ is $p$, $l = 2$ is $d$, ...), now needs to be enlarged to specify the total angular momentum as well. This is done by adding a subscript indicating the value of $j$. For $l = 0$, the value of $j$ must be $\frac{1}{2}$; for $l = 1$, its value is $\frac{1}{2}$ or $\frac{3}{2}$, while for $l = 2$, the value of $j$ is $\frac{3}{2}$ or $\frac{5}{2}$, and so on. Thus the states of the atom, including the total angular momentum, are $1s_{1/2}$, $2s_{1/2}$, $2p_{1/2}$, $2p_{3/2}$, and so on. Equation (11.111) shows, for example, that the $2p_{1/2}$ and $2p_{3/2}$ states have different energies when the spin-orbit interaction is included.

As we will discuss in the next chapter, this labeling can also be extended to multielectron atoms. Multielectron atoms that are quite similar to the single-electron hydrogen atom include the alkali elements, such as sodium. For sodium in the ground state, 10 electrons fill up the $1s$, $2s$, and $2p$ energy states, while the 11th electron is in the $3s$ level. Although the $3s$ electron tends to reside in a shell that is outside the other 10 electrons, the radial wave functions shown in Fig. 10.5 reveal that its wave function penetrates inside the electron cloud formed by the inner electrons. It is thus only partially shielded from the nucleus with its $Z = 11$ positive charge, and its energy is reduced from the $n = 3$ value (10.34) for hydrogen. Unlike the $l = 0$ states, the wave function of the $3p$ electron vanishes at the origin. It thus doesn't "see" the nucleus as much as does the $3s$ electron, and consequently its energy is not reduced as much. Thus the degeneracy between the $3s$ and $3p$ energy states that is present in hydrogen is broken in the sodium atom. The spin-orbit interaction then adds a fine structure to the sodium energy levels. In particular, the difference in energy between the $3p_{1/2}$ and $3p_{3/2}$ states is responsible for the two closely spaced yellow lines in the spectrum, known as the sodium D lines, which are produced when the atom makes a transition from the $3p$ to the $3s$ state (see Fig. 11.7).

## The Darwin Term

If we evaluate (11.111) for $l = 0$, we obtain a finite result. Of course, in a state with zero orbital angular momentum, there cannot be any spin-orbit interaction. The finite result arises because the expectation value (11.110) of $1/r^3$ in the hydrogenic wave functions has a $1/l$ dependence that cancels the factor of $l$ from the eigenvalue of $2\hat{\mathbf{L}} \cdot \hat{\mathbf{S}}$ for a state with $j = l + \frac{1}{2}$. In fact, if we were to evaluate the expectation value (11.110) using more exact relativistic wave functions from the Dirac equation, we would find that there is actually no spin-orbit contribution for $l = 0$, as you would expect physically. However, a perturbative solution to the Dirac equation shows that there does exist an additional interaction that we have not included in our discussion of relativistic perturbations.

The Dirac equation is a four-component wave equation, as compared with the two-component spinors that we introduced in Chapter 2 to represent the spin states of spin-$\frac{1}{2}$ particles. When one reduces the equation to an effective Schrödinger-like equation by eliminating the lower two components, one finds in addition to the perturbations (11.82) and (11.93) an additional perturbation of the form

$$\hat{H}_D = \frac{-1}{8m_e^2 c^2}\left[\hat{\mathbf{p}}\cdot,\left[\hat{\mathbf{p}}, V(|\hat{\mathbf{r}}|)\right]\right] \tag{11.112}$$

where the momentum operators are dotted with each other.[10] Thus $\hat{H}_D$ is rotationally invariant like $\hat{H}_K$, and we can calculate the first-order energy shift by

---

[10] For example, see R. Shankar, *Principles of Quantum Mechanics,* Plenum, New York, 1980, pp. 581–586. Other references for learning about the Dirac equation include J. D. Bjorken and S. D. Drell, *Relativistic Quantum Mechenics*, McGraw-Hill, New York, 1964, and J. J. Sakurai, *Advanced Quantum Mechanics*, Addison-Wesley, Reading, Mass., 1967. This latter book is highly recommended for its excellent discussion of the physics associated with the Dirac equation, although it does use a somewhat old-fashioned *ict* metric.

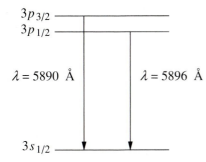

**FIGURE 11.7**
The spin-orbit splitting of the $3p_{1/2}$ and $3p_{3/2}$ levels leads to a fine structure that is responsible for the sodium D lines. The energy difference between the $3s$ and $3p$ levels results from the fact that the potential energy experienced by the $n = 3$ electron is not a pure $-e^2/r$ Coulomb potential.

$$E_D^{(1)} = \langle n, l, m | \frac{-1}{8m_e^2 c^2} \left[ \hat{\mathbf{p}} \cdot, \left[ \hat{\mathbf{p}}, V(|\hat{\mathbf{r}}|) \right] \right] | n, l, m \rangle$$

$$= \int d^3r \, |R_{n,l}|^2 |Y_{l,m}|^2 \frac{\hbar^2}{8m_e^2 c^2} \nabla^2 V$$

$$= \int d^3r \, |R_{n,l}|^2 |Y_{l,m}|^2 \frac{Z e^2 \hbar^2 \pi}{2 m_e^2 c^2} \delta^3(\mathbf{r}) \tag{11.113}$$

Since only $l = 0$ states are nonzero at the origin, the Darwin term contributes only for $s$ states. In fact, the magnitude of this contribution

$$\langle n, 0, 0 | \hat{H}_D | n, 0, 0 \rangle = \frac{m_e c^2 Z^4 \alpha^4}{2n^3} \tag{11.114}$$

turns out to be exactly the same as the spurious $l = 0$ contribution from (11.111) for the spin-orbit interaction.

Why does the Dirac equation have four components instead of two? Any quantum mechanical relativistic description of particles must include the anti-particles as well as the particles—in this case positrons as well as electrons. *Each* of these particles is a spin-$\frac{1}{2}$ particle, and thus we end up with a four-component equation. Why must the positrons be included in our treatment? One way to see this is to go back to the energy-time uncertainty relation (4.62) and note that for time intervals

$$\Delta t \sim \frac{\hbar}{m_e c^2} \tag{11.115}$$

the uncertainty in the energy $\Delta E \sim 2m_e c^2$, which is sufficient to create an electron-positron pair. Thus, in addition to an amplitude for the hydrogen atom to be an electron and a proton, there is an amplitude for the atom to be an electron, a proton, *and* an electron-positron pair. In fact, you can see that for sufficiently short time intervals, the atom can be teeming with activity with many pairs of electrons and positrons, and even particle-antiparticle pairs of heavier particles as well. It is this sort of behavior that makes quantum field theory a complicated many-particle theory.

In an atom containing electron-positron pairs in addition to the usual electron and proton, the concept of a simple potential energy of interaction between the electron and the proton must break down. This breakdown occurs on the distance scale

$$\Delta r \sim c \Delta t \sim \frac{\hbar}{m_e c} \tag{11.116}$$

roughly the Compton wavelength of the electron, which becomes effectively the electron's charge radius. Note that (11.116) is a factor of $\alpha$ smaller than the Bohr radius $a_0$ of the atom. It is interesting to note that if we replace the potential energy with a smeared average over this distance scale, we obtain

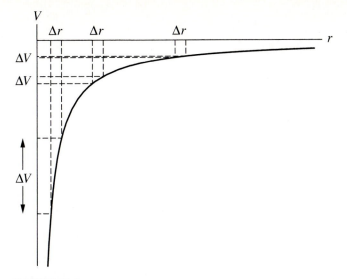

**FIGURE 11.8**
Fluctuations on the distance scale $\hbar/m_e c$ produce a significant change in the potential energy near the origin.

$$\overline{V} = V(r) + \overline{\Delta \mathbf{r}} \cdot \nabla V + \frac{1}{2} \sum_{i,j} \overline{\Delta x_i \Delta x_j} \frac{\partial^2 V}{\partial x_i \partial x_j} + \cdots$$

$$= V(r) + \frac{1}{6} \Delta r^2 \nabla^2 V + \cdots = V(r) + \frac{1}{6} \left( \frac{\hbar}{m_e c} \right)^2 \nabla^2 V + \cdots \quad (11.117)$$

where we have assumed that the vector displacements average to zero and that there is spherical symmetry. Equation (11.117) yields the same form for the perturbing Hamiltonian that appears in the middle of (11.113), except the factor of $\frac{1}{8}$ is replaced by $\frac{1}{6}$. As Fig. 11.8 shows, fluctuations on the distance scale (11.116) for the Coulomb potential only have a substantial effect near the origin, and that is why only $s$ states are affected.

## 11.7 THE ENERGY LEVELS OF HYDROGEN, INCLUDING FINE STRUCTURE, THE LAMB SHIFT, AND HYPERFINE SPLITTING

Adding the energy shifts (11.89), (11.111), and (11.114) together, we obtain

$$E_K^{(1)} + E_{S-O}^{(1)} + E_D^{(1)} = E_{n,j}^{(1)} = -\frac{m_e c^2 (Z\alpha)^4}{2n^3} \left( \frac{1}{j + \frac{1}{2}} - \frac{3}{4n} \right) \quad (11.118)$$

Notice that the magnitude of the total energy shift is of order $(Z\alpha)^2$ times the unperturbed energy (10.34) of the atom. In particular, for hydrogen ($Z = 1$), the energy shift is roughly $10^{-5}$ as large as the unperturbed energy. Thus the perturbations do indeed contribute a *fine structure* to the energy levels—hence the

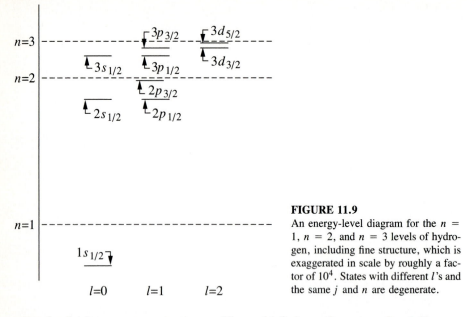

**FIGURE 11.9**
An energy-level diagram for the $n = 1$, $n = 2$, and $n = 3$ levels of hydrogen, including fine structure, which is exaggerated in scale by roughly a factor of $10^4$. States with different $l$'s and the same $j$ and $n$ are degenerate.

name for the fine-structure constant $\alpha$. Figure 11.9 shows the energy-level diagram for hydrogen, including this fine structure. Also note that although each of the individual energy shifts (11.89), (11.111), and (11.114) depends on the value of $l$, the total shift does not. This surprising degeneracy is actually maintained to all orders in the relativistic perturbation when the Dirac equation with a Coulomb potential is solved exactly.[11]

In 1947 W. E. Lamb and R. C. Retherford observed a very small energy difference between the $2s_{1/2}$ and the $2p_{1/2}$ levels through the absorption of microwave radiation with a frequency of 1058 MHz, corresponding to an energy splitting of $4.4 \times 10^{-6}$ eV (see Fig. 11.10b).[12] This *Lamb shift*, which is of the order $m_e c^2 (Z\alpha)^4 \alpha \log \alpha$, can be explained by quantum electrodynamics in terms of the interaction of the electron with the quantized electromagnetic field.[13] The

---

[11] The exact energy eigenvalues for the Dirac equation with a Coulomb potential are given by

$$E_{n,j} = m_e c^2 \left\{ \left[ 1 + \left( \frac{Z\alpha}{n - (j + \frac{1}{2}) + \sqrt{(j + \frac{1}{2})^2 - (Z\alpha)^2}} \right)^2 \right]^{-1/2} - 1 \right\}$$

[12] W. E. Lamb and R. C. Retherford, *Phys. Rev.* **72**, 241 (1947); **86**, 1014 (1952). This latter paper contains their most precise results. Lamb received the Nobel prize in 1953 for this work.

[13] We examine the quantized electromagnetic field in Chapter 14. However, we will not attempt to work out the value of the Lamb shift, which is itself a taxing problem. For an interesting discussion of the difficulties that this calculation presented to R. P. Feynman and H. Bethe, two of the more clever physicists at performing calculations, see Feynman's Nobel prize speech in *Nobel Lectures—Physics*, vol. III, Elsevier Publications, New York, 1972.

Lamb shift has been measured to five significant figures, providing one of the most sensitive tests of quantum electrodynamics (QED). Note that the magnitude of the Lamb shift is roughly $10^{-6}$ of the spacing between levels that produce the Balmer series. Thus, measuring the shift itself with an accuracy of one part in $10^5$ by detecting the difference in wavelength of visible photons emitted as the atom makes transitions from higher energy states to the $2s_{1/2}$ or $2p_{1/2}$ states would require a resolution of 1 part in $10^{11}$! The main reason that we can isolate these QED corrections experimentally with such precision is the fortunate degeneracy, apart from these QED effects, of the $2s_{1/2}$ and $2p_{1/2}$ states. In essence, the experiment of Lamb and Retherford is a sensitive null test in that any absorption at the appropriate microwave frequency is an indication of an energy splitting that cannot be explained through purely relativistic effects.

In our discussion of the perturbations for the hydrogen atom, we have so far neglected the effect of the proton's spin. As we discussed in Chapter 5, the proton's intrinsic magnetic moment interacts with the electron's magnetic moment, leading to a hyperfine interaction. When we include the proton's spin degrees of freedom as well as those of the electron, the ground state has a four-fold degeneracy, which is split into two energy levels by the hyperfine interaction, as indicated in Fig. 11.10a. When the atom makes a transition between these two levels, it emits a photon with a frequency of 1420 MHz, or a wavelength of 21 cm. This energy splitting, which is on the order of $(m_e/m_p)\alpha^4 m_e c^2$, is a factor of $(m_e/m_p)$ smaller than the fine structure—hence the name *hyperfine structure*. This factor of $(m_e/m_p)$ enters because the magnetic moment of the proton is smaller than

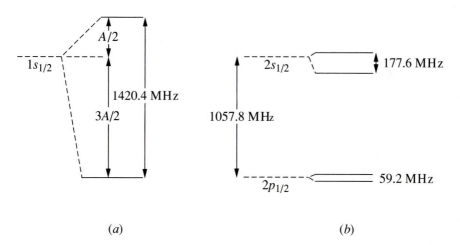

(a)                                                    (b)

**FIGURE 11.10**

The hyperfine splitting of (a) the $n = 1$ and (b) the $n = 2$ energy levels of hydrogen. The Lamb shift is the 1057.8 MHz splitting between the $2s_{1/2}$ and $2p_{1/2}$ states of hydrogen, which is due to quantum electrodynamic effects. Without these QED effects, states of different $l$ and the same $j$ would be degenerate, as shown in Fig. 11.7. The hyperfine splitting of the $2p_{3/2}$ level is not shown because the fine-structure splitting between the $j = \frac{3}{2}$ and $j = \frac{1}{2}$ levels is roughly ten times larger than the Lamb shift.

that of the electron by $(m_e/m_p)$. As Fig. 11.10$b$ shows, this hyperfine structure occurs in excited states of the atom as well, producing splittings equal to 24 MHz for the $2p_{3/2}$ level, 178 MHz for the $2s_{1/2}$ level, and 59 MHz for the $2p_{1/2}$ level. However, the form of the Hamiltonian is not as simple as (5.9) for states with orbital angular momentum $l \neq 0$.[14]

## 11.8   THE ZEEMAN EFFECT IN HYDROGEN

In Section 11.4 we examined the Stark effect, which is produced when a hydrogen atom is placed in an external electric field. In 1896 Zeeman observed the splitting of the spectral lines of the light emitted by an atom placed in an external magnetic field. To analyze this Zeeman effect in hydrogen, we take

$$\hat{H}_B = -\boldsymbol{\hat{\mu}} \cdot \mathbf{B} = -\left(\frac{-e}{2m_e c}\hat{\mathbf{L}} + \frac{-e}{m_e c}\hat{\mathbf{S}}\right) \cdot \mathbf{B} \tag{11.119}$$

as the form for the interaction Hamiltonian, where the first term in parentheses is the magnetic moment operator due to orbital motion [see (1.2)], while the second term is that due to intrinsic spin for the electron. We have neglected the contribution of the proton to (11.119) because of the small magnitude of the proton's magnetic moment. Equation (11.119) includes the dominant part of the magnetic interaction for a one-electron atom if the applied field is not extremely strong.[15] If we orient our coordinate axes so that the magnetic field points in the $z$ direction, the perturbing Hamiltonian becomes

$$\hat{H}_B = \frac{eB}{2m_e c}(\hat{L}_z + 2\hat{S}_z) \tag{11.120}$$

Let's consider the effect of an external magnetic field on the $n = 2$ states of hydrogen. For magnetic fields on the order of $10^4$ gauss, the magnitude $e\hbar B/m_e c$ of $\hat{H}_B$ is comparable with that of spin-orbit energy (11.111) for $n = 2$. Thus for magnetic fields with a strength of a few thousand gauss or less, we can treat $\hat{H}_B$ as a perturbation to the Hamiltonian of the hydrogen atom *including* spin-orbit coupling. Since $\hat{H}_B$ may be written as

$$\hat{H}_B = \frac{eB}{2m_e c}(\hat{J}_z + \hat{S}_z) \tag{11.121}$$

which clearly commutes with $\hat{J}_z$ as well as $\hat{\mathbf{L}}^2$, we can calculate the first-order shift in the energy as the expectation value of (11.121) in the states (11.108):

$$E_B^{(1)} = \frac{eB}{2m_e c}\langle j = l \pm \tfrac{1}{2}, m_j | (\hat{J}_z + \hat{S}_z) | j = l \pm \tfrac{1}{2}, m_j \rangle \tag{11.122}$$

---

[14] For a derivation of the full hyperfine Hamiltonian, see S. Gasiorowicz, *Quantum Physics,* Wiley, New York, 1974. For a calculation of these hyperfine splittings, see H. A. Bethe and E. E. Salpeter, *Quantum Mechanics of One- and Two-Electron Atoms,* Springer-Verlag, Berlin, 1957, Section 22.
[15] The full Hamiltonian in a magnetic field, ignoring intrinsic spin, is derived in Appendix E.

Of course, the expectation value of $J_z$ in these states is just its eigenvalue $m_j\hbar$, but to evaluate $\langle S_z \rangle$, we must use the explicit form (11.108) for the states:

$$\langle S_z \rangle = \frac{\hbar}{2}\left(\frac{l \pm m_j + \frac{1}{2}}{2l + 1} - \frac{l \mp m_j + \frac{1}{2}}{2l + 1}\right) = \pm \frac{m_j \hbar}{2l + 1} \qquad (11.123)$$

Hence

$$E_B^{(1)} = \frac{e\hbar B}{2m_e c} m_j \left(1 \pm \frac{1}{2l + 1}\right) \qquad (11.124)$$

Notice that we can express this energy shift compactly in the form

$$E_B^{(1)} = \frac{g(j, l)e\hbar B}{2m_e c} m_j \qquad (11.125)$$

which is reminiscent of the energy of a particle of spin $j$ in an external magnetic field with a $g$ factor

$$g(j = l \pm \tfrac{1}{2}, l) = \left(1 \pm \frac{1}{2l + 1}\right) \qquad (11.126)$$

known as the Landé $g$ factor.

Figure 11.11 shows the splitting of the $1s_{1/2}$, $2p_{1/2}$, and $2p_{3/2}$ states in an external magnetic field. Notice that the Landé $g$ factor is 2 for the $s_{1/2}$ states, $\frac{2}{3}$ for the $p_{1/2}$ states, and $\frac{4}{3}$ for the $p_{3/2}$ states. In Chapter 14 we will see how selection rules for electromagnetic transitions arise. The allowed electric-dipole transitions ($\Delta m_j = 0, \pm 1$) are indicated in the figure along with the corresponding spectrum. It is interesting to compare these results with what the spectrum would look like if the electron did not have intrinsic spin (see Problem 11.11).

## 11.9  SUMMARY

To analyze a system using time-independent perturbation theory, we express the full Hamiltonian for the system in the form

$$\hat{H} = \hat{H}_0 + \hat{H}_1 \qquad (11.127)$$

where the eigenstates and eigenvalues of the unperturbed Hamiltonian $\hat{H}_0$ are given by

$$\hat{H}_0|\varphi_n^{(0)}\rangle = E_n^{(0)}|\varphi_n^{(0)}\rangle \qquad (11.128)$$

The perturbing Hamiltonian $\hat{H}_1$ may arise from external perturbations such as those that come from applying electric fields—the Stark effect (Sections 11.2, 11.4, 11.5)—or those that come from applying magnetic fields—the Zeeman effect (Section 11.8)—to the system, or from internal perturbations such as those causing the fine structure of the hydrogen atom (Section 11.6). If the state $|\varphi_n^{(0)}\rangle$ is not degenerate, the first-order and second-order corrections to the energy are

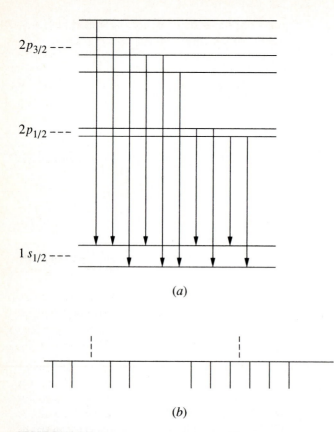

(a)

(b)

**FIGURE 11.11**
(a) The Zeeman effect for the 1s and 2p levels of hydrogen in a weak external magnetic field, showing the allowed electric dipole transitions. (b) A schematic diagram of the resulting spectrum. The dashed lines show the fine structure that is present in the absence of an external magnetic field.

given by

$$E_n^{(1)} = \langle \varphi_n^{(0)} | \hat{H}_1 | \varphi_n^{(0)} \rangle \tag{11.129}$$

and

$$E_n^{(2)} = \sum_{k \neq n} \frac{|\langle \varphi_k^{(0)} | \hat{H}_1 | \varphi_n^{(0)} \rangle|^2}{E_n^{(0)} - E_k^{(0)}} \tag{11.130}$$

When the unperturbed energy eigenstates are degenerate, formula (11.129) [as well as (11.130)] does *not* apply. Rather, the first-order corrections to the energy are the eigenvalues of the eigenvalue equation for the operator $\hat{H}_1$ using the degenerate eigenstates of $\hat{H}_0$ as a basis (see Section 11.3). Often we can take advantage of a symmetry of the perturbing Hamiltonian $\hat{H}_1$ to reduce the size of the degenerate subspace in which we need to work. In particular, if $[\hat{H}_1, \hat{A}] = 0$ (where $\hat{A}$ may be the generator of a symmetry operation for $\hat{H}_1$), only states that have both the same energy $E_n^{(0)}$ and the same eigenvalue $a$ of the operator $\hat{A}$ are mixed together by the perturbation. See Sections 11.4 and 11.6 for illustrations.

## PROBLEMS

**11.1.** Consider a perturbation $\hat{H}_1 = b\hat{x}^4$ to the simple harmonic oscillator Hamiltonian

$$\hat{H}_0 = \frac{\hat{p}_x^2}{2m} + \frac{1}{2}m\omega^2\hat{x}^2$$

This is an example of an anharmonic oscillator, one with a nonlinear restoring force.

(a) Show that the first-order shift in the energy is given by

$$E_n^{(1)} = \frac{3\hbar^2 b}{4m^2\omega^2}\left(1 + 2n + 2n^2\right)$$

(b) Argue that no matter how small $b$ is, the perturbation expansion will break down for some sufficiently large $n$. What is the physical reason?

**11.2.** Use the series expansion for the exponential in (11.41),

$$|\psi_n\rangle = \hat{T}(q|\mathbf{E}|/m\omega^2)|n\rangle = e^{-iq|\mathbf{E}|\hat{p}_x/m\omega^2\hbar}|n\rangle$$

to evaluate the first-order correction to the state of the harmonic oscillator due to an applied electric field. Compare your results with the perturbative result (11.23).

**11.3.** For the simple harmonic oscillator, for which

$$\hat{H}_0 = \frac{\hat{p}_x^2}{2m} + \frac{1}{2}m\omega^2\hat{x}^2$$

take the perturbing Hamiltonian to be

$$\hat{H}_1 = \frac{1}{2}m\omega_1^2\hat{x}^2$$

where $\omega_1 \ll \omega$. Calculate the energy shifts through second order and compare with the exact eigenvalues.

**11.4.** Calculate the first-order shift to the energy of the ground state and first-excited state of a particle of mass $m$ in the one-dimensional infinite square well

$$V(x) = \begin{cases} 0 & 0 < x < L \\ \infty & \text{elsewhere} \end{cases}$$

of (a) the constant perturbation $\hat{H}_1 = V_1$ and (b) the linearly increasing perturbation $\hat{H}_1 = \varepsilon E_1^{(0)}\hat{x}/L$, where $E_1^{(0)}$ is the unperturbed energy of the ground state and $\varepsilon \ll 1$.

**11.5.** (a) Calculate the exact energy eigenstates of the Hamiltonian (11.7) of the ammonia molecule in an external electric field.

(b) Assuming that $\mu_e|\mathbf{E}| \ll A$, use perturbation theory to determine the first-order correction to the unperturbed eigenstates $|I\rangle$ and $|II\rangle$ and compare with the results of (a).

**11.6.** The spin Hamiltonian for a spin-$\frac{1}{2}$ particle in an external magnetic field is

$$\hat{H} = -\hat{\boldsymbol{\mu}} \cdot \mathbf{B} = -\frac{gq}{2mc}\hat{\mathbf{S}} \cdot \mathbf{B}$$

Take $\mathbf{B} = B_0\mathbf{k} + B_2\mathbf{j}$, with $B_2 \ll B_0$. Determine the energy eigenvalues exactly and compare with the results of perturbation theory through second order in $B_2/B_0$.

**11.7.** Assume that the proton is a uniformly charged sphere of radius $R$.

(a) Show for the hydrogen atom that the potential energy of the electron in the field of the proton is given by

$$
V(r) = \begin{cases}
-\dfrac{3e^2}{2R^3}\left(R^2 - \dfrac{1}{3}r^2\right) & r < R \\[4mm]
-\dfrac{e^2}{r} & r > R
\end{cases}
$$

*Hint:* Use Gauss's law and remember that the potential energy $V(r)$ must be continuous.

(b) Calculate the energy shift for the $1s$ and $2p$ states of hydrogen if the potential energy in (a) is used. What effect does this shift have upon the Lyman $\alpha$ wavelength? *Suggestion:* You can use the fact that $R \ll a_0$ to simplify the integrand *before* evaluating the integrals.

**11.8.** Use the form of the $Y_{l,m}(\theta, \phi)$'s to verify that $\langle n, l', m'|\hat{z}|n, l, m\rangle = 0$ for $m \neq m'$.

**11.9.** The spin Hamiltonian of a spin-1 ion in a crystal is given by

$$
\hat{H} = \frac{a}{\hbar^2}\hat{S}_z^2 + \frac{b}{\hbar^2}(\hat{S}_x^2 - \hat{S}_y^2)
$$

Assume $b \ll a$ and treat

$$
\frac{b}{\hbar^2}(\hat{S}_x^2 - \hat{S}_y^2)
$$

as a perturbation. Calculate the unperturbed energies and the first-order corrections using perturbation theory. Beware of the degeneracy. Compare your perturbative results with the exact eigenvalues.

**11.10.** For the two-dimensional harmonic oscillator, the unperturbed Hamiltonian is given by

$$
\hat{H}_0 = \frac{\hat{p}_x^2}{2m} + \frac{1}{2}m\omega^2\hat{x}^2 + \frac{\hat{p}_y^2}{2m} + \frac{1}{2}m\omega^2\hat{y}^2
$$

Determine the first-order energy shifts to the ground state and the degenerate first-excited states due to the perturbation

$$
\hat{H}_1 = 2b\hat{x}\hat{y}
$$

**11.11.** (a) Determine how the energy levels of the hydrogen atom for the $1s$ and $2p$ states would appear in the absence of any intrinsic spin for the electron, with the only contribution to the fine structure coming form the relativistic correction to the kinetic energy.

(b) What happens to these energy levels if the atom is placed in an external magnetic field?

(c) What is the resulting spectrum?

**11.12.** Show for a general potential energy $V(r)$ that the form of the spin-orbit Hamiltonian (11.93) becomes

$$
\hat{H}_{\text{S-O}} = \frac{1}{2m_e^2 c^2|\hat{\mathbf{r}}|}\frac{d\hat{V}}{dr}\hat{\mathbf{L}}\cdot\hat{\mathbf{S}}
$$

*Suggestion:* Start with (11.92).

**11.13.** Obtain the states (11.107a) and (11.107b).

**11.14.** Determine the effect of an external magnetic field on the energy levels of the $n = 2$ states of hydrogen when the applied magnetic field $B$ has a magnitude much greater than $10^4$ gauss, in which case the spin-orbit interaction may be neglected as a first approximation. This is the Paschen-Bach effect.

The following four problems provide us with some techniques for evaluating some of the hydrogen-atom expectation values that we have used in this chapter. These "tricks" are given by R. Shankar, *Principles of Quantum Mechanics*.

**11.15.** In order to evaluate $\langle 1/r \rangle$ consider $\gamma/r$ as a perturbation for the hydrogenic atom, where we can think of $\gamma$ as some "small" constant. The first-order shift in the energy is given by

$$E_n^{(1)} = \left\langle \frac{\gamma}{r} \right\rangle_{n,l,m}$$

which is clearly linear in $\gamma$.

(a) First show that the exact eigenvalues are given by

$$E_n = -\frac{\mu(Ze^2 - \gamma)^2}{2\hbar^2 n^2}$$

*Suggestion:* Examine (10.32).

(b) Since $E_n = E_n^{(0)} + E_n^{(1)} + E_n^{(2)} + \cdots$, we can obtain $E_n^{(1)}$ either by explicitly finding the contribution to $E_n$ that is linear in $\gamma$, or, more generally, noting that

$$E_n^{(1)} = \gamma \left( \frac{dE_n}{d\gamma} \right)_{\gamma=0}$$

since $E_n^{(0)}$ is of course independent of $\gamma$ and the higher-order terms in the expansion are at least of order $\gamma^2$. In this way show that

$$\left\langle \frac{1}{r} \right\rangle_{n,l,m} = \frac{\mu Ze^2}{\hbar^2 n^2} = \frac{Z\mu c\alpha}{\hbar n^2} = \frac{Z}{a_0 n^2}$$

**11.16.** (a) Treat $\gamma \mathbf{p}^2/2\mu$ as a perturbation for the hydrogen-like atom and, using the techniques of Problem 11.15, show that

$$\left\langle \frac{\mathbf{p}^2}{2\mu} \right\rangle_{n,l,m} = \frac{\mu c^2 Z^2 \alpha^2}{2n^2}$$

(b) Use the results of (a) and 11.15 to show that for the hydrogen atom

$$\langle K \rangle = -\tfrac{1}{2} \langle V \rangle$$

in agreement with the virial theorem in quantum mechanics. See Problem 10.12.

**11.17.** In order to evaluate $\langle 1/r^2 \rangle$ take $\gamma/r^2$ as a perturbation for the hydrogen atom. Here again we can obtain the exact solution since the perturbation modifies the centrifugal potential in (10.12) as follows:

$$\frac{l(l+1)\hbar^2}{2\mu r^2} + \frac{\gamma}{r^2} = \frac{l'(l'+1)\hbar^2}{2\mu r^2}$$

Thus the exact energy is given by

$$E = -\frac{\mu c^2 Z^2 \alpha^2}{2(n_r + l' + 1)^2}$$

Show that

$$\left\langle \frac{\gamma}{r^2} \right\rangle_{n,l,m} = E^{(1)}_{n,l,m} = \gamma \left( \frac{dE}{d\gamma} \right)_{\gamma=0} = \gamma \left( \frac{dE}{dl'} \right)_{l'=l} \left( \frac{dl'}{d\gamma} \right)_{l'=l} = \frac{\gamma Z^2}{n^3 a_0^2 (l + \frac{1}{2})}$$

**11.18.** We cannot use the techniques of Problem 11.15 to evaluate $\langle 1/r^3 \rangle$, since there is no term in the Coulomb Hamiltonian that involves $1/r^3$. However, use the fact that

$$\langle n, l, m | [\hat{H}_0, \hat{p}_r] | n, l, m \rangle = 0$$

where $\hat{p}_r$ is radial momentum operator introduced in (9.92) and $\hat{H}_0$ is the unperturbed hydrogenic Hamiltonian (11.80), to show that

$$\left\langle \frac{1}{r^3} \right\rangle_{n,l,m} = \frac{Z}{a_0 l(l + 1)} \left\langle \frac{1}{r^2} \right\rangle_{n,l,m} = \frac{Z^3}{a_0^3 n^3 l(l + 1)(l + \frac{1}{2})}$$

# 12

## IDENTICAL PARTICLES

In any discussion of multielectron atoms, molecules, solids, nuclei, or elementary particles, we face systems that involve identical particles. As we will discuss in this chapter, the truly indistinguishable nature of identical particles within quantum mechanics has profound consequences for the way the physical world behaves.

## 12.1 INDISTINGUISHABLE PARTICLES IN QUANTUM MECHANICS

As far as we can tell, all electrons are identical. They all have the same mass, the same charge, and the same intrinsic spin. There are no additional properties, such as color, that allow us to distinguish one electron from another. Yet within classical mechanics, identical particles are, in principle, distinguishable. You don't have to paint one of them red and one of them green to be able to tell two identical particles apart. If at some initial time you specify the positions and the velocities $(\mathbf{r}_1, \mathbf{v}_1)$ and $(\mathbf{r}_2, \mathbf{v}_2)$ of two interacting particles, you can calculate their positions and velocities at all later times. The particles follow well-defined trajectories, so you don't need to actually observe the particles to be sure which is which when you find one of the particles at a later time. In any case, within classical theory you would, in principle, be permitted to make measurements of the particles' positions and velocities without influencing their motions so that you could actually follow the trajectories of the two particles and thus keep track of them.

   Life in the real world is different, at least on the microscopic level. As we have seen in Chapter 8, in many microscopic situations there is no well-defined trajectory that a particle follows. The particle has amplitudes to take all paths. Or in the language of wave functions, each of the particles may have an amplitude to be at a variety of overlapping positions, as indicated in Fig. 12.1, so we cannot be sure

**FIGURE 12.1**
A schematic diagram indicating the position probability distribution for two particles. Since these distributions overlap, there is no way to be sure which particle we have detected if we make a measurement of the particle's position and the two particles are identical.

which of the particles we have found if we make a subsequent measurement of the particle's position. Moreover, any attempt to keep track of the particle by measuring its position is bound to change fundamentally the particle's quantum state.

With these considerations in mind, let's see what types of states are allowed for a pair of identical particles. We specify a two-particle state by

$$|a, b\rangle = |a\rangle_1 \otimes |b\rangle_2 \tag{12.1}$$

where a single-particle state such as $|a\rangle_1$ specifies the state of particle 1 and $|b\rangle_2$ specifies the state of particle 2.

We introduce the exchange operator $\hat{P}_{12}$, which is defined by

$$\hat{P}_{12}|a, b\rangle = |b, a\rangle \tag{12.2a}$$

or

$$\hat{P}_{12}(|a\rangle_1 \otimes |b\rangle_2) = |b\rangle_1 \otimes |a\rangle_2 \tag{12.2b}$$

As an example, the effect of the exchange operator on the state $|\mathbf{r}_1, +\mathbf{z}\rangle_1 \otimes |\mathbf{r}_2, -\mathbf{z}\rangle_2$, which has particle 1 at position $\mathbf{r}_1$ with $S_z = \hbar/2$ and particle 2 at position $\mathbf{r}_2$ with $S_z = -\hbar/2$, is to produce the state $|\mathbf{r}_2, -\mathbf{z}\rangle_1 \otimes |\mathbf{r}_1, +\mathbf{z}\rangle_2$, which has particle 1 at position $\mathbf{r}_2$ with $S_z = -\hbar/2$ and particle 2 at position $\mathbf{r}_1$ with $S_z = \hbar/2$ (see Fig. 12.2). The exchange operator interchanges the particles, switching the subscript labels 1 and 2 on the states. Since for any physical state of two identical particles we cannot tell if we have exchanged the particles, the "exchanged" state must be the same physical state and therefore can differ from the initial state by at most an overall phase:

$$\hat{P}_{12}|\psi\rangle = e^{i\delta}|\psi\rangle = \lambda|\psi\rangle \tag{12.3}$$

Thus the allowed physical states are eigenstates of the exchange operator with eigenvalue $\lambda$. Applying the exchange operator twice yields the identity operator. Therefore

$$\hat{P}_{12}^2|\psi\rangle = \lambda^2|\psi\rangle = |\psi\rangle \tag{12.4}$$

which shows that $\lambda^2 = 1$, or $\lambda = \pm 1$ are the two allowed eigenvalues.[1]

---

[1] There are exceptions to this rule in two-dimensional systems. See the article "Anyons" by F. Wilczek, *Scientific American*, May 1991, p. 58.

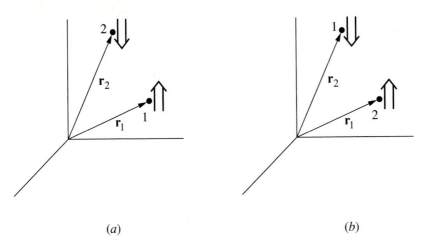

(a)                                                    (b)

**FIGURE 12.2**
The effect of the exchange operator on a state of two spin-$\frac{1}{2}$ particles as shown in (a) is to exchange
both the positions and the spins (indicated by the double arrow), as shown in (b).

Clearly, if the two identical particles are each in the same state $|a\rangle$, they are
in an eigenstate of the exchange operator with eigenvalue $\lambda = 1$:

$$\hat{P}_{12}|a, a\rangle = |a, a\rangle \tag{12.5}$$

indicating that the state is symmetric under exchange. If $b \neq a$, we can find the
linear combinations of the two states $|a, b\rangle$ and $|b, a\rangle$ that are eigenstates of the
exchange operator. The matrix representation of the exchange operator using these
states as a basis is given by

$$\hat{P}_{12} \rightarrow \begin{pmatrix} \langle a, b|\hat{P}_{12}|a, b\rangle & \langle a, b|\hat{P}_{12}|b, a\rangle \\ \langle b, a|\hat{P}_{12}|a, b\rangle & \langle b, a|\hat{P}_{12}|b, a\rangle \end{pmatrix} = \begin{pmatrix} 0 & 1 \\ 1 & 0 \end{pmatrix} \tag{12.6}$$

where we have used action of the exchange operator as given in (12.2) and
assumed that the two states $|a, b\rangle$ and $|b, a\rangle$ are normalizable and orthogonal.
Thus the condition that the eigenvalue equation (12.3) has a nontrivial solution is
given by

$$\begin{vmatrix} -\lambda & 1 \\ 1 & -\lambda \end{vmatrix} = 0 \tag{12.7}$$

which also yields $\lambda = \pm 1$ as before. Substituting the eigenvalues into the eigenvalue equation, we find that the eigenstates corresponding to these eigenvalues
are given by

$$|\psi_S\rangle = \frac{1}{\sqrt{2}}|a, b\rangle + \frac{1}{\sqrt{2}}|b, a\rangle \qquad \lambda = 1 \tag{12.8a}$$

$$|\psi_A\rangle = \frac{1}{\sqrt{2}}|a, b\rangle - \frac{1}{\sqrt{2}}|b, a\rangle \qquad \lambda = -1 \tag{12.8b}$$

where the subscripts $S$ and $A$ indicate that these two eigenstates are symmetric and antisymmetric, respectively, under the interchange of the two particles. Notice that two identical particles must be in *either* the state $|\psi_S\rangle$ *or* the state $|\psi_A\rangle$, but they cannot be in a superposition of these states, for then exchanging the two particles does not lead to a state that differs from the initial state by an overall phase:

$$\hat{P}_{12}\big(c_S|\psi_S\rangle + c_A|\psi_A\rangle\big) = c_S|\psi_S\rangle - c_A|\psi_A\rangle \tag{12.9}$$

Thus the particles must make a choice between $|\psi_S\rangle$ and $|\psi_A\rangle$. In fact, it turns out that Nature makes the choice for them in a strikingly comprehensive way:

Particles with an integral intrinsic spin, $s = 0, 1, 2, \ldots$, are found to be only in symmetric states and are called *bosons*; these particles obey Bose-Einstein statistics.[2] Examples of such particles include fundamental elementary particles such as photons, gluons, the $W_\pm$ and $Z_0$ intermediate vector bosons, and the graviton—particles that mediate the electromagnetic, strong, weak, and gravitational interactions, respectively—as well as composite particles such as pions and nuclei such as $He^4$.

Particles with half-integral intrinsic spin, $s = \frac{1}{2}, \frac{3}{2}, \frac{5}{2}, \ldots$, are found to be only in antisymmetric states and are called *fermions*; these particles obey Fermi-Dirac statistics. Examples of such particles include fundamental elementary particles such as electrons, muons, neutrinos, and quarks, as well as composite particles such as protons, neutrons, and nuclei such as $He^3$.

At the level of nonrelativistic quantum mechanics, this relationship between the intrinsic spin of the particle and the exchange symmetry of the quantum state is a law of nature—often referred to as the spin-statistics theorem—that we must accept as a given. We can take comfort in the fact that this spin-statistics theorem can be shown to be a necessary consequence of relativistic quantum field theory.[3] In Chapter 14 we consider the fully relativistic quantum field theory for photons, and we can then see why, as an example, photons must indeed be bosons.

---

[2] The symmetry requirement on the allowed quantum states of identical bosons leads to a statistical distribution function for an ensemble of $N$ identical bosons in thermal equilibrium at a temperature $T$ that is different from the classical Boltzmann distribution function. In particular, the number of bosons in a particular quantum state with energy $E$ is given by

$$n(E) = \frac{1}{e^\alpha e^{E/kT} - 1}$$

where the value of $\alpha$ is chosen so as to ensure that the total number of particles is indeed $N$. On the other hand, the antisymmetry requirement on the allowed quantum states for an ensemble of $N$ identical fermions leads to the distribution function

$$n(E) = \frac{1}{e^\alpha e^{E/kT} + 1}$$

Note: $n(E)$ can be very large for the Bose-Einstein distribution, while for the Fermi-Dirac distribution $n(E) \leq 1$. For a derivation of these quantum distribution functions, see, for example, F. Reif, *Fundamentals of Statistical and Thermal Physics*, McGraw-Hill, New York, 1965, Chapter 9.

[3] A comprehensive but advanced discussion is given by R. Streater and A. S. Wightman, *PCT, Spin and Statistics, and All That*, W. A. Benjamin, New York, 1964.

Finally, as we saw in (12.5), if two identical particles are in the same state, the state is necessarily symmetric under interchange, and therefore such a state cannot be occupied by fermions. Thus two electrons, two spin-$\frac{1}{2}$ particles, cannot occupy the same state—a statement of the Pauli exclusion principle. We will see how this principle plays a fundamental role in determining the structure of atoms and molecules.

## 12.2 THE HELIUM ATOM

As an interesting example of a system containing two identical fermions, we start with the Hamiltonian

$$\hat{H} = \frac{\hat{\mathbf{p}}_1^2}{2m_e} + \frac{\hat{\mathbf{p}}_2^2}{2m_e} - \frac{Ze^2}{|\hat{\mathbf{r}}_1|} - \frac{Ze^2}{|\hat{\mathbf{r}}_2|} + \frac{e^2}{|\hat{\mathbf{r}}_1 - \hat{\mathbf{r}}_2|} \qquad (12.10)$$

which includes the principle electrostatic interactions between the nucleus and the electrons in the helium atom when we take $Z = 2$ for the charge on the nucleus and ignore the contribution of the kinetic energy of the nucleus to the energy of the atom (see Fig. 12.3). One approach is to treat

$$\hat{H}_0 = \frac{\hat{\mathbf{p}}_1^2}{2m_e} - \frac{Ze^2}{|\hat{\mathbf{r}}_1|} + \frac{\hat{\mathbf{p}}_2^2}{2m_e} - \frac{Ze^2}{|\hat{\mathbf{r}}_2|} \qquad (12.11)$$

as the unperturbed Hamiltonian and the Coulomb energy of repulsion of the two electrons

$$\hat{H}_1 = \frac{e^2}{|\hat{\mathbf{r}}_1 - \hat{\mathbf{r}}_2|} \qquad (12.12)$$

as a perturbation. Although we do not expect the perturbation to be much smaller than the interaction of the electrons with the nucleus, nonetheless breaking the Hamiltonian into (12.11) and (12.12) is an attractive option. Since $\hat{\mathbf{r}}_1$ and $\hat{\mathbf{p}}_1$ commute with $\hat{\mathbf{r}}_2$ and $\hat{\mathbf{p}}_2$, the unperturbed Hamiltonian is just the sum of two independent Coulomb Hamiltonians. Thus we can express the eigenstates of the unperturbed Hamiltonian as simultaneous hydrogenic eigenstates $|n_1, l_1, m_1\rangle_1 \otimes |n_2, l_2, m_2\rangle_2$, which we know well. On the other hand, the full Hamiltonian (12.10) is just too complicated to solve directly, and so we must resort to approx-

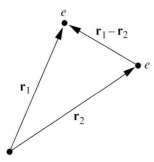

**FIGURE 12.3**
The positions $\mathbf{r}_1$ and $\mathbf{r}_2$ of the two electrons with respect to the nucleus in the helium atom.

imation methods. Moreover, this perturbative approach has much to teach us, both qualitatively and quantitatively, about the important effects of the identical nature of the electrons on the spectrum of the helium atom.

## The Ground State

Let's start with the ground state of $\hat{H}_0$, in which each of the particles is in the lowest-energy state

$$|1, 0, 0\rangle_1 \otimes |1, 0, 0\rangle_2 = |1, 0, 0\rangle_1 |1, 0, 0\rangle_2, \tag{12.13}$$

where on the right-hand side we have dispensed with the direct-product symbol, just as we did in our earlier discussion of two-particle spin states in Chapter 5. Although this state is clearly symmetric under interchange of the two particles, we have not yet specified the spin states of the two electrons. There is only one way to do this and make the *total* state of the two particles antisymmetric; namely, the spin state of the two particles must be

$$\frac{1}{\sqrt{2}} \left( |+\mathbf{z}\rangle_1 |-\mathbf{z}\rangle_2 - |-\mathbf{z}\rangle_1 |+\mathbf{z}\rangle_2 \right) \tag{12.14}$$

which is antisymmetric under exchange of the spins of the two particles. Thus the ground state of the two-electron system is given by

$$|1s, 1s\rangle = |1, 0, 0\rangle_1 |1, 0, 0\rangle_2 \frac{1}{\sqrt{2}} \left( |+\mathbf{z}\rangle_1 |-\mathbf{z}\rangle_2 - |-\mathbf{z}\rangle_1 |+\mathbf{z}\rangle_2 \right) \tag{12.15}$$

where the label $1s, 1s$ on the overall ket indicates that each of the electrons is in the $n = 1$, $l = 0$ state of the hydrogenic Hamiltonian. Recall from (5.31) that the spin state (12.14) is just the state

$$|0, 0\rangle = \frac{1}{\sqrt{2}} \left( |+\mathbf{z}\rangle_1 |-\mathbf{z}\rangle_2 - |-\mathbf{z}\rangle_1 |+\mathbf{z}\rangle_2 \right) \tag{12.16}$$

where

$$(\hat{\mathbf{S}}_1 + \hat{\mathbf{S}}_2)^2 |0, 0\rangle = 0 \tag{12.17a}$$

$$(\hat{S}_{1z} + \hat{S}_{2z}) |0, 0\rangle = 0 \tag{12.17b}$$

The ground state of the two-electron system must be a total-spin-0 state, even though the Hamiltonian (12.10) itself does not depend on spin. The spectroscopic notation[4] for this state is $^1S_0$. From (10.34), the energy of the unperturbed ground

---

[4] It is conventional to label the *total* spin $S$, the *total* orbital angular momentum $L$, and the *total* angular momentum $J$ of atomic states in the form $^{2S+1}L_J$, where the orbital angular momentum label is given in the usual spectroscopic notation: $L = 0$ is $S$, $L = 1$ is $P$, $L = 2$ is $D$, and so on. Note the $2S + 1$ superscript gives the spin multiplicity for each state: $S = 0$ corresponds to a single total-spin state, while for $S = 1$ there is the usual triplet of spin-1 states. For atoms with more than two electrons, the values of total spin will, of course, differ from these values.

state is

$$E^{(0)}_{1s,1s} = 2\left(-\tfrac{1}{2}m_e c^2 Z^2 \alpha^2\right) = -8(13.6 \text{ eV}) = -108.8 \text{ eV} \qquad (12.18)$$

where we have put $Z = 2$ to determine a numerical value.

The first-order shift in the ground-state energy is given by

$$E^{(1)}_{1s,1s} = \langle 1s, 1s| \frac{e^2}{|\hat{\mathbf{r}}_1 - \hat{\mathbf{r}}_2|} |1s, 1s\rangle \qquad (12.19)$$

Since the perturbing Hamiltonian depends on the position operators $\hat{\mathbf{r}}_1$ and $\hat{\mathbf{r}}_2$, it is natural to evaluate (12.19) in position space. For the ground state, the spin-independent part of the state is symmetric under exchange of the two particles, and we need to ask how we express a symmetric spatial state $|\psi_S\rangle$ of two identical particles in position space. We first write

$$|\psi_S\rangle = \frac{1}{2} \int d^3 r_1\, d^3 r_2 \left(\frac{1}{\sqrt{2}} |\mathbf{r}_1, \mathbf{r}_2\rangle + \frac{1}{\sqrt{2}} |\mathbf{r}_2, \mathbf{r}_1\rangle\right)$$

$$\times \left(\frac{1}{\sqrt{2}} \langle \mathbf{r}_1, \mathbf{r}_2|\psi_S\rangle + \frac{1}{\sqrt{2}} \langle \mathbf{r}_2, \mathbf{r}_1|\psi_S\rangle\right) \qquad (12.20)$$

where we have been careful to use the appropriate two-particle position states

$$\frac{1}{\sqrt{2}}|\mathbf{r}_1, \mathbf{r}_2\rangle + \frac{1}{\sqrt{2}}|\mathbf{r}_2, \mathbf{r}_1\rangle = \frac{1}{\sqrt{2}}|\mathbf{r}_1\rangle_1 |\mathbf{r}_2\rangle_2 + \frac{1}{\sqrt{2}}|\mathbf{r}_2\rangle_1 |\mathbf{r}_1\rangle_2 \qquad (12.21)$$

for identical particles that are symmetric under interchange of the two identical particles. The factor of $\frac{1}{2}$ in front of the right-hand side of (12.20) is necessary because when we integrate over all values of $\mathbf{r}_1$ *and* $\mathbf{r}_2$, we count each two-particle position state twice. However, for a symmetric state

$$\langle \mathbf{r}_1, \mathbf{r}_2|\psi_S\rangle = \langle \mathbf{r}_2, \mathbf{r}_1|\psi_S\rangle \qquad (12.22)$$

and thus

$$|\psi_S\rangle = \frac{1}{2} \int d^3 r_1\, d^3 r_2 \left(|\mathbf{r}_1, \mathbf{r}_2\rangle\langle \mathbf{r}_1, \mathbf{r}_2|\psi_S\rangle + |\mathbf{r}_2, \mathbf{r}_1\rangle\langle \mathbf{r}_1, \mathbf{r}_2|\psi_S\rangle\right)$$

$$= \int d^3 r_1\, d^3 r_2 \, |\mathbf{r}_1, \mathbf{r}_2\rangle\langle \mathbf{r}_1, \mathbf{r}_2|\psi_S\rangle \qquad (12.23)$$

where in the last step we have interchanged the dummy variables $\mathbf{r}_1 \leftrightarrow \mathbf{r}_2$ in the second term and taken advantage of (12.22). Notice that the result (12.23) is exactly what we would have obtained had we just inserted two-particle position states $|\mathbf{r}_1, \mathbf{r}_2\rangle$ for nonidentical particles. You can verify that a similar result holds for an antisymmetric spatial state $|\psi_A\rangle$. See Problem 12.1.

Using (12.23), we now see that (12.19) becomes

$$E^{(1)}_{1s,1s} = \iint d^3 r_1\, d^3 r_2 |\langle \mathbf{r}_1|1, 0, 0\rangle|^2 |\langle \mathbf{r}_2|1, 0, 0\rangle|^2 \frac{e^2}{|\mathbf{r}_1 - \mathbf{r}_2|} \qquad (12.24)$$

Since $|\langle \mathbf{r}_1|1, 0, 0\rangle|^2$ is the probability density of finding an electron at $\mathbf{r}_1$,

$$e|\langle \mathbf{r}_1|1, 0, 0\rangle|^2 = \rho(\mathbf{r}_1) \tag{12.25}$$

has the form of the charge density due to this electron. The electric potential at $\mathbf{r}_2$ produced by this charge density is

$$\int d^3 r_1 \frac{\rho(\mathbf{r}_1)}{|\mathbf{r}_1 - \mathbf{r}_2|} \tag{12.26}$$

Therefore the energy of interaction of this charge density with the charge density

$$e|\langle \mathbf{r}_2|1, 0, 0\rangle|^2 = \rho(\mathbf{r}_2) \tag{12.27}$$

of the other electron is the usual result from electrostatics

$$E_{1s,1s}^{(1)} = \int\!\!\int d^3 r_1\, d^3 r_2 \frac{\rho(\mathbf{r}_1)\rho(\mathbf{r}_2)}{|\mathbf{r}_1 - \mathbf{r}_2|} \tag{12.28}$$

which gives us a nice, physical way to interpret (12.24).

Using the wave function $\langle \mathbf{r}|1, 0, 0\rangle = R_{1,0}Y_{0,0} = (1/\sqrt{\pi})(Z/a_0)^{3/2}e^{-Zr/a_0}$, with $R_{1,0}$ from (10.43) and $Y_{0,0} = 1/\sqrt{4\pi}$, the energy shift (12.24) becomes

$$E_{1s,1s}^{(1)} = \left[\frac{1}{\pi}\left(\frac{Z}{a_0}\right)^3\right]^2 e^2 \int_0^\infty r_1^2\, dr_1\, e^{-2Zr_1/a_0} \int_0^\infty r_2^2\, dr_2\, e^{-2Zr_2/a_0}$$
$$\times \int d\Omega_1 \int d\Omega_2 \frac{1}{|\mathbf{r}_1 - \mathbf{r}_2|} \tag{12.29}$$

Evaluating the integrals in (12.29) is relatively straightforward, since there is no angular dependence in the ground-state wave function. In particular, we have the freedom to choose the $z$ axis of the dummy variable $\mathbf{r}_1$ to point in the direction of $\mathbf{r}_2$ so that $\mathbf{r}_1 \cdot \mathbf{r}_2 = r_1 r_2 \cos\theta_2$, giving us an exact differential for the $\theta_2$ integral. Then

$$\int d\Omega_2 \frac{1}{|\mathbf{r}_1 - \mathbf{r}_2|} = \int_0^{2\pi} d\phi_2 \int_0^\pi \sin\theta_2\, d\theta_2 \frac{1}{(r_1^2 + r_2^2 - 2r_1 r_2 \cos\theta_2)^{1/2}}$$
$$= -2\pi \frac{1}{2r_1 r_2}\left[(r_1^2 + r_2^2 - 2r_1 r_2 \cos\theta_2)^{1/2}\right]_{\theta_2=0}^{\theta_2=\pi}$$
$$= \frac{\pi}{r_1 r_2}(r_1 + r_2 - |r_1 - r_2|) \tag{12.30}$$

Since there is no angular dependence left in the integral (12.29), we can use $\int d\Omega_1 = 4\pi$ to do the remaining angular integrals. Finally, doing the radial integrals, we obtain

$$E_{1s,1s}^{(1)} = \left[\left(\frac{Z}{a_0}\right)^3\right]^2 4e^2 \int_0^\infty dr_1\, r_1 e^{-2Zr_1/a_0} \int_0^\infty dr_2\, r_2 e^{-2Zr_2/a_0}(r_1 + r_2 - |r_1 - r_2|)$$

$$= \left[\left(\frac{Z}{a_0}\right)^3\right]^2 4e^2 \int_0^\infty dr_1\, r_1 e^{-2Zr_1/a_0}$$

$$\times \left(2 \int_0^{r_1} dr_2\, r_2^2 e^{-2Zr_2/a_0} + 2r_1 \int_{r_1}^\infty dr_2\, r_2\, e^{-2Zr_2/a_0}\right)$$

$$= \tfrac{5}{8} Z m_e c^2 \alpha^2 = 34.0 \text{ eV} \tag{12.31}$$

where we have replaced the Bohr radius with $a_0 = \hbar/m_e c\alpha$, ignoring reduced-mass effects, and put $Z = 2$ to obtain a numerical result.

Adding the first-order shift in the energy to the unperturbed value, we find that the ground-state energy of the helium atom is given by

$$E_{1s,1s} \cong E_{1s,1s}^{(0)} + E_{1s,1s}^{(1)} = -108.8 \text{ eV} + 34.0 \text{ eV} = -74.8 \text{ eV} \tag{12.32}$$

This is to be compared with the experimental value $E_{\exp} = -79.0$ eV. Thus there is a sizeable discrepancy between our perturbative calculation and experiment. This is actually not surprising, since we have no reason to expect that the Coulomb repulsion between the two electrons in the helium atom, which we treated as a perturbation, should be much less than the energy of attraction of the electrons to the nucleus. In fact, considering the size of the first-order shift (12.31), we haven't done badly at all in this primitive attempt at a perturbative solution. After a discussion of the excited states, we will examine an alternative method of determining the energies with higher accuracy.

## The Excited States

Let's turn our attention to the first-excited states, in which one of the particles is in the state $|1,0,0\rangle$ while the other is in one of the four states $|2, l, m\rangle$, which are all degenerate eigenstates of the single-particle hydrogenic Hamiltonian. Taking the spin states of the two electrons into account, we can construct a number of two-electron states of the form (12.8b) that are antisymmetric under exchange of the two particles:

$$\frac{1}{\sqrt{2}}\big(|1,0,0,+\mathbf{z}\rangle_1|2,l,m,+\mathbf{z}\rangle_2 - |2,l,m,+\mathbf{z}\rangle_1|1,0,0,+\mathbf{z}\rangle_2\big)$$

$$= \frac{1}{\sqrt{2}}\big(|1,0,0\rangle_1|2,l,m\rangle_2 - |2,l,m\rangle_1|1,0,0\rangle_2\big)|+\mathbf{z}\rangle_1|+\mathbf{z}\rangle_2 \tag{12.33a}$$

$$\frac{1}{\sqrt{2}}\big(|1,0,0,-\mathbf{z}\rangle_1|2,l,m,-\mathbf{z}\rangle_2 - |2,l,m,-\mathbf{z}\rangle_1|1,0,0,-\mathbf{z}\rangle_2\big)$$

$$= \frac{1}{\sqrt{2}}\big(|1,0,0\rangle_1|2,l,m\rangle_2 - |2,l,m\rangle_1|1,0,0\rangle_2\big)|-\mathbf{z}\rangle_1|-\mathbf{z}\rangle_2 \tag{12.33b}$$

$$\frac{1}{\sqrt{2}}\big(|1,0,0,+\mathbf{z}\rangle_1|2,l,m,-\mathbf{z}\rangle_2 - |2,l,m,-\mathbf{z}\rangle_1|1,0,0,+\mathbf{z}\rangle_2\big) \quad (12.33c)$$

$$\frac{1}{\sqrt{2}}\big(|1,0,0,-\mathbf{z}\rangle_1|2,l,m,+\mathbf{z}\rangle_2 - |2,l,m,+\mathbf{z}\rangle_1|1,0,0,-\mathbf{z}\rangle_2\big) \quad (12.33d)$$

The unperturbed energy of each of these states is given by

$$E^{(0)}_{1s,2s\text{ or }2p} = -\frac{1}{2}m_ec^2Z^2\alpha^2\left(1 + \frac{1}{2^2}\right) = -68.0 \text{ eV} \quad (12.34)$$

where the subscript label shows that one of the electrons has $n = 1$, $l = 0$, while the other has $n = 2$ and either $l = 0$ or $l = 1$. As before, for helium we have set $Z = 2$ to obtain a numerical value.

Unlike (12.33a) and (12.33b), which have total-spin states $|1,1\rangle$ and $|1,-1\rangle$, respectively, the states (12.33c) and (12.33d) are not eigenstates of total spin. However, we have not yet taken into account the effect of the perturbing Hamiltonian (12.12). Since the states (12.33) are all degenerate, we must find the proper linear combinations that diagonalize $\hat{H}_1$. First, note that since the two electrons are identical, $\hat{H}_1$ as well as $\hat{H}_0$ must commute with the exchange operator. Although we are required to form total eigenstates of the two electrons that are antisymmetric with respect to exchange, we have not taken advantage of the full exchange symmetry of the Hamiltonian. In particular, we can express the exchange operator

$$\hat{P}_{12} = \hat{P}^{\text{space}}_{12}\hat{P}^{\text{spin}}_{12} \quad (12.35)$$

where the operator $\hat{P}^{\text{spin}}_{12}$ exchanges the spin states of the two electrons and $\hat{P}^{\text{space}}_{12}$ exchanges the spatial states. Since

$$\big[\hat{H}_1, \hat{P}^{\text{space}}_{12}\big] = \big[\hat{H}_1, \hat{P}^{\text{spin}}_{12}\big] = 0 \quad (12.36)$$

we can diagonalize the interacting Hamiltonian $\hat{H}_1$ by choosing states that are eigenstates of *both* $\hat{P}^{\text{spin}}_{12}$ and $\hat{P}^{\text{space}}_{12}$, provided we are careful to choose states that are overall antisymmetric under complete exchange.

The states (12.33a) and (12.33b) already satisfy this requirement. They are both symmetric under exchange of the spins of the electrons and antisymmetric under exchange of the spatial states. From the states (12.33c) and (12.33d) we can choose two combinations that are completely antisymmetric under exchange. One combination has a symmetric spin state and an antisymmetric spatial state

$$\frac{1}{\sqrt{2}}\big(|1,0,0\rangle_1|2,l,m\rangle_2 - |2,l,m\rangle_1|1,0,0\rangle_2\big)\frac{1}{\sqrt{2}}\big(|+\mathbf{z}\rangle_1|-\mathbf{z}\rangle_2 + |-\mathbf{z}\rangle_1|+\mathbf{z}\rangle_2\big)$$

$$(12.37)$$

where the spin state is just the total-spin-1 state $|1,0\rangle$, and the other has an antisymmetric spin state and a symmetric spatial state

$$\frac{1}{\sqrt{2}}\big(|1,0,0\rangle_1|2,l,m\rangle_2 + |2,l,m\rangle_1|1,0,0\rangle_2\big)\frac{1}{\sqrt{2}}\big(|+\mathbf{z}\rangle_1|-\mathbf{z}\rangle_2 - |-\mathbf{z}\rangle_1|+\mathbf{z}\rangle_2\big)$$

$$(12.38)$$

where the spin state is the total-spin-0 state $|0,0\rangle$. We can therefore condense our notation and express the excited states in the form

$$\frac{1}{\sqrt{2}}\big(|1,0,0\rangle_1|2,l,m\rangle_2 - |2,l,m\rangle_1|1,0,0\rangle_2\big)|1,m_s\rangle \qquad (12.39)$$

with $m_s$ taking in the values 1, 0, and $-1$, and

$$\frac{1}{\sqrt{2}}\big(|1,0,0\rangle_1|2,l,m\rangle_2 + |2,l,m\rangle_1|1,0,0\rangle_2\big)|0,0\rangle \qquad (12.40)$$

We also might have been led to select these particular combinations of states by noting that

$$\big[\hat{H}, \hat{\mathbf{S}}_1 + \hat{\mathbf{S}}_2\big] = 0 \qquad (12.41)$$

and thus we can find eigenstates of the Hamiltonian that are eigenstates of total spin.

What is the effect of the perturbing Hamiltonian on the energy of these states? At first glance, the problem of evaluating the first-order shifts still seems to be a large one, since there are four different degenerate spatial states ($l = 0$, $m = 0$ and $l = 1$, $m = \pm 1$ and 0) for each total-spin state. However, there is an additional symmetry of $\hat{H}_1$ that we have not yet utilized. Notice that if we *rotate* the positions $\mathbf{r}_1$ and $\mathbf{r}_2$ of *both* particles in (12.12), $\hat{H}_1$ doesn't change. The generator of position rotations for the two-electron system is just the *total* orbital angular momentum operator

$$\hat{\mathbf{L}} = \hat{\mathbf{r}}_1 \times \hat{\mathbf{p}}_1 + \hat{\mathbf{r}}_2 \times \hat{\mathbf{p}}_2 \qquad (12.42)$$

Thus the perturbing Hamiltonian commutes with the total orbital angular momentum. Since one of the particles in the states (12.39) and (12.40) is in a state with zero orbital angular momentum, these states are already eigenstates with a total orbital angular momentum $l$ and $z$-component $m$. Therefore we can calculate the first-order shift simply as

$$E^{(1)} = \frac{1}{2}\big(_2\langle 1,0,0|_1\langle 2,l,m| \pm {}_2\langle 2,l,m|_1\langle 1,0,0|\big)\frac{e^2}{|\hat{\mathbf{r}}_1 - \hat{\mathbf{r}}_2|}$$

$$\times \big(|1,0,0\rangle_1|2,l,m\rangle_2 \pm |2,l,m\rangle_1|1,0,0\rangle_2\big) \qquad (12.43)$$

Evaluating this expectation value in position space, we find

$$E^{(1)} = \frac{1}{2}\iint d^3r_1\,d^3r_2\,\big(\langle 1,0,0|\mathbf{r}_1\rangle\langle 2,l,m|\mathbf{r}_2\rangle \pm \langle 1,0,0|\mathbf{r}_2\rangle\langle 2,l,m|\mathbf{r}_1\rangle\big)\frac{e^2}{|\mathbf{r}_1 - \mathbf{r}_2|}$$

$$\times \big(\langle \mathbf{r}_1|1,0,0\rangle\langle \mathbf{r}_2|2,l,m\rangle \pm \langle \mathbf{r}_1|2,l,m\rangle\langle \mathbf{r}_2|1,0,0\rangle\big)$$

$$= \iint d^3r_1 \, d^3r_2 \, |\langle 1, 0, 0|\mathbf{r}_1\rangle|^2 |\langle 2, l, m|\mathbf{r}_2\rangle|^2 \frac{e^2}{|\mathbf{r}_1 - \mathbf{r}_2|}$$

$$\pm \iint d^3r_1 \, d^3r_2 \, \langle 1, 0, 0|\mathbf{r}_1\rangle\langle 2, l, m|\mathbf{r}_2\rangle \frac{e^2}{|\mathbf{r}_1 - \mathbf{r}_2|} \langle \mathbf{r}_1|2, l, m\rangle\langle \mathbf{r}_2|1, 0, 0\rangle$$

$$(12.44)$$

where in the last step we have made use of the symmetry of perturbation under $\mathbf{r}_1 \leftrightarrow \mathbf{r}_2$. Thus we can express the first-order shift in the energy of the first-excited states in the form

$$E^{(1)} = J \pm K \qquad (12.45)$$

Notice that the $+$ and $-$ signs in (12.44) are correlated with the value of the total spin so that the two-particle state is antisymmetric under exchange. The $J$ term, which is manifestly positive, is similar to the expression (12.24) that we obtained when we calculated the shift of the ground-state energy. As in (12.28), we can describe this term as the charge density of one of the electrons interacting with the charge density of the other.

The $K$ term, however, is pure quantum mechanics. There is no classical interpretation that we can assign to this term. It arises, as we have seen, because of the identical nature of the particles. We can argue that $K$ must be positive. Note that if we put $\mathbf{r}_1 = \mathbf{r}_2$ in the wave functions in the first line of (12.44), we find that the antisymmetric wave functions vanish, while the symmetric wave functions add together constructively. Thus the electrons in the antisymmetric spatial state tend to avoid each other in space, which should lower their energy due to Coulomb repulsion relative to the electrons in the symmetric spatial state, in which the electrons prefer to be close together. Thus the existence of such an exchange term produces an energy shift of the total-spin states: the energy of the triplet of spin-1 states is shifted by $J - K$, while the singlet spin-0 state is shifted in energy by $J + K$. Provided $K$ is positive, the energy of the spin-1 states will be lower than the spin-0 state, as we have argued physically should be true.

If we evaluate the integrals in (12.44), we find[5]

$$E^{(1)}_{1s,2s} = J_{1s,2s} \pm K_{1s,2s} = 11.4 \text{ eV} \pm 1.2 \text{ eV} \qquad (12.46a)$$

and

$$E^{(1)}_{1s,2p} = J_{1s,2p} \pm K_{1s,2p} = 13.2 \text{ eV} \pm 0.9 \text{ eV} \qquad (12.46b)$$

Adding these first-order corrections to (12.34), we obtain $-56.6$ eV $\pm$ 1.2 eV for the $1s, 2s$ states and $-54.8$ eV $\pm$ 0.9 eV for the $1s, 2p$ states. The observed values, shown in Figure 12.4, are $-58.8$ eV $\pm$ 0.4 eV for the $1s, 2s$ states and $-57.9$ eV $\pm$ 0.1 eV for the $1s, 2p$ states. Thus there is almost a 1-eV energy difference between the $^3S_1$ and $^1S_0$ states. Not surprisingly, as for the ground

---

[5] See J. L. Powell and B. Crasemann, *Quantum Mechanics*, Addison-Wesley, Reading, Mass., 1961, p. 457.

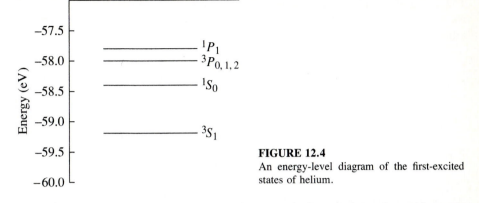

**FIGURE 12.4**
An energy-level diagram of the first-excited states of helium.

state, the agreement between our first-order perturbative results and experiment is not excellent. Nonetheless, our results show two striking features: not only have spin-dependent energy splittings been generated from a Hamiltonian that did not involve the spins of the particles at all, but the magnitude of this triplet-singlet splitting is much larger than is generated from the spin-spin interactions of the magnetic moments of the two electrons (see Problem 12.3). A similar mechanism is presumably responsible for the large spin-spin interaction that aligns the spins in a ferromagnet. However, it is much more difficult to calculate these effects for ferromagnetic materials.

## The Variational Method

In practice, detailed calculations of the energy levels of helium are carried out using the variational method. This is a simple but powerful technique that can be used in a variety of problems. We start with the expectation value of the energy in an arbitrary state $|\psi\rangle$:

$$\langle E \rangle = \langle \psi | \hat{H} | \psi \rangle \tag{12.47}$$

where we have assumed that $\langle \psi | \psi \rangle = 1$. Although in practice we are not able to determine the exact eigenvalues $E_n$ and corresponding eigenstates $|E_n\rangle$, in principle we can express $|\psi\rangle$ as the superposition

$$|\psi\rangle = \sum_n c_n |E_n\rangle \tag{12.48}$$

Then

$$\langle E \rangle = \sum_n |c_n|^2 E_n \geq \sum_n |c_n|^2 E_0 = E_0 \tag{12.49}$$

where we have assumed that $E_n \geq E_0$, with $E_0$ the exact ground-state energy. Thus for any state $|\psi\rangle$

$$E_0 \leq \langle \psi | \hat{H} | \psi \rangle \tag{12.50}$$

The key to the variational method is to choose a trial state $|\psi(\alpha_1, \alpha_2, \alpha_3, \ldots)\rangle$, which depends on parameters $\alpha_1, \alpha_2, \alpha_3, \ldots$, and then vary the parameters to minimize $\langle E \rangle$ for this state. In this way we can zero in on the ground-state energy.

We now use the variational method to determine the ground-state energy of helium. In choosing our trial state, we must keep in mind two goals: we want to pick a state that is not too far away from the exact state and one for which we can actually evaluate $\langle E \rangle$. For helium, a good starting choice is

$$|\psi\rangle = |1, 0, 0(\tilde{Z})\rangle_1 |1, 0, 0(\tilde{Z})\rangle_2 \qquad (12.51)$$

where the state $|1, 0, 0(\tilde{Z})\rangle$ is the single-particle ground state of a hydrogenic atom with charge $\tilde{Z}$, which we will take as the variational parameter. In position space

$$\langle \mathbf{r} | 1, 0, 0(\tilde{Z}) \rangle = \frac{1}{\sqrt{\pi}} \left( \frac{\tilde{Z}}{a_0} \right)^{3/2} e^{-\tilde{Z} r / a_0} \qquad (12.52)$$

In evaluating $\langle E \rangle = \langle \psi | \hat{H} | \psi \rangle$, it is convenient to group the terms in the Hamiltonian as follows:

$$\begin{aligned}
\hat{H} &= \frac{\hat{\mathbf{p}}_1^2}{2m_e} + \frac{\hat{\mathbf{p}}_2^2}{2m_e} - \frac{Ze^2}{|\hat{\mathbf{r}}_1|} - \frac{Ze^2}{|\hat{\mathbf{r}}_2|} + \frac{e^2}{|\hat{\mathbf{r}}_1 - \hat{\mathbf{r}}_2|} \\
&= \left[ \frac{\hat{\mathbf{p}}_1^2}{2m_e} - \frac{\tilde{Z}e^2}{|\hat{\mathbf{r}}_1|} + \frac{\hat{\mathbf{p}}_2^2}{2m_e} - \frac{\tilde{Z}e^2}{|\hat{\mathbf{r}}_2|} \right] + \left[ \frac{(\tilde{Z} - Z)e^2}{|\hat{\mathbf{r}}_1|} + \frac{(\tilde{Z} - Z)e^2}{|\hat{\mathbf{r}}_2|} \right] + \frac{e^2}{|\hat{\mathbf{r}}_1 - \hat{\mathbf{r}}_2|}
\end{aligned}$$

$$(12.53)$$

where the term in the first brackets is just the sum of two hydrogenic Hamiltonians with charge $\tilde{Z}$, whose expectation value is straightforward to determine for the state (12.52).[6] Then

$$\begin{aligned}
\langle E \rangle &= \tfrac{1}{2} m_e c^2 \alpha^2 (-2\tilde{Z}^2 + 4\tilde{Z}(\tilde{Z} - Z) + \tfrac{5}{4}\tilde{Z}) \\
&= \tfrac{1}{2} m_e c^2 \alpha^2 (2\tilde{Z}^2 - 4\tilde{Z}Z + \tfrac{5}{4}\tilde{Z})
\end{aligned} \qquad (12.54)$$

Setting

$$\frac{\partial \langle E \rangle}{\partial \tilde{Z}} = 0 \qquad (12.55)$$

we find the value of $\tilde{Z}$ that minimizes the energy:

$$\tilde{Z} = Z - \tfrac{5}{16} \qquad (12.56)$$

Substituting this result into $\langle E \rangle$, we obtain

$$E_0 \leq -\tfrac{1}{2} m_e c^2 \alpha^2 \left[ 2\left(Z - \tfrac{5}{16}\right)^2 \right] = -77.4 \text{ eV} \qquad (12.57)$$

---

[6] Note: The Hamiltonian itself does *not* depend on $\tilde{Z}$. We have added *and* subtracted terms such as $\tilde{Z}e^2/|\hat{\mathbf{r}}_1|$ to make the evaluation of $\langle \psi | \hat{H} | \psi \rangle$ as easy as possible. For the variational method to work, the variational parameter *cannot* be a parameter appearing in $\hat{H}$.

which is much closer to the observed value $E_{\text{exp}} = -79.0\,\text{eV}$ than our earlier first-order perturbative result (12.32). Notice that (12.56) has the simple interpretation that each electron is partially screened from the nucleus by the presence of the other electron in the atom. As before, we have put $Z = 2$ in (12.57) to obtain the numerical result.

The variational method for the ground state of helium can be improved with more complicated trial wave functions. In fact, Pekeris has used a wave function involving 1075 parameters to obtain numerically an estimate for the ground-state energy that agrees with the experimental results to within the experimental errors.[7] In general, the reason for working so hard to obtain such good agreement is that then one can use the wave function that has been deduced to calculate other quantities, such as the lifetimes of excited states.

We can also estimate the energies of these excited states and obtain the corresponding wave functions using the variational method. Note in (12.48) that if we choose a trial state $|\psi\rangle$ that does not have any of the true ground state in it, that is, if it is orthogonal to the ground state:

$$\langle E_0|\psi\rangle = 0 \tag{12.58}$$

then the expectation value of the energy in this trial state satisfies $\langle E\rangle \geq E_1$, and we can obtain an estimate of the energy $E_1$ by minimizing this expectation value. Sometimes it is easy to satisfy the condition (12.58). For example, if the excited state has nonzero orbital angular momentum, choosing a trial wave function involving the appropriate spherical harmonic for the angular dependence automatically guarantees that this state is orthogonal to the ground state. Or, if necessary, we can determine an excited-energy trial state that is orthogonal to the trial state $|\psi\rangle$ that we obtained from estimating the ground-state energy. We choose a new trial state $|\phi\rangle$ and then explicitly construct a state that is orthogonal to the trial ground state $|\psi\rangle$:

$$|\phi\rangle \rightarrow \frac{|\phi\rangle - |\psi\rangle\langle\psi|\phi\rangle}{\sqrt{1 - |\langle\psi|\phi\rangle|^2}} \tag{12.59}$$

However, since we are using trial state $|\psi\rangle$ rather than the true ground state in this superposition, we should not expect minimizing the expectation value of the energy in this state to yield an upper bound for $E_1$.

## 12.3  MULTIELECTRON ATOMS AND THE PERIODIC TABLE

Let's turn our attention to multielectron atoms. Even if we neglect the contribution of the kinetic energy of the nucleus (with charge $Ze$), then, as for helium, the energy eigenvalue equation for the Hamiltonian

$$\hat{H} = \sum_{i=1}^{Z} \frac{\hat{\mathbf{p}}_i^2}{2m_e} - \sum_{i=1}^{Z} \frac{Ze^2}{|\hat{\mathbf{r}}_i|} + \sum_{i<j} \frac{e^2}{|\hat{\mathbf{r}}_i - \hat{\mathbf{r}}_j|} \tag{12.60}$$

---

[7] C. L. Pekeris, *Phys. Rev.* **115**, 1216 (1959).

is too complicated to solve exactly. Of course, if we could neglect the contribution of the Coulomb repulsion between the individual electrons, the solution would be straightforward. The Hamiltonian would then just be the sum of $Z$ individual Coulomb Hamiltonians (with charge $Ze$ on the nucleus for each), and the allowed eigenstates could be formed from a direct product of these individual Coulomb eigenstates, provided we require that the total state, including spin, is antisymmetric under exchange of each pair of electrons. However, there is no reason to expect that we can treat the Coulomb repulsion of the electrons as a perturbation, as we attempted for helium. In particular, the typical distance separating the electrons in the atom should be of the same size as their distance from the nucleus, and although the size of the mutual interaction between each pair of electrons should be smaller than the interaction of the electrons with the nucleus because of the factor of $Z$ in the nuclear charge, there are $Z(Z - 1)/2$ different pairings of the $Z$ electrons to take into account. Thus, even for modest $Z$ we should expect the mutual interaction term of all the electrons in the atom to be comparable in size to the interaction of these electrons with the nucleus.

We clearly need an alternative way of dealing with the mutual interaction of the electrons. One approach, first used by Hartree, is to treat each of the electrons as moving *independently* in a spherically symmetric potential energy $V(r)$ due to the nucleus and the other $Z - 1$ electrons. This potential energy should have the form

$$V(r) = \begin{cases} -\dfrac{Ze^2}{r} & r \to 0 \\[2mm] -\dfrac{e^2}{r} & r \to \infty \end{cases} \qquad (12.61)$$

because for small $r$ the electron sees only the nucleus while for large $r$ the nucleus is shielded by the other $Z - 1$ electrons. How do we determine the form for this potential energy, since it depends on the charge distribution of the electrons, which is itself determined by solving the Schrödinger equation? One approach is to guess a reasonable form for $V(r)$ and then solve the Schrödinger equation (numerically) to determine the wave functions. As in (12.25), we can use these wave functions to determine a charge distribution, which can then be used to determine the potential energy that each electron experiences. We can continue with this procedure until we obtain a self-consistent solution—namely, the potential energy that we determine from charge distribution of the electrons is the same, to a given accuracy, as the potential energy that we used to determine the wave functions that yielded this charge distribution.

We can label the energy eigenstates by the three quantum numbers $n$, $l$, and $m$, just as we did for the hydrogen atom. However, since the potential energy does not have a simple $1/r$ dependence, states with a particular $n$ and different $l$ do not have the same energy. In particular, states with lower $l$ should have lower energy, since the lowering of the centrifugal barrier with lower $l$ permits these wave functions to penetrate more deeply inside the charge distribution formed by the other electrons and thus "see," at least at smaller $r$, the full attractive potential energy of the nucleus with its charge $Ze$. Thus the "accidental" degeneracy of the hydrogen atom disappears. The independence of the energy on $m$ persists, however, since the potential energy is taken to be spherically symmetric.

Let's discuss the ground-state electronic configuration of the elements, especially those of low $Z$. Start with the first row of the periodic table, shown in Fig. 12.5. Hydrogen has $n = 1$ and $l = 0$ for the ground state. We call this a $1s$ electron configuration. For helium, as we saw in Section 12.2, we can put the two electrons both in the $1s$ state, which we now denote by $1s^2$ as a shorthand notation for $1s1s$. Of course, the spin state must be a total-spin-singlet state in order to make the total state antisymmetric under exchange. These two electrons fill the $n = 1$ shell. It takes more than 19 eV to excite one of the electrons to the $n = 2$ level and 24.6 eV to ionize the atom by removing one of the electrons. Helium is exceptionally stable and, correspondingly, not chemically active.[8]

The next element in the periodic table, lithium, has three electrons, but we cannot add this third electron to the $1s$ level, for then two of the electrons would be in the same state, since there are only two possible spin states, spin up and spin down, for each of the electrons. Thus the total state could not be completely antisymmetric under exchange of any pair of the electrons. One of the electrons must therefore be in the next highest energy state, the $2s$ state, which has lower energy than the $2p$ state. We label this electron configuration by $1s^2 2s$. It takes only 5.4 eV to ionize lithium and thus lithium is chemically quite active. The ground state of the next element in the periodic table, beryllium, is $1s^2 2s^2$, while boron with five electrons is in the $1s^2 2s^2 2p$ state. The next five elements—C, N, O, F, and Ne—fill up the $2p$ level, which can accommodate 6 electrons, since for $l = 1$, we can have $m = 1, 0$, and $-1$ and two possible intrinsic spin states for each of these orbital states. Figure 12.6 shows the ionization energy for each of the elements. Notice that as $Z$ increases within a shell, the electrons are pulled in towards the nucleus and the ionization energy increases. Neon, like helium, has a completely filled shell and a high ionization energy of 21.6 eV. The charge density for a closed shell

$$\sum_{m=-l}^{l} e|R_{n,l}(r)|^2 |Y_{l,m}(\theta, \phi)|^2 \tag{12.62}$$

is spherically symmetric since

$$\sum_{m=-l}^{l} |Y_{l,m}(\theta, \phi)|^2 = \frac{2l + 1}{4\pi} \tag{12.63}$$

Thus the electronic charge effectively shields the nuclear charge. Although exciting one of the electrons to an excited state would change this situation, the energy gap between the $n = 3$ and $n = 2$ levels is sufficiently large that the atom has little affinity for other electrons and is chemically quite inert.

---

[8] See also the discussion at the end of Section 12.4.

Periodic table of the elements, including electronic configurations. Each cell lists the element symbol with atomic number, the electronic configuration beyond the closed-shell core, and the ground-state term symbol.

| I | II | | | | | | | | | | | III | IV | V | VI | VII | VIII |
|---|---|---|---|---|---|---|---|---|---|---|---|---|---|---|---|---|---|
| $\mathrm{H}^{1}$ $1s^1$ $^2S_{1/2}$ | | | | | | | | | | | | | | | | | $\mathrm{He}^{2}$ $1s^2$ $^1S_0$ |
| $\mathrm{Li}^{3}$ $2s^1$ $^2S_{1/2}$ | $\mathrm{Be}^{4}$ $2s^2$ $^1S_0$ | | | | | | | | | | | $\mathrm{B}^{5}$ $2p^1$ $^2P_{1/2}$ | $\mathrm{C}^{6}$ $2p^2$ $^3P_0$ | $\mathrm{N}^{7}$ $2p^3$ $^4S_{3/2}$ | $\mathrm{O}^{8}$ $2p^4$ $^3P_2$ | $\mathrm{F}^{9}$ $2p^5$ $^2P_{3/2}$ | $\mathrm{Ne}^{10}$ $2p^6$ $^1S_0$ |
| $\mathrm{Na}^{11}$ $3s^1$ $^2S_{1/2}$ | $\mathrm{Mg}^{12}$ $3s^2$ $^1S_0$ | | | | | | | | | | | $\mathrm{Al}^{13}$ $3p^1$ $^2P_{1/2}$ | $\mathrm{Si}^{14}$ $3p^2$ $^3P_0$ | $\mathrm{P}^{15}$ $3p^3$ $^4S_{3/2}$ | $\mathrm{S}^{16}$ $3p^4$ $^3P_2$ | $\mathrm{Cl}^{17}$ $3p^5$ $^2P_{3/2}$ | $\mathrm{Ar}^{18}$ $3p^6$ $^1S_0$ |
| $\mathrm{K}^{19}$ $4s^1$ $^2S_{1/2}$ | $\mathrm{Ca}^{20}$ $4s^2$ $^1S_0$ | $\mathrm{Sc}^{21}$ $3d^1$ $^2D_{3/2}$ | $\mathrm{Ti}^{22}$ $3d^2$ $^3F_2$ | $\mathrm{V}^{23}$ $3d^3$ $^4F_{3/2}$ | $\mathrm{Cr}^{24}$ $4s^13d^5$ $^7S_3$ | $\mathrm{Mn}^{25}$ $3d^5$ $^6S_{5/2}$ | $\mathrm{Fe}^{26}$ $3d^6$ $^5D_4$ | $\mathrm{Co}^{27}$ $3d^7$ $^4F_{9/2}$ | $\mathrm{Ni}^{28}$ $3d^8$ $^3F_4$ | $\mathrm{Cu}^{29}$ $4s^13d^{10}$ $^2S_{1/2}$ | $\mathrm{Zn}^{30}$ $3d^{10}$ $^1S_0$ | $\mathrm{Ga}^{31}$ $4p^1$ $^2P_{1/2}$ | $\mathrm{Ge}^{32}$ $4p^2$ $^3P_0$ | $\mathrm{As}^{33}$ $4p^3$ $^4S_{3/2}$ | $\mathrm{Se}^{34}$ $4p^4$ $^3P_2$ | $\mathrm{Br}^{35}$ $4p^5$ $^2P_{3/2}$ | $\mathrm{Kr}^{36}$ $4p^6$ $^1S_0$ |
| $\mathrm{Rb}^{37}$ $5s^1$ $^2S_{1/2}$ | $\mathrm{Sr}^{38}$ $5s^2$ $^1S_0$ | $\mathrm{Y}^{39}$ $4d^1$ $^2D_{3/2}$ | $\mathrm{Zr}^{40}$ $4d^2$ $^3F_2$ | $\mathrm{Nb}^{41}$ $5s^14d^4$ $^6D_{1/2}$ | $\mathrm{Mo}^{42}$ $5s^14d^5$ $^7S_3$ | $\mathrm{Tc}^{43}$ $5s^14d^6$ $^6D_{9/2}$ | $\mathrm{Ru}^{44}$ $5s^14d^7$ $^5F_5$ | $\mathrm{Rh}^{45}$ $5s^14d^8$ $^4F_{9/2}$ | $\mathrm{Pd}^{46}$ $5s^04d^{10}$ $^1S_0$ | $\mathrm{Ag}^{47}$ $5s^14d^{10}$ $^2S_{1/2}$ | $\mathrm{Cd}^{48}$ $4d^{10}$ $^1S_0$ | $\mathrm{In}^{49}$ $5p^1$ $^2P_{1/2}$ | $\mathrm{Sn}^{50}$ $5p^2$ $^3P_0$ | $\mathrm{Sb}^{51}$ $5p^3$ $^4S_{3/2}$ | $\mathrm{Te}^{52}$ $5p^4$ $^3P_2$ | $\mathrm{I}^{53}$ $5p^5$ $^2P_{3/2}$ | $\mathrm{Xe}^{54}$ $5p^6$ $^1S_0$ |
| $\mathrm{Cs}^{55}$ $6s^1$ $^2S_{1/2}$ | $\mathrm{Ba}^{56}$ $6s^2$ $^1S_0$ | 57–71 Rare Earths | $\mathrm{Hf}^{72}$ $5d^2$ $^3F_2$ | $\mathrm{Ta}^{73}$ $5d^3$ $^4F_{3/2}$ | $\mathrm{W}^{74}$ $5d^4$ $^5D_0$ | $\mathrm{Re}^{75}$ $5d^5$ $^6S_{5/2}$ | $\mathrm{Os}^{76}$ $5d^6$ $^5D_4$ | $\mathrm{Ir}^{77}$ $5d^7$ $^4F_{9/2}$ | $\mathrm{Pt}^{78}$ $6s^15d^9$ $^3D_3$ | $\mathrm{Au}^{79}$ $6s^15d^{10}$ $^2S_{1/2}$ | $\mathrm{Hg}^{80}$ $5d^{10}$ $^1S_0$ | $\mathrm{Tl}^{81}$ $6p^1$ $^2P_{1/2}$ | $\mathrm{Pb}^{82}$ $6p^2$ $^3P_0$ | $\mathrm{Bi}^{83}$ $6p^3$ $^4S_{3/2}$ | $\mathrm{Po}^{84}$ $6p^4$ $^3P_2$ | $\mathrm{At}^{85}$ $6p^5$ $^2P_{3/2}$ | $\mathrm{Rn}^{86}$ $6p^6$ $^1S_0$ |
| $\mathrm{Fr}^{87}$ $7s^1$ $^2S_{1/2}$ | $\mathrm{Ra}^{88}$ $7s^2$ $^1S_0$ | 89–103 Actinides | $\mathrm{Rf}^{104}$ $5f^{14}6d^2$ | $\mathrm{Ha}^{105}$ | | | | | | | | | | | | | |

Rare earths (Lanthanides)

| $\mathrm{La}^{57}$ $5d^1$ $^2D_{3/2}$ | $\mathrm{Ce}^{58}$ $4f^2$ $^3H_4$ | $\mathrm{Pr}^{59}$ $4f^3$ $^4I_{9/2}$ | $\mathrm{Nd}^{60}$ $4f^4$ $^5I_4$ | $\mathrm{Pm}^{61}$ $4f^5$ $^6H_{5/2}$ | $\mathrm{Sm}^{62}$ $4f^6$ $^7F_0$ | $\mathrm{Eu}^{63}$ $4f^7$ $^8S_{7/2}$ | $\mathrm{Gd}^{64}$ $5d^14f^7$ $^9D_2$ | $\mathrm{Tb}^{65}$ $6s^14f^9$ $^6H_{15/2}$ | $\mathrm{Dy}^{66}$ $4f^{10}$ $^5I_8$ | $\mathrm{Ho}^{67}$ $4f^{11}$ $^4I_{15/2}$ | $\mathrm{Er}^{68}$ $4f^{12}$ $^3H_6$ | $\mathrm{Tm}^{69}$ $4f^{13}$ $^2F_{7/2}$ | $\mathrm{Yb}^{70}$ $4f^{14}$ $^1S_0$ | $\mathrm{Lu}^{71}$ $5d^14f^{14}$ $^2D_{3/2}$ |
|---|---|---|---|---|---|---|---|---|---|---|---|---|---|---|

Actinides

| $\mathrm{Ac}^{89}$ $6d^1$ $^2D_{3/2}$ | $\mathrm{Th}^{90}$ $6d^2$ $^3F_2$ | $\mathrm{Pa}^{91}$ $6d^15f^2$ $^4K_{11/2}$ | $\mathrm{U}^{92}$ $6d^15f^3$ $^5L_6$ | $\mathrm{Np}^{93}$ $6d^15f^4$ $^6L_{11/2}$ | $\mathrm{Pu}^{94}$ $5f^6$ $^7F_0$ | $\mathrm{Am}^{95}$ $5f^7$ $^8S_{7/2}$ | $\mathrm{Cm}^{96}$ $6d^15f^7$ $^9D_2$ | $\mathrm{Bk}^{97}$ $6d^15f^8$ $^6G_{15/2}$ | $\mathrm{Cf}^{98}$ $5f^{10}$ $^5I_8$ | $\mathrm{Es}^{99}$ $5f^{11}$ $^4I_{15/2}$ | $\mathrm{Fm}^{100}$ $5f^{12}$ $^3H_6$ | $\mathrm{Md}^{101}$ $5f^{13}$ $^2F_{7/2}$ | $\mathrm{No}^{102}$ $5f^{14}$ $^1S_0$ | $\mathrm{Lr}^{103}$ $5f^{14}6d^1$ $^2D_{3/2}$ |
|---|---|---|---|---|---|---|---|---|---|---|---|---|---|---|

**FIGURE 12.5**

The periodic table of the elements, including the electronic configurations (from *Phys. Lett.* **B 239**).

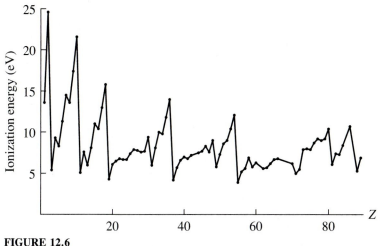

**FIGURE 12.6**
The ionization energy of the elements.

The third row of the periodic table starts with sodium. After filling the $n = 2$ shell, the eleventh electron must go into the $n = 3$ level. Again, the $3s$ level lies lower than the $3p$ level and thus the electron configuration is $1s^22s^22p^63s$. Since this last electron is in a new shell, it is primarily shielded from the nucleus by the inner ten electrons and thus, like lithium, has a low ionization energy (5.1 eV) and a high chemical activity. Both lithium and sodium are alkali metals, which are quite reactive chemically. As an example, if fluorine is present, sodium sees a natural home for its "extra" electron; it can donate it to fluorine completing the $n = 2$ shell for that element. These ions then bind together through electrostatic attraction, forming an ionic bond. In moving from sodium to argon along the third row of the periodic table, the $s$ and the $p$ levels fill up just as they did in going from lithium to neon along the second row, and thus the chemical properties of these elements are strikingly similar to the ones above them in the periodic table.[9] In particular, chlorine, like fluorine, is a halogen that needs just one electron to complete a shell—the $3p$ shell in this case. Chlorine can take that electron from sodium, producing sodium chloride, or ordinary salt. Argon, like neon, has a closed shell and is relatively inert; it is one of the noble gases.

The next row in the periodic table contains some surprises. Instead of filling the $3d$ level next, as you might expect based on the energy levels of hydrogen, the $4s$ electrons penetrate the electron cloud of the inner electrons and have their energy pulled down below that of the $3d$ level. In fact, there is very little separation between the $4s$ and $3d$ levels. By the time the $3d$ level is filled with four electrons, the interaction between the electrons raises the energy of the $4s$ level so that it is slightly above the $3d$ level. Thus chromium has an outer shell electron configuration of $4s^13d^5$ instead of $4s^23d^4$. The next few elements—manganese,

---

[9] The ionization energies are slightly less, since $n = 3$ instead of 2.

iron, cobalt, and nickel—have a filled $4s$ level with electronic configurations ranging from $4s^2 3d^5$ to $4s^2 3d^8$. The chemical properties of these elements are all similar. Since the average radius of the $3d$ electrons is somewhat less than the $4s$ electrons, the outer $4s$ electrons tend to shield the inner $3d$ electrons from outside influences.[10] At copper, which is $3d^{10} 4s^1$, the pattern shifts again, with one of the electrons from the $4s$ level shifting to the $3d$ level. However, the two configurations $4s^2 3d^9$ and $4s^1 3d^{10}$ are so close together in energy that copper can behave as if it has one or two valence electrons depending on its chemical environment. Finally, after both the $4s$ and $3d$ levels are filled, the $4p$ level fills, repeating the pattern of the previous two rows.

One of the triumphs of quantum mechanics is that it provides us with a detailed understanding of the physics responsible for the periodicity in the chemical properties of the elements.

## 12.4 COVALENT BONDING

In our discussion of chemical activity of the elements, we have indicated that certain elements can bind ionically together through electrostatic attraction after they have transferred an electron from one of the elements to the other. There is another type of bond, the covalent bond, in which the elements actually share rather than exchange their electrons. This sort of bonding is pure quantum mechanics in action. To see how it arises, we first consider as a specific case the positively charged hydrogen molecule ion, where a single electron is shared by two protons; then we consider the more prototypical case of the hydrogen molecule, where two electrons are shared. We will see that the identical nature of the electrons plays a crucial role in this covalent bonding.

### The Hydrogen Molecule Ion

Although molecules, even simple diatomic molecules, are complex systems with many degrees of freedom, fortunately there are approximations that we can make that make the problem of determining the bound states of molecules a reasonably tractable one. In particular, recall from Section 9.7 that the energy of vibration of the nuclei of a diatomic molecule is on the order of $(m_e/m_N)^{1/2}$ smaller than the electronic energy of the molecule. Thus the typical period of the motion of electrons in the molecule is much shorter than that of the nuclei, and we can neglect the motion of the nuclei as a first approximation, part of the Born-Oppenheimer approximation. For a diatomic molecule we then consider the behavior of electrons under the influence of two *fixed* nuclei separated by a distance $R$. In fact, we can take $R$ as a parameter that also appears in the wave function and that we can adjust using the variational method to determine the value of $R$ that minimizes

---

[10] This effect is even more pronounced in the sixth and seventh rows of the periodic table. The $4f$ electrons do not fill up until after the $6s$ shell. Thus the rare earth elements with $Z$ ranging from 57 to 71 have similar chemical properties. Also, the $5f$ electrons do not fill until after the $7s$ shell, leading to very similar chemical properties for the actinides, $Z = 89$ to $Z = 103$.

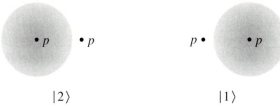

$|2\rangle$ $|1\rangle$

**FIGURE 12.7**
A schematic representation of the two states used as a basis for a variational calculation of the ground state of the hydrogen molecule ion. In $|1\rangle$ the electron combines with one of the protons to form a hydrogen atom in its ground state, while in $|2\rangle$ it combines with the other proton, again forming hydrogen in the ground state.

the energy of the molecule, thus determining both the energy and the size of the molecule.

A natural trial wave function for the hydrogen molecule ion $H_2^+$ is determined by first considering the lowest-energy state of the system when the two protons are widely separated. Then there are clearly two possible states: the electron is attached either to one of the protons, forming a hydrogen atom in the ground state, or the electron is attached to the other proton, again in the ground state of a hydrogen atom. These two states are indicated in Fig. 12.7. In terms of the coordinates shown in Fig. 12.8, the corresponding position-space wave functions are given by

$$\langle \mathbf{r}|1\rangle = \frac{1}{\sqrt{\pi a_0^3}} e^{-|\mathbf{r}-\mathbf{R}/2|/a_0} \tag{12.64a}$$

$$\langle \mathbf{r}|2\rangle = \frac{1}{\sqrt{\pi a_0^3}} e^{-|\mathbf{r}+\mathbf{R}/2|/a_0} \tag{12.64b}$$

There are, of course, many other possible states of the system that we are neglecting in which, for example, the electron is in an excited state of the hydrogen atom.

What is the proper linear combination of states $|1\rangle$ and $|2\rangle$ to use for the variational method? Here, as was the case for the N atom in the ammonia molecule that we treated as a two-state system in Chapter 4, there is an amplitude for the electron attached to one of the protons to jump to the other proton. This amplitude means that the matrix representation of the Hamiltonian

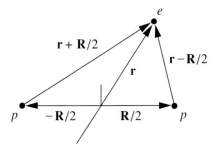

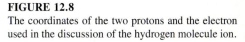

**FIGURE 12.8**
The coordinates of the two protons and the electron used in the discussion of the hydrogen molecule ion.

$$\hat{H} = \frac{\hat{\mathbf{p}}^2}{2m_e} - \frac{e^2}{|\hat{\mathbf{r}} - \mathbf{R}/2|} - \frac{e^2}{|\hat{\mathbf{r}} + \mathbf{R}/2|} + \frac{e^2}{R} \tag{12.65}$$

using the states $|1\rangle$ and $|2\rangle$ as a basis will have off-diagonal matrix elements, similar to what we assumed in our treatment of the ammonia molecule. Here, because of the relative simplicity of the $H_2^+$ ion, we can actually calculate the matrix elements:

$$H_{11} = \langle 1| \left( \frac{\hat{\mathbf{p}}^2}{2m_e} - \frac{e^2}{|\hat{\mathbf{r}} - \mathbf{R}/2|} \right) |1\rangle - \langle 1| \frac{e^2}{|\hat{\mathbf{r}} + \mathbf{R}/2|} |1\rangle + \frac{e^2}{R} \langle 1|1\rangle$$

$$= E_1 - \int d^3r \, \frac{e^2}{|\mathbf{r} + \mathbf{R}/2|} |\langle \mathbf{r}|1\rangle|^2 + \frac{e^2}{R} \tag{12.66}$$

where $E_1$ is the ground-state energy of the hydrogen atom;

$$H_{22} = \langle 2| \left( \frac{\hat{\mathbf{p}}^2}{2m_e} - \frac{e^2}{|\hat{\mathbf{r}} + \mathbf{R}/2|} \right) |2\rangle - \langle 2| \frac{e^2}{|\hat{\mathbf{r}} - \mathbf{R}/2|} |2\rangle + \frac{e^2}{R} \langle 2|2\rangle$$

$$= E_1 - \int d^3r \, \frac{e^2}{|\mathbf{r} - \mathbf{R}/2|} |\langle \mathbf{r}|2\rangle|^2 + \frac{e^2}{R} = H_{11} \tag{12.67}$$

where the last step follows from the symmetry of the two configurations;

$$H_{12} = \langle 1| \left( \frac{\hat{\mathbf{p}}^2}{2m_e} - \frac{e^2}{|\hat{\mathbf{r}} + \mathbf{R}/2|} \right) |2\rangle - \langle 1| \frac{e^2}{|\hat{\mathbf{r}} - \mathbf{R}/2|} |2\rangle + \frac{e^2}{R} \langle 1|2\rangle$$

$$= \left( E_1 + \frac{e^2}{R} \right) \langle 1|2\rangle - \int d^3r \, \langle 1|\mathbf{r}\rangle \frac{e^2}{|\mathbf{r} - \mathbf{R}/2|} \langle \mathbf{r}|2\rangle \tag{12.68}$$

where the amplitude

$$\langle 1|2\rangle = \int d^3r \, \langle 1|\mathbf{r}\rangle \langle \mathbf{r}|2\rangle \tag{12.69}$$

is called the *overlap integral*. Note that the nonvanishing of the off-diagonal matrix element depends on the states $|1\rangle$ and $|2\rangle$ overlapping in space. Since the wave functions are real, the off-diagonal matrix elements are equal:

$$H_{21} = \langle 2| \left( \frac{\hat{\mathbf{p}}^2}{2m_e} - \frac{e^2}{|\hat{\mathbf{r}} - \mathbf{R}/2|} \right) |1\rangle - \langle 2| \frac{e^2}{|\hat{\mathbf{r}} + \mathbf{R}/2|} |1\rangle + \frac{e^2}{R} \langle 2|1\rangle$$

$$= \left( E_1 + \frac{e^2}{R} \right) \langle 2|1\rangle - \int d^3r \, \langle 2|\mathbf{r}\rangle \frac{e^2}{|\mathbf{r} + \mathbf{R}/2|} \langle \mathbf{r}|1\rangle = H_{12} \tag{12.70}$$

Because $H_{11} = H_{22}$ as well as $H_{12} = H_{21}$, the linear combinations of the states $|1\rangle$ and $|2\rangle$ that diagonalize the Hamiltonian are

$$|\pm\rangle = \frac{1}{\sqrt{2 \pm 2\langle 1|2\rangle}} (|1\rangle \pm |2\rangle) \tag{12.71}$$

where the overall factor is needed to normalize the states because the basis states are not orthogonal (see Problem 12.8). Note from the form of the wave functions (12.64) that the wave function $\langle \mathbf{r}|+\rangle$ has even parity while the wave function $\langle \mathbf{r}|-\rangle$ has odd parity. We could have selected the two states $|+\rangle$ and $|-\rangle$ initially as the proper linear combinations from the inversion symmetry of the Hamiltonian (12.65). Figure 12.9 shows a sketch of the wave functions, and Fig. 12.10 shows the expectation values of the energies

$$E_{\pm} = \frac{1}{1 \pm \langle 1|2\rangle}\left(H_{11} \pm H_{12}\right) \tag{12.72}$$

plotted as a function of $R$. Only the even parity state has a minimum, which occurs

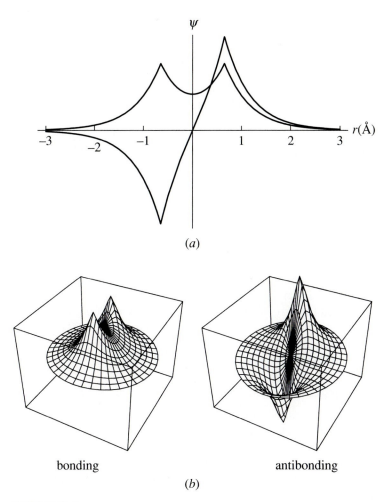

(a)

bonding          antibonding

(b)

**FIGURE 12.9**
The wave functions for the bonding and antibonding orbital of the hydrogen molecule ion plotted (a) along the axis connecting the two protons and (b) in two dimensions.

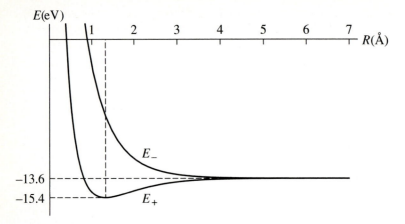

**FIGURE 12.10**
The energies of the bonding and antibonding orbitals as a function of the interproton separation $R$.

for a separation of 1.3 Å for the protons, corresponding to a binding energy of 1.8 eV. Thus the state $|+\rangle$ is referred to as a bonding molecular orbital, while the state $|-\rangle$ is called an antibonding molecular orbital. These molecular orbitals are linear combinations of atomic orbitals. Note from Fig. 12.9 that only for the bonding orbital is the electron shared *between* the two protons. For the antibonding orbital, on the other hand, there is a node in the wave function midway between the two protons where the potential energy is quite negative, and thus the electron in this state doesn't benefit from the full attraction of the two protons.

The experimental separation between the protons for the hydrogen molecule ion is 1.06 Å, with a binding energy of 2.8 eV. The reason for the lack of better agreement between our variational results and experiment resides in our choice of trial wave function. In particular, notice that as $R \to 0$, the system reduces to He$^+$, while our trial wave function remains the $1s$ ground state of hydrogen. Thus we should not be surprised that we have underestimated the size of the binding energy and overestimated the size of the molecule. One way of doing better is to use a trial wave function with an effective charge $Z$ (as in Section 12.2) as well as the interproton distance $R$ as parameters. Nonetheless, our solution demonstrates the key qualitative features of the binding.

## Muon-Catalyzed Fusion

Note that the interproton separation in the hydrogen molecule ion is on the order of the Bohr radius $a_0$ of the hydrogen atom. This length scale enters in the trial wave functions (12.64) that we used in our variational approach to the molecule. Since $a_0 = \hbar/\mu c \alpha$, the size of the molecule depends on the reduced mass $\mu$ of corresponding atom. In particular, suppose we were to replace the electron in the hydrogen molecule ion with a muon. Then the reduced mass is given by $m_\mu m_p/(m_\mu + m_p)$ instead of $m_e m_p/(m_e + m_p)$. Since the reduced mass for the muonic atom is roughly a factor of $m_\mu/m_e$ larger than the reduced mass of hydrogen, the interproton separation in a muonic hydrogen molecule ion should be a factor of $m_e/m_\mu$ smaller than for the molecule with an electron generating the binding. Replacing the electron with a muon produces a much smaller molecule, just as replacing the electron with a muon in the hydrogen atom produces a much smaller atom.

Suppose that the nuclei of this diatomic molecule consist of two deuterons instead of two protons. The small size of the muonic molecule means that the deuterons have a significantly greater probability of being close together than is the case when an electron is responsible for the binding of the molecule. In fact, for this muonic molecule nuclear reactions such as

$$d + d \rightarrow t + p + 4.0 \text{ MeV} \tag{12.73}$$

have been observed to take place. This is a typical fusion reaction of the sort that is contemplated as a plentiful energy source (using the deuterium naturally present in sea water). However, whereas the attempts to generate fusion commercially depend on thermonuclear reactions generated by heating the deuterons to sufficiently high temperatures that they have a significant chance of overcoming their Coulomb repulsion and being close enough together (on the order of a fermi) to permit the reaction (12.73) to occur, the quantum mechanical binding in the muonic molecule does this at ordinary room temperature. Thus this muon catalysis can be described as a form of "cold fusion." Unfortunately, muons themselves are not freely available (except in cosmic rays) and thus an accelerator is required to generate the energy required to produce muons, for example, in the form of muon-antimuon pairs. The only way for reactions such as (12.73) to be a net generator of energy is for each muon to form a large number of muonic molecules in which it catalyzes a nuclear reaction before the muon decays in roughly 2.2 microseconds.[11] So far this has not been feasible, but work is still in progress.

## The Hydrogen Molecule

We are now ready to turn our attention to the hydrogen molecule, $H_2$, where we must examine the effect of the identical nature of the two electrons on the binding of the molecule. Although we won't spell out the details here,[12] we can use an approach similar to the one we used for the hydrogen molecule ion to understand the binding of the hydrogen molecule. For $H_2$, each hydrogen atom in the molecule supplies, of course, one electron. For each of the electrons to be in the same region of space between the two protons, the spatial state must be symmetric under exchange of the two particles, and consequently the total-spin state must be a spin-0 state. With two electrons being shared instead of one, the molecule has a binding energy of 4.7 eV, as compared with 2.8eV for $H_2{}^+$, despite the extra Coulomb repulsion of the electrons. The interproton separation for the

---

[11] L. Alvarez et al., *Phys. Rev.* **105**, 1127 (1957). For an entertaining account of Luis Alvarez's discovery of muon catalyzed fusion, see *Nobel Lectures—Physics,* vol. II, Elsevier Publication, New York, 1969.

[12] For a nice discussion, see S. Gasiorowicz, *Quantum Physics,* Wiley, New York, 1974.

molecule is 0.7 Å, as compared with 1.3 Å for the ion. Binding only occurs for the total-spin-0 state; the total-spin-1 state, corresponding to an antisymmetric spatial state in which the electrons, in general, do not reside together in the region between the two protons, does not exhibit a minimum in the total energy for any separation of the protons. Thus repulsion rather than binding occurs in this case.

We can now also see why, for example, a hydrogen atom and a helium atom do not bind together to form a molecule. The two electrons in the ground state of helium are both in the same spatial $1s$ state, and therefore their total-spin state is the singlet spin-0 state. We say these two electrons in helium are paired together. The electron in the hydrogen atom cannot form a covalent bond and pair up with either one of these electrons, since this would either mean that three electrons are in the same spatial state (which is forbidden) or else one of the electrons would have to be excited to the $2s$ state of helium, which is energetically quite costly. On the other hand, as we have seen in the hydrogen molecule, if the electron from hydrogen interacts with an electron from helium in a total-spin-1 state, repulsion occurs between the two atoms. This is the reason helium is an inert element without affinity for other atoms, unlike an element with an unpaired outer electron that can pair up with an electron from another atom to form a covalent bond. Moreover, once two unpaired electrons from different atoms have paired up to form a covalent bond, an electron from a third atom cannot pair up with either one of them. Thus we see why the chemical forces saturate.

## 12.5 CONCLUSION

Most of our attention in this chapter has been devoted to systems containing identical fermions. The requirement that the state of the system be antisymmetric under the exchange of any two identical fermions means, in particular, that two identical fermions cannot occupy the same state (the Pauli principle). For identical bosons, on the other hand, the state of the system must be symmetric under exchange of any two of the particles. Consequently, it is possible, and in fact preferred, to have many identical bosons in the same state. In Chapter 14 we will see an example when we discuss how a laser operates. Other examples in which many bosons condense to the ground state at sufficiently low temperatures (Bose-Einstein condensation) include superconductivity and superfluidity. Like the laser, these phenomena are interesting and exciting macroscopic manifestations of purely quantum behavior.

## PROBLEMS

**12.1.** Verify for an antisymmetric spatial state under exchange of two particles that

$$|\psi_A\rangle = \frac{1}{2} \int d^3r_1 \, d^3r_2 \left( \frac{1}{\sqrt{2}}|\mathbf{r}_1, \mathbf{r}_2\rangle - \frac{1}{\sqrt{2}}|\mathbf{r}_2, \mathbf{r}_1\rangle \right)\left( \frac{1}{\sqrt{2}}\langle\mathbf{r}_1, \mathbf{r}_2|\psi_A\rangle - \frac{1}{\sqrt{2}}\langle\mathbf{r}_2, \mathbf{r}_1|\psi_A\rangle \right)$$

$$= \int d^3r_1 \, d^3r_2 \, |\mathbf{r}_1, \mathbf{r}_2\rangle\langle\mathbf{r}_1, \mathbf{r}_2|\psi_A\rangle$$

**12.2.** Two identical, noninteracting spin-$\frac{1}{2}$ particles of mass $m$ are in the one-dimensional harmonic oscillator for which the Hamiltonian is

$$\hat{H} = \frac{\hat{p}_{1x}^2}{2m} + \frac{1}{2}m\omega^2\hat{x}_1^2 + \frac{\hat{p}_{2x}^2}{2m} + \frac{1}{2}m\omega^2\hat{x}_2^2$$

(a) Determine the ground-state and first-excited-state kets and corresponding energies when the two particles are in a total-spin-0 state. What are the lowest energy states and corresponding kets for the particles if they are in a total-spin-1 state?

(b) Suppose the two particles interact with a potential energy of interaction

$$V(|x_1 - x_2|) = \begin{cases} -V_0 & |x_1 - x_2| < a \\ 0 & \text{elsewhere} \end{cases}$$

Argue what the effect will be on the energies that you determined in (a), that is, whether the energy of each state moves up, moves down, or remains unchanged.

**12.3.** Obtain an order-of-magnitude estimate for the singlet-triplet splitting of the energy levels of the two electrons in helium due to a direct spin-spin interaction. *Suggestion:* Compare with the magnitude of the hyperfine interaction in hydrogen as discussed in Chapter 5.

**12.4.** Use the variational principle to estimate the ground-state energy for the one-dimensional anharmonic oscillator

$$\hat{H} = \frac{\hat{p}_x^2}{2m} + b\hat{x}^4$$

Compare your result with the exact result

$$E_0 = 1.060b^{1/3}\left(\frac{\hbar^2}{2m}\right)^{2/3}$$

**12.5.** For the delta function potential well

$$\frac{2m}{\hbar^2}V(x) = -\frac{\lambda}{b}\delta(x)$$

use a Guassian wave function as a trial wave function to obtain an upper bound for the ground-state energy. Compare with the result of Problem 6.16.

**12.6.** Consider the one-dimensional motion of particle of mass $m$ in a uniform gravitational field above an impenetrable plane. Take the potential energy to be infinite at the plane and locate the plane at $z = 0$.

(a) Plot the potential energy of the particle. What is the Hamiltonian? Sketch roughly the ground-state wave function.

(b) Use an appropriate trial wave function to estimate the ground-state energy.

(c) What is the average position $\langle z \rangle$ of the particle above the plane?

**12.7.** A muon and a proton are bound together in the ground state of a muonic "hydrogen atom." As this atom diffuses around, it bumps into a deuteron. What is the incentive for the muon to jump from the proton to the deuteron, forming another muonic atom, but this time with the deuteron instead of the proton as the nucleus? Explain.

**12.8.** Show that the linear combinations of states that diagonalize the Hamiltonian of the hydrogen molecule ion are given by (12.71). Verify that these states are properly normalized and that the corresponding energy expectation values are given by (12.72). *Note:* If $|\psi\rangle$ is not normalized, then $\langle E \rangle = \langle \psi|\hat{H}|\psi\rangle/\langle\psi|\psi\rangle$.

# CHAPTER
# 13

# SCATTERING

How do we learn about the nature of the fundamental interactions on a microscopic level? Solving the Schrödinger equation for the hydrogen atom yields an infinite set of energy levels that gives us proof that the Coulomb interaction is the predominant interaction between an electron and a proton on a distance scale on the order of angstroms. However, for the two-body problem in nuclear physics, as we discussed in Chapter 10, there is a single bound state, so we must resort to scattering techniques to learn about the nature of the nuclear force. After all, it was through a scattering experiment that Rutherford first discovered the very existence of the nucleus within the atom. Subsequently, scattering has played a major role in helping us learn about nuclear physics and particle physics, as well as more about atomic physics. After introducing the concept of the scattering cross section, we will use the Born approximation and the partial wave expansion to calculate the cross section in quantum mechanics. These two approaches are in a sense complementary to each other: the Born approximation works best at high energies and the partial wave expansion has its greatest utility at low energies.

## 13.1 THE ASYMPTOTIC WAVE FUNCTION
## AND THE DIFFERENTIAL CROSS SECTION

In a typical scattering experiment, a beam of particles, often from an accelerator, is projected at a fixed target composed of other particles. In Rutherford's experiment, the incident particles were $\alpha$ particles, He$^4$ nuclei, which were emitted in radioactive decay, while the target consisted of gold atoms in the form of a thin gold foil. A schematic diagram of this experiment is given in Fig. 13.1. The angular distribution of the scattered $\alpha$ particles provided clear evidence of the existence of a relatively massive gold nucleus. More recently, experiments done

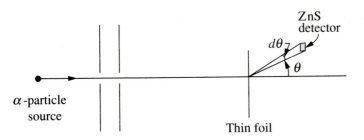

**FIGURE 13.1**
A schematic diagram of the Rutherford scattering experiment.

at the Stanford Linear Accelerator Center with high-energy electrons, accelerated in the two-mile-long accelerator to 20 GeV, as the incident projectiles and target protons in the form of liquid hydrogen revealed that protons were actually composed of fractionally charged constituents, which are called quarks. A convenient way to describe such experiments is to picture the incident particle, initially far from the target, at least on a microscopic scale, as an essentially free particle that interacts with the target only when it is within the range of the potential energy of interaction $V(r)$, which we will take to be spherically symmetric. Just as a comet on a hyperbolic orbit can be deflected by its gravitational interaction with the sun, so too can the incident projectiles in these scattering experiments be deflected by their interaction with the target particle. Since the particles interacting through this potential energy $V(r)$ are not bound, the energy of the incident particle can take on a continuum of different values, just as it did when we analyzed the free particle in one dimension in Section 6.6. As we saw there, physical states can be formed from the superposition of these continuum states in the form of a wave packet.

In practice, this wave packet generally has a sharp peak in momentum space about some incident momentum $\mathbf{p}_0$, and consequently the wave packet in position space is quite broad. In fact, we will assume that it is sufficiently broad that we can treat it as a plane wave for the purposes of analyzing the experiment. This is generally the way one-dimensional scattering is discussed within wave mechanics. Specifically, as we discussed in Section 6.10, one considers a potential energy function such as the potential barrier shown in Fig. 13.2. Outside the range $a$ of the potential, we can express the wave function as a plane wave:

$$\psi(x) = \begin{cases} Ae^{ikx} + Be^{-ikx} & x < 0 \\ Ce^{ikx} & x > a \end{cases} \tag{13.1}$$

where $k = \sqrt{2mE/\hbar^2}$. The time dependence of such an energy eigenfunction is the usual $e^{-iEt/\hbar}$. This is of course a stationary state. However, by superposing these energy eigenstates together, we can produce a wave packet with time dependence such that the incident wave packet alone approaches the barrier, and after interaction with the barrier there is an amplitude for the wave packet to be reflected and an amplitude for it to be transmitted, as depicted in Fig. 6.10.

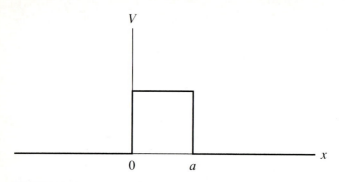

**FIGURE 13.2**
A potential barrier of width $a$ in a one-dimensional scattering experiment.

Examining the stationary states to determine these amplitudes is analogous to doing a scattering experiment with water waves in a pond in which the source of the waves is not a single stone thrown into the pond (which would generate a wave packet) but a harmonic source that continually beats up and down in the water at a steady frequency.

What is the analogue of (13.1) in three dimensions? We take the incident wave to be traveling along the $z$ axis, with the target located at the origin. Far from the target the asymptotic wave function should include this incident wave together with an outgoing wave produced by interaction with the potential:

$$\psi \xrightarrow[r \to \infty]{} A e^{ikz} + \text{(outgoing wave)} \tag{13.2}$$

(see Fig. 13.3). Outside the range of the potential, where the particle detectors are located, the outgoing spherical wave must be a solution to the Schrödinger equation with $V = 0$. Thus, we start with the differential equation (10.4) for the radial wave function $u = Rr$ with $V = 0$,

$$-\frac{\hbar^2}{2\mu} \frac{d^2 u}{dr^2} + \frac{l(l + 1)\hbar^2}{2\mu r^2} u = E u \tag{13.3}$$

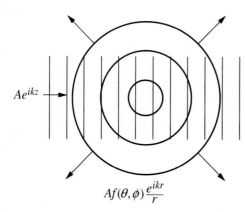

$A e^{ikz}$

$A f(\theta, \phi) \dfrac{e^{ikr}}{r}$

**FIGURE 13.3**
A schematic diagram of a three-dimensional scattering experiment indicating the incident plane wave and the outgoing spherical wave.

For large $r$ we can neglect the centrifugal barrier term and obtain, for any $l$, the equation

$$-\frac{\hbar^2}{2\mu}\frac{d^2u}{dr^2} = Eu \tag{13.4}$$

which has the two solutions

$$u = e^{ikr} \quad \text{and} \quad u = e^{-ikr} \tag{13.5}$$

with

$$k = \sqrt{\frac{2\mu E}{\hbar^2}} \tag{13.6}$$

If we attach the time dependence $e^{-iEt/\hbar}$ to each of these solutions, we see that only $e^{ikr}$ corresponds to the outgoing wave, since as time increases, $r$ must also increase to keep the phase constant. This outgoing wave is the type that we expect to be generated by the interaction of the incident wave with the potential. Since $R = u/r$, this suggests that we express the asymptotic wave function (13.2) in the form

$$\psi \xrightarrow[r\to\infty]{} Ae^{ikz} + Af(\theta, \phi)\frac{e^{ikr}}{r} = \psi_{\text{inc}} + \psi_{\text{sc}} \tag{13.7}$$

where the function $f(\theta, \phi)$ allows for angular dependence in the outgoing scattered wave. After all, we have no reason to expect that the outgoing scattered wave should have the same amplitude in the forward direction ($\theta = 0$) as at other angles. Also, the spherical harmonics that appear in the energy eigenfunctions $\langle \mathbf{r}|E, l, m\rangle = R(r)Y_{l,m}(\theta, \phi)$ can add up to produce, in principle, any angular dependence.

How do we relate this asymptotic wave function (13.7) to what is measured in the laboratory? Let's return to the way the experiments are actually performed. As in Section 6.10, the incident flux of particles can be related to a probability current, whose form follows from the Schrödinger equation in position space,

$$-\frac{\hbar^2}{2\mu}\nabla^2\psi + V\psi = i\hbar\frac{\partial\psi}{\partial t} \tag{13.8}$$

We start with the probability density

$$|\langle \mathbf{r}|\psi\rangle|^2 = \psi^*(\mathbf{r})\psi(\mathbf{r}) \tag{13.9}$$

By evaluating the time derivative of this probability density and taking advantage of the form of the Schrödinger equation (13.8), we find that we can write

$$\frac{\partial}{\partial t}(\psi^*\psi) + \boldsymbol{\nabla}\cdot\mathbf{j} = 0 \tag{13.10}$$

where

$$\mathbf{j} = \frac{\hbar}{2\mu i}(\psi^*\boldsymbol{\nabla}\psi - \psi\boldsymbol{\nabla}\psi^*) \tag{13.11}$$

The dimensions of the probability current $\mathbf{j}$ are probability per unit area per unit time. Equation (13.10) expresses conservation of probability in the form of a local conservation law, since it implies that

$$\frac{d}{dt} \int d^3r\, \psi^*\psi = -\int d^3r\, \boldsymbol{\nabla} \cdot \mathbf{j} = -\oint dS\, \mathbf{n} \cdot \mathbf{j} \qquad (13.12)$$

where in the last step we have used Gauss's theorem to convert the volume integral to a closed surface integral over the surface enclosing the volume. Note that if the integral over the surface of the dot product of the probability current $\mathbf{j}$ with outward normal $\mathbf{n}$ to the surface is positive, there is a net outflow of probability from the volume, and consequently the probability of finding the particle in the volume decreases.

In Section 6.10 we saw that the one-dimensional probability current for the wave function (13.1) is given by

$$j = \frac{\hbar k}{m}(|A|^2 - |B|^2) \qquad x < 0 \qquad (13.13a)$$

$$j = \frac{\hbar k}{m}|C|^2 \qquad x > a \qquad (13.13b)$$

Thus the reflection coefficient can be calculated from

$$R = \frac{j_{\text{ref}}}{j_{\text{inc}}} = \frac{(\hbar k/m)|B|^2}{(\hbar k/m)|A|^2} = \frac{|B|^2}{|A|^2} \qquad (13.14)$$

while the transmission coefficient is given by

$$T = \frac{j_{\text{tran}}}{j_{\text{inc}}} = \frac{(\hbar k/m)|C|^2}{(\hbar k/m)|A|^2} = \frac{|C|^2}{|A|^2} \qquad (13.15)$$

In one dimension, conservation of probability requires that $R + T = 1$.

In three dimensions, there is more than just reflection and transmission. The experimentalist counts the number of particles scattered through angles $\theta$ and $\phi$ that enter a detector that subtends a certain solid angle, as indicated in Fig. 13.4. In particular, the differential cross section $d\sigma$ is defined by

$$d\sigma = \frac{d\sigma}{d\Omega}d\Omega = \frac{\text{number of particles scattered into } d\Omega \text{ per second}}{\text{number of particles incident per unit area per second}}$$

$$(13.16)$$

Note that the dimensions of cross section are area. We can think of it as an effective area that the target presents to the incident flux of particles. Also note from (13.16) that if we multiply this incident flux, which is the number of incident particles per unit area per unit time, by the differential cross sectional area $d\sigma$, we obtain the number of particles scattered into the solid angle $d\Omega$ per unit time. The total cross section $\sigma$ is obtained by integrating over all solid angle:

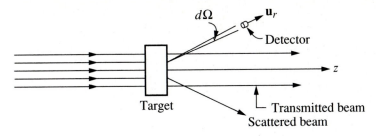

**FIGURE 13.4**
An experimental setup characteristic of a three-dimensional scattering experiment. The detector sub-tends a solid angle $d\Omega$.

$$\sigma = \int \frac{d\sigma}{d\Omega} d\Omega \qquad (13.17)$$

This total area may have a simple physical significance in some cases. For example, in a nuclear scattering experiment with neutrons as the projectiles and a nucleus as the target, the total cross section is on the order of the size of the nucleus, since the nuclear force is short range and neutrons that strike the nucleus are likely to interact with it. On the other hand, if neutrinos are the projectiles and the target is a nucleus, the cross section is many orders of magnitude smaller. This is not because the nucleus has suddenly shrunk in size but because neutrinos interact so weakly with the nucleus that most of the neutrinos pass right through the nucleus without scattering at all.

The flux of particles in a scattering experiment is proportional to the probability current. If we use the asymptotic wave function (13.7) to calculate the probability current, we find that the current does not simply break up into two pieces consisting of an incident current and a scattered current. Unlike the one-dimensional case (13.13), there is *interference* between the incident wave and the scattered wave. This interference in the forward direction ($\theta = 0$) is, in fact, responsible for the reduction of flux in the forward direction from its incident value and therefore is necessary for scattering to occur at all.[1] Nonetheless, in practice it is possible to calculate probability flux entering a detector that is not directly in the forward direction by using the scattered outgoing wave function alone. In the actual experiments, the incident beam is limited in the transverse direction by the sides of the beam tube, if by nothing else. If we call this transverse dimension $b$, then at a distance $r$ from the target the incident wave is present only for angles $\theta \lesssim b/r$ (see Fig. 13.5). Since the particle detectors are located at large distances from the target ($r \rightarrow \infty$), the scattered wave will be the only wave of interest unless the detector is placed essentially at $\theta = 0$, namely, in the incident beam.

---

[1] See also the optical theorem in Section 13.4.

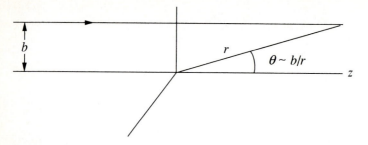

**FIGURE 13.5**
If the incident "plane" wave extends over a distance $b$ transverse to the $z$ axis, a distant detector located a distance $r$ from the origin and at an angle $\theta \gtrsim b/r$ will *not* be in the path of the incident beam. As $r \to \infty$, such a detector would only detect the incident beam as $\theta \to 0$.

To determine the probability current that flows into a detector that subtends a solid angle $d\Omega$, we calculate

$$\mathbf{j}_{sc} = \frac{\hbar}{2\mu i}(\psi_{sc}^* \boldsymbol{\nabla}\psi_{sc} - \psi_{sc}\boldsymbol{\nabla}\psi_{sc}^*) \tag{13.18}$$

where the asymptotic form of the scattered wave function is given by

$$\psi_{sc} \xrightarrow[r\to\infty]{} Af(\theta, \phi)\frac{e^{ikr}}{r} \tag{13.19}$$

The scattered probability current is then given by (see Problem 13.2)

$$\mathbf{j}_{sc} \xrightarrow[r\to\infty]{} \frac{\hbar k}{\mu r^2}|A|^2|f|^2\mathbf{u}_r \tag{13.20}$$

where the unit vector $\mathbf{u}_r$ shows that this current is radially directed. The probability flow (probability per unit time) into a solid angle $d\Omega$ is determined by taking the dot product of the scattered probability current (probability per unit area per unit time) with the area $\mathbf{u}_r r^2\,d\Omega$ covered by an infinitesimal detector:

$$\mathbf{j}_{sc}\cdot\mathbf{u}_r r^2\,d\Omega = \frac{\hbar k}{\mu}|A|^2|f|^2\,d\Omega \tag{13.21}$$

Since the incident probability flux, which is directed along the $z$ axis, is given by

$$j_{inc} = \frac{\hbar k}{\mu}|A|^2 \tag{13.22}$$

the differential cross section that follows from the definition (13.16) is

$$\frac{d\sigma}{d\Omega}d\Omega = \frac{\mathbf{j}_{sc}\cdot\mathbf{u}_r r^2\,d\Omega}{j_{inc}} = |f|^2\,d\Omega \tag{13.23}$$

Thus

$$\frac{d\sigma}{d\Omega} = |f(\theta, \phi)|^2 \tag{13.24}$$

It is this differential cross section that replaces the reflection coefficient $R$ used to describe one-dimensional scattering. The function $f(\theta, \phi)$ is often referred to as the *scattering amplitude*.

## 13.2 THE BORN APPROXIMATION

A particularly good approach for calculating the scattering amplitude when the energy of the incident beam is large in comparison with the magnitude of the potential energy is the Born approximation. We begin by expressing the position-space energy eigenvalue equation

$$\left(-\frac{\hbar^2}{2\mu}\nabla^2 + V\right)\psi(\mathbf{r}) = E\psi(\mathbf{r}) \tag{13.25}$$

in the form

$$\left(\nabla^2 + k^2\right)\psi(\mathbf{r}) = \frac{2\mu}{\hbar^2}V(r)\psi(\mathbf{r}) \tag{13.26}$$

with $k$ given by (13.6). The incoming plane wave $Ae^{ikz}$ is a solution to the equation

$$\left(\nabla^2 + k^2\right)\psi(\mathbf{r}) = 0 \tag{13.27}$$

It is convenient to express the formal solution to (13.26) in the form

$$\psi(\mathbf{r}) = Ae^{ikz} + \int d^3r'\, G(\mathbf{r}, \mathbf{r}')\frac{2\mu}{\hbar^2}V(\mathbf{r}')\psi(\mathbf{r}') \tag{13.28}$$

The function $G(\mathbf{r}, \mathbf{r}')$ is called the Green's function of the differential equation (13.26) and itself satisfies the differential equation

$$\left(\nabla^2 + k^2\right)G(\mathbf{r}, \mathbf{r}') = \delta^3(\mathbf{r} - \mathbf{r}') \tag{13.29}$$

That (13.28) is a solution to (13.26) can be verified by applying the differential operator $\nabla^2 + k^2$ to (13.28). It is only a formal solution since the wave function $\psi$ appears on the right-hand side of this equation as well as on the left-hand side. Nonetheless, we will see that (13.28) provides a useful route for determining the wave function in an iterative procedure known as the Born approximation.

We first determine the Green's function, making sure that the solution (13.28) satisfies the appropriate boundary conditions. As indicated by the argument of the delta function, the Green's function itself depends only on the difference of the vectors $\mathbf{r}$ and $\mathbf{r}'$. With this in mind, let's first set $\mathbf{r}'$ to zero in (13.29) and determine the solution to the equation

$$\left(\nabla^2 + k^2\right)G(\mathbf{r}, 0) = \delta^3(\mathbf{r}) \tag{13.30}$$

Given the spherical symmetry of this differential equation, we naturally search for a solution in spherical coordinates. Notice that, except at the origin, the Green's function is a solution to (13.27). We saw in Section 13.1 that a solution to the Schrödinger equation for a free particle in the form of an outgoing wave is given by

$$G(\mathbf{r}, 0) = C \frac{e^{ikr}}{r} \tag{13.31}$$

where $C$ is some constant. In fact, (13.31) is actually a solution to (13.30) even at the origin, provided we choose the value of $C$ properly. Recall from (10.9) that

$$\nabla^2 \frac{1}{r} = -4\pi\delta^3(\mathbf{r}) \tag{13.32}$$

Note that

$$\nabla^2 \frac{e^{ikr}}{r} = \nabla \cdot \left( \nabla \frac{e^{ikr}}{r} \right) = \nabla \cdot (\nabla e^{ikr}) \frac{1}{r} + \nabla \cdot \left( \nabla \frac{1}{r} \right) e^{ikr}$$

$$= \frac{1}{r} \nabla^2 e^{ikr} + 2\nabla e^{ikr} \cdot \left( \nabla \frac{1}{r} \right) + e^{ikr} \nabla^2 \frac{1}{r}$$

$$= \frac{1}{r} \left( -k^2 + \frac{2ik}{r} \right) e^{ikr} - 2ik \frac{e^{ikr}}{r^2} - 4\pi\delta^3(\mathbf{r})e^{ikr}$$

$$= -k^2 \frac{e^{ikr}}{r} - 4\pi\delta^3(\mathbf{r})e^{ikr} \tag{13.33}$$

where the evaluation of the action of $\nabla^2$ on $e^{ikr}$ has been carried out using

$$\nabla^2 = \frac{\partial^2}{\partial r^2} + \frac{2}{r} \frac{\partial}{\partial r} + \text{angular derivatives} \tag{13.34}$$

Thus

$$(\nabla^2 + k^2)C \frac{e^{ikr}}{r} = -4\pi C \delta^3(\mathbf{r}) \tag{13.35}$$

and consequently

$$G(\mathbf{r}, 0) = -\frac{e^{ikr}}{4\pi r} \tag{13.36}$$

Therefore

$$G(\mathbf{r}, \mathbf{r}') = -\frac{e^{ik|\mathbf{r}-\mathbf{r}'|}}{4\pi|\mathbf{r} - \mathbf{r}'|} \tag{13.37}$$

and

$$\psi(\mathbf{r}) = A e^{ikz} - \frac{\mu}{2\pi\hbar^2} \int d^3r' \frac{e^{ik|\mathbf{r}-\mathbf{r}'|}}{|\mathbf{r} - \mathbf{r}'|} V(\mathbf{r}')\psi(\mathbf{r}') \tag{13.38}$$

Generally, the range of the $\mathbf{r}'$ integral is limited by the range of the potential energy $V(\mathbf{r}')$ to a microscopic distance (see Fig. 13.6). In order to determine the

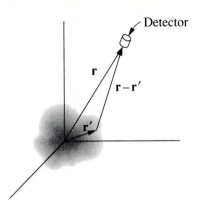

**FIGURE 13.6**
The $\mathbf{r}'$ integration is restricted to lie within the range of the potential energy, which is indicated by the shaded region, while the particle is detected at $\mathbf{r}$.

scattering amplitude, we need to examine the behavior of the wave function as $r \to \infty$. In this case, since $r \gg |\mathbf{r}'|$, we can approximate

$$|\mathbf{r} - \mathbf{r}'| = (r^2 - 2\mathbf{r} \cdot \mathbf{r}' + r'^2)^{1/2} = r\left(1 - 2\mathbf{u}_r \cdot \frac{\mathbf{r}'}{r} + \frac{r'^2}{r^2}\right)^{1/2} \xrightarrow[r \to \infty]{} r\left(1 - \mathbf{u}_r \cdot \frac{\mathbf{r}'}{r}\right)$$

(13.39)

where we have neglected terms of order $(r'/r)^2$. Thus in (13.38) we can make the replacement

$$\frac{1}{|\mathbf{r} - \mathbf{r}'|} \xrightarrow[r \to \infty]{} \frac{1}{r}\left(1 + \mathbf{u}_r \cdot \frac{\mathbf{r}'}{r}\right) \xrightarrow[r \to \infty]{} \frac{1}{r}$$

(13.40)

since the terms that we are neglecting make a vanishing contribution to the integral relative to the one that we have retained as $r \to \infty$. However, within the exponent all that matters is the value of the phase $kr$ modulo $2\pi$, no matter how large $r$ is. Thus we need to retain both the first two terms in the expansion (13.39), namely,

$$e^{ik|\mathbf{r} - \mathbf{r}'|} \xrightarrow[r \to \infty]{} e^{ikr(1 - \mathbf{u}_r \cdot \frac{\mathbf{r}'}{r})} = e^{ikr} e^{-i\mathbf{k}_f \cdot \mathbf{r}'}$$

(13.41)

where

$$\mathbf{k}_f = k\mathbf{u}_r$$

(13.42)

points in the direction of the outgoing scattered wave. Again, the terms of order $(r'/r)^2$ can be safely neglected relative to the two terms that we have retained. With these approximations, (13.38) in the asymptotic limit becomes

$$\psi(\mathbf{r}) \xrightarrow[r \to \infty]{} A e^{ikz} - \frac{\mu e^{ikr}}{2\pi\hbar^2 r} \int d^3r' \, e^{-i\mathbf{k}_f \cdot \mathbf{r}'} V(\mathbf{r}')\psi(\mathbf{r}')$$

(13.43)

In retrospect, we can now see why we made the choices we did in deriving (13.43). Clearly, we chose the particular solution $Ae^{ikz}$ to (13.27) to match up with the boundary condition that our wave function include the correct incident wave. Similarly, in deriving the Green's function (13.37), we discarded the incoming spherical wave $Ce^{-ikr}/r$ because of the physical requirement that the potential generate only such outgoing waves. In practice it may be possible to do scattering with incoming spherical waves,[2] but most experiments are similar to the approach described earlier in which an incoming plane wave generates an outgoing spherical wave upon interaction with the target.

We are now ready for the Born approximation. As we remarked earlier, (13.43) is an integral equation that involves the wave function $\psi$ on the right-hand side within the integral, as well as on the left-hand side. If the potential energy $V$ were set to zero, the solution for $\psi$ would be simply $\psi = Ae^{ikz}$. This suggests that if magnitude of the potential energy is small compared with the energy $E$, we can replace the wave function $\psi(\mathbf{r}')$ within the integral with that of the incident wave. Then

$$\psi(\mathbf{r}) \xrightarrow[r\to\infty]{} Ae^{ikz} - \frac{\mu}{2\pi\hbar^2} \frac{e^{ikr}}{r} \int d^3r' \, e^{-i\mathbf{k}_f\cdot\mathbf{r}'} V(\mathbf{r}') Ae^{ikz'} \tag{13.44}$$

Comparing this equation with general asymptotic expression (13.7) reveals that the scattering amplitude is given in the Born approximation by

$$f(\theta, \phi) = -\frac{\mu}{2\pi\hbar^2} \int d^3r' \, e^{-i\mathbf{k}_f\cdot\mathbf{r}'} V(\mathbf{r}') e^{ikz'}$$

$$= -\frac{\mu}{2\pi\hbar^2} \int d^3r' \, e^{-i\mathbf{k}_f\cdot\mathbf{r}'} V(\mathbf{r}') e^{i\mathbf{k}_i\cdot\mathbf{r}'}$$

$$= -\frac{\mu}{2\pi\hbar^2} \int d^3r' \, V(\mathbf{r}') e^{i\mathbf{q}\cdot\mathbf{r}'} \tag{13.45}$$

where we have introduced the vector $\mathbf{k}_i$ with magnitude $k$ directed along the $z$ axis, the direction of the incident wave. Note that the vector $\hbar\mathbf{q} = \hbar\mathbf{k}_i - \hbar\mathbf{k}_f = \mathbf{p}_i - \mathbf{p}_f$ is just the momentum transferred from the incident beam to the target during the scattering process and that the scattering amplitude is, up to an overall constant, just the Fourier transform of the potential energy with respect to $\mathbf{q}$. This Born approximation can be considered as the first in a series of approximations arising from an iterative procedure in which the wave function determined by the previous iteration, such as (13.44), is then substituted for the exact wave function on the right-hand side of (13.43).

A rough estimate of the range of validity of the Born approximation can be made by noting that since we replace $\psi(\mathbf{r}')$ by $\psi_{inc}$ in (13.43), we want $|\psi_{sc}/\psi_{inc}| \ll 1$ within the range of the potential (where $V ( 0)$ , that is, in the

---

[2] An example might be using laser light to implode a pellet of deuterium in an attempt to generate thermonuclear fusion reactions.

vicinity of the origin. Comparing (13.7) with (13.44), we see that

$$\psi_{sc}(\mathbf{r}) = -\frac{\mu}{2\pi\hbar^2} \int d^3r' \frac{e^{ik|\mathbf{r}-\mathbf{r}'|}}{|\mathbf{r}-\mathbf{r}'|} V(\mathbf{r}')\psi(\mathbf{r}') \qquad (13.46)$$

Thus since $\psi_{inc}(0) = A$,

$$\left|\frac{\psi_{sc}(0)}{\psi_{inc}(0)}\right| = \left|-\frac{\mu}{2\pi\hbar^2} \int d^3r' \frac{e^{ikr'}}{r'} V(\mathbf{r}')e^{ikz'}\right| \qquad (13.47)$$

where we have replaced the exact wave function on the right-hand side of (13.46) with the incident wave function in accord with the Born approximation. If the potential energy is spherically symmetric, $V(\mathbf{r}') = V(r')$, we can carry out the angular integrals, and the condition for the validity of the Born approximation becomes

$$\left|\frac{\psi_{sc}(0)}{\psi_{inc}(0)}\right| = \left|\frac{\mu}{2\pi\hbar^2} \int_0^\infty dr' \int_0^{2\pi} d\phi' \int_0^\pi \sin\theta' d\theta' \, r' e^{ikr'} V(r') e^{ikr'\cos\theta'}\right|$$

$$= \left|\frac{2\mu}{\hbar^2 k} \int_0^\infty dr' \, e^{ikr'} V(r') \sin kr'\right| \ll 1 \qquad (13.48)$$

At *high energies* ($k \to \infty$), the exponential and the sine in (13.48) oscillate rapidly and cut off the integral for $r' \gtrsim 1/k$. The condition (13.48) becomes in this case

$$\left|\frac{2\mu}{\hbar^2 k} \int_0^{1/k} dr' \, V(r')kr'\right| \sim \frac{\mu V_0}{\hbar^2 k} \int_0^{1/k} dr' \, kr' \sim \frac{\mu V_0}{\hbar^2 k^2} \ll 1 \qquad (13.49a)$$

or

$$\frac{V_0}{E} \ll 1 \qquad (13.49b)$$

as we argued earlier. The Born approximation may also be valid at low energies, but under more restrictive conditions (see Problem 13.3).

## 13.3   AN EXAMPLE OF THE BORN APPROXIMATION: THE YUKAWA POTENTIAL

Let's evaluate the scattering amplitude for the potential energy

$$V(r) = g\frac{e^{-m_0 r}}{r} \qquad (13.50)$$

known as the Yukawa potential. With the appropriate choice for the values of $g$ and $m_0$, this potential could represent the short-range potential energy between two nucleons, say a neutron and a proton. Or if we choose $g = Z_1 Z_2 e^2$ and $m_0 = 0$, the potential reduces to the Coulomb potential energy between a projectile with charge $Z_1e$ and a nucleus with charge $Z_2e$, as in Rutherford scattering. However, the Coulomb potential does not vanish fast enough for large $r$ to ensure that (13.7) is an asymptotic solution to the Schrödinger equation in this case, and thus our formalism is not appropriate for a pure Coulomb interaction. Nonetheless, we can consider the factor $e^{-m_0 r}$ as a mathematically convenient way to introduce the screening that actually occurs in Rutherford scattering, where the electrons within the atom shield the incident $\alpha$ particle from the nucleus until the $\alpha$ particle penetrates the electron cloud. Recall that the size of the atom is on the order of an angstrom, while the size of the nucleus is between $10^4$ and $10^5$ times smaller. From (13.45)

$$f(\theta, \phi) = -\frac{\mu g}{2\pi\hbar^2} \int d^3r' \frac{e^{-m_0 r'}}{r'} e^{i\mathbf{q}\cdot\mathbf{r}'} \tag{13.51}$$

In order to carry out the integrals, it is convenient to choose our dummy integration variables so that the $z'$ axis is parallel to $\mathbf{q}$. Then $\mathbf{q}\cdot\mathbf{r}' = qz' = qr'\cos\theta'$, where $\theta'$ is the usual polar angle in spherical coordinates. Thus

$$f = -\frac{\mu g}{2\pi\hbar^2} \int_0^\infty dr' \int_0^{2\pi} d\phi' \int_0^\pi d\theta' \sin\theta' r' e^{-m_0 r'} e^{iqr'\cos\theta'} \tag{13.52}$$

Carrying out the angular integrals first, we obtain

$$f = \frac{\mu g i}{\hbar^2 q} \int_0^\infty dr' \, e^{-m_0 r'} \left( e^{iqr'} - e^{-iqr'} \right)$$

$$= \frac{-2\mu g}{\hbar^2 (m_0^2 + q^2)} \tag{13.53}$$

Note that

$$q^2 = (\mathbf{k}_i - \mathbf{k}_f)^2$$
$$= (k^2 - 2\mathbf{k}_i \cdot \mathbf{k}_f + k^2)$$
$$= 2k^2(1 - \cos\theta) = 4k^2 \sin^2\frac{\theta}{2} \tag{13.54}$$

and therefore $f = f(\theta)$ only. The angle $\theta$ is shown in Fig. 13.7. Thus

$$\frac{d\sigma}{d\Omega} = |f(\theta)|^2 = \frac{4\mu^2 g^2}{\hbar^4 [m_0^2 + 4k^2 \sin^2(\theta/2)]^2} \tag{13.55}$$

The differential cross section depends only on the angle $\theta$ and not on the angle $\phi$ because of azimuthal symmetry under rotations about the $z$ axis for a spherical potential.

We now specialize to the case of Coulomb scattering. At energies and/or angles such that $4k^2 \sin^2(\theta/2) \gg m_0^2$, the expression (13.55) for the differential cross section reduces to

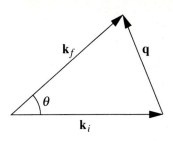

**FIGURE 13.7**
The incident wave vector $\mathbf{k}_i$, the scattered wave vector $\mathbf{k}_f$, and the vector $\mathbf{q}$.

$$\frac{d\sigma}{d\Omega} = \frac{\mu^2(Z_1Z_2e^2)^2}{\hbar^4 4k^4 \sin^4(\theta/2)}$$

$$= \frac{(Z_1Z_2e^2)^2}{16E^2 \sin^4(\theta/2)} \tag{13.56}$$

the famous result for Rutherford scattering. Interestingly, the dependence on Planck's constant has disappeared entirely. This differential cross section (13.56) is in complete agreement with that obtained from a classical analysis of Coulomb scattering, as well as with that obtained from an exact solution using quantum mechanics. Without this rather fortunate agreement between the classical and quantum results, the historical development of quantum mechanics would almost certainly have been quite different. It was the agreement between (13.56) and experiment that led Rutherford, *before* the advent of quantum mechanics, to the nuclear model of the atom. In classical physics, this model could not be stable as the electrons in their classical orbits radiated away their energy and spiraled into the nucleus. Bohr first addressed these problems with his own model, with its stationary states and discrete energies, in the crucial transition between classical and quantum mechanics.

## 13.4   THE PARTIAL WAVE EXPANSION

In general, the Born approximation is a high-energy approximation for calculating the differential cross section. There is another approach, known as the partial wave expansion, that is most useful at low energies and is therefore somewhat complementary to the Born approximation.

As we have seen in Rutherford scattering, if the potential energy is spherically symmetric, the scattering amplitude is a function of $\theta$ only:

$$f(\theta, \phi) = f(\theta) \tag{13.57}$$

We begin by writing

$$f(\theta) = \sum_{l=0}^{\infty}(2l + 1)a_l(k)P_l(\cos\theta) \tag{13.58}$$

as a superposition of the partial waves, where

$$P_l(\cos\theta) = \sqrt{\frac{4\pi}{2l+1}}\, Y_{l,0} \tag{13.59}$$

is a Legendre polynomial (see Problem 9.17). In a sense, all we are saying in (13.58) is that we can write any function of $\theta$ as a superposition of Legendre polynomials. After all, these functions form a complete set. The rationale for expressing the coefficients in the expansion in the form (13.58) will become clear shortly. For now, let us note that the value of the coefficient $a_l(k)$ will, in general, depend on the value of the energy, a dependence exhibited explicitly by indicating that the partial wave is a function of $k$.

Although in principle we have traded $f(\theta)$ for an infinite set of partial waves, the utility of (13.58) comes from the fact that at low energies only a few of the $a_l(k)$ are significantly different from zero. To see why, we first give a heuristic semiclassical argument. In general, a beam of particles incident on a target in a scattering experiment with a broad spectrum of impact parameters consists of many different orbital angular momenta, as can be seen by evaluating $\mathbf{r} \times \mathbf{p}$ for each impact parameter (see Fig. 13.8). However, if the potential energy has a finite range $a$, scattering will occur only for those impact parameters that are less than $a$. Thus for interaction there is a maximum angular momentum, whose value is roughly given by $\hbar l_{\max} = ap = a(\hbar k)$, or $l_{\max} = ak$. The lower the energy and the smaller the value of $k$, the fewer angular momentum states can interact with the target.

To start our partial wave analysis in quantum mechanics, we express the incident plane wave in the form (see Problem 13.8):

$$e^{ikz} = e^{ikr\cos\theta} = \sum_{l=0}^{\infty} i^l(2l+1)j_l(kr)P_l(\cos\theta) \tag{13.60}$$

which can be considered as a special case of the more general expansion

$$\psi(\mathbf{r}) = \sum_l c_l R_l(r)Y_{l,0}(\theta) \tag{13.61}$$

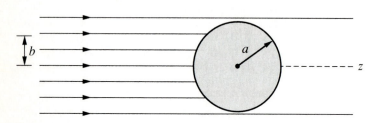

**FIGURE 13.8**
A classical representation of the incident flux in terms of particles following well-defined trajectories. Those particles with impact parameter $b$ possess orbital angular momentum $|\mathbf{L}| = |\mathbf{r} \times \mathbf{p}| = b\hbar k$. Only particles with impact parameters less than or equal to the range $a$ of the potential energy would interact with the target.

Only the $Y_{l,0}$'s enter because the plane wave (13.60) is independent of $\phi$. Since the plane wave is the wave function of a free particle, the radial functions must be spherical Bessel functions. We must discard the spherical Neumann functions in the expansion because they blow up at the origin, as we discussed in Chapter 10. Clearly, the plane wave is finite at the origin. Note that the appearance of all angular momenta in this expansion is consistent with a picture of a plane wave, which is infinite in extent, as having all impact parameters.

What can we say about the asymptotic form of the full wave function (13.7), including the scattered wave, in terms of partial waves? Since we are searching for a solution of the Schrödinger equation for $r \to \infty$, a region in which $V = 0$ but one that excludes the origin, we must include both spherical Bessel and Neumann functions in our general expression for the function $R_l(r)$ in (13.61):

$$\psi(\mathbf{r}) \xrightarrow[r \to \infty]{} \sum_l [A_l j_l(kr) + B_l \eta_l(kr)] \, P_l(\cos \theta) \tag{13.62}$$

The asymptotic expressions for the Bessel functions themselves are given by

$$j_l(kr) \xrightarrow[r \to \infty]{} \frac{\sin(kr - l\pi/2)}{kr} \qquad \eta_l(kr) \xrightarrow[r \to \infty]{} -\frac{\cos(kr - l\pi/2)}{kr} \tag{13.63}$$

Substituting these forms into (13.60) and (13.62), we obtain

$$e^{ikz} \xrightarrow[r \to \infty]{} \sum_{l=0}^{\infty} i^l (2l + 1) \frac{\sin(kr - l\pi/2)}{kr} P_l(\cos \theta) \tag{13.64}$$

for the incident wave and

$$\psi(\mathbf{r}) \xrightarrow[r \to \infty]{} \sum_l \left[ A_l \frac{\sin(kr - l\pi/2)}{kr} - B_l \frac{\cos(kr - l\pi/2)}{kr} \right] P_l(\cos \theta)$$

$$= \sum_{l=0}^{\infty} C_l \frac{\sin[kr - l\pi/2 + \delta_l(k)]}{kr} P_l(\cos \theta) \tag{13.65}$$

for the complete wave function, where in the last step we have combined the sine and cosine into a sine function with its phase shifted by $\delta_l(k)$. Comparing (13.64) and (13.65), we see that the effect of the potential is to introduce a phase shift in the asymptotic wave function. Figure 13.9 shows qualitatively how this happens. We can express (13.65) in the form

$$\psi \xrightarrow[r \to \infty]{} \sum_{l=0}^{\infty} C_l \frac{e^{i(kr - l\pi/2 + \delta_l)} - e^{-i(kr - l\pi/2 + \delta_l)}}{2ikr} P_l(\cos \theta) \tag{13.66}$$

which contains both incoming and outgoing spherical waves. What is the source of these incoming spherical waves? They must be due to the presence of the incident plane wave in the full asymptotic wave function. If we rewrite (13.64) as

$$e^{ikz} \xrightarrow[r \to \infty]{} \sum_{l=0}^{\infty} i^l (2l + 1) \frac{e^{i(kr - l\pi/2)} - e^{-i(kr - l\pi/2)}}{2ikr} P_l(\cos \theta) \tag{13.67}$$

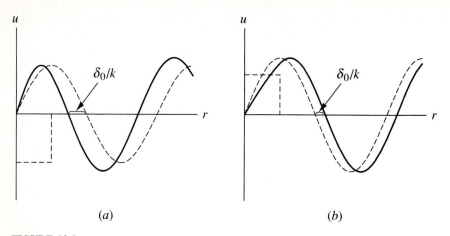

**FIGURE 13.9**
A depiction of how the potential energy affects the phase of a wave. ($a$) A potential well (an attractive potential) produces a positive phase shift ($\delta_0 > 0$) for the radial function $u = rR$ while in ($b$) a potential barrier (a repulsive potential) generates a negative phase shift ($\delta_0 < 0$). The dashed curve shows $u$ when $V = 0$ in each case.

we see these incoming spherical waves explicitly. Since[3]

$$\psi - e^{ikz} \xrightarrow[r \to \infty]{} f(\theta)\frac{e^{ikr}}{r} \tag{13.68}$$

which is an outgoing spherical wave only, the incoming spherical waves must cancel if we subtract (13.67) from (13.66), which implies that

$$C_l = (2l + 1)e^{il\pi/2}e^{i\delta_l} \tag{13.69}$$

With this result, we see that

$$f(\theta) = \sum_{l=0}^{\infty}(2l + 1)\frac{1}{2ik}(e^{2i\delta_l} - 1)P_l(\cos\theta)$$

$$= \sum_{l=0}^{\infty}(2l + 1)\frac{e^{i\delta_l}}{k}\sin\delta_l P_l(\cos\theta) \tag{13.70}$$

Thus, comparing (13.58) and (13.70), we find

$$a_l(k) = \frac{e^{i\delta_l}}{k}\sin\delta_l \tag{13.71}$$

Determining the scattering amplitude through a decomposition into partial waves is equivalent to determining the phase shift for each of these partial waves.

---

[3] In this section we have set the amplitude $A$ of the incident plane wave equal to unity for mathematical simplicity. Note that the differential cross section (13.24) is independent of $A$.

In order to determine the total cross section

$$\sigma = \int d\Omega \, \frac{d\sigma}{d\Omega} = \int d\Omega \, |f(\theta)|^2 \tag{13.72}$$

we take advantage of (13.59) and the orthogonality of the spherical harmonics, namely,

$$\int d\Omega \, Y^*_{l,m}(\theta, \phi) Y_{l',m'}(\theta, \phi) = \delta_{l,l'}\delta_{m,m'} \tag{13.73}$$

to do the integral over the solid angle:

$$\sigma = \frac{4\pi}{k^2} \sum_{l=0}^{\infty} (2l + 1) \sin^2 \delta_l \tag{13.74}$$

Comparing this result with the expression (13.70) for the scattering amplitude, we see that

$$\sigma = \frac{4\pi}{k} \, \text{Im} \, f(0), \tag{13.75}$$

where we have taken advantage of the fact that the Legendre polynomials satisfy $P_l(1) = 1$. Equation (13.75) is known as the *optical theorem* and is a reflection of the fact that, as we discussed earlier, the very existence of scattering requires scattering in the forward direction in order to interfere with the incident wave and reduce the probability current in the forward direction.

Finally, it is common to express (13.74) as

$$\sigma = \sum_{l=0}^{\infty} \sigma_l \tag{13.76}$$

where

$$\sigma_l = \frac{4\pi}{k^2} (2l + 1) \sin^2 \delta_l \tag{13.77}$$

the $l$th partial cross section, is the contribution to the total cross section by the $l$th partial wave. Note that the maximum value for the $l$th partial cross section occurs when the phase shift $\delta_l = \pi/2$.

## 13.5   EXAMPLES OF PHASE-SHIFT ANALYSIS

### Hard-Sphere Scattering

We first analyze the scattering from the repulsive potential

$$V(r) = \begin{cases} \infty & r < a \\ 0 & r > a \end{cases} \tag{13.78}$$

which characterizes a very hard (impenetrable) sphere. Our earlier discussion suggests that at sufficiently low energy the $l = 0$ partial wave dominates the expansion (13.70). Determining the phase shift is particular easy for S-wave scattering, since when $l = 0$ the radial equation simplifies considerably with the elimination of the centrifugal barrier $l(l + 1)\hbar^2/2\mu r^2$. Outside the sphere, the function $u = Rr$ satisfies the free-particle equation

$$-\frac{\hbar^2}{2\mu}\frac{d^2u}{dr^2} = Eu \qquad r > a \qquad (13.79)$$

Rather than write the solution to (13.79) in the form $u = B\cos kr + C\sin kr$ (or in the form $u = Be^{ikr} + Ce^{-ikr}$), we are guided by asymptotic form for the radial wave function in (13.65) to write

$$u = C\sin(kr + \delta_0) \qquad r > a \qquad (13.80)$$

where, as usual, $k = \sqrt{2\mu E/\hbar^2}$. Figure 13.10 shows a plot of $u$. The boundary condition $u(a) = 0$ determines the S-wave phase shift:

$$C\sin(ka + \delta_0) = 0 \qquad \text{or} \qquad \delta_0 = -ka \qquad (13.81)$$

Thus the S-wave total cross section is given by

$$\sigma_{l=0} = \frac{4\pi}{k^2}\sin^2\delta_0 = \frac{4\pi}{k^2}\sin^2 ka \qquad (13.82)$$

Problem 13.10 shows that the higher partial waves, such as the P-wave, can be neglected relative to the S-wave for hard-sphere scattering at low energy. Thus

$$\sigma \xrightarrow[ka\to 0]{} \sigma_{l=0} \xrightarrow[ka\to 0]{} \frac{4\pi}{k^2}(ka)^2 = 4\pi a^2 \qquad (13.83)$$

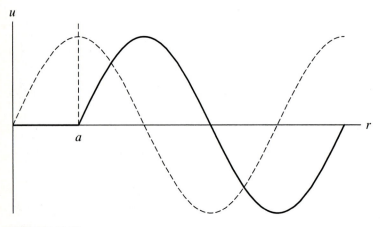

**FIGURE 13.10**
A plot of the wave function $u = rR$ for S-wave scattering from a hard sphere showing the phase shift $\delta_0 = -ka$. The dashed curve shows $u$ when $V = 0$.

Notice the cross section is indeed an area, but in this case the area is *four* times the classical cross section $\pi a^2$ that the sphere presents in the form of a disk that blocks the incident plane wave. Of course, low-energy scattering corresponds to a very long wavelength for the incident wave, and thus we should not expect to obtain the classical result. However, even at high energies and very short wavelengths we cannot completely avoid diffraction effects. We give a heuristic argument. At high energies, many partial waves, up to $l_{\max} = ka$, should contribute to the scattering. Therefore

$$\sigma \xrightarrow[ka \gg 1]{} \sum_{l=0}^{ka} \frac{4\pi}{k^2}(2l + 1)\sin^2 \delta_l \tag{13.84}$$

With so many $l$ values contributing, we assume we can replace $\sin^2 \delta_l$ by its average value, $\frac{1}{2}$, in the sum.[4] Then

$$\sigma \xrightarrow[ka \gg 1]{} \sum_{l=0}^{ka} \frac{2\pi}{k^2}(2l + 1) \xrightarrow[ka \gg 1]{} 2\pi a^2 \tag{13.85}$$

Why do we obtain twice the classical result, even at high energy? For hard-sphere scattering, partial waves with impact parameters less than $a$ must be deflected. However, in order to produce a "shadow" behind the sphere, there must be scattering in the forward direction (recall the optical theorem) to produce destructive interference with the incident plane wave. In fact, the interference is not completely destructive and the shadow has a bright spot (known in optics as Poisson's bright spot[5]) in the forward direction.

## S-Wave Scattering for the Finite Potential Well

As another example of the determination of the phase shift at low energy, we examine scattering from an attractive potential, namely, the spherical well

$$V(r) = \begin{cases} -V_0 & r < a \\ 0 & r > a \end{cases} \tag{13.86}$$

In terms of the function $u = Rr$, the energy eigenvalue equation becomes

$$-\frac{\hbar^2}{2\mu}\frac{d^2 u}{dr^2} - V_0 u = Eu \qquad r < a \tag{13.87a}$$

$$-\frac{\hbar^2}{2\mu}\frac{d^2 u}{dr^2} = Eu \qquad r > a \tag{13.87b}$$

---

[4] For a detailed analysis, see J. J. Sakurai, *Modern Quantum Mechanics*, Benjamin-Cummings, Menlo Park, Calif., 1985, p. 408.

[5] Ironically, Poisson, who supported a corpuscular theory for light, refused to believe Fresnel's prediction that a bright spot would occur in the shadow of an illuminated disk, until it was experimentally verified.

Equation (13.87$a$) can be written as

$$\frac{d^2u}{dr^2} = -k_0^2 u \qquad k_0 = \sqrt{\frac{2\mu}{\hbar^2}(E + V_0)} \qquad r < a \qquad (13.88a)$$

while equation (13.87$b$) is the usual

$$\frac{d^2u}{dr^2} = -k^2 u \qquad k = \sqrt{\frac{2\mu}{\hbar^2}E} \qquad r > a \qquad (13.88b)$$

The solution to (13.88$a$) that satisfies the boundary condition $u(0) = 0$ is

$$u = A \sin k_0 r \qquad r < a \qquad (13.89a)$$

As before, we write the solution outside the well in the form

$$u = C \sin(kr + \delta_0) \qquad r > a \qquad (13.89b)$$

allowing explicitly for the appearance of a phase shift. The finite spherical well is especially nice because we can determine analytically the wave function everywhere, both inside and outside the well.[6]

Making sure that $u$ is continuous and has a continuous first derivative at $r = a$, we obtain

$$A \sin k_0 a = C \sin(ka + \delta_0) \qquad (13.90a)$$

$$A k_0 \cos k_0 a = C k \cos(ka + \delta_0) \qquad (13.90b)$$

Dividing these two equations, we obtain

$$\tan(ka + \delta_0) = \frac{k}{k_0} \tan k_0 a = \frac{ka}{k_0 a} \tan k_0 a \qquad (13.91)$$

This equation for the S-wave phase shift simplifies considerably at sufficiently low energy, that is, $ka \to 0$. Since $ka/k_0 a \ll 1$, as long as $\tan k_0 a$ is not too large, the right-hand side of (13.91) is much less than one and we can replace the tangent of a small quantity with the quantity itself. Thus

$$ka + \delta_0 \cong \frac{ka}{k_0 a} \tan k_0 a, \qquad (13.92a)$$

or

$$\delta_0 \cong ka\left(\frac{\tan k_0 a}{k_0 a} - 1\right) \qquad (13.92b)$$

---

[6] For other potential energies for which an analytic solution is not so easy to determine, we can still solve the energy eigenvalue equation by integrating the Schrödinger equation numerically outwards from the origin. Comparison of the numerical solution with the asymptotic form of the radial wave function that appears in (13.65) permits a determination of the phase shift.

From (13.77),

$$\sigma_{l=0} = \frac{4\pi}{k^2} \sin^2 \delta_0 \cong \frac{4\pi}{k^2} \left[ ka \left( \frac{\tan k_0 a}{k_0 a} - 1 \right) \right]^2$$

$$= 4\pi a^2 \left( \frac{\tan k_0 a}{k_0 a} - 1 \right)^2 \qquad (13.93)$$

Since

$$k_0 a = \sqrt{k^2 a^2 + \frac{2\mu V_0 a^2}{\hbar^2}} \qquad (13.94)$$

then for sufficiently small $ka$

$$k_0 a \cong \sqrt{\frac{2\mu V_0 a^2}{\hbar^2}} \qquad (13.95)$$

and the total (S-wave) cross section is independent of energy.

## Resonances

There is a significant exception to this independence of the cross section on energy. Suppose that the quantity $\sqrt{2\mu V_0 a^2/\hbar^2}$ is slightly less than $\pi/2$. Then as the energy increases, $k_0 a$, as given in (13.94), can reach the value of $\pi/2$. In this case, $\tan k_0 a$ is infinite, and therefore we can no longer assume that the right-hand side of (13.91) is small, even for small $ka$. In fact, at the value of the energy when $k_0 a = \pi/2$, $\tan(ka + \delta_0) = \infty$ and, consequently, $ka + \delta_0 = \pi/2$, which implies $\delta_0 \cong \pi/2$ since we are presuming $ka \ll 1$.[7] Thus

$$\sigma_{l=0} = \frac{4\pi}{k^2} \sin^2 \delta_0 \cong \frac{4\pi}{k^2} = 4\pi a^2 \left( \frac{1}{k^2 a^2} \right) \qquad (13.96)$$

Here we see a pronounced dependence of the total cross section on energy. Also notice that the magnitude of the total cross section is much larger than that given in (13.93). Instead of being a small quantity, the phase shift $\delta_0 = \pi/2$.

What is causing this unusual behavior? If you return to our discussion of the bound states of the finite spherical well in Chapter 10, you will recognize the condition

$$\sqrt{\frac{2\mu V_0 a^2}{\hbar^2}} \cong \frac{\pi}{2} \qquad (13.97)$$

---

[7] Since the phase shift $\delta_0$ starts at zero, we are assuming that the condition $\delta_0 = \pi/2$ is the first resonance condition that we reach as this phase shift grows.

as the condition that the well has a bound state at zero energy. Thus for a potential well satisfying (13.97), the energy of the scattering system is essentially the same as the energy of the bound state. In this case, the incident particle in a scattering experiment would like to form a bound state in the well. Since the system has a small but positive energy, the bound state isn't stable. However, pulling in the wave function in an attempt to form such a bound system dramatically changes the scattering behavior.

Although it is more difficult to determine the phase shifts and cross sections in such a nice analytic form as (13.92) for higher partial waves, it is easy to see physically why these higher partial waves may exhibit resonant behavior, with large "bumps" in the partial cross sections. These bumps arise when the phase shift goes through odd multiples of $\pi/2$. Figure 13.11 shows a plot of the effective potential energy

$$V_{\text{eff}}(r) = V(r) + \frac{l(l+1)\hbar^2}{2\mu r^2} \tag{13.98}$$

for the spherical well. A particle with energy $E$ greater than zero but less than the height of the barrier can tunnel through the barrier and form a metastable bound state in the well. This state is metastable (and not stable) because a particle "trapped" inside the well can also tunnel out. Thus if the energy of the beam in a scattering experiment is tuned to one of the energies of these metastable states, there is an enhanced tendency for the particle to get stuck in the well. The system then loses track of the mechanism by which the bound state was formed, that is, in particular it may lose track of the direction of the incident beam. When this metastable state decays, it emits the particle with the characteristic angular distribution for that particular decay mode.

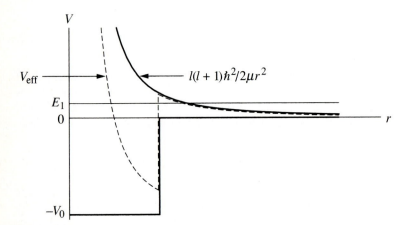

**FIGURE 13.11**
The centrifugal barrier combines with the potential well to produce an effective potential that can produce a metastable state, as indicated. If the energy of the incident beam coincides with the energy of one of these metastable states, a resonance in the scattering cross section can occur.

A convenient way to parametrize the behavior of the phase shift in the vicinity of a resonance leads to the famous Breit-Wigner formula. We assume the phase shift $\delta_l$ of the $l$th partial wave goes through $\pi/2$ at an energy $E_0$:

$$\delta_l(E_0) = \frac{\pi}{2} \tag{13.99}$$

We next make a Taylor series expansion of $\cot \delta_l$ in the vicinity of the resonant energy:

$$\cot \delta_l(E) = \cot \delta_l(E_0) + \left(\frac{d \cot \delta_l}{dE}\right)_{E=E_0} (E - E_0) + \cdots$$

$$= -\left(\frac{1}{\sin^2 \delta_l} \frac{d\delta_l}{dE}\right)_{E=E_0} (E - E_0) + \cdots \tag{13.100}$$

Defining

$$\left(\frac{d\delta_l(E)}{dE}\right)_{E=E_0} = \frac{2}{\Gamma} \tag{13.101}$$

we obtain

$$\cot \delta_l(E) = -\frac{2}{\Gamma}(E - E_0) + \cdots \tag{13.102}$$

Finally, we can express the function $a_l(k)$ (see 13.71) as

$$a_l(k) = \frac{e^{i\delta_l}}{k} \sin \delta_l$$

$$= \frac{1}{k} \frac{1}{\cot \delta_l - i}$$

$$\cong \frac{1}{k} \frac{1}{-(2/\Gamma)(E - E_0) - i} = -\frac{1}{k} \frac{\Gamma/2}{(E - E_0) + i\Gamma/2} \tag{13.103}$$

Repeating the steps leading to (13.74), we find that the total cross section for the $l$th partial wave in the vicinity of a resonance is thus given by

$$\sigma_l \cong \frac{4\pi}{k^2}(2l + 1)\frac{\Gamma^2/4}{(E - E_0)^2 + \Gamma^2/4} \tag{13.104}$$

As an example, Fig. 13.12$a$ shows a strong resonance in $\pi^+ - p$ scattering with a peak at roughly 190 MeV of incident pion kinetic energy. This resonance, known as the $\Delta(1232)$ because the center-of-mass energy of the resonance at the peak is 1232 MeV, has a full width at half maximum $\Gamma$ of 110–120 MeV. Figure 13.12$b$ shows the P-wave phase shift, which reaches $\pi/2$ at the resonance peak. The resonance is thus formed in the $l = 1$ channel. In fact, the intrinsic spin of the $\Delta(1232)$ is $j = \frac{3}{2}$. From (13.103), we expect $\sigma_{l=1} = 12\pi/k^2$

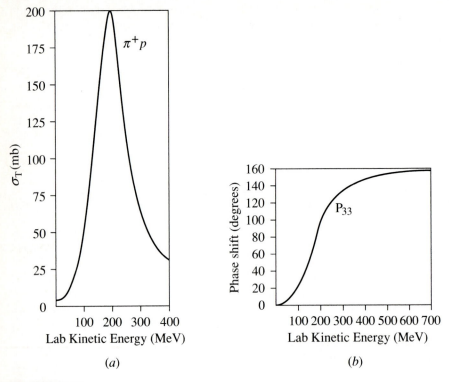

**FIGURE 13.12**
(a) The total cross section for $\pi^+ - p$ scattering with pion kinetic energies up to 400 MeV *(From W. R. Frazer, Elementary Particles, Prentice-Hall, Englewood Cliffs, N. J., 1966.)* (b) The P-wave phase shift for $\pi^+ - p$ scattering. *(From L. D. Roper, R. M. Wright, and B. T. Feld, Phys. Rev. 138, B190, 1965.)*

when $E = E_0$. However, when we add the orbital angular momentum $l = 1$ of the pion-proton system to an intrinsic spin $s = \frac{1}{2}$ of the proton, we can form a total angular momentum $j = \frac{1}{2}$ (two states) as well as $j = \frac{3}{2}$ (four states). Thus of the six total angular momentum states that are generated (presumably incoherently) in the collision, only the four $j = \frac{3}{2}$ states can produce the resonance. If we assume the scattering cross section for the $j = \frac{1}{2}$ states is negligible in comparison with the $j = \frac{3}{2}$ resonance scattering in the vicinity of the peak, the $\pi^+ - p$ total cross section at the resonance should be $\frac{4}{6}$ of the P-wave peak cross section, that is $8\pi/k^2$, which is about 190 mb, in good agreement with the observed value.

Finally, we return to our discussion of S-wave scattering for the finite potential well and ask what happens if we increase the energy of the beam so that the phase shift $\delta_0 \to \pi$, as illustrated in Fig. 13.13. Then the wave function outside the well,

$$u = C \sin(kr + \pi) = -C \sin kr \qquad (13.105)$$

is the same as the wave function outside the well with zero phase shift [see

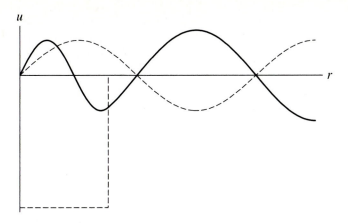

**FIGURE 13.13**
The wave function $u(r)$ for S-wave scattering for a potential well at an energy such that the phase shift $\delta_0 = \pi$. The dashed curve shows $u$ when $V = 0$.

(13.89b)] up to an *overall* phase. Moreover, since $\sin \delta_0 = 0$, the S-wave partial cross section vanishes ($\sigma_{l=0} = 0$). This effect, known as the Ramsauer-Townsend effect,[8] is clearly seen in the very low scattering cross section from noble gases at about 0.7 eV. These rare-gas atoms have a potential well with a sharply defined range, and it is possible at low energies to have $\delta_0 = \pi$, with all other phase shifts negligible, leading to essentially perfect transmission of the incident wave.

## 13.6  SUMMARY

The differential scattering cross section is given by

$$\frac{d\sigma}{d\Omega} = |f(\theta, \phi)|^2 \tag{13.106}$$

where the scattering amplitude $f(\theta, \phi)$ determines the angular dependence in the asymptotic form for the solution to the Schrödinger equation:

$$\psi \xrightarrow[r \to \infty]{} A e^{ikz} + A f(\theta, \phi) \frac{e^{ikr}}{r} \tag{13.107}$$

At "high" energies, the scattering amplitude can be determined through the Born approximation

$$f(\theta, \phi) = -\frac{\mu}{2\pi\hbar^2} \int d^3\mathbf{r}' \, V(r') e^{i\mathbf{q}\cdot\mathbf{r}'} \tag{13.108}$$

---

[8] Unfortunately, a different Townsend.

which is (up to constants) the Fourier transform of the potential energy with respect to the vector $\mathbf{q} = \mathbf{k}_i - \mathbf{k}_f$.

At "low" energies, on the other hand, the scattering amplitude for scattering from a central potential can be determined from the partial wave expansion

$$f(\theta) = \sum_{l=0}^{\infty} (2l + 1) \frac{e^{i\delta_l}}{k} \sin \delta_l P_l(\cos \theta) \tag{13.109}$$

leading to a total cross section

$$\sigma = \frac{4\pi}{k^2} \sum_{l=0}^{\infty} (2l + 1) \sin^2 \delta_l \tag{13.110}$$

The phase shifts $\delta_l$, which are generated when the particular partial waves interact with the potential, enter in the asymptotic expression for the wave function

$$\psi(\mathbf{r}) \xrightarrow[r \to \infty]{} \sum_{l=0}^{\infty} C_l \frac{\sin [kr - l\pi/2 + \delta_l(k)]}{kr} P_l(\cos \theta) \tag{13.111}$$

A useful way to determine the phase shifts is to solve the Schrödinger equation, either numerically or analytically, for the radial wave function $u = rR$, which obeys the one-dimensional Schrödinger equation

$$-\frac{\hbar^2}{2\mu} \frac{d^2 u}{dr^2} + \frac{l(l + 1)\hbar^2}{2\mu r^2} u + V(r)u = Eu \tag{13.112}$$

and compare the asymptotic form of the solution with (13.111).

## PROBLEMS

**13.1.** Use the three-dimensional time-dependent Schrödinger equation

$$-\frac{\hbar^2}{2\mu} \nabla^2 \psi + V\psi = i\hbar \frac{\partial \psi}{\partial t}$$

to establish that the probability density $\psi^*(\mathbf{r}, t)\psi(\mathbf{r}, t)$ obeys the local conservation law

$$\frac{\partial}{\partial t}(\psi^*\psi) + \nabla \cdot \mathbf{j} = 0$$

where

$$\mathbf{j} = \frac{\hbar}{2\mu i}(\psi^* \nabla \psi - \psi \nabla \psi^*)$$

What would happen to your derivation if the potential energy $V$ were imaginary? Is probability conserved? Explain. In nonrelativistic quantum mechanics, such an imaginary potential energy can be used, for example, to account for particle absorption in interactions with the nucleus.

**13.2.** Evaluate the probability current for the scattered wave

$$\psi_{sc} \underset{r \to \infty}{\longrightarrow} Af(\theta, \phi)\frac{e^{ikr}}{r}$$

and show that

$$\mathbf{j}_{sc} \underset{r \to \infty}{\longrightarrow} \frac{\hbar k}{\mu r^2}|A|^2|f|^2\mathbf{u}_r$$

where $\mathbf{u}_r$ is a unit vector in the direction of the radius.

**13.3.** Show that at low energies ($k \to 0$) the requirement (13.48) for the validity of the Born approximation becomes

$$\left|\frac{2\mu}{\hbar^2 k}\int_0^\infty dr' V(r')kr'\right| \sim \frac{\mu V_0 a^2}{\hbar^2} \ll 1$$

where $V_0$ is the order of magnitude of the potential energy, $a$ is the range of the potential, and we have neglected constants of order unity. By comparing this result with (13.49), argue that if the Born approximation is valid at low energies, it works at high energies as well.

**13.4.** Using the Born approximation to determine the one-dimensional reflection coefficient $R$ for a potential energy $V(x)$ that vanishes everywhere except in the vicinity of the origin:

(a) Show that we can write the solution to the one-dimensional Schrödinger equation in the form

$$\psi(x) = Ae^{ikx} + \int dx' G(x, x')\frac{2m}{\hbar^2}V(x')\psi(x')$$

where

$$\frac{\partial^2}{\partial x^2}G(x, x') + k^2 G(x, x') = \delta(x - x')$$

(b) Since $G$ satisfies a second-order differential equation, $G$ must be a continuous function and, in particular, it must be continuous at $x = x'$. By integrating the differential equation for $G$ from just below to just above $x = x'$, show that the first derivative of $G$ is discontinuous at $x = x'$ and that it satisfies

$$\left(\frac{\partial G}{\partial x}\right)_{x=x'_+} - \left(\frac{\partial G}{\partial x}\right)_{x=x'_-} = 1$$

Then show that one solution for $G$ is given by

$$G = \begin{cases} \dfrac{1}{2ik}e^{ik(x-x')} & x > x' \\[2mm] \dfrac{1}{2ik}e^{-ik(x-x')} & x < x' \end{cases}$$

(c) Substitute this expression for $G$ into the equation for $\psi$ in (a). Show that in the Born approximation

$$\psi \xrightarrow[x \to -\infty]{} Ae^{ikx} + Ae^{-ikx} \int_{-\infty}^{\infty} dx' \frac{e^{2ikx'}}{2ik} \frac{2m}{\hbar^2} V(x')$$

and that consequently

$$R = \left| \frac{m}{ik\hbar^2} \int_{-\infty}^{\infty} dx' \, e^{2ikx'} V(x') \right|^2$$

(d) For the potential barrier

$$V(x) = \begin{cases} V_0 & 0 < x < a \\ 0 & \text{elsewhere} \end{cases}$$

the exact reflection coefficient is given by $R = 1 - T$ with

$$T = \frac{1}{1 + [V_0^2/4E(E - V_0)] \sin^2 \sqrt{(2m/\hbar^2)(E - V_0)}a}$$

Show that the exact result for R in the limit $V_0/E \ll 1$ agrees with the result of the Born approximation.

**13.5.** In the initial Rutherford scattering experiment Geiger and Marsden used $\alpha$ particles with an energy of 5 MeV. Choose a reasonable value for $m_0$ and determine the range of angles for which (13.56) should be valid. *Suggestion:* Write $m_0 = 1/a$, where $a$ is a characteristic screening length. How large should $a$ be? *Note:* Interaction with the *electrons* in the atom produces a deflection on the order of $10^{-4}$ radians.

**13.6.** Use the Born approximation to determine the differential cross section for the potential energy

$$V = \frac{C}{r^2}$$

where $C$ is a constant, corresponding to a $1/r^3$ force. *Note:* The result depends on $\hbar$, so it is not the same as the classical result.

**13.7.** Use the Born approximation to determine the differential cross section for the potential energy

$$V(r) = V_0 e^{-r/a}$$

**13.8.** Our goal is to establish (13.60), which, with the aid of (13.59), can be put in the form

$$e^{ikr\cos\theta} = \sum_{l=0}^{\infty} i^l \sqrt{4\pi(2l + 1)} \, j_l(kr) Y_{l,0}(\theta)$$

(a) Explain why the expansion of the plane wave must be of the form

$$e^{ikr\cos\theta} = \sum_{l=0}^{\infty} c_l j_l(kr) Y_{l,0}(\theta)$$

(b) Use the fact that

$$|l, 0\rangle = \frac{1}{\sqrt{(2l)!}} \left(\frac{\hat{L}_-}{\hbar}\right)^l |l, l\rangle$$

(see Problem 9.16) to show

$$c_l j_l(kr) = \frac{1}{\sqrt{(2l)!}} \int d\Omega \, Y_{l,l}^* \left[ \left(\frac{1}{i} e^{i\phi} \left(i \frac{\partial}{\partial\theta} - \cot\theta \frac{\partial}{\partial\phi}\right)\right)^l e^{ikr\cos\theta} \right]$$

$$= \frac{1}{\sqrt{(2l)!}} \int d\Omega \, Y_{l,l}^* \left[ (-1)^l e^{il\phi} \sin^l\theta \frac{d^l}{d(\cos\theta)^l} e^{ikr\cos\theta} \right]$$

where the last step follows from the explicit form (9.142) for the raising operator in position space.

(c) Use the explicit form (9.146) for $Y_{l,l}(\theta, \phi)$ to express this result in the form

$$c_l j_l(kr) = (ikr)^l \frac{2^l l!}{\sqrt{(2l)!}} \sqrt{\frac{4\pi}{(2l+1)!}} \int d\Omega \, |Y_{l,l}(\theta, \phi)|^2 e^{ikr\cos\theta}$$

(d) Finally, isolate $c_l$ by evaluating this expression as $r \to 0$. *Hint:* For small $r$, $j_l(kr)$ behaves as $(kr)^l/(2l+1)!!$, where $(2l+1)!! = (2l+1)(2l-1)\ldots 5 \cdot 3 \cdot 1$.

**13.9.** A particle is scattered by a spherically symmetric potential at sufficiently low energy that the phase shifts $\delta_l = 0$ for $l > 1$ (that is, only $\delta_0$ and $\delta_1$ are nonzero). Show that the differential cross section has the form

$$\frac{d\sigma}{d\Omega} = A + B\cos\theta + C\cos^2\theta$$

and determine $A$, $B$, and $C$ in terms of the phase shifts. Determine the total cross section $\sigma$ in terms of $A$, $B$, and $C$.

**13.10.** Evaluate the P-wave phase shift $\delta_1$ for scattering from a hard sphere, for which the potential energy is given by

$$V(r) = \begin{cases} \infty & r < a \\ 0 & r > a \end{cases}$$

Express your result in terms of $j_1(ka)$ and $\eta_1(ka)$. Use the leading behavior of $j_1(\rho)$ and $\eta_1(\rho)$ for small $\rho$ to show that $\delta_1 \to -(ka)^3/3$ as $ka \to 0$, and thus $\delta_1$ can indeed be neglected in comparison to $\delta_0$ [see (13.81)] at sufficiently low energy.

**13.11.** Compare the Born approximation result for the total cross section for scattering from the potential well

$$V(r) = \begin{cases} -V_0 & r < a \\ 0 & r > a \end{cases}$$

with that obtained by using S-wave phase shift analysis. Using the condition for the validity of the Born approximation at low energy (see Problem 13.3), show that the two approaches are in agreement when the Born approximation is valid.

**13.12.** Consider the spherically symmetric potential energy $2\mu V(r)/\hbar^2 = \gamma\delta(r - a)$, where $\gamma$ is a constant and $\delta(r - a)$ is a Dirac delta function that vanishes everywhere except on the spherical surface specified by $r = a$.

(a) Show that the S-wave phase shift $\delta_0$ for scattering from this potential satisfies the equation

$$\tan(ka + \delta_0) = \frac{\tan ka}{1 + (\gamma/k)\tan ka}$$

(b) Evaluate the phase shift in the low-energy limit and show that the total cross section for S-wave scattering is

$$\sigma \cong 4\pi a^2 \left(\frac{\gamma a}{1 + \gamma a}\right)^2$$

**13.13** (a) Determine the differential cross section $d\sigma/d\Omega$ in the Born approximation for scattering from the potential energy $2\mu V(r)/\hbar^2 = \gamma\delta(r - a)$ (see Problem 13.12). Show the explicit dependence of $d\sigma/d\Omega$ on $\theta$.

(b) Evaluate $d\sigma/d\Omega$ in the low-energy limit. Show that the differential cross section is isotropic. What is the total cross section?

(c) Use the condition for the validity of the Born approximation at low energy (see Problem 13.3) to establish that your result in (b) for the total cross section agrees with that given in Problem 13.12 in the appropriate limit.

# CHAPTER
# 14

# PHOTONS
# AND ATOMS

In this chapter we turn our attention to a quantum treatment of the electromagnetic field. After analyzing the Aharonov-Bohm effect, which demonstrates the unusual role played by the vector potential in quantum mechanics, we use the vector potential to show that the Hamiltonian for the electromagnetic field can be expressed as a collection of harmonic oscillators. The raising and lowering operators for these oscillators turn out to be creation and annihilation operators for photons, the quanta of the electromagnetic field. This quantum theory of the electromagnetic field is then used to determine the lifetimes of excited states of the hydrogen atom using time-dependent perturbation theory.

## 14.1   THE AHARONOV-BOHM EFFECT

Within classical physics, the vector potential $\mathbf{A}$ is simply an auxiliary field that is introduced to help determine the physical electromagnetic fields $\mathbf{E}$ and $\mathbf{B}$. In particular, Gauss's law for magnetism,

$$\nabla \cdot \mathbf{B} = 0 \tag{14.1}$$

implies that we can write

$$\mathbf{B} = \nabla \times \mathbf{A} \tag{14.2}$$

since the divergence of a curl vanishes. Moreover, when expressed in terms of the vector potential, Faraday's law,

$$\nabla \times \mathbf{E} + \frac{1}{c} \frac{\partial \mathbf{B}}{\partial t} = 0 \tag{14.3}$$

becomes

$$\nabla \times \left( \mathbf{E} + \frac{1}{c} \frac{\partial \mathbf{A}}{\partial t} \right) = 0 \tag{14.4}$$

which implies

$$\mathbf{E} + \frac{1}{c} \frac{\partial \mathbf{A}}{\partial t} = -\nabla \varphi \tag{14.5}$$

since the curl of a gradient vanishes.

We can always alter the function $\mathbf{A}$ by adding to it a gradient of a scalar function $\chi$:

$$\mathbf{A} \rightarrow \mathbf{A} + \nabla \chi \tag{14.6a}$$

This transformation does not affect the magnetic field (14.2), and the electric field (14.5) will also be unaffected provided

$$\varphi \rightarrow \varphi - \frac{1}{c} \frac{\partial \chi}{\partial t} \tag{14.6b}$$

as well. The transformation specified by (14.6) is known as a *gauge transformation*. Although the potentials $\varphi$ and $\mathbf{A}$ are altered by a gauge transformation, the "physical" electric and magnetic fields are not.

We can see the special role the vector potential plays in nonrelativistic quantum mechanics by considering the Aharonov-Bohm effect. As background, first consider a long solenoid carrying a current. The magnetic field *inside* the solenoid is uniform and has the magnitude $B_0$. From the definition (14.2) of the vector potential, we find

$$\int (\nabla \times \mathbf{A}) \cdot d\mathbf{S} = \int \mathbf{B} \cdot d\mathbf{S} \tag{14.7}$$

for the flux of the magnetic field through any surface $\mathbf{S}$. We can take advantage of Stokes's theorem to convert the surface integral on the left-hand side of (14.7) to a closed line integral:

$$\oint \mathbf{A} \cdot d\mathbf{r} = \int \mathbf{B} \cdot d\mathbf{S} \tag{14.8}$$

For the solenoid we take as our path a circle of radius $\rho$ centered on the axis of the solenoid, as shown in Fig. 14.1. From the azimuthal symmetry of the solenoid, the magnitude of the azimuthal component of $\mathbf{A}$ must be the same everywhere

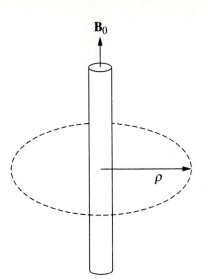

**FIGURE 14.1**
A line integral for evaluating the vector potential for a solenoid.

along the path. Thus we find for a circular path of radius $\rho$ that is less than the radius $R$ of the solenoid

$$\oint \mathbf{A} \cdot d\mathbf{r} = A2\pi\rho = B_0\pi\rho^2 \qquad \rho < R \tag{14.9}$$

or

$$\mathbf{A} = \left(\frac{B_0\rho}{2}\right)\mathbf{u}_\phi \qquad \rho < R \tag{14.10}$$

Outside the solenoid, the integral for the magnetic flux is given by

$$\int \mathbf{B} \cdot d\mathbf{S} = B_0\pi R^2 \qquad \rho > R \tag{14.11}$$

since the magnetic field vanishes outside a long solenoid. Thus from (14.8) we find

$$\mathbf{A} = \left(\frac{B_0 R^2}{2\rho}\right)\mathbf{u}_\phi \qquad \rho > R \tag{14.12}$$

We can check our results (14.10) and (14.12) by using the gradient in cylindrical coordinates,

$$\nabla = \mathbf{u}_\rho \frac{\partial}{\partial\rho} + \mathbf{u}_\phi \frac{1}{\rho}\frac{\partial}{\partial\phi} + \mathbf{u}_z \frac{\partial}{\partial z} \tag{14.13}$$

to evaluate the curl of the vector potential and to verify that it yields a uniform magnetic field $B_0$ within the solenoid and zero field outside the solenoid.

Let's now reconsider the famous double-slit experiment with an additional feature. Suppose that directly behind the barrier between the two slits we insert a small solenoid, as indicated in Fig. 14.2. Recall that the intensity at an arbitrary point $P$ on the screen arises from the interference between the amplitude $\psi_1$ for the particle starting at the source point $S$ to arrive at $P$ after passing through one of the slits and the amplitude $\psi_2$ for it to arrive at $P$ after passing through the other slit. Of course, as we saw in Chapter 8, there are many neighboring paths for both paths 1 and 2 that have essentially the same phase and therefore contribute coherently when evaluating the path integral (8.28).

Surprisingly, the phase for each path that contributes to the path integral is modified by the presence of the solenoid, even though the magnetic field may vanish at all points along the path. According to (E.6), the Lagrangian for a particle of charge $q$ picks up an additional term $q\mathbf{A} \cdot \mathbf{v}/c$, and thus the amplitude to take path 1 is modified by

$$\psi_1 \rightarrow \psi_1 \exp\left[ i\left(\frac{q}{\hbar c}\right)\int_{t_0}^{t'} \mathbf{A} \cdot \mathbf{v}\, dt \right] \tag{14.14}$$

where $t_0$ is the initial time at which it leaves the source and $t'$ is the final time when it reaches the point $P$. Since $\mathbf{v}\, dt = d\mathbf{r}$, we can express (14.14) as

$$\psi_1 \rightarrow \psi_1 \exp\left[ i\left(\frac{q}{\hbar c}\right) \int_{\text{path 1}} \mathbf{A} \cdot d\mathbf{r} \right] \tag{14.15}$$

while the corresponding expression for the amplitude to take path 2 is modified by

$$\psi_2 \rightarrow \psi_2 \exp\left[ i\left(\frac{q}{\hbar c}\right) \int_{\text{path 2}} \mathbf{A} \cdot d\mathbf{r} \right] \tag{14.16}$$

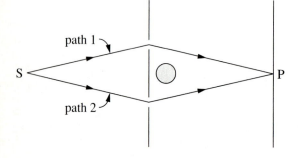

**FIGURE 14.2**
The double-slit experiment with a long solenoid inserted behind the barrier. The closed contour formed by following path 1 from the source $S$ to the point $P$ on the distant screen and then back to source along path 2 includes the magnetic flux of the solenoid.

Thus the amplitude to reach the point $P$ by passing through either of the slits is given by

$$\psi_1 + \psi_2 \rightarrow$$

$$\psi_1 \exp\left[i\left(\frac{q}{\hbar c}\right) \int_{\text{path 1}} \mathbf{A} \cdot d\mathbf{r}\right] + \psi_2 \exp\left[i\left(\frac{q}{\hbar c}\right) \int_{\text{path 2}} \mathbf{A} \cdot d\mathbf{r}\right]$$

$$= \exp\left[i\left(\frac{q}{\hbar c}\right) \int_{\text{path 2}} \mathbf{A} \cdot d\mathbf{r}\right]\left\{\psi_1 \exp\left[i\left(\frac{q}{\hbar c}\right)\left(\int_{\text{path 1}} \mathbf{A} \cdot d\mathbf{r} - \int_{\text{path 2}} \mathbf{A} \cdot d\mathbf{r}\right)\right] + \psi_2\right\}$$

$$= \exp\left[i\left(\frac{q}{\hbar c}\right) \int_{\text{path 2}} \mathbf{A} \cdot d\mathbf{r}\right]\left\{\psi_1 \exp\left[i\left(\frac{q}{\hbar c}\right)\oint \mathbf{A} \cdot d\mathbf{r}\right] + \psi_2\right\} \qquad (14.17)$$

In the last step we find that the *relative* phase between $\psi_1$ and $\psi_2$ is proportional to the *closed* line integral of the vector potential going from the source to the point $P$ along path 1 and *back* to the source from point $P$ along path 2. Taking advantage of (14.8), we see that

$$\psi_1 + \psi_2 \rightarrow \exp\left[i\left(\frac{q}{\hbar c}\right) \int_{\text{path 2}} \mathbf{A} \cdot d\mathbf{r}\right]\left\{\psi_1 \exp\left[i\left(\frac{q}{\hbar c}\right)\int \mathbf{B} \cdot d\mathbf{S}\right] + \psi_2\right\} \qquad (14.18)$$

where the relative phase has now been expressed in terms of the flux of the magnetic field through the closed path. The presence of this relative phase will cause a shift in the interference pattern as the magnetic field in the solenoid varies. For example, when

$$\frac{q}{\hbar c} \int \mathbf{B} \cdot d\mathbf{S} = 2n\pi \qquad n = 0, 1, 2, \ldots \qquad (14.19a)$$

the pattern will be the same as without the magnetic field present, while when

$$\frac{q}{\hbar c} \int \mathbf{B} \cdot d\mathbf{S} = (2n + 1)\pi \qquad n = 0, 1, 2, \ldots \qquad (14.19b)$$

the position of the minima and the maxima in the pattern will be interchanged.

This is a rather startling result. Classically, we would expect that the particle must follow either path 1 or path 2. Along each of these paths the magnetic field $\mathbf{B}$ vanishes everywhere. How then does the charged particle *know* about the magnetic field within the solenoid? While the classical particle responds to the magnetic field only where the particle is, that is, locally, the quantum particle takes both paths. Since the solenoid produces a vector potential that changes the phase for each of the paths, in some sense we might say that the particle compares the phase that it has picked up along the two different paths and responds directly to the phase difference. Notice that this relative phase difference depends on the magnetic flux passing through the surface bounded by the paths and not on the vector potential itself. Thus the phase difference is a gauge invariant quantity that may in fact be measured.[1] Even though the phase difference depends on the magnetic field **B** and not on vector potential **A** directly, the Aharonov-Bohm effect suggests that the particle learns about the magnetic field by responding to the vector potential along the path.

## 14.2 THE HAMILTONIAN FOR THE ELECTROMAGNETIC FIELD

Given the results of the preceding section, we should not be surprised that our discussion of the quantum mechanics of the electromagnetic field starts with the vector potential. We have already taken advantage of two of Maxwell's equations, (14.1) and (14.3), to introduce the scalar potential $\varphi$ and the vector potential **A**. Expressed in terms of these potentials, the remaining two equations, Gauss's law,

$$\mathbf{\nabla} \cdot \mathbf{E} = 4\pi\rho \tag{14.20}$$

and Ampere's law,

$$\mathbf{\nabla} \times \mathbf{B} = \frac{4\pi}{c}\mathbf{j} + \frac{1}{c}\frac{\partial \mathbf{E}}{\partial t} \tag{14.21}$$

are given by

$$-\nabla^2\varphi - \frac{1}{c}\frac{\partial}{\partial t}\mathbf{\nabla} \cdot \mathbf{A} = 4\pi\rho \tag{14.22}$$

and

$$\mathbf{\nabla}(\mathbf{\nabla} \cdot \mathbf{A}) - \nabla^2\mathbf{A} = \frac{4\pi}{c}\mathbf{j} + \frac{1}{c}\frac{\partial}{\partial t}\left(-\mathbf{\nabla}\varphi - \frac{1}{c}\frac{\partial \mathbf{A}}{\partial t}\right) \tag{14.23}$$

respectively.

---

[1] A number of experiments confirming the prediction (14.19) have been carried out. The first was done by R. G. Chambers, *Phys. Rev. Lett.* **5**, 3 (1960). A recent experiment is that of A. Tonomura et al., *Phys. Rev. Lett.* **48**, 1443 (1982). Many people found Y. Aharonov and D. Bohm's 1959 paper (*Phys. Rev.* **115**, 485 (1959)) difficult to believe, which surprised Bohm, for he knew it would be much more surprising if the experiments did not confirm their prediction, since that would mean that quantum mechanics itself was wrong.

A physically transparent gauge for analyzing the electromagnetic field is the Coulomb gauge, which takes advantage of our freedom to make gauge transformations to impose the constraint

$$\nabla \cdot \mathbf{A} = 0 \tag{14.24}$$

on the vector potential. The reason for calling this gauge the Coulomb gauge becomes apparent when we make the replacement (14.24) in (14.22). Then the scalar potential satisfies the equation

$$\nabla^2 \varphi = -4\pi\rho \tag{14.25}$$

for which the solution can be expressed as

$$\varphi(\mathbf{r}, t) = \int d^3r' \, \frac{\rho(\mathbf{r}', t)}{|\mathbf{r} - \mathbf{r}'|} \tag{14.26}$$

This is just the usual expression for the scalar potential arising from a charge distribution $\rho$ in electrostatics. Notice, however, that we have not restricted ourselves to static charge distributions and, in fact, the value of the scalar potential at the position $\mathbf{r}$ at time $t$ is determined by the charge distribution $\rho(\mathbf{r}', t)$ at the same time $t$. Thus there are no retardation effects arising from the finite time $|\mathbf{r} - \mathbf{r}'|/c$ it takes for a change in the charge distribution at $\mathbf{r}'$ to produce a change in the field at $\mathbf{r}$. The absence of these effects in the scalar potential in this gauge emphasizes that physical effects depend directly on the electric and magnetic fields and not on the potentials, which can always be altered by a gauge transformation.

In Appendix E we find that the nonrelativistic Hamiltonian for a particle of mass $m$ and charge $q$ interacting with the electromagnetic field is given by

$$H = \frac{(\mathbf{p} - q\mathbf{A}/c)^2}{2m} + q\varphi \tag{14.27}$$

If we set $\mathbf{A} = 0$, the remaining terms are the kinetic energy and electrostatic potential energy that we have included in our initial treatment of systems such as the hydrogen atom. We will examine the effect of the interaction of the charged particle with the vector potential in the next few sections. First, however, we need to examine the Hamiltonian for the free electromagnetic field, that is, the field energy in the absence of charges and currents ($\rho = 0$, $\mathbf{j} = 0$). In this case, the electromagnetic field energy is given by

$$H_{\text{E\&M}} = \frac{1}{8\pi} \int d^3r \, (\mathbf{E}^2 + \mathbf{B}^2) \tag{14.28}$$

See Problem 14.1. According to (14.26), with no charges, $\varphi = 0$, and from (14.5), $\mathbf{E} = -(1/c)(\partial \mathbf{A}/\partial t)$. Therefore,

$$H_{\text{E\&M}} = \frac{1}{8\pi} \int d^3r \left[ \left( -\frac{1}{c} \frac{\partial \mathbf{A}}{\partial t} \right)^2 + (\nabla \times \mathbf{A})^2 \right] \tag{14.29}$$

Note that we have not put a hat on the Hamiltonian because we are treating the electromagnetic field as a classical field. In fact, our goal of this section is to begin to see how we can make the transition from a classical theory to a fully quantum treatment of the electromagnetic field.[2]

Without charges and currents, the equation of motion (14.23) for the vector potential in the Coulomb gauge is given by

$$\frac{1}{c^2}\frac{\partial^2 \mathbf{A}}{\partial t^2} - \nabla^2 \mathbf{A} = 0 \tag{14.30}$$

A specific solution to this wave equation is given by the plane wave

$$\mathbf{A} = \mathbf{A}_0 e^{i(\mathbf{k}\cdot\mathbf{r}-\omega t)} \tag{14.31}$$

with $\omega = kc$. In addition, the Coulomb gauge condition (14.24) imposes the constraint

$$\nabla \cdot \mathbf{A} = i\mathbf{k} \cdot \mathbf{A}_0 e^{i(\mathbf{k}\cdot\mathbf{r}-\omega t)} = 0 \tag{14.32}$$

Thus $\mathbf{k} \cdot \mathbf{A}_0 = 0$, indicating that the wave is transverse to the direction of propagation, which is in the direction of $\mathbf{k}$. For this reason, the Coloumb gauge is often called the transverse gauge as well.

What is the most general solution to the wave equation (14.30)? It can be obtained by superposing all the different plane wave solutions. In general, this superposition takes the form of an *integral* over all possible values of $\mathbf{k}$. However, in our discussion of the quantum properties of the electromagnetic field it is somewhat easier conceptually to impose boundary conditions on the solutions to the wave equation that dictate that the allowed values of $\mathbf{k}$ take on discrete rather than continuous values and the superposition is in the form of a *sum* rather than an integral. One convenient way to do this is to work in a cubic box of length $L$ on a side subject to *periodic boundary conditions*. For example, we require

$$e^{ik_x x} = e^{ik_x(x+L)} \tag{14.33}$$

The condition (14.33) and the corresponding conditions imposed on $k_y$ and $k_z$ are satisfied provided

$$k_x L = 2\pi n_x \qquad k_y L = 2\pi n_y$$
$$k_z L = 2\pi n_z \qquad n_x, n_y, n_z = 0, \pm 1, \pm 2, \ldots \tag{14.34}$$

Notice, for example, that as $n_x$ takes on all positive and negative integral values, $k_x$ runs from $-\infty$ to $\infty$. Moreover, the separation $\Delta k_x$ between *adjacent* modes is given by $\Delta k_x = 2\pi/L$. Thus as $L \to \infty$ and the volume $V$ of the box in which we are working approaches infinity, the allowed values of $k_x$ approach the continuum that we would have expected had we chosen to work in the infinite volume limit

---

[2] Our approach follows that of J. J. Sakurai, *Advanced Quantum Mechanics*, Addison-Wesley, Reading, Mass., 1967.

initially.[3] It should be emphasized that this discrete set of solutions is a result of our having imposed certain boundary conditions on the solutions to the wave equation. We are still doing strictly classical physics.

With **k** taking on discrete values, the most general solution to the wave equation is given by

$$\mathbf{A}(\mathbf{r}, t) = \sum_{\mathbf{k}, \lambda} \left( c_{\mathbf{k}, \lambda} \, \boldsymbol{\varepsilon}(\mathbf{k}, \lambda) \frac{e^{i(\mathbf{k} \cdot \mathbf{r} - \omega t)}}{\sqrt{V}} + c_{\mathbf{k}, \lambda}^* \, \boldsymbol{\varepsilon}(\mathbf{k}, \lambda) \frac{e^{-i(\mathbf{k} \cdot \mathbf{r} - \omega t)}}{\sqrt{V}} \right) \tag{14.35}$$

The vectors $\boldsymbol{\varepsilon}(\mathbf{k}, \lambda)$ are unit vectors indicating the direction, or polarization, of the vector potential for each value of **k**. Note that the requirement that **A** satisfy the gauge condition (14.24) reduces to the condition that

$$\mathbf{k} \cdot \boldsymbol{\varepsilon}(\mathbf{k}, \lambda) = 0 \tag{14.36}$$

Thus the polarization vector $\boldsymbol{\varepsilon}(\mathbf{k}, \lambda)$ is perpendicular to the direction of **k**. For any particular **k**, there are two linearly independent vectors that satisfy this condition. We indicate these two vectors by having $\lambda$ take on the values 1 and 2 in the sum over $\lambda$. For example, if **k** points in the $z$ direction, we can choose $\boldsymbol{\varepsilon}(\mathbf{k}, 1)$ to be a unit vector in the $x$ direction and $\boldsymbol{\varepsilon}(\mathbf{k}, 2)$ to be a unit vector in the $y$ direction. Since **k** may point in an arbitrary direction, the unit vectors $\boldsymbol{\varepsilon}(\mathbf{k}, 1)$ and $\boldsymbol{\varepsilon}(\mathbf{k}, 2)$ will not, in general, lie along the $x$ and $y$ axes, respectively. It is, however, convenient to choose the set of vectors $(\boldsymbol{\varepsilon}(\mathbf{k}, 1), \boldsymbol{\varepsilon}(\mathbf{k}, 2), \mathbf{k}/|\mathbf{k}|)$ as a right-handed set of orthogonal unit vectors. The factors $c_{\mathbf{k}, \lambda}$ in (14.35) can be considered as coefficients in the expansion allowing for arbitrary amplitudes for each of the plane waves. We have added a term involving $c_{\mathbf{k}, \lambda}^*$ to ensure that the classical vector potential is a real field.

We are now ready to evaluate the energy (14.29) of the electromagnetic field. A typical term comes from the electric field energy, which is given by

$$\frac{1}{8\pi} \int d^3 r \frac{1}{c} \frac{\partial \mathbf{A}}{\partial t} \cdot \frac{1}{c} \frac{\partial \mathbf{A}}{\partial t}$$

$$= \frac{1}{8\pi} \int d^3 r \sum_{\mathbf{k}, \lambda} \left( \frac{-i\omega}{c} c_{\mathbf{k}, \lambda} \, \boldsymbol{\varepsilon}(\mathbf{k}, \lambda) \frac{e^{i(\mathbf{k} \cdot \mathbf{r} - \omega t)}}{\sqrt{V}} + \frac{i\omega}{c} c_{\mathbf{k}, \lambda}^* \boldsymbol{\varepsilon}(\mathbf{k}, \lambda) \frac{e^{-i(\mathbf{k} \cdot \mathbf{r} - \omega t)}}{\sqrt{V}} \right)$$

$$\cdot \sum_{\mathbf{k}', \lambda'} \left( \frac{-i\omega'}{c} c_{\mathbf{k}', \lambda'} \, \boldsymbol{\varepsilon}(\mathbf{k}', \lambda') \frac{e^{i(\mathbf{k}' \cdot \mathbf{r} - \omega' t)}}{\sqrt{V}} + \frac{i\omega'}{c} c_{\mathbf{k}', \lambda'}^* \, \boldsymbol{\varepsilon}(\mathbf{k}', \lambda') \frac{e^{-i(\mathbf{k}' \cdot \mathbf{r} - \omega' t)}}{\sqrt{V}} \right) \tag{14.37}$$

---

[3] Working inside a box may seem quite unphysical, especially a cubic box, which we have chosen for mathematical convenience. However, we will see that any physically measurable quantity is independent of the volume of the box, and thus we can let $L \to \infty$ without changing any of our results. Given that the volume of the universe itself may be bounded, the idea of introducing such a box into our calculations may not seem so strange after all. Moreover, if your universe is similar to that of a creature living on the two-dimensional surface of a balloon, the periodic boundary conditions that we are imposing may seem almost natural as well. We can take comfort in the fact that the effects we are calculating do not depend on either the size or the shape of the box.

Note that we have to sum over all possible values of **k** and $\lambda$ twice, each independently of the other, since the vector potential appears twice. In evaluating (14.37), it is convenient to first carry out the integral over all space. Here we can take advantage of the orthonormality relation

$$\int d^3r \frac{e^{i\mathbf{k}\cdot\mathbf{r}}}{\sqrt{V}} \frac{e^{-i\mathbf{k}'\cdot\mathbf{r}}}{\sqrt{V}} = \delta_{\mathbf{k},\mathbf{k}'} \tag{14.38}$$

(see Problem 14.2). You can now see why we inserted $1/\sqrt{V}$ factors explicitly in the expansion. Thus a sample term from (14.37) is given by

$$\frac{1}{8\pi} \int d^3r \sum_{\mathbf{k},\lambda} \left( \frac{-i\omega}{c} c_{\mathbf{k},\lambda} \boldsymbol{\varepsilon}(\mathbf{k}, \lambda) \frac{e^{i(\mathbf{k}\cdot\mathbf{r}-\omega t)}}{\sqrt{V}} \right) \cdot \sum_{\mathbf{k}',\lambda'} \left( \frac{i\omega'}{c} c^*_{\mathbf{k}',\lambda'} \boldsymbol{\varepsilon}(\mathbf{k}', \lambda') \frac{e^{-i(\mathbf{k}'\cdot\mathbf{r}-\omega' t)}}{\sqrt{V}} \right)$$

$$= \frac{1}{8\pi} \sum_{\mathbf{k},\lambda} \sum_{\mathbf{k}',\lambda'} \left( \frac{-i\omega}{c} c_{\mathbf{k},\lambda}(t) \boldsymbol{\varepsilon}(\mathbf{k}, \lambda) \right) \cdot \left( \frac{i\omega'}{c} c^*_{\mathbf{k}',\lambda'}(t) \boldsymbol{\varepsilon}(\mathbf{k}', \lambda') \right) \delta_{\mathbf{k},\mathbf{k}'}$$

$$= \frac{1}{8\pi} \sum_{\mathbf{k},\lambda} \sum_{\lambda'} \frac{\omega^2}{c^2} c^*_{\mathbf{k},\lambda'} c_{\mathbf{k},\lambda} \boldsymbol{\varepsilon}(\mathbf{k}, \lambda) \cdot \boldsymbol{\varepsilon}(\mathbf{k}, \lambda')$$

$$= \frac{1}{8\pi} \sum_{\mathbf{k},\lambda} \sum_{\lambda'} \frac{\omega^2}{c^2} c^*_{\mathbf{k},\lambda'} c_{\mathbf{k},\lambda} \delta_{\lambda,\lambda'} = \frac{1}{8\pi} \sum_{\mathbf{k},\lambda} \frac{\omega^2}{c^2} c^*_{\mathbf{k},\lambda} c_{\mathbf{k},\lambda} \tag{14.39}$$

There are actually four such terms present in (14.37), two of which are time dependent. However, when we add the terms arising from evaluating the magnetic field energy to those arising from calculating the electric field energy, we find that the *total* energy is simply

$$H = \frac{1}{2\pi} \sum_{\mathbf{k},\lambda} \frac{\omega^2}{c^2} c^*_{\mathbf{k},\lambda} c_{\mathbf{k},\lambda} \tag{14.40}$$

The time-dependent pieces from the electric and magnetic field energies have canceled, just as we would expect, since the total Hamiltonian for a closed system should be time independent.

Unless your name is Dirac, this Hamiltonian may not look familiar. However, following Dirac, we can make the underlying physics more apparent by the following nifty change of variables:

$$q_{\mathbf{k},\lambda} = \frac{1}{c\sqrt{4\pi}}(c_{\mathbf{k},\lambda} + c^*_{\mathbf{k},\lambda}) \qquad p_{\mathbf{k},\lambda} = \frac{-i\omega}{c\sqrt{4\pi}}(c_{\mathbf{k},\lambda} - c^*_{\mathbf{k},\lambda}) \tag{14.41}$$

in which case we find

$$H = \frac{1}{2\pi} \sum_{\mathbf{k},\lambda} \frac{\omega^2}{c^2} \left( c\sqrt{4\pi} \frac{(\omega q_{\mathbf{k},\lambda} - i p_{\mathbf{k},\lambda})}{2\omega} \right) \left( c\sqrt{4\pi} \frac{(\omega q_{\mathbf{k},\lambda} + i p_{\mathbf{k},\lambda})}{2\omega} \right)$$

$$= \sum_{\mathbf{k},\lambda} \left( \frac{p^2_{\mathbf{k},\lambda}}{2} + \frac{1}{2}\omega^2 q^2_{\mathbf{k},\lambda} \right) \tag{14.42}$$

Thus we see that formally the electromagnetic field can be considered as a collection of independent harmonic oscillators. This fact is often used as the starting point for a derivation of Planck's blackbody spectrum. In that approach, the electromagnetic energy density is determined as the number of modes (oscillators) in a particular frequency range multiplied by the average energy of each oscillator. The key ingredient in resolving the classical Rayleigh ultraviolet catastrophe, which arises from giving each oscillator an average energy of $kT$, is to restrict the allowed energies of each oscillator to the discrete values that we determined in Chapter 7.

## 14.3   QUANTIZING THE RADIATION FIELD

We are now ready to "turn on" quantum mechanics in our treatment of the electromagnetic field. We assume that the variables $q_{k,\lambda}$ and $p_{k,\lambda}$ should be *operators* obeying the commutation relations

$$\left[\hat{q}_{\mathbf{k},\lambda}, \hat{p}_{\mathbf{k}',\lambda'}\right] = i\hbar\delta_{\mathbf{k},\mathbf{k}'}\delta_{\lambda,\lambda'} \tag{14.43}$$

just as for the three-dimensional harmonic oscillator for which

$$\hat{H} = \frac{\hat{p}_x^2}{2m} + \frac{1}{2}m\omega^2\hat{x}^2 + \frac{\hat{p}_y^2}{2m} + \frac{1}{2}m\omega^2\hat{y}^2 + \frac{\hat{p}_z^2}{2m} + \frac{1}{2}m\omega^2\hat{z}^2 \tag{14.44}$$

with $\left[\hat{x}, \hat{p}_x\right] = \left[\hat{y}, \hat{p}_y\right] = \left[\hat{z}, \hat{p}_z\right] = i\hbar$, and $\left[\hat{x}, \hat{p}_y\right] = 0$, and so on. The commutation relations (14.43) seem a natural step in our treatment of the Hamiltonian (14.42), although, unlike (14.44), (14.42) is an *infinite* collection of independent oscillators.[4] Moreover, from the relations (14.41) relating the variables $q_{\mathbf{k},\lambda}$ and $p_{\mathbf{k},\lambda}$ to the coefficients $c_{\mathbf{k},\lambda}$ and $c_{\mathbf{k},\lambda}^*$ in the expansion of the vector potential, we see that if $\hat{q}_{\mathbf{k},\lambda}$ and $\hat{p}_{\mathbf{k},\lambda}$ are operators, so are the $c_{\mathbf{k},\lambda}$'s.

The natural operators for analyzing the harmonic oscillator are the raising and lowering operators, which for the Hamiltonian (14.42) are given by

$$\hat{a}_{\mathbf{k},\lambda} = \sqrt{\frac{\omega}{2\hbar}}\left(\hat{q}_{\mathbf{k},\lambda} + \frac{i}{\omega}\hat{p}_{\mathbf{k},\lambda}\right) \qquad \hat{a}_{\mathbf{k},\lambda}^{\dagger} = \sqrt{\frac{\omega}{2\hbar}}\left(\hat{q}_{\mathbf{k},\lambda} - \frac{i}{\omega}\hat{p}_{\mathbf{k},\lambda}\right) \tag{14.45}$$

In terms of these operators,

$$\hat{q}_{\mathbf{k},\lambda} = \sqrt{\frac{\hbar}{2\omega}}\left(\hat{a}_{\mathbf{k},\lambda} + \hat{a}_{\mathbf{k},\lambda}^{\dagger}\right) \qquad \hat{p}_{\mathbf{k},\lambda} = -i\sqrt{\frac{\hbar\omega}{2}}\left(\hat{a}_{\mathbf{k},\lambda} - \hat{a}_{\mathbf{k},\lambda}^{\dagger}\right) \tag{14.46}$$

A comparison of these equations with (14.41) suggests the replacements

$$c_{\mathbf{k},\lambda} \rightarrow c\sqrt{\frac{2\pi\hbar}{\omega}}\,\hat{a}_{\mathbf{k},\lambda} \qquad c_{\mathbf{k},\lambda}^* \rightarrow c\sqrt{\frac{2\pi\hbar}{\omega}}\,\hat{a}_{\mathbf{k},\lambda}^{\dagger} \tag{14.47}$$

---

[4] A more rigorous way to introduce these commutation relations is to start with a field Lagrangian that yields the equations of motion and then to postulate commutation relations for the generalized field coordinates and the corresponding momenta. See, for example, R. Shankar, *Principles of Quantum Mechanics*, Plenum, New York, 1980, p. 516.

and therefore the classical vector potential **A** has been replaced by the *Hermitian operator* $\hat{\mathbf{A}}$,

$$\hat{\mathbf{A}} = \sum_{\mathbf{k},\lambda} c\sqrt{\frac{2\pi\hbar}{\omega}}\left(\hat{a}_{\mathbf{k},\lambda}\boldsymbol{\varepsilon}(\mathbf{k},\lambda)\frac{e^{i(\mathbf{k}\cdot\mathbf{r}-\omega t)}}{\sqrt{V}} + \hat{a}^{\dagger}_{\mathbf{k},\lambda}\boldsymbol{\varepsilon}(\mathbf{k},\lambda)\frac{e^{-i(\mathbf{k}\cdot\mathbf{r}-\omega t)}}{\sqrt{V}}\right) \quad (14.48)$$

Since the vector potential is now an operator, both the electric and magnetic fields are operators as well. If we use this expression (14.48) for the vector potential to evaluate the Hamiltonian (14.29), which is now also an operator, we obtain

$$\hat{H} = \tfrac{1}{2}\sum_{\mathbf{k},\lambda}\hbar\omega\left(\hat{a}_{\mathbf{k},\lambda}\hat{a}^{\dagger}_{\mathbf{k},\lambda} + \hat{a}^{\dagger}_{\mathbf{k},\lambda}\hat{a}_{\mathbf{k},\lambda}\right)$$

$$= \sum_{\mathbf{k},\lambda}\hbar\omega\left(\hat{a}^{\dagger}_{\mathbf{k},\lambda}\hat{a}_{\mathbf{k},\lambda} + \tfrac{1}{2}\right) \quad (14.49)$$

where in the last step we have taken advantage of the commutation relations

$$\left[\hat{a}_{\mathbf{k},\lambda}, \hat{a}^{\dagger}_{\mathbf{k}',\lambda'}\right] = \delta_{\mathbf{k},\mathbf{k}'}\delta_{\lambda,\lambda'} \quad (14.50)$$

which follow from (14.43) and from the definitions (14.45) of the raising and lowering operators. The reason that the Hamiltonian (14.49) cannot simply be obtained from the expression (14.40) for the energy of the field using the replacements (14.47) is that in working out (14.40) we assumed that the $c_{\mathbf{k},\lambda}$'s and $c^{*}_{\mathbf{k},\lambda}$'s were numbers, not operators that do not commute; thus we did not keep track of the order in which these numbers appeared in evaluating the Hamiltonian. The right way to derive (14.49) is to go back to the beginning and use the expansion (14.48) for the vector potential operator together with the commutation relations (14.50) from the start in evaluating the Hamiltonian

$$\hat{H} = \frac{1}{8\pi}\int d^{3}r\left\{\left(\frac{1}{c}\frac{\partial\hat{\mathbf{A}}}{\partial t}\right)^{2} + \left(\boldsymbol{\nabla}\times\hat{\mathbf{A}}\right)^{2}\right\} \quad (14.51)$$

In nonrelativistic quantum mechanics we are accustomed to replacing classical variables such as the position and momentum by operators. Now we see in a quantum treatment of the electromagnetic field that the field itself becomes an operator, an operator that annihilates and creates photons, as we will show in the next section. This transition from a classical to a quantum field theory represents a conceptual revolution in the way we think about fields. It also indicates the way that quantum mechanics and relativity, the two major cornerstones in the way we view the physical world that have originated in the twentieth century, are joined together in the form of a relativistic quantum field.

## 14.4 THE PROPERTIES OF PHOTONS

The lowest-energy state of the Hamiltonian (14.49) is called the vacuum state and is denoted by the ket $|0\rangle$. This is a state such that for all values of **k** and $\lambda$

$$\hat{a}_{\mathbf{k},\lambda}|0\rangle = 0 \quad (14.52)$$

We can consider the ground state as a direct product of lowest-energy states for each of the individual independent harmonic oscillators that comprise the Hamiltonian:

$$|0\rangle = |0_{\mathbf{k}_1,\lambda_1}\rangle \otimes |0_{\mathbf{k}_2,\lambda_2}\rangle \otimes |0_{\mathbf{k}_3,\lambda_3}\rangle \otimes \cdots \qquad (14.53)$$

The energy of this ground state is determined by

$$\hat{H}|0\rangle = \sum_{\mathbf{k},\lambda} \hbar\omega \left(\hat{a}_{\mathbf{k},\lambda}^{\dagger} \hat{a}_{\mathbf{k},\lambda} + \tfrac{1}{2}\right)|0\rangle$$

$$= \tfrac{1}{2} \sum_{\mathbf{k},\lambda} \hbar\omega |0\rangle \qquad (14.54)$$

Thus the ground-state energy $E_0$ is the sum of the zero-point energies of each of the harmonic oscillators:

$$E_0 = \tfrac{1}{2} \sum_{\mathbf{k},\lambda} \hbar\omega \qquad (14.55)$$

Unless there is some cutoff in the theory that limits the number of oscillators with arbitrarily high frequencies, this sum diverges because there are an infinite number of such oscillators. Nonetheless, it is convenient to treat $E_0$ as if it were finite. We will see that it is only differences in energy that matter in any case.[5]

If we apply the raising operator $\hat{a}_{\mathbf{k},\lambda}^{\dagger}$ for one of these oscillators to the ground state, we obtain the state

$$\hat{a}_{\mathbf{k},\lambda}^{\dagger}|0\rangle = |0_{\mathbf{k}_1,\lambda_1}\rangle \otimes |0_{\mathbf{k}_2,\lambda_2}\rangle \otimes \cdots \hat{a}_{\mathbf{k},\lambda}^{\dagger}|0_{\mathbf{k},\lambda}\rangle \otimes \cdots$$

$$= |0_{\mathbf{k}_1,\lambda_1}\rangle \otimes |0_{\mathbf{k}_2,\lambda_2}\rangle \otimes \cdots |1_{\mathbf{k},\lambda}\rangle \otimes \cdots \qquad (14.56)$$

For simplicity, we denote this whole state by

$$|1_{\mathbf{k},\lambda}\rangle = \hat{a}_{\mathbf{k},\lambda}^{\dagger}|0\rangle \qquad (14.57)$$

with the understanding that each oscillator except the one specified by the vector $\mathbf{k}$ and the polarization state $\lambda$ is in the ground state. The energy of this state is determined by letting the Hamiltonian act on it:

$$\hat{H}|1_{\mathbf{k},\lambda}\rangle = \sum_{\mathbf{k}',\lambda'} \hbar\omega' \left(\hat{a}_{\mathbf{k}',\lambda'}^{\dagger} \hat{a}_{\mathbf{k}',\lambda'} + \tfrac{1}{2}\right)|1_{\mathbf{k},\lambda}\rangle \qquad (14.58)$$

Since

$$\tfrac{1}{2} \sum_{\mathbf{k}',\lambda'} \hbar\omega' = E_0 \qquad (14.59)$$

---

[5] An interesting manifestation of the zero-point field energy is the Casimir effect, in which two neutral conducting plates attract each other because of vacuum fluctuations. See, for example, C. Itzykson and J. -B. Zuber, *Quantum Field Theory,* McGraw-Hill, New York, 1980, p. 138.

and

$$\hat{a}_{\mathbf{k},\lambda}^{\dagger}\hat{a}_{\mathbf{k},\lambda}|1_{\mathbf{k},\lambda}\rangle = |1_{\mathbf{k},\lambda}\rangle \tag{14.60}$$

while

$$\hat{a}_{\mathbf{k}',\lambda'}^{\dagger}\hat{a}_{\mathbf{k}',\lambda'}|1_{\mathbf{k},\lambda}\rangle = 0 \qquad \mathbf{k}' \neq \mathbf{k} \qquad \lambda' \neq \lambda \tag{14.61}$$

then

$$\hat{H}|1_{\mathbf{k},\lambda}\rangle = (E_0 + \hbar\omega)|1_{\mathbf{k},\lambda}\rangle \tag{14.62}$$

Thus the energy added to the system by the action of the operator $\hat{a}_{\mathbf{k},\lambda}^{\dagger}$ on the ground state is $\hbar\omega$.

We can see that we have added momentum to the system as well. Even classically we know that the electromagnetic field carries momentum. In fact, the direction of the momentum of the field is the same as the Poynting vector $\mathbf{S}_P = (c/4\pi)\mathbf{E} \times \mathbf{B}$, which gives the field energy per unit area per unit time (see Problem 14.1). If we place a black disk in front of a light source, as illustrated in Fig. 14.3$a$, the disk will recoil as well as heat up as it absorbs momentum and energy from the field. We construct the momentum operator for the electromagnetic field by expressing the electric and magnetic field operators in terms of the vector potential:

$$\hat{\mathbf{P}} = \frac{1}{4\pi c}\int d^3r\,\hat{\mathbf{E}} \times \hat{\mathbf{B}}$$

$$= \frac{1}{4\pi c}\int d^3r\left(-\frac{1}{c}\frac{\partial\hat{\mathbf{A}}}{\partial t}\right) \times \left(\boldsymbol{\nabla} \times \hat{\mathbf{A}}\right) \tag{14.63}$$

If we substitute the expansion (14.48) for the vector potential into this expression for the momentum operator for the electromagnetic field, we find

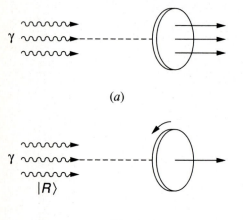

(a)

(b)

**FIGURE 14.3**
(a) A black disk recoils as it absorbs photons.
(b) The rate of rotation increases as it absorbs circularly polarized photons.

$$\hat{\mathbf{P}} = \tfrac{1}{2} \sum_{\mathbf{k},\lambda} \hbar\mathbf{k}(\hat{a}^\dagger_{\mathbf{k},\lambda}\hat{a}_{\mathbf{k},\lambda} + \hat{a}_{\mathbf{k},\lambda}\hat{a}^\dagger_{\mathbf{k},\lambda})$$

$$= \sum_{\mathbf{k},\lambda} \hbar\mathbf{k}\left(\hat{a}^\dagger_{\mathbf{k},\lambda}\hat{a}_{\mathbf{k},\lambda} + \tfrac{1}{2}\right) = \sum_{\mathbf{k},\lambda} \hbar\mathbf{k}\hat{a}^\dagger_{\mathbf{k},\lambda}\hat{a}_{\mathbf{k},\lambda} \qquad (14.64)$$

where in the last step we have used the fact that

$$\sum_{\mathbf{k},\lambda} \hbar\mathbf{k} = 0 \qquad (14.65)$$

since for every value of $\mathbf{k}$ in the sum there is a $-\mathbf{k}$ to cancel it. Thus, as expected, the vacuum state has no momentum:

$$\hat{\mathbf{P}}|0\rangle = 0 \qquad (14.66)$$

Applying the momentum operator (14.64) to the state $|1_{\mathbf{k},\lambda}\rangle$, we obtain

$$\hat{\mathbf{P}}|1_{\mathbf{k},\lambda}\rangle = \sum_{\mathbf{k}',\lambda'} \hbar\mathbf{k}'\hat{a}^\dagger_{\mathbf{k}',\lambda'}\hat{a}_{\mathbf{k}',\lambda'}|1_{\mathbf{k},\lambda}\rangle = \hbar\mathbf{k}|1_{\mathbf{k},\lambda}\rangle \qquad (14.67)$$

Thus the state $|1_{\mathbf{k},\lambda}\rangle$ has additional momentum $\hbar\mathbf{k}$ as well as additional energy $\hbar\omega$ in comparison with the vacuum state. Since $\omega = |\mathbf{k}|c$, the additional energy and momentum are related by $E = pc$, as expected for a particle like the photon that moves at the speed of light. We can create a state with $n_{\mathbf{k},\lambda}$ photons, each with momentum $\hbar\mathbf{k}$ and polarization $\lambda$, by acting $n_{\mathbf{k},\lambda}$ times with the creation operator:

$$|n_{\mathbf{k},\lambda}\rangle = \frac{(\hat{a}^\dagger_{\mathbf{k},\lambda})^{n_{\mathbf{k},\lambda}}}{\sqrt{n_{\mathbf{k},\lambda}!}}|0\rangle \qquad (14.68)$$

Recall from the commutation relations of the raising and lowering operators that

$$\hat{a}^\dagger_{\mathbf{k},\lambda}|n_{\mathbf{k},\lambda}\rangle = \sqrt{n_{\mathbf{k},\lambda} + 1}|n_{\mathbf{k},\lambda} + 1\rangle \qquad (14.69a)$$

and

$$\hat{a}_{\mathbf{k},\lambda}|n_{\mathbf{k},\lambda}\rangle = \sqrt{n_{\mathbf{k},\lambda}}|n_{\mathbf{k},\lambda} - 1\rangle \qquad (14.69b)$$

Thus it is appropriate to call the operator $\hat{a}^\dagger_{\mathbf{k},\lambda}$ a *creation operator* for a photon, since it increases the number of photons in a state by one. Similarly, we call $\hat{a}_{\mathbf{k},\lambda}$ an *annihilation operator,* since it reduces the photon content of a state by one.

As we discussed in Chapter 2, the $\lambda$ label on the photon state indicates its polarization. For example, for a photon traveling in the $z$ direction, the single-photon state $|1_{\mathbf{k},1}\rangle$ is an $x$-polarized photon, while the state $|1_{\mathbf{k},2}\rangle$ is a $y$-polarized photon. In particular, the right-circularly polarized state is given by

$$|R\rangle = \frac{1}{\sqrt{2}}\big(|1_{\mathbf{k},1}\rangle + i|1_{\mathbf{k},2}\rangle\big) = \frac{1}{\sqrt{2}}\left(\hat{a}^\dagger_{\mathbf{k},1}|0\rangle + i\hat{a}^\dagger_{\mathbf{k},2}|0\rangle\right) \qquad (14.70a)$$

while the left-circularly polarized state is given by

$$|L\rangle = \frac{1}{\sqrt{2}}\big(|1_{\mathbf{k},1}\rangle - i|1_{\mathbf{k},2}\rangle\big) = \frac{1}{\sqrt{2}}\left(\hat{a}^\dagger_{\mathbf{k},1}|0\rangle - i\hat{a}^\dagger_{\mathbf{k},2}|0\rangle\right) \qquad (14.70b)$$

Using the quantum theory of the electromagnetic field, we can check that these states do correspond to eigenstates of angular momentum with eigenvalues $\hbar$ and $-\hbar$, respectively, along the direction of the momentum of the photon by verifying that

$$\frac{\hat{\mathbf{J}} \cdot \mathbf{k}}{|\mathbf{k}|}\left[\frac{1}{\sqrt{2}}\left(\hat{a}_{\mathbf{k},1}^{\dagger}|0\rangle + i\,\hat{a}_{\mathbf{k},2}^{\dagger}|0\rangle\right)\right] = \hbar\left[\frac{1}{\sqrt{2}}\left(\hat{a}_{\mathbf{k},1}^{\dagger}|0\rangle + i\,\hat{a}_{\mathbf{k},2}^{\dagger}|0\rangle\right)\right] \qquad (14.71a)$$

and

$$\frac{\hat{\mathbf{J}} \cdot \mathbf{k}}{|\mathbf{k}|}\left[\frac{1}{\sqrt{2}}\left(\hat{a}_{\mathbf{k},1}^{\dagger}|0\rangle - i\,\hat{a}_{\mathbf{k},2}^{\dagger}|0\rangle\right)\right] = -\hbar\left[\frac{1}{\sqrt{2}}\left(\hat{a}_{\mathbf{k},1}^{\dagger}|0\rangle - i\,\hat{a}_{\mathbf{k},2}^{\dagger}|0\rangle\right)\right] \qquad (14.71b)$$

where the angular momentum operator for the electromagnetic field is given by

$$\hat{\mathbf{J}} = \int d^{3}r\,\mathbf{r} \times \frac{(\hat{\mathbf{E}} \times \hat{\mathbf{B}})}{4\pi c} \qquad (14.72)$$

The photon has an intrinsic spin of one, as we also deduced from the behavior of the photon polarization states under rotations in Section 2.7. The classical physicist knows there is angular momentum in the electromagnetic field from the expression (14.72) (without the hats). For example, the disk shown in Fig. 14.3b will start to spin about its axis if the electromagnetic field incident on the disk is circularly polarized. But of course it is pure quantum mechanics that this angular momentum is quantized in units of $\hbar$.

Based on our discussion in Section 12.1 on the connection between the intrinsic spin of a particle and its statistics, we expect that the spin-1 photon should be a boson. This is confirmed by (14.68), which shows that there can be more than one photon with momentum $\hbar\mathbf{k}$ and polarization $\lambda$ in the same state. This connection between spin and statistics actually entered our theory when we chose to make the creation and annihilation operators for photons obey *commutation* relations. In order to see the effect of an alternative way of turning on quantum mechanics in a field theory—namely, *anticommutation* relations—that limits the number of particles that can be in the same state to one, see Problem 14.5.

## 14.5   THE HAMILTONIAN OF THE ATOM AND THE ELECTROMAGNETIC FIELD

We are now ready to consider the interaction of photons and atoms. If you examine expression (14.48) for the vector potential operator $\hat{\mathbf{A}}(\mathbf{r}, t)$, you will notice that both the position $\mathbf{r}$ and the time $t$ enter as parameters specifying the field. We emphasized in our discussion of the energy-time uncertainty relation in nonrelativistic quantum mechanics that $t$ is not an operator but rather a parameter used to specify the state. Not surprisingly, in relativistic quantum field theory, position and time enter on an equal footing. In particular, the position $\mathbf{r}$ is no longer an operator but a parameter that, for example, we integrate over to express the

Hamiltonian operator (14.51) in terms of annihilation and creation operators. In a fully relativistic treatment of charged particles such as electrons and positrons, there is also another field—a function of $\mathbf{r}$ and $t$ called a Dirac field—that is a superposition of creation and annihilation operators for electrons and positrons. In determining the full Hamiltonian for the interactions of charged particles with photons, the position $\mathbf{r}$ of the Dirac field is integrated over in the same way it is integrated in determining the energy of the electromagnetic field.

In the approach that we will follow in this chapter, we will not use quantum field theory for the charged particles. It isn't necessary, since our goal is to treat the interactions of photons with atoms, and the physics of the atoms is essentially nonrelativistic in nature. Photons, on the other hand, are inherently relativistic and thus including them requires a quantized electromagnetic field. If you look back to the derivation of the Hamiltonian (14.27) in Appendix E, you will see that the vector potential that enters the Hamiltonian in the form $\mathbf{p} - q\mathbf{A}(\mathbf{r}, t)/c$ is actually evaluated at the position of the charged particle. Thus in order to treat the interaction of the vector potential operator $\hat{\mathbf{A}}(\mathbf{r}, t)$ with charged particles in a manner that is self-consistent with the way that we treat it in evaluating the electromagnetic field energy, we need to use the position-space representation of the Hamiltonian for the charged particles so that the position $\mathbf{r}$ is treated as a variable rather than an operator.

For simplicity, we concentrate on the Hamiltonian of the hydrogen atom, including the Hamiltonian of the electromagnetic field. The Hamiltonian in the center-of-mass frame of the atom is given by[6]

---

[6] The Coulomb interaction of the charged particles actually arises from that portion of the electromagnetic field energy due to the scalar potential $\varphi$ in the Coulomb gauge. The electric field energy is given by

$$\frac{1}{8\pi} \int d^3r\, \mathbf{E}^2 = \frac{1}{8\pi} \int d^3r \left( -\boldsymbol{\nabla}\varphi - \frac{1}{c}\frac{\partial \mathbf{A}}{\partial t} \right)^2$$

$$= \frac{1}{8\pi} \int d^3r \left[ \boldsymbol{\nabla}\varphi \cdot \boldsymbol{\nabla}\varphi + 2\boldsymbol{\nabla}\varphi \cdot \frac{1}{c}\frac{\partial \mathbf{A}}{\partial t} + \left( \frac{1}{c}\frac{\partial A}{\partial t} \right)^2 \right]$$

$$= \frac{1}{8\pi} \int d^3r \left[ -\varphi\nabla^2\varphi - 2\varphi\frac{1}{c}\frac{\partial \boldsymbol{\nabla} \cdot \mathbf{A}}{\partial t} + \left( \frac{1}{c}\frac{\partial \mathbf{A}}{\partial t} \right)^2 \right]$$

where in the last step we have performed two integrations by parts and assumed that the fields vanish at infinity so that there is no contribution at the end points of the integration. Note that the middle term in the brackets vanishes since $\boldsymbol{\nabla} \cdot \mathbf{A} = 0$. Finally, taking advantage of Gauss's law (14.26) in the Coulomb gauge, we can write this expression for the electric field energy as

$$\frac{1}{2} \int d^3r\, \rho\varphi + \frac{1}{8\pi} \int d^3r \left( \frac{1}{c}\frac{\partial A}{\partial t} \right)^2 = \frac{1}{2} \int d^3r\, \rho(\mathbf{r}) \int d^3r' \frac{\rho(\mathbf{r}')}{|\mathbf{r} - \mathbf{r}'|} + \frac{1}{8\pi} \int d^3r \left( \frac{1}{c}\frac{\partial \mathbf{A}}{\partial t} \right)^2$$

For hydrogen $\rho = -e\delta^3(\mathbf{r} - \mathbf{r}_e) + e\delta^3(\mathbf{r} - \mathbf{r}_p)$, so that the first term yields the Coulomb energy $-e^2/|\mathbf{r}_e - \mathbf{r}_p|$ of interaction between the electron and the proton, as well as the (infinite) self-energy of the particles, which we are neglecting.

$$\hat{H} \rightarrow \frac{1}{2m_e}\left(\frac{\hbar}{i}\boldsymbol{\nabla} + \frac{e\hat{\mathbf{A}}}{c}\right)^2 + \frac{1}{2m_p}\left(-\frac{\hbar}{i}\boldsymbol{\nabla}\right)^2 - \frac{e^2}{r} + \frac{1}{8\pi}\int d^3r\left[\left(\frac{1}{c}\frac{\partial\hat{\mathbf{A}}}{\partial t}\right)^2 + \left(\boldsymbol{\nabla}\times\hat{\mathbf{A}}\right)^2\right]$$

$$(14.73)$$

where the arrow indicates that the atom part of the Hamiltonian is given in position space. We have neglected the interaction of the proton with $\mathbf{A}$ because $m_p \gg m_e$. If we wish to include the interaction of the intrinsic spin of the electron with the magnetic field, we would obtain an additional contribution to the Hamiltonian of the form

$$\frac{g_e e}{2m_e c}\hat{\mathbf{S}}_e \cdot \boldsymbol{\nabla}\times\hat{\mathbf{A}} \tag{14.74}$$

where again we have neglected the interaction of the proton's magnetic moment because of the large mass of the proton.

Unfortunately, we are not able to determine the exact energy eigenstates and eigenvalues of the full Hamiltonian (14.73). The system is just too complex. Thus we must resort to a perturbative approach. We express the whole Hamiltonian in the form

$$\hat{H} = \hat{H}_0 + \hat{H}_1 \tag{14.75}$$

where

$$\hat{H}_0 \rightarrow -\frac{\hbar^2}{2\mu}\nabla^2 - \frac{e^2}{r} + \frac{1}{8\pi}\int d^3r\left[\left(\frac{1}{c}\frac{\partial\hat{\mathbf{A}}}{\partial t}\right)^2 + \left(\boldsymbol{\nabla}\times\hat{\mathbf{A}}\right)^2\right] \tag{14.76}$$

is the sum of two Hamiltonians: the Hamiltonian for the hydrogen atom without interaction with the vector potential (see Section 10.2) and the free Hamiltonian for the electromagnetic field, which we examined in the preceding section. Thus we know the eigenstates and eigenvalues of $\hat{H}_0$. The perturbing Hamiltonian $\hat{H}_1$ is the remainder of the full Hamiltonian (14.73):

$$\hat{H}_1 \rightarrow \frac{e}{2m_e c}\hat{\mathbf{A}}\cdot\frac{\hbar}{i}\boldsymbol{\nabla} + \frac{e}{2m_e c}\frac{\hbar}{i}\boldsymbol{\nabla}\cdot\hat{\mathbf{A}} + \frac{e^2}{2m_e c^2}\hat{\mathbf{A}}^2 \tag{14.77}$$

The gradient operator in position space acts on a wave function, say for the initial state of the atom. Since

$$\boldsymbol{\nabla}\cdot\hat{\mathbf{A}}\psi = (\boldsymbol{\nabla}\cdot\hat{\mathbf{A}})\psi + \hat{\mathbf{A}}\cdot\boldsymbol{\nabla}\psi = \hat{\mathbf{A}}\cdot\boldsymbol{\nabla}\psi \tag{14.78}$$

because $\boldsymbol{\nabla}\cdot\hat{\mathbf{A}} = 0$ in the Coulomb gauge, we can safely move the gradient through the vector potential operator in the second piece of the interaction Hamiltonian (14.77). Therefore, this Hamiltonian simplifies to

$$\hat{H}_1 \rightarrow \frac{e}{m_e c}\hat{\mathbf{A}}\cdot\frac{\hbar}{i}\boldsymbol{\nabla} + \frac{e^2}{2m_e c^2}\hat{\mathbf{A}}^2 \tag{14.79}$$

## 14.6 TIME-DEPENDENT PERTURBATION THEORY

Our goal is to work out how a state such as an excited state of the atom evolves in time. However, since we are not able to determine the eigenstates and eigenvalues of the full Hamiltonian $\hat{H} = \hat{H}_0 + \hat{H}_1$, we cannot determine the time dependence of the full system by expressing an arbitrary state as a superposition of energy eigenstates,

$$|\psi(0)\rangle = \sum_E |E\rangle\langle E|\psi(0)\rangle \tag{14.80}$$

and then taking advantage of the time-development operator (4.9):

$$|\psi(t)\rangle = e^{-i\hat{H}t/\hbar}|\psi(0)\rangle = \sum_E |E\rangle\langle E|\psi(0)\rangle e^{-iEt/\hbar} \tag{14.81}$$

Of course this procedure would also fail if the Hamiltonian itself were time dependent, as would happen, for example, if $\hat{H}_1$ were to vary in time: $\hat{H}_1 = \hat{H}_1(t)$. Such a situation arises if the system is subject to an external influence that changes with time, such as the spin system in a classical, external, time-dependent magnetic field, like the one we examined in Section 4.4. In order to handle cases such as these (see the example of this section and Problems 14.6 through 14.8) as well as deal with time evolution for the Hamiltonian (14.73), we resort to the techniques of time-dependent perturbation theory.

We begin by expressing an arbitrary state at the initial time $t = 0$ as a superposition of the eigenstates that we know, the eigenstates of $\hat{H}_0$:

$$|\psi(0)\rangle = \sum_n |E_n^{(0)}\rangle\langle E_n^{(0)}|\psi(0)\rangle = \sum_n c_n(0)|E_n^{(0)}\rangle \tag{14.82}$$

We then write the time dependence in the form

$$|\psi(t)\rangle = \sum_n c_n(t)e^{-iE_n^{(0)}t/\hbar}|E_n^{(0)}\rangle \tag{14.83}$$

If the Hamiltonian were only to consist of $\hat{H}_0$, the $c_n$ would be time independent. Thus if $\hat{H}_1$ is "small," we expect that the time dependence of the $c_n(t)$ can be handled perturbatively. We will first demonstrate how this works using techniques similar to those that we used in Chapter 11, and then we will show how we can obtain the results to all orders in a more compact, elegant manner (that is especially appropriate for quantum field theory) utilizing what is termed the interaction picture.

We obtain the equations governing the time evolution of the $c_n(t)$ by substituting (14.83) into the Schrödinger equation

$$\hat{H}|\psi(t)\rangle = i\hbar\frac{d}{dt}|\psi(t)\rangle \tag{14.84}$$

yielding

$$\sum_n c_n(t) e^{-iE_n^{(0)}t/\hbar} (\hat{H}_0 + \hat{H}_1)|E_n^{(0)}\rangle = i\hbar \sum_n \left[ \frac{d}{dt} c_n(t) - \frac{i E_n^{(0)}}{\hbar} c_n(t) \right] e^{-iE_n^{(0)}t/\hbar} |E_n^{(0)}\rangle$$

(14.85)

Taking the inner product with the bra $\langle E_f^{(0)}|$ and using

$$\hat{H}_0|E_n^{(0)}\rangle = E_n^{(0)}|E_n^{(0)}\rangle$$

(14.86)

on the left-hand side of (14.85), this equation becomes

$$\frac{d}{dt} c_f(t) = -\frac{i}{\hbar} \sum_n c_n(t) e^{i(E_f^{(0)} - E_n^{(0)})t/\hbar} \langle E_f^{(0)}|\hat{H}_1|E_n^{(0)}\rangle$$

(14.87)

We will assume that at time $t = 0$ the system is in an eigenstate $|E_i^{(0)}\rangle$ of $\hat{H}_0$. Thus the initial conditions for the coupled differential equations (14.87) are

$$c_n(0) = \delta_{ni}$$

(14.88)

As in Chapter 11, we insert a parameter $\lambda$ in the Hamiltonian to keep track of the order of smallness ($\hat{H}_1 \to \lambda\hat{H}_1$) and expand the coefficients $c_n(t)$ in a power series in $\lambda$:

$$c_n(t) = c_n^{(0)} + \lambda c_n^{(1)} + \lambda^2 c_n^{(2)} + \cdots$$

(14.89)

Making these substitutions on both sides of (14.87), we obtain

$$\frac{d}{dt} \left( c_f^{(0)} + \lambda c_f^{(1)} + \lambda^2 c_f^{(2)} + \cdots \right)$$

$$= -\frac{i}{\hbar} \sum_n (c_n^{(0)} + \lambda c_n^{(1)} + \lambda^2 c_n^{(2)} + \cdots) e^{i(E_f^{(0)} - E_n^{(0)})t/\hbar} \langle E_f^{(0)}|\lambda\hat{H}_1|E_n^{(0)}\rangle$$

(14.90)

The only term of order $\lambda^0$ is on the left-hand side of (14.90), indicating that

$$\frac{d}{dt} c_f^{(0)} = 0$$

(14.91)

The initial condition (14.88) is satisfied provided

$$c_f^{(0)} = \delta_{fi} \quad \text{and} \quad c_n^{(k)}(0) = 0 \quad \text{for} \quad k \geq 1$$

(14.92)

Collecting the $\lambda^1$ terms in (14.90), we obtain

$$\frac{d}{dt} c_f^{(1)}(t) = -\frac{i}{\hbar} \sum_n c_n^{(0)}(t) e^{i(E_f^{(0)} - E_n^{(0)})t/\hbar} \langle E_f^{(0)}|\hat{H}_1|E_n^{(0)}\rangle$$

$$= -\frac{i}{\hbar} e^{i(E_f^{(0)} - E_i^{(0)})t/\hbar} \langle E_f^{(0)}|\hat{H}_1|E_i^{(0)}\rangle$$

(14.93)

which can be integrated to yield

$$c_f^{(1)} = -\frac{i}{\hbar} \int_0^t dt' \, e^{i(E_f^{(0)} - E_i^{(0)})t'/\hbar} \langle E_f^{(0)}|\hat{H}_1(t')|E_i^{(0)}\rangle$$

(14.94)

In (14.94) we have allowed explicitly for the possibility that $\hat{H}_1$ depends on time. Combining (14.92) and (14.94), we see that through first order

$$c_f(t) = \delta_{fi} - \frac{i}{\hbar} \int_0^t dt' \, e^{i(E_f^{(0)} - E_i^{(0)})t'/\hbar} \langle E_f^{(0)}|\hat{H}_1(t')|E_i^{(0)}\rangle + \cdots \qquad (14.95)$$

## An Example

Let's consider a simple example. Equation (11.27) gives the Hamiltonian for a charge $q$ in a one-dimensional harmonic oscillator in a classical (not quantized) electric field, which we will take here to have the time-dependence $|\mathbf{E}| = |\mathbf{E}_0|e^{-t/\tau}$, such as would arise if the oscillator were situated between the plates of a discharging parallel-plate capacitor. Thus the Hamiltonian is given by

$$\hat{H} = \frac{\hat{p}_x^2}{2m} + \frac{1}{2}m\omega^2\hat{x}^2 - q\hat{x}|\mathbf{E}_0|e^{-t/\tau} \qquad (14.96)$$

We choose

$$\hat{H}_0 = \frac{\hat{p}_x^2}{2m} + \frac{1}{2}m\omega^2\hat{x}^2 \qquad (14.97)$$

and

$$\hat{H}_1(t) = -q\hat{x}|\mathbf{E}_0|e^{-t/\tau} \qquad (14.98)$$

Suppose that at $t = 0$ the oscillator is in the $n = 0$ ground state. What is the probability that the oscillator is in an excited state at $t = \infty$? From (14.95)

$$c_n(\infty) = \frac{iq|\mathbf{E}_0|}{\hbar} \int_0^\infty dt' \, e^{in\omega t'} e^{-t'/\tau} \langle n|\hat{x}|0\rangle \qquad n \neq 0 \qquad (14.99)$$

Expressing the position operator in terms of raising and lowering operators, we find

$$c_n(\infty) = \frac{iq|\mathbf{E}_0|}{\hbar} \sqrt{\frac{\hbar}{2m\omega}} \int_0^\infty dt' \, e^{in\omega t'} e^{-t'/\tau} \langle n|(\hat{a} + \hat{a}^\dagger)|0\rangle$$

$$= \frac{iq|\mathbf{E}_0|}{\hbar} \sqrt{\frac{\hbar}{2m\omega}} \int_0^\infty dt' \, e^{in\omega t'} e^{-t'/\tau} \langle n|1\rangle \qquad (14.100)$$

Thus only $c_1(\infty)$ is nonzero:

$$c_1(\infty) = \frac{iq|\mathbf{E}_0|}{\hbar} \sqrt{\frac{\hbar}{2m\omega}} \int_0^\infty dt' \, e^{i\omega t'} e^{-t/\tau}$$

$$= \frac{iq|\mathbf{E}_0|}{\hbar} \sqrt{\frac{\hbar}{2m\omega}} \frac{\tau}{1 - i\omega\tau} \qquad (14.101)$$

The probability of the oscillator making the transition from the ground state to the first-excited state is given by

$$|c_1(\infty)|^2 = \frac{(q|\mathbf{E}_0|\tau)^2}{2m\hbar\omega} \frac{1}{1 + \omega^2\tau^2} \tag{14.102}$$

This result is our first hint of how a selection rule might arise: Since the position operator in the electric dipole Hamiltonian (14.98) involves a single raising and a single lowering operator, only transitions in which the quantum number $n$ of the oscillator changes by 1 are permitted through first order in perturbation theory.

## The Schrödinger Picture

Before going on, it is instructive to rephrase our discussion of time-dependent perturbation theory using the interaction picture. First, let's summarize time development in the familiar Schrödinger "picture" that we have used so far in our discussion of time evolution. In this picture, states evolve in time according to

$$|\psi_S(t)\rangle = \hat{U}_S(t)|\psi_S(0)\rangle \tag{14.103}$$

where

$$\hat{U}_S(t) = e^{-i\hat{H}t/\hbar} \tag{14.104}$$

when the Hamiltonian $\hat{H}$ is independent of time. In general, as shown in (4.7), the time-development operator satisfies the differential equation

$$i\hbar\frac{d}{dt}\hat{U}_S(t) = \hat{H}\hat{U}_S(t) \tag{14.105}$$

Note that we have added a subscript $S$ to the states and to the time-development operator to distinguish them from states and operators in other pictures. According to (4.16), the expectation value of an operator $\hat{O}_S$ is given by

$$\frac{d\langle\psi_S(t)|\hat{O}_S|\psi_S(t)\rangle}{dt} = \frac{i}{\hbar}\langle\psi_S(t)|[\hat{H}, \hat{O}_S]|\psi_S(t)\rangle \tag{14.106}$$

where we are assuming that the operator $\hat{O}_S$ does not itself depend explicitly on time.

## The Heisenberg Picture

An alternative to the Schrödinger picture for describing time evolution is the Heisenberg picture. In the Heisenberg picture, it is the states that are constant in time:

$$|\psi_H(t)\rangle = \hat{U}_S^\dagger(t)|\psi_S(t)\rangle = e^{i\hat{H}t/\hbar}|\psi_S(t)\rangle = |\psi_S(0)\rangle \tag{14.107}$$

On the other hand, an expectation value can be written as

$$\langle\psi_S(t)|\hat{O}_S|\psi_S(t)\rangle = \langle\psi_S(0)|e^{i\hat{H}t/\hbar}\hat{O}_S e^{-i\hat{H}t/\hbar}|\psi_S(0)\rangle$$
$$= \langle\psi_H|e^{i\hat{H}t/\hbar}\hat{O}_S e^{-i\hat{H}t/\hbar}|\psi_H\rangle \tag{14.108}$$

which suggests that we define the operator $\hat{O}_H$ in the Heisenberg picture by

$$\hat{O}_H = e^{i\hat{H}t/\hbar}\hat{O}_S e^{-i\hat{H}t/\hbar} \tag{14.109}$$

so that

$$\langle\psi_S(t)|\hat{O}_S|\psi_S(t)\rangle = \langle\psi_H(t)|\hat{O}_H(t)|\psi_H(t)\rangle \tag{14.110}$$

Notice that

$$\frac{d\hat{O}_H}{dt} = \frac{d\,e^{i\hat{H}t/\hbar}}{dt}\hat{O}_S e^{-i\hat{H}t/\hbar} + e^{i\hat{H}t/\hbar}\hat{O}_S\frac{d\,e^{-i\hat{H}t/\hbar}}{dt}$$

$$= \frac{i}{\hbar}\left(e^{i\hat{H}t/\hbar}\hat{H}\hat{O}_S e^{-i\hat{H}t/\hbar} - e^{i\hat{H}t/\hbar}\hat{O}_S\hat{H}e^{-i\hat{H}t/\hbar}\right)$$

$$= \frac{i}{\hbar}[\hat{H},\hat{O}_H] \tag{14.111}$$

where in the last step we have taken advantage of the fact the Hamiltonian commutes with the time-development operator.[7] Thus in the Heisenberg picture the state vectors are fixed in time while the operators carry all the time dependence.

As an illustration, consider the harmonic oscillator, for which the Hamiltonian is given by

$$\hat{H} = \frac{\hat{p}_x^2}{2m} + \frac{1}{2}m\omega^2\hat{x}^2 = \hbar\omega\left(\hat{a}^\dagger\hat{a} + \frac{1}{2}\right) \tag{14.112}$$

We will take operators without subscripts to be operators in the Schrödinger picture. We can use (14.111) to determine how the operators evolve in time in the Heisenberg picture. In particular, the lowering operator in the Heisenberg picture satisfies

$$\frac{d\hat{a}_H}{dt} = \frac{i}{\hbar}[\hat{H},\hat{a}_H]$$

$$= \frac{i}{\hbar}e^{i\hat{H}t/\hbar}[H,\hat{a}]e^{-i\hat{H}t/\hbar}$$

$$= \frac{i}{\hbar}e^{i\hat{H}t/\hbar}(-\hbar\omega\hat{a})e^{-i\hat{H}t/\hbar} = -i\omega\hat{a}_H \tag{14.113}$$

The solution to this differential equation is

$$\hat{a}_H(t) = \hat{a}_H(0)e^{-i\omega t} = \hat{a}e^{-i\omega t} \tag{14.114}$$

where the last step follows since the two pictures coincide at $t = 0$. Similarly, we find that

$$\hat{a}_H^\dagger(t) = \hat{a}_H^\dagger(0)e^{i\omega t} = \hat{a}^\dagger e^{i\omega t} \tag{14.115}$$

---

[7] Note that the Hamiltonian is the same in the two pictures.

Notice that if we express the Hamiltonian in terms of the operators in the Heisenberg picture, the time dependence cancels out as it must because the Hamiltonian itself is independent of time:

$$\hat{H} = \hbar\omega\left(\hat{a}_H^\dagger(t)\hat{a}_H(t) + \tfrac{1}{2}\right)$$

$$= \hbar\omega\left(\hat{a}_H^\dagger(0)\hat{a}_H(0) + \tfrac{1}{2}\right) = \hbar\omega\left(\hat{a}^\dagger\hat{a} + \tfrac{1}{2}\right) \tag{14.116}$$

Also note that the commutator of the raising and lowering operators in the Heisenberg picture is given by

$$\left[\hat{a}_H(t), \hat{a}_H^\dagger(t)\right] = 1 \tag{14.117}$$

provided we are careful to evaluate the commutator at *equal times*. See also Problem 14.9.

Although we have phrased our whole discussion of nonrelativistic quantum mechanics to this point in the Schrödinger picture, the Heisenberg picture is a natural picture for quantum field theory. In fact, we slipped into this picture naturally when determining the vector potential operator, which in (14.48) varies in time.

### The Interaction Picture

We are now ready for an intermediate picture in which *both* the operators and the states carry some of the time dependence. We presume that we can break up the total Hamiltonian into two parts in the Schrödinger picture: $\hat{H} = \hat{H}_0 + \hat{H}_1$. We define the state in the interaction picture by

$$|\psi_I(t)\rangle = e^{i\hat{H}_0 t/\hbar}|\psi_S(t)\rangle \tag{14.118}$$

Note that

$$i\hbar\frac{d}{dt}|\psi_I(t)\rangle = -\hat{H}_0 e^{i\hat{H}_0 t/\hbar}|\psi_S(t)\rangle + e^{i\hat{H}_0 t/\hbar}i\hbar\frac{d}{dt}|\psi_S(t)\rangle$$

$$= -\hat{H}_0 e^{i\hat{H}_0 t/\hbar}|\psi_S(t)\rangle + e^{i\hat{H}_0 t/\hbar}\left(\hat{H}_0 + \hat{H}_1\right)|\psi_S(t)\rangle$$

$$= e^{i\hat{H}_0 t/\hbar}\hat{H}_1|\psi_S(t)\rangle = e^{i\hat{H}_0 t/\hbar}\hat{H}_1 e^{-i\hat{H}_0 t/\hbar}|\psi_I(t)\rangle \tag{14.119}$$

Time dependence of the state in the interaction picture is governed by $\hat{H}_1$.

If we examine the expectation value of an operator, we find

$$\langle\psi_S(t)|\hat{O}_S|\psi_S(t)\rangle = \langle\psi_I(t)|e^{i\hat{H}_0 t/\hbar}\hat{O}_S e^{-i\hat{H}_0 t/\hbar}|\psi_I(t)\rangle \tag{14.120}$$

which suggests that we define an operator in the interaction picture by

$$\hat{O}_I = e^{i\hat{H}_0 t/\hbar}\hat{O}_S e^{-i\hat{H}_0 t/\hbar} \tag{14.121}$$

Presuming again that the operator $\hat{O}_S$ in the Schrödinger picture does not depend on time, the time dependence of the operator $\hat{O}_I$ in the interaction picture is governed by

$$\frac{d\hat{O}_I}{dt} = \frac{i}{\hbar} \left( e^{i\hat{H}_0 t/\hbar} \hat{H}_0 \hat{O}_S e^{-i\hat{H}_0 t/\hbar} - e^{i\hat{H}_0 t/\hbar} \hat{O}_S \hat{H}_0 e^{-i\hat{H}_0 t/\hbar} \right)$$

$$= \frac{i}{\hbar} [\hat{H}_0, \hat{O}_I] \tag{14.122}$$

Thus the time development of operators in the interaction picture is determined by $\hat{H}_0$. From the definition (14.121) of operators in the interaction picture, we see that the Hamiltonian $\hat{H}_0$ is the same in both these pictures. On the other hand, $\hat{H}_1$ and

$$\hat{H}_{1I} = e^{i\hat{H}_0 t/\hbar} \hat{H}_1 e^{-i\hat{H}_0 t/\hbar} \tag{14.123}$$

differ, since $\hat{H}_0$ and $\hat{H}_1$ do not in general commute. Consequently, even if $\hat{H}_0$ and $\hat{H}_1$ are both time independent, $\hat{H}_{1I}$ will, in general, be time dependent. The time evolution (14.119) of the states in the interaction picture can be expressed conveniently in terms of $\hat{H}_{1I}$ as

$$i\hbar \frac{d}{dt} |\psi_I(t)\rangle = \hat{H}_{1I} |\psi_I(t)\rangle \tag{14.124}$$

Equations (14.122) and (14.124) are the fundamental equations governing time dependence in the interaction picture. In this picture the time evolution of the operators is determined by $\hat{H}_0$ and time evolution of the states is determined by $\hat{H}_1$. The interaction picture is intermediate between the Schrödinger picture, in which the states carry all the time dependence, and the Heisenberg picture, in which the operators carry the time dependence. It is convenient to express the solution to (14.124) in terms of a time-development operator in the interaction picture:

$$|\psi_I(t)\rangle = \hat{U}_I(t) |\psi_I(0)\rangle \tag{14.125}$$

Because $\hat{H}_{1I}$ depends on time, the time-development operator in the interaction picture is *not* simply given by $e^{-i\hat{H}_{1I} t/\hbar}$. However, from (14.124) we see that the time-development operator in the interaction picture satisfies the equation

$$i\hbar \frac{d}{dt} \hat{U}_I(t) = \hat{H}_{1I} \hat{U}_I(t) \tag{14.126}$$

and thus we can at least write a formal solution in the form

$$\hat{U}_I(t) = 1 - \frac{i}{\hbar} \int_0^t dt' \, \hat{H}_{1I}(t') \hat{U}_I(t') \tag{14.127}$$

as can be verified by substituting this expression into (14.126) and noting that $\hat{U}_I(0) = 1$, the appropriate initial condition. We can obtain a perturbative solution by iteration:

$$\hat{U}_I(t) = 1 - \frac{i}{\hbar} \int_0^t dt' \, \hat{H}_{1I}(t') \left[ 1 - \frac{i}{\hbar} \int_0^{t'} dt'' \, \hat{H}_{1I}(t'') \hat{U}_I(t'') \right]$$

$$= 1 - \frac{i}{\hbar} \int_0^t dt' \, \hat{H}_{1I}(t') + \left( -\frac{i}{\hbar} \right)^2 \int_0^t dt' \, \hat{H}_{1I}(t') \int_0^{t'} dt'' \, \hat{H}_{1I}(t'') + \cdots$$

$$(14.128)$$

If, as before, we assume that the perturbation is small and insert a parameter $\lambda$ in $\hat{H}_1$ to keep track of the order of smallness, the expansion (14.128) can be considered as a power series in $\lambda$. We hope that the series converges sufficiently rapidly that retaining the first few terms in the series gives a good approximation for the time development of the system.

Finally, let's return to our initial problem of determining the $c_n(t)$ in (14.83). Comparing (14.118), which defines the state $|\psi_I(t)\rangle$ in the interaction picture, with (14.83), we see that

$$|\psi_I(t)\rangle = e^{i\hat{H}_0 t/\hbar} |\psi_S(t)\rangle$$

$$= e^{i\hat{H}_0 t/\hbar} \sum_n c_n(t) e^{-iE_n^{(0)} t/\hbar} |E_n^{(0)}\rangle$$

$$= \sum_n c_n(t) |E_n^{(0)}\rangle \tag{14.129}$$

Thus

$$c_f(t) = \langle E_f^{(0)} | \psi_I(t) \rangle \tag{14.130}$$

If, as in (14.88), we choose the initial state to be an energy eigenstate of $\hat{H}_0$, namely, $|\psi_I(0)\rangle = |E_i^{(0)}\rangle$, then using the expansion (14.128) we find

$$\langle E_f^{(0)} | \hat{U}_I(t) | E_i^{(0)} \rangle = \langle E_f^{(0)} | E_i^{(0)} \rangle - \frac{i}{\hbar} \int_0^t dt' \, \langle E_f^{(0)} | \hat{H}_{1I}(t') | E_i^{(0)} \rangle + \cdots$$

$$= \delta_{fi} - \frac{i}{\hbar} \int_0^t dt' \, \langle E_f^{(0)} | e^{i\hat{H}_0 t'/\hbar} \hat{H}_1 e^{-i\hat{H}_0 t'/\hbar} | E_i^{(0)} \rangle + \cdots$$

$$= \delta_{fi} - \frac{i}{\hbar} \int_0^t dt' \, e^{i(E_f^{(0)} - E_i^{(0)})t'/\hbar} \langle E_f^{(0)} | \hat{H}_1 | E_i^{(0)} \rangle + \cdots$$

$$(14.131)$$

in agreement with (14.95). Thus the probability of making a transition from the $|E_i^{(0)}\rangle$ to the state $|E_f^{(0)}\rangle$ is given by

$$|c_f(t) e^{-iE_f^{(0)} t/\hbar}|^2 = |c_f(t)|^2 = |\langle E_f^{(0)} | \hat{U}_I(t) | E_i^{(0)} \rangle|^2 \tag{14.132}$$

## 14.7 FERMI'S GOLDEN RULE

We are ready to examine the time evolution of an excited state of an atom. We take the initial state to be an energy eigenstate $|i\rangle = |n_i, l_i, m_i\rangle \otimes |0\rangle$, where the atom is in the state $|n_i, l_i, m_i\rangle$, with no photons present. The final state is $|n_f, l_f, m_f\rangle \otimes |1_{\mathbf{k},\lambda}\rangle$, where the atom is in the state $|n_f, l_f, m_f\rangle$ and a photon with momentum $\hbar \mathbf{k}$ and polarization $\lambda$ has been emitted. For example, we might be interested in calculating the lifetime for a hydrogen atom in the $2p$ state to emit a photon and make a transition to the $1s$ ground state. The Hamiltonian $\hat{H}_1$ is given by (14.79). Since the total Hamiltonian (14.73) is the Hamiltonian for a closed system, with no external sources or sinks of energy, it must be independent of time. Thus we can use the expansion (14.48) for the vector potential evaluated at a particular time, say $t = 0$,

$$\hat{\mathbf{A}}(\mathbf{r}) = \sum_{\mathbf{k},\lambda} c \sqrt{\frac{2\pi\hbar}{\omega}} \left[ \hat{a}_{\mathbf{k},\lambda} \boldsymbol{\epsilon}(\mathbf{k}, \lambda) \frac{e^{i\mathbf{k}\cdot\mathbf{r}}}{\sqrt{V}} + \hat{a}^{\dagger}_{\mathbf{k},\lambda} \boldsymbol{\epsilon}(\mathbf{k}, \lambda) \frac{e^{-i\mathbf{k}\cdot\mathbf{r}}}{\sqrt{V}} \right] \qquad (14.133)$$

to express the Hamiltonian in terms of the annihilation and creation operators for photons. The amplitude to find the system in the state $|n_f, l_f, m_f\rangle \otimes |1_{\mathbf{k},\lambda}\rangle$ at time $t$ is given by

$$\langle 1_{\mathbf{k},\lambda}| \otimes \langle n_f, l_f, m_f|\hat{U}_I(t)|n_i, l_i, m_i\rangle \otimes |0\rangle \qquad (14.134)$$

or more simply

$$\langle f|\hat{U}_I(t)|i\rangle \qquad (14.135)$$

where it is understood that the initial and final states are eigenstates of $\hat{H}_0$.

Since the $\hat{H}_1$ in (14.79) is independent of time in the Schrödinger picture, evaluation of the time integral in (14.131) is straightforward. Defining

$$\eta = (E_f^{(0)} - E_i^{(0)})/\hbar \qquad (14.136)$$

we obtain

$$\int_0^t dt' \, e^{i\eta t'} = \frac{1}{i\eta}(e^{i\eta t} - 1) = \frac{e^{i\eta t/2}}{(\eta/2)} \sin(\eta t/2) \qquad (14.137)$$

Thus the probability of making a transition is given by

$$|\langle f|\hat{U}_I(t)|i\rangle|^2 = \frac{1}{\hbar^2} \frac{\sin^2(\eta t/2)}{(\eta/2)^2} |\langle f|\hat{H}_1|i\rangle|^2 \qquad (14.138)$$

Figure 14.4 shows

$$\frac{\sin^2(\eta t/2)}{(\eta/2)^2} \qquad (14.139)$$

plotted as a function of $\eta$. There is clearly a nonzero probability of making transitions to states with energy such that $\eta \approx 0$, that is, to states such that

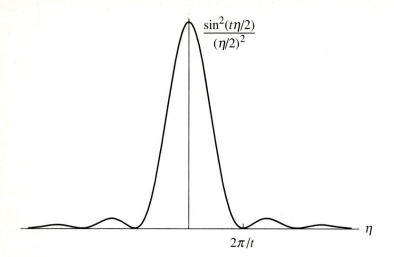

**FIGURE 14.4**

A plot of $\dfrac{\sin^2(t\eta/2)}{(\eta/2)^2}$ versus $\eta$.

$E_f^{(0)} \neq E_i^{(0)}$. For an atom in an excited state making a transition to another state with the emission of a photon, this means

$$E_{n_f} + \hbar\omega \neq E_{n_i} \qquad (14.140)$$

where the energies $E_{n_i}$ and $E_{n_f}$ are the initial and final energies, respectively, of the atom. (For the hydrogen atom, $E_n = -\mu c^2 \alpha^2/2n^2$.)

How then does conservation of energy arise? Notice that the first zero of (14.139) occurs when $t\eta/2 = \pi$, or $\eta = 2\pi/t$. Also notice that as $\eta \to 0$, $(\sin^2 \eta t/2)/(\eta/2)^2 \to t^2$. Thus as $t$ increases, the function (14.139) becomes narrower and higher, and the probability of making a transition to a state that doesn't conserve energy decreases. In fact, for large $t$ this sounds like a Dirac delta function, except that the area of the central peak, which is roughly the height times the width, is growing like $t$. Indeed, we can make use of the representation of a Dirac delta function

$$\lim_{t\to\infty} \frac{1}{\pi} \frac{\sin^2(t\eta/2)}{t(\eta/2)^2} = \delta\left(\frac{\eta}{2}\right) \qquad (14.141)$$

to express the transition probability as

$$\lim_{t\to\infty} |\langle f|\hat{U}_I(t)|i\rangle|^2 = \frac{\pi t \delta(\eta/2)}{\hbar^2}|\langle f|\hat{H}_1|i\rangle|^2 \qquad (14.142)$$

Thus in the large-$t$ limit we see energy conservation appearing. For finite times, we should not expect strict energy conservation in any case. After all, the energy-time uncertainty relation $\Delta E \Delta t \geq \hbar/2$ implies that if the evolutionary time $\Delta t$ for the system is finite, the energy of the system is uncertain by $\Delta E \geq \hbar/2\Delta t$. In our example, the time $t$ is such an evolutionary time. Thus, as (14.138) shows, we should expect to find transitions to states with a spread in energy $\Delta E$, where $\Delta E t \sim \hbar$. In practice, the evolutionary time $t$ imposed by the experimental set-up makes the large-time limit the appropriate one. For example, detecting nonconservation of energy by one part in $10^6$ for photons emitted in a transition in the visible part of the spectrum would require that the time $t$ be on the order of $10^{-9}$ seconds. In a typical experiment in which atoms are excited in a discharge tube, we do not know when the atom was actually excited to this precision, let alone interrupting the time evolution of the system within this time period. Such interruption can occur naturally, however, such as when the atom is de-excited by colliding with other atoms in the discharge tube. Although, in principle, such collisions could shorten the "natural" evolutionary time for the system, leading to a spread in energy $\Delta E$ in the photons emitted, the natural time scale for such collisions is actually large in comparison with the natural lifetime.[8]

Neglecting effects such as those due to collisions, we can safely calculate the transition probability in the large-time limit using (14.142). At second glance, the appearance of the energy-conserving Dirac delta function in this equation may now seem disturbing. After all, when $\eta = 0$, the probability of making a transition appears to be infinite. However, the transition probability (14.142) is not physically significant. It is not possible to observe a transition to a particular final state involving a photon with a particular $\mathbf{k}$. Any detector that we use to detect photons counts photons within a range of angles, which are determined by the solid angle subtended by the detector. Also, there is always some energy resolution for the detector; it detects photons within a range of energies. In order to compare (14.142) with experiment, we need to calculate

$$\sum_{\mathbf{k}}^{\mathbf{k}+\Delta\mathbf{k}} |\langle f|\hat{U}_I(t)|i\rangle|^2 \tag{14.143}$$

where the range of $\Delta\mathbf{k}$ included in the sum over final photon states is determined by the resolution of the detector.

## The Density of States

How many photon states are there between $\mathbf{k}$ and $\mathbf{k} + \Delta\mathbf{k}$? Recall that we are working in a cubic box of volume $V = L^3$ and that the allowed photon states are discrete, as indicated by (14.34). In order to count the density of states, we set up a lattice with each state represented by a point in the lattice with the integer

---

[8] The natural lifetime of the excited state sets an evolutionary time for the system that is on the order of $10^{-9}$ seconds, as we will calculate in the next section, and therefore from the energy-time uncertainty relation we should expect to see a spread in the energy of the emitted photons that is on the order of 1 part in $10^6$.

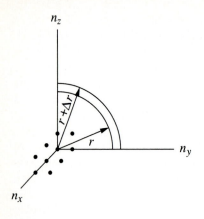

**FIGURE 14.5**
Each allowed value of **k** corresponds to a point in the $(n_x, n_y, n_z)$ lattice. For the sake of clarity, only a few of the points are shown. The number of points between $k$ and $k + \Delta k$ is given by the volume of the *spherical* shell between $r$ and $r + \Delta r$.

coordinates $(n_x, n_y, n_z)$, as shown in Fig. 14.5. Each point on the lattice labels a photon state.[9] Since the radius in the lattice is given by

$$r = \sqrt{n_x^2 + n_y^2 + n_z^2}$$

$$= \frac{L}{2\pi} \sqrt{k_x^2 + k_y^2 + k_z^2} = \frac{L}{2\pi} k \qquad (14.144)$$

and there is one state per unit volume in this lattice, the number of states between $r$ and $r + \Delta r$ is given by the volume of the spherical shell between these two radii,

$$4\pi r^2 \Delta r = 4\pi \left(\frac{L}{2\pi}\right)^3 k^2 \Delta k \qquad (14.145)$$

assuming $\Delta r \ll r$. Notice that each positive and negative integer on the lattice corresponds to a different photon state, and therefore we have to count the number of states in the whole spherical shell between $r$ and $r + \Delta r$. The fraction of the states in a solid angle $\Delta\Omega$ is just given by

$$\left(\frac{L}{2\pi}\right)^3 k^2 \Delta k \, \Delta\Omega \qquad (14.146)$$

As $V$ becomes large, the spacing between photon states becomes small, approaching a continuum as $V \to \infty$. In this case, the number of states between

---

[9] Strictly, there are two photon states for each value of **k**, taking into account photon polarization. To get the total transition rate at the end we must sum over these different polarizations.

$k$ and $k + dk$ and between $\Omega$ and $\Omega + d\Omega$ is given by[10]

$$\left(\frac{L}{2\pi}\right)^3 k^2 \, dk \, d\Omega = \frac{V}{(2\pi)^3} \, d^3k \tag{14.147}$$

Consequently, the sum in (14.143) is replaced by an integral:

$$\sum_{k}^{k+\Delta k} |\langle f|\hat{U}_I(t)|i\rangle|^2 = \int \frac{V \, d^3k}{(2\pi)^3} |\langle f|\hat{U}_I(t)|i\rangle|^2 \tag{14.148}$$

where the limits of the integral are determined by the resolution of the detector. Using $E = \hbar\omega = \hbar k c$, we can write the number of states (14.147) as

$$\frac{V}{(2\pi)^3} \frac{\omega^2}{\hbar c^3} \, d\Omega \, dE = \rho \, d\Omega \, dE \tag{14.149}$$

where we have labeled by $\rho \, d\Omega$ the *density of states* with photon energy between $E$ and $E + dE$. Notice that $\rho \, d\Omega$ is proportional to $\omega^2$, that is, the (angular) frequency squared. Incidentally, our notation is somewhat unconventional because $\rho \, d\Omega$ is often just called $\rho$ for historical reasons.

## The Transition Rate

We are now ready to determine the transition probability to a set of photon states that are covered by the detector:

$$\lim_{t\to\infty} \sum_{k}^{k+\Delta k} |\langle f|\hat{U}_I(t)|i\rangle|^2 = \int dE \int d\Omega \frac{\pi t \delta(\eta/2)}{\hbar^2} |\langle f|\hat{H}_1|i\rangle|^2 \rho \tag{14.150}$$

where the angular and energy limits of the integration are determined by the resolution of the detector. Since

$$\delta\left(\frac{\eta}{2}\right) = \delta\left(\frac{E_{n_f} + E - E_{n_i}}{2\hbar}\right)$$

$$= 2\hbar\delta\left[E - (E_{n_i} - E_{n_f})\right] \tag{14.151}$$

the *transition probability per unit time* into a solid angle $d\Omega$ that we obtain by carrying out the energy integration in (14.150) and dividing the probability of

---

[10] Incidentally, if we make the substitution $\mathbf{p} = \hbar\mathbf{k}$, we see that the number of states between $\mathbf{p}$ and $\mathbf{p} + d\mathbf{p}$ is given by

$$\frac{V \, d^3p}{(2\pi)^3\hbar^3} = \frac{V \, d^3p}{h^3}$$

showing that each state occupies a volume $h^3$ in phase space.

making a transition by the total time $t$ is simply given by

$$dR = \frac{2\pi}{\hbar}|\langle f|\hat{H}_1|i\rangle|^2 \rho \, d\Omega \tag{14.152}$$

Integrating over all solid angles and summing over the two polarization states of the photon, we obtain for the total transition probability per unit time to a particular final state of the atom

$$R = \sum_\lambda \int \frac{2\pi}{\hbar}|\langle f|\hat{H}_1|i\rangle|^2 \rho \, d\Omega \tag{14.153}$$

The result (14.153) is often referred to as *Fermi's Golden Rule*. Note that $R$, the transition probability per unit time, is independent of time. Since probability of decay in time $dt$ is given by $R \, dt$, the probability of the atom decaying in the next time interval $dt$ does not increase the longer the atom survives. For a sample of $N$ atoms excited at $t = 0$, the number $dN$ that decay in time $dt$ is given by

$$dN = -NR \, dt \tag{14.154}$$

which can be integrated to yield

$$N(t) = N(0)e^{-Rt}$$
$$= N(0)e^{-t/\tau} \tag{14.155}$$

Thus the lifetime $\tau$ for this decay is given by $\tau = 1/R$.[11]

## 14.8 SPONTANEOUS EMISSION

In order to determine the lifetime for spontaneous emission of an excited state of the hydrogen atom, we need to calculate the matrix element in (14.153). We use the interaction Hamiltonian

$$\hat{H}_1 \rightarrow \frac{e}{m_e c}\hat{\mathbf{A}} \cdot \frac{\hbar}{i}\nabla + \frac{e^2}{2m_e c^2}\hat{\mathbf{A}}^2 \tag{14.156}$$

In evaluating the matrix element

$$\langle 1_{\mathbf{k},\lambda}| \otimes \langle n_f, l_f, m_f|\hat{H}_1|n_i, l_i, m_i\rangle \otimes |0\rangle \tag{14.157}$$

the photon part is the easiest. The only term in the expansion (14.133) for the vector potential in terms of annihilation and creation operators that contributes to this matrix element is when the particular $\hat{a}_{\mathbf{k},\lambda}^\dagger$ that creates the final-state photon acts on the photon vacuum state $|0\rangle$. Note that the piece of the Hamiltonian involving $\hat{\mathbf{A}}^2$ cannot contribute to this matrix element, since it involves terms such as $\hat{a}_{\mathbf{k},\lambda}^\dagger \hat{a}_{\mathbf{k}',\lambda'}^\dagger$, which changes the number of photons in the initial state by two. Here we are calculating the amplitude for emission of a single photon. The

---

[11] Fortunately, more complicated systems such as human beings need not conform to this behavior.

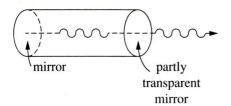

**FIGURE 14.6**

A resonant cavity formed by two reflecting surfaces traps photons with a particular wave vector **k**. These photons then stimulate emission of other atoms in the cavity that have been pumped into the excited state.

mirror

partly transparent mirror

reason that two-photon emission is less likely than single-photon emission is that the rate for single-photon emission is proportional to $e^2$ (from the square of the matrix element), while the rate for two-photon emission is proportional to $e^4$. We can use $\hbar$ and $c$ to express these factors in terms of the fine-structure constant $\alpha = e^2/\hbar c \cong 1/137$. Thus the rate for single-photon emission is on the order of $\alpha$, while the two-photon emission rate is on the order of $\alpha^2$.

Using the interaction Hamiltonian (14.156), we find

$$\langle 1_{\mathbf{k},\lambda}| \otimes \langle n_f, l_f, m_f|\hat{H}_1|n_i, l_i, m_i\rangle \otimes |0\rangle$$

$$= \frac{e}{m_e c}c\sqrt{\frac{2\pi\hbar}{\omega V}}\int d^3r\, \psi^*_{n_f,l_f,m_f}e^{-i\mathbf{k}\cdot\mathbf{r}}\boldsymbol{\varepsilon}(\mathbf{k},\lambda)\cdot\frac{\hbar}{i}\boldsymbol{\nabla}\psi_{n_i,l_i,m_i}\left(\langle 1_{\mathbf{k},\lambda}|\hat{a}^\dagger_{\mathbf{k},\lambda}|0\rangle\right)$$

$$\tag{14.158}$$

Note that

$$\langle 1_{\mathbf{k},\lambda}|\hat{a}^\dagger_{\mathbf{k},\lambda}|0\rangle = \langle 1_{\mathbf{k},\lambda}|1_{\mathbf{k},\lambda}\rangle = 1 \tag{14.159}$$

Interestingly, for emission in the presence of the $n_{\mathbf{k},\lambda}$ photons, each with momentum $\hbar\mathbf{k}$ and polarization $\lambda$, we would have

$$\langle n_{\mathbf{k},\lambda}+1|\hat{a}^\dagger_{\mathbf{k},\lambda}|n_{\mathbf{k},\lambda}\rangle = \sqrt{n_{\mathbf{k},\lambda}+1}\langle n_{\mathbf{k},\lambda}+1|n_{\mathbf{k},\lambda}+1\rangle = \sqrt{n_{\mathbf{k},\lambda}+1} \tag{14.160}$$

Thus the transition rate for *stimulated emission* is a factor of $n_{\mathbf{k},\lambda}+1$ larger than for spontaneous emission. This is the key to the operation of the laser. A resonant cavity is set up that traps photons with a particular **k**, say between two reflecting surfaces, as shown in Fig. 14.6. Then subsequent decays of excited atoms are more likely to be into states with a particular type of photon if there are already photons of this type present.

## The Electric Dipole Approximation

How large is the argument $\mathbf{k}\cdot\mathbf{r}$ of the exponential in (14.158)? The size of $\mathbf{r}$ is on the order of the size of the atom, that is, on the order of $a_0$, the Bohr radius. The wavelength $\lambda$ for transitions in the visible is 4000–7000 Å, while even for Lyman $\alpha$, the $n = 2$ to $n = 1$ transition, the wavelength is on the order of 1200 Å. Since $k = 2\pi/\lambda$, $\mathbf{k}\cdot\mathbf{r} \ll 1$, and it is a good approximation to use the series expansion

$$e^{-i\mathbf{k}\cdot\mathbf{r}} = 1 - i\mathbf{k}\cdot\mathbf{r} + \frac{(-i\mathbf{k}\cdot\mathbf{r})^2}{2!} + \cdots \tag{14.161a}$$

and replace the exponential with the first term:

$$e^{-i\mathbf{k}\cdot\mathbf{r}} \rightarrow 1 \qquad (14.161b)$$

This straightforward mathematical approximation leads to the electric dipole approximation, in which the effective interaction Hamiltonian in position space responsible for the transition becomes

$$\hat{H}_1 \rightarrow -\boldsymbol{\mu}_e \cdot \hat{\mathbf{E}} \qquad (14.162)$$

To establish this result, it is convenient to use a small trick. For the Hamiltonian

$$\hat{H}_0 = \frac{\hat{\mathbf{p}}^2}{2\mu} - \frac{e^2}{|\hat{\mathbf{r}}|} \qquad (14.163)$$

the commutator of the Hamiltonian and the position operator is given by

$$\begin{aligned}
\left[\hat{H}_0, \hat{x}_i\right] &= \left[\frac{\hat{\mathbf{p}}^2}{2\mu}, \hat{x}_i\right] \\
&= \sum_j \left[\frac{\hat{p}_j \hat{p}_j}{2\mu}, \hat{x}_i\right] = \frac{1}{2\mu} \sum_j \left(\hat{p}_j[\hat{p}_j, \hat{x}_i] + [\hat{p}_j, \hat{x}_i]\hat{p}_j\right) \\
&= -\frac{1}{\mu} \sum_j \hat{p}_j i\hbar \delta_{i,j} = -\frac{i\hbar}{\mu} \hat{p}_i \qquad (14.164)
\end{aligned}$$

Thus

$$\begin{aligned}
\langle n_f, l_f, m_f | \hat{p}_i | n_i, l_i, m_i \rangle &= \frac{i\mu}{\hbar} \langle n_f, l_f, m_f | [\hat{H}_0, \hat{x}_i] | n_i, l_i, m_i \rangle \\
&= \frac{i\mu}{\hbar} \left(E_{n_f} - E_{n_i}\right) \langle n_f, l_f, m_f | \hat{x}_i | n_i, l_i, m_i \rangle \\
&= -i\mu\omega \langle n_f, l_f, m_f | \hat{x}_i | n_i, l_i, m_i \rangle \qquad (14.165)
\end{aligned}$$

or

$$\langle n_f, l_f, m_f | \hat{\mathbf{p}} | n_i, l_i, m_i \rangle = -i\mu\omega \langle n_f, l_f, m_f | \hat{\mathbf{r}} | n_i, l_i, m_i \rangle \qquad (14.166)$$

where $\hbar\omega = E_{n_i} - E_{n_f}$ is the energy carried by the photon. Using this result, we see that the matrix element of the interaction Hamiltonian taken between the initial atom state and the final atom-photon state is the same as that generated by the replacement

$$\frac{e}{m_e c} \hat{\mathbf{A}} \cdot \frac{\hbar}{i} \boldsymbol{\nabla} \rightarrow \frac{-e}{c} \frac{\partial \hat{\mathbf{A}}}{\partial t} \cdot \mathbf{r} = e\mathbf{r} \cdot \hat{\mathbf{E}} = -\boldsymbol{\mu}_e \cdot \hat{\mathbf{E}} \qquad (14.167)$$

in the limit that we make the replacement of the exponential (14.161) by one.[12]

---

[12] In calculating the time derivative in (14.167), we can take the time dependence as given in (14.48), since the operators in the interaction picture evolve according to $\hat{H}_0$. We then evaluate the whole Hamiltonian at $t = 0$ to determine the form of $\hat{H}_1$. Note the radius vector $\mathbf{r}$ points from the proton to the electron and thus the dipole moment of the electron-proton system is given by $\boldsymbol{\mu}_e = -e\mathbf{r}$. Also, we have replaced the reduced mass in (14.166) by the electron mass.

The transition rate (14.152) can now be expressed as

$$dR = \frac{2\pi}{\hbar} \left( c\sqrt{\frac{2\pi\hbar}{\omega V}} \right)^2 e^2 \frac{\omega^2}{c^2} \left| \int d^3r \, \psi^*_{n_f,l_f,m_f} \, \mathbf{r} \psi_{n_i,l_i,m_i} \cdot \boldsymbol{\varepsilon}(\mathbf{k}, \lambda) \right|^2 \frac{V\omega^2 \, d\Omega}{(2\pi)^3 c^3 \hbar}$$

$$= \frac{\alpha\omega^3}{2\pi c^2} \left| \int d^3r \, \psi^*_{n_f,l_f,m_f} \, \mathbf{r} \psi_{n_i,l_i,m_i} \cdot \boldsymbol{\varepsilon}(\mathbf{k}, \lambda) \right|^2 d\Omega \qquad (14.168)$$

where in the last step we have introduced the fine-structure constant $\alpha$.

## The Lifetime of the 2p State of Hydrogen

Our problem now reduces to evaluating the matrix element of the position operator between the initial and final atom states. Note that the components of the position vector $\mathbf{r}$ can be put in the following form:

$$-\frac{1}{\sqrt{2}}(x + iy) = -\frac{1}{\sqrt{2}} r e^{i\phi} \sin\theta = \sqrt{\frac{4\pi}{3}} r Y_{1,1} \qquad (14.169a)$$

$$\frac{1}{\sqrt{2}}(x - iy) = \frac{1}{\sqrt{2}} r e^{-i\phi} \sin\theta = \sqrt{\frac{4\pi}{3}} r Y_{1,-1} \qquad (14.169b)$$

$$z = r \cos\theta = \sqrt{\frac{4\pi}{3}} r Y_{1,0} \qquad (14.169c)$$

In terms of these components

$$\mathbf{r} \cdot \boldsymbol{\varepsilon} = \left( \frac{\varepsilon_x + i\varepsilon_y}{\sqrt{2}} \right) \left( \frac{x - iy}{\sqrt{2}} \right) + \left( \frac{\varepsilon_x - i\varepsilon_y}{\sqrt{2}} \right) \left( \frac{x + iy}{\sqrt{2}} \right) + \varepsilon_z z$$

$$= r\sqrt{\frac{4\pi}{3}} \left( \frac{\varepsilon_x + i\varepsilon_y}{\sqrt{2}} Y_{1,-1} - \frac{\varepsilon_x - i\varepsilon_y}{\sqrt{2}} Y_{1,1} + \varepsilon_z Y_{1,0} \right) \qquad (14.170)$$

Thus

$$\int d^3r \, \psi^*_{n_f,l_f,m_f} \, \mathbf{r} \cdot \boldsymbol{\varepsilon}(\mathbf{k}, \lambda) \psi_{n_i,l_i,m_i}$$

$$= \int d^3r \, \psi^*_{n_f,l_f,m_f} \, r\sqrt{\frac{4\pi}{3}} \left( \frac{\varepsilon_x + i\varepsilon_y}{\sqrt{2}} Y_{1,-1} - \frac{\varepsilon_x - i\varepsilon_y}{\sqrt{2}} Y_{1,1} + \varepsilon_z Y_{1,0} \right) \psi_{n_i,l_i,m_i}$$

$$\qquad (14.171)$$

In order to calculate the lifetime of the 2p state of the hydrogen atom, we must evaluate the matrix element

$$\int d^3r \, \psi_{1,0,0}^* \, \mathbf{r} \cdot \boldsymbol{\varepsilon}(\mathbf{k}, \lambda) \psi_{2,1,m_i}$$

$$= \int d^3r \, R_{1,0}^* Y_{0,0}^* \, r \sqrt{\frac{4\pi}{3}} \left( \frac{\varepsilon_x + i\varepsilon_y}{\sqrt{2}} \, Y_{1,-1} - \frac{\varepsilon_x - i\varepsilon_y}{\sqrt{2}} \, Y_{1,1} + \varepsilon_z Y_{1,0} \right) R_{2,1} Y_{1,m_i}$$

$$(14.172)$$

Note that $Y_{0,0} = 1/\sqrt{4\pi}$ is a constant, while from (14.169) we see that $Y_{1,1} = -Y_{1,-1}^*$. Thus in carrying out the angular integrals in (14.172), we can take advantage of the orthogonality relation

$$\int d\Omega \, Y_{1,m}^* Y_{1,m_i} = \delta_{m,m_i} \tag{14.173}$$

to express (14.172) as

$$\int d^3r \, \psi_{1,0,0}^* \mathbf{r} \cdot \boldsymbol{\varepsilon}(\mathbf{k}, \lambda) \psi_{2,1,m_i}$$

$$= \sqrt{\frac{1}{3}} \left( -\frac{\varepsilon_x + i\varepsilon_y}{\sqrt{2}} \delta_{m_i,1} + \frac{\varepsilon_x - i\varepsilon_y}{\sqrt{2}} \delta_{m_i,-1} + \varepsilon_z \delta_{m_i,0} \right) \int_0^\infty dr \, r^2 R_{1,0}^* r R_{2,1}$$

$$(14.174)$$

Notice that if, for example, the atom is initially in a state with $m_i = 1$, then the only nonvanishing piece of the matrix element contributes a term proportional to $\varepsilon_x + i\varepsilon_y$. If the photon is propagating in the $z$ direction, this corresponds to a right-circularly polarized photon state, as discussed in Sections 2.7 and 14.4. Thus the angular momentum of the initial state is carried off in the intrinsic spin of the emitted photon in this particular case. In general, a photon emitted in some other direction carries the angular momentum of the atom in a combination of both orbital and spin angular momentum.

Using the radial wave functions (10.43) and (10.44b), we find

$$\int_0^\infty dr \, r^2 R_{1,0}^* r R_{2,1} = \sqrt{\frac{3}{2}} \frac{2^8}{3^5} a_0 \tag{14.175}$$

Thus the absolute square of (14.174) is given by

$$\frac{1}{3} \left( \frac{\varepsilon_x^2 + \varepsilon_y^2}{2} \delta_{m_i,1} + \frac{\varepsilon_x^2 + \varepsilon_y^2}{2} \delta_{m_i,-1} + \varepsilon_z^2 \delta_{m_i,0} \right) \frac{2^{15}}{3^9} a_0^2 \tag{14.176}$$

We can determine the transition rate separately for each of the three possible $m_i$. If, as is commonly the case, the atom is initially unpolarized and thus each of the three $m_i$ values occur with equal probability, we can obtain the transition rate by averaging over these different values of $m_i$:

$$\frac{1}{3} \sum_{m_i} \left| \int d^3r \, \psi_{1,0,0}^* \mathbf{r} \cdot \boldsymbol{\varepsilon}(\mathbf{k}, \lambda) \psi_{2,1,m_i} \right|^2 = \frac{2^{15}}{3^{10}} a_0^2 \frac{1}{3} \left( \varepsilon_x^2 + \varepsilon_y^2 + \varepsilon_z^2 \right) = \frac{2^{15}}{3^{11}} a_0^2$$

$$(14.177)$$

The transition rate is independent of the direction of $\boldsymbol{\varepsilon}$ when we average over the initial $m$ values. The total transition rate is obtained by integrating over all possible directions in which the photon is emitted *and* summing over the two possible polarization states for each photon:

$$R_{2p \to 1s} = \sum_{\lambda} \int d\Omega \, \frac{\alpha \omega^3}{2\pi c^2} \frac{2^{15}}{3^{11}} a_0^2 = \frac{\alpha \omega^3}{c^2} \frac{2^{17}}{3^{11}} a_0^2 \qquad (14.178)$$

Since

$$\hbar\omega = E_{2p} - E_{1s} = \frac{1}{2} m_e c^2 \alpha^2 \left(1 - \frac{1}{2^2}\right)$$

$$= \tfrac{3}{8} m_e c^2 \alpha^2 \qquad (14.179)$$

we can express the transition rate in the form

$$R_{2p \to 1s} = \left(\frac{2}{3}\right)^8 \alpha^5 \frac{m_e c^2}{\hbar}$$

$$= 0.6 \times 10^9 \ \text{s}^{-1} \qquad (14.180)$$

The lifetime for the transmission is therefore

$$\tau_{2p \to 1s} = \frac{1}{R_{2p \to 1s}} = 1.6 \times 10^{-9} \ \text{s} \qquad (14.181)$$

Our ability to calculate this lifetime from first principles is one of the triumphs of quantum mechanics. Not only can quantum mechanics make detailed predictions about the energy levels of the hydrogen atom, it can also predict the rates for the transitions between these levels,[13] as well as the angular distributions of the photon emitted in the decay. In fact, using quantum mechanics, we are able to predict the results of any measurement that the experimentalist can make on the hydrogen atom.

## Magnetic Dipole and Electric Quadrupole Transitions

A transition in which the matrix element of the electric dipole Hamiltonian between the initial and final states is nonzero is known as an electric dipole transition. However, if we evaluate the probability of a $3d$ state of hydrogen making a transition to the $1s$ ground state using this Hamiltonian, we find that the matrix element vanishes. We can see the reason by evaluating the angular part of the matrix element $\int d^3r \, \psi_{1,0,0}^* \mathbf{r} \cdot \boldsymbol{\varepsilon}(\mathbf{k}, \lambda)\psi_{3,2,m_i}$, which involves integrals of the form

$$\int d\Omega \, Y_{0,0}^* Y_{1,m} Y_{2,m_i} = \frac{1}{\sqrt{4\pi}} \int d\Omega \, Y_{1,m} Y_{2,m_i} \qquad (14.182)$$

---

[13] The agreement between (14.180) and experiment is, of course, excellent.

which vanishes since

$$\int d\Omega \, Y_{1,m}^* Y_{2,m_i} = 0 \tag{14.183}$$

Although an electric dipole transition from the $3d$ state to the $2p$ state is allowed, and a subsequent transition to the ground state through a second electric dipole transition is also permitted, (14.183) shows that a direct electric dipole transition between the $3d$ state and the ground state is not possible.[14] This raises the question: Is any direct transition with the emission of a single photon from the $3d$ state to the $1s$ state allowed?

In fact, the Hamiltonian (14.156) contains not just the electric dipole Hamiltonian but higher multipole contributions as well. To see how such contributions arise, consider retaining the next term in the series expansion (14.161$a$) of the exponential $e^{-i\mathbf{k}\cdot\mathbf{r}}$ in evaluating the matrix element. This leads to $-i\mathbf{k}\cdot\mathbf{r}\boldsymbol{\varepsilon}\cdot(\hbar\nabla/i)$ sandwiched between the initial and final atom wave functions, instead of just $\boldsymbol{\varepsilon}\cdot(\hbar\nabla/i)$ in (14.158). To see the physical significance of this term, it is helpful to express the operator version as

$$-i\mathbf{k}\cdot\hat{\mathbf{r}}\boldsymbol{\varepsilon}\cdot\hat{\mathbf{p}} = -\frac{i}{2}\left(\mathbf{k}\cdot\hat{\mathbf{r}}\boldsymbol{\varepsilon}\cdot\hat{\mathbf{p}} + \mathbf{k}\cdot\hat{\mathbf{p}}\boldsymbol{\varepsilon}\cdot\hat{\mathbf{r}}\right) - \frac{i}{2}\left(\mathbf{k}\cdot\hat{\mathbf{r}}\boldsymbol{\varepsilon}\cdot\hat{\mathbf{p}} - \mathbf{k}\cdot\hat{\mathbf{p}}\boldsymbol{\varepsilon}\cdot\hat{\mathbf{r}}\right) \tag{14.184}$$

The second term on the right-hand side of (14.184) can be rewritten as

$$-\frac{i}{2}\left(\mathbf{k}\cdot\hat{\mathbf{r}}\boldsymbol{\varepsilon}\cdot\hat{\mathbf{p}} - \mathbf{k}\cdot\hat{\mathbf{p}}\boldsymbol{\varepsilon}\cdot\hat{\mathbf{r}}\right) = -\frac{i}{2}\left(\mathbf{k}\times\boldsymbol{\varepsilon}\cdot\hat{\mathbf{r}}\times\hat{\mathbf{p}}\right) \tag{14.185}$$

We immediately recognize the orbital angular momentum operator $\hat{\mathbf{L}} = \hat{\mathbf{r}}\times\hat{\mathbf{p}}$. Thus if the electric dipole matrix element vanishes, following the type of argument that we used to obtain the electric dipole Hamiltonian in (14.167), we find that

$$\frac{e}{m_e c}\hat{\mathbf{A}}\cdot\frac{\hbar}{i}\nabla \rightarrow \frac{e}{2m_e c}\hat{\mathbf{B}}\cdot\hat{\mathbf{L}} \tag{14.186}$$

which is of the form

$$\hat{H}_1 = -\hat{\boldsymbol{\mu}}\cdot\hat{\mathbf{B}} \tag{14.187}$$

where

$$\hat{\boldsymbol{\mu}} = \frac{-e}{2m_e c}\hat{\mathbf{L}} \tag{14.188}$$

---

[14] In the general case of a transition from the state $|n_i, l_i, m_i\rangle$ to the state $|n_f, l_f, m_f\rangle$, the matrix element (14.171) is proportional to $\int d\Omega \, Y_{l_f,m_f}^* Y_{1,m} Y_{l_i,m_i}$. The product of the two $Y_{l,m}$'s can be expressed as $Y_{1,m} Y_{l_i,m_i} = \sum_{L,M} c_{L,M} Y_{L,M}$, where $L = l_i + 1, l_i, l_i - 1$, assuming $l_i \langle\ 0$. After all, the $Y_{l,m}$'s do form a complete set, and the values of $L$ are determined by the addition of angular momenta, as discussed in Appendix B. From the orthogonality of the $Y_{l,m}$'s, this result shows that $l_f = l_i + 1, l_i, l_i - 1$. However, under parity $Y_{l,m} \rightarrow (-1)^l Y_{l,m}$ (see Problem 9.15), and thus for $\int d\Omega \, Y_{l_f,m_f}^* Y_{1,m} Y_{l_i,m_i}$ to be an even function (so that the integral is nonzero), $l_f$ cannot equal $l_i$, and therefore $\Delta l = l_f - l_i = \pm 1$.

This expression for the magnetic moment of a particle with charge $-e$ is the familiar result with which we started our discussion of the interaction of a magnetic moment in an external magnetic field in Chapter 1. See (1.2). The Hamiltonian (14.187) contributes to what are called magnetic dipole transitions. Since we have included the second term in the expansion for the exponential, we expect the size of the matrix element to be of order $ka_0$ smaller than for an electric dipole transition and the transition rate to be of order $(ka_0)^2$ smaller. This suggests that the atomic transition rate for a magnetic dipole transition should be on the order of $10^6$ times smaller than for electric dipole transitions, and the corresponding lifetime should be on the order of $10^6$ times longer.

So far in our discussion of magnetic dipole transitions, we have neglected the part of the magnetic dipole Hamiltonian that depends on the intrinsic spin of the electron:

$$\frac{e g_e}{2 m_e c} \hat{\mathbf{S}} \cdot \hat{\mathbf{B}} \tag{14.189}$$

In a nonrelativistic approach to the quantum mechanics of the electron, we must still put this term into the Hamiltonian by hand. In a fully relativistic approach in which the electron is treated on the same basis as the photon and the Dirac relativistic "wave function" becomes a quantum field operator, the intrinsic spin part of the interaction enters naturally. In any case, we can now see why we neglected (14.189) in analyzing an allowed electric dipole transition. The spin magnetic dipole Hamiltonian (14.189) would make a contribution of the same size as the magnetic dipole Hamiltonian (14.186), which is due to orbital angular momentum.

Actually, you can show that even the magnetic dipole Hamiltonian cannot contribute to a direct transition from the $3d$ to the $1s$ state. However, so far we have neglected entirely the "symmetric" piece of the operator (14.184), that is, the first term on the right-hand side. This term leads to an electric quadrupole Hamiltonian that does connect the $3d$ and $1s$ states (see Problem 14.13). Thus the transition rate to go directly between these two states with the emission of a single photon should be roughly a million times smaller than for a typical electric dipole transition.

## 14.9 HIGHER-ORDER PROCESSES AND FEYNMAN DIAGRAMS

Using the procedures outlined in the preceding sections, you can calculate the transition rate for any first-order process involving the interactions of photons and atoms. In particular, it is straightforward to calculate the differential cross section for the photoelectric effect in hydrogen, where a photon ionizes the atom, kicking out a "free" electron. For the photoelectric effect, the final-state wave function for a sufficiently energetic electron can be taken to be a momentum eigenfunction, instead of the bound-state wave functions that we used in determining the lifetime of the $2p$ state (see Problem 14.12).

It is interesting to consider, at least conceptually, how you would calculate the cross section for a higher-order process such as photon-atom scattering. In this case, we are interested in the matrix element

$$\langle f|\hat{H}_1|i\rangle = \langle 1_{\mathbf{k}_f, \lambda_f}| \otimes \langle n_f, l_f, m_f|\hat{H}_1|n_i, l_i, m_i\rangle \otimes |1_{\mathbf{k}_i, \lambda_i}\rangle \tag{14.190}$$

Now the

$$\frac{e^2}{2m_e c^2}\hat{\mathbf{A}}^2 \tag{14.191}$$

part of $\hat{H}_1$ can contribute to lowest order. The operator $\hat{\mathbf{A}}^2$ contains the appropriate photon operators $\hat{a}^\dagger_{\mathbf{k}_f, \lambda_f} \hat{a}_{\mathbf{k}_i, \lambda_i}$ to annihilate the initial photon and create the appropriate final photon. Thus (14.191) can contribute to photon-atom scattering in first order (the operator $\hat{H}_1$ acting once) in the expansion (14.128) for the time-development operator.

There is another higher-order pathway that contributes to the amplitude for photon-atom scattering as well. Consider the second-order contribution to the transition amplitude

$$\left(-\frac{i}{\hbar}\right)^2 \int_0^t dt' \int_0^{t'} dt'' \langle f|\hat{H}_{1I}(t')\hat{H}_{1I}(t'')|i\rangle \tag{14.192}$$

There is a nonzero contribution to (14.192) arising from the other part of the interaction Hamiltonian,

$$\frac{e}{m_e c}\hat{\mathbf{A}} \cdot \frac{\hbar}{i}\mathbf{\nabla} \tag{14.193}$$

In practice, we evaluate (14.192) by inserting a complete set of energy eigenstates $|I\rangle$ of $\hat{H}_0$:

$$\left(-\frac{i}{\hbar}\right)^2 \int_0^t dt' \int_0^{t'} dt'' \sum_I \langle f|\hat{H}_{1I}(t')|I\rangle\langle I|\hat{H}_{1I}(t'')|i\rangle \tag{14.194}$$

Notice that the operator (14.193) acts twice, once at $t'$ and once at $t''$. Also note that $t' > t''$.

There are two different ways in which the operator (14.193) can contribute to the photon-atom scattering amplitude given in (14.194). One possibility is for the operator (14.193) to act at $t''$ to annihilate the incident photon, with the atom making a transition to some intermediate state $|I\rangle$, and then the operator acts again at $t'$ to create the final photon. We denote this transition amplitude graphically by the diagram in Fig. 14.7a. It is also possible for the operator (14.193) to act at $t''$ to create the final photon, and then when the operator acts again at $t'$ to annihilate the incident photon. This amplitude is represented graphically by Fig. 14.7b. It might seem strange that the final-state photon can be emitted before

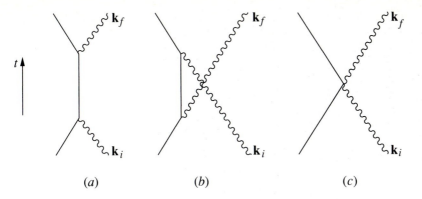

**FIGURE 14.7**
Time-ordered diagrams for calculating the amplitude for photon-atom scattering. In ($a$) the incident photon is absorbed first with a transition to an intermediate atom state $|i\rangle$, and then at a later time the final-state photon is emitted. In ($b$) the final-state photon is emitted first with a transition to an intermediate state consisting of the atom and the incident photon and the final photon, and then at a later time the incident photon is absorbed. In ($c$) the first-order amplitude due to the $\hat{\mathbf{A}}^2$ term in the interaction Hamiltonian in which both the incident photon is absorbed and the final-state photon created at the same time. This latter diagram is often referred to as a seagull diagram.

the incident photon is absorbed. In particular, this means that for $t'' < t < t'$ both photons exist, as indicated in Fig. 14.7$b$. However, these intermediate states in Fig. 14.7 need not conserve energy; energy is only conserved when you wait for a long period of time, as the experimentalist does when observing the incident and scattered photons.

The graphical pictures shown in Fig. 14.7 for the quantum mechanical transition amplitudes are known as Feynman diagrams. The diagrams are a convenient way of keeping track of the terms in the perturbative expansion (14.131) for the transition amplitude. Once you see how the analysis works, what the rules relating the amplitude to the diagram are, you can learn to write these amplitudes simply by constructing the possible diagrams that can contribute to a particular process. The diagrams are just a shorthand device that provide a powerful tool for calculating these amplitudes.

The Feynman diagrams that arise in a full treatment within quantum electrodynamics (QED) of photon-electron scattering are actually easier to evaluate than those for photon-atom scattering, basically because the electron is not as complicated as the atom with all its bound states. In QED you treat the field that creates and annihilates electrons on the same footing as the vector potential that creates and annihilates photons. QED is the best theory of any kind in its agreement between theory and experiment. With it, quantities such as the $g$ factor of the electron have been determined to better than nine significant figures. Feynman has noted that if the distance between Los Angeles and New York were measured to this precision, it would be accurate to the thickness of a human hair. It is this sort of agreement, and the lack of *any* significant disagreement between theory and experiment, that has caused Feynman to describe quantum electrodynamics as the "jewel of physics—our proudest possession."[15]

## PROBLEMS

**14.1.** By taking the dot product of Ampere's law, (14.21), with $\mathbf{E}$ and using the vector identity

$$\nabla \cdot (\mathbf{E} \times \mathbf{B}) = \mathbf{B} \cdot (\nabla \times \mathbf{E}) - \mathbf{E} \cdot (\nabla \times \mathbf{B})$$

show that energy conservation follows from Maxwell's equations in the form

$$\frac{\partial u}{\partial t} + \nabla \cdot \mathbf{S}_P = -\mathbf{J} \cdot \mathbf{E}$$

where

$$u = \frac{1}{8\pi}(\mathbf{E}^2 + \mathbf{B}^2) \quad \text{and} \quad \mathbf{S}_P = \frac{c}{4\pi}(\mathbf{E} \times \mathbf{B})$$

Discuss the physical significance of each term in this equation. *Suggestion:* Integrate the equation over an arbitrary volume and use Gauss's theorem to make the physical significance more transparent.

**14.2.** Show that

$$\int d^3 r \left(\frac{e^{i\mathbf{k}\cdot\mathbf{r}}}{\sqrt{V}}\right)^* \left(\frac{e^{i\mathbf{k}'\cdot\mathbf{r}}}{\sqrt{V}}\right) = \delta_{\mathbf{k},\mathbf{k}'}$$

given periodic boundary conditions. See (14.34).

**14.3.** The nonrelativistic wave equation is the Schrödinger equation

$$i\hbar\frac{\partial \psi}{\partial t} = \frac{1}{2m}\left(\frac{\hbar}{i}\nabla\right)^2 \psi + V\psi$$

For a *relativistic* free particle, for which $E^2 = \mathbf{p}^2 c^2 + m^2 c^4$, a natural wave equation is

$$\left(i\hbar\frac{\partial}{\partial t}\right)^2 \psi = c^2\left(\frac{\hbar}{i}\nabla\right)^2 \psi + m^2 c^4 \psi$$

or

$$\nabla^2 \psi - \frac{1}{c^2}\frac{\partial^2 \psi}{\partial t^2} - \left(\frac{mc}{\hbar}\right)^2 \psi = 0$$

which is called the Klein-Gordon equation. Use this equation to show that there is a local conservation law of the form

$$\frac{\partial \rho}{\partial t} + \nabla \cdot \mathbf{j} = 0 \quad \text{with} \quad \mathbf{j} = \frac{\hbar}{2mi}(\psi^*\nabla\psi - \psi\nabla\psi^*)$$

---

[15] R. P. Feynman, *QED—The Strange Theory of Light and Matter,* Princeton University Press, Princeton, N.J., 1985. Although this book is intended for someone with no familiarity with quantum mechanics, it is nonetheless an excellent place to start your reading about quantum electrodynamics.

Determine the form of $\rho(\mathbf{r}, t)$. From this form for $\rho$, give an argument for why the Klein-Gordon equation is not a good candidate for a one-particle relativistic wave equation in place of the Schrödinger equation, for which $\rho = \psi^*\psi$.

**14.4.** The resolution of the problem outlined in Problem 14.3 is to treat the solution to the Klein-Gordon equation as a quantum field.

(a) Verify that if we write

$$\hat{\varphi}(\mathbf{r}, t) = \sum_{\mathbf{k}} c\sqrt{\frac{\hbar}{2\omega}} \left( \hat{a}_{\mathbf{k}} \frac{e^{i(\mathbf{k}\cdot\mathbf{r}-\omega t)}}{\sqrt{V}} + \hat{a}_{\mathbf{k}}^{\dagger} \frac{e^{-i(\mathbf{k}\cdot\mathbf{r}-\omega t)}}{\sqrt{V}} \right)$$

then $\hat{\varphi}$ is a solution to the Klein-Gordon equation

$$\nabla^2\hat{\varphi} - \frac{1}{c^2}\frac{\partial^2\hat{\varphi}}{\partial t^2} - \left(\frac{mc}{\hbar}\right)^2\hat{\varphi} = 0$$

provided $\omega = \sqrt{\mathbf{k}^2 + (mc/\hbar)^2}\, c$.

(b) One can show that the Hamiltonian for this system is given by

$$\hat{H} = \frac{1}{2}\int d^3r \left[ \left(\frac{1}{c}\frac{\partial\hat{\varphi}}{\partial t}\right)^2 + \nabla\hat{\varphi}\cdot\nabla\hat{\varphi} + \left(\frac{mc}{\hbar}\right)^2\hat{\varphi}^2 \right]$$

Show that if $[\hat{a}_{\mathbf{k}}, \hat{a}_{\mathbf{k}'}] = 0$, $[\hat{a}_{\mathbf{k}}^{\dagger}, \hat{a}_{\mathbf{k}'}^{\dagger}] = 0$, and $[\hat{a}_{\mathbf{k}}, \hat{a}_{\mathbf{k}'}^{\dagger}] = \delta_{\mathbf{k},\mathbf{k}'}$, then the Hamiltonian becomes

$$\hat{H} = \sum_{\mathbf{k}} \hbar\omega\left(\hat{a}_{\mathbf{k}}^{\dagger}\hat{a}_{\mathbf{k}} + \tfrac{1}{2}\right)$$

Argue that the field $\hat{\varphi}$ creates and annihilates (spin-0) particles of mass $m$ and energy $E = \sqrt{\mathbf{p}^2c^2 + m^2c^4}$ and that these particles are indeed bosons; that is, it is possible to put more than a single particle in a state with momentum $\mathbf{p}$.

**14.5.** In order to see why the particles created by a scalar field must be bosons, consider an alternative procedure for quantizing this field. Try writing

$$\hat{\varphi}(\mathbf{r}, t) = \sum_{\mathbf{k}} c\sqrt{\frac{\hbar}{2\omega}} \left( \hat{b}_{\mathbf{k}} \frac{e^{i(\mathbf{k}\cdot\mathbf{r}-\omega t)}}{\sqrt{V}} + \hat{b}_{\mathbf{k}}^{\dagger} \frac{e^{-i(\mathbf{k}\cdot\mathbf{r}-\omega t)}}{\sqrt{V}} \right)$$

where the annihilation operators $\hat{b}_{\mathbf{k}}$ and the creation operators $\hat{b}_{\mathbf{k}}^{\dagger}$ obey the *anticommutation* relations

$$\{\hat{b}_{\mathbf{k}}, \hat{b}_{\mathbf{k}'}^{\dagger}\} = 0 \qquad \{\hat{b}_{\mathbf{k}}, \hat{b}_{\mathbf{k}'}\} = 0 \qquad \{\hat{b}_{\mathbf{k}}^{\dagger}, \hat{b}_{\mathbf{k}'}^{\dagger}\} = 0$$

where the anticommutator is defined by

$$\{\hat{A}, \hat{B}\} \equiv \hat{A}\hat{B} + \hat{B}\hat{A}$$

(a) Show that these anticommutation relations require that there cannot be more than one particle in a state:

$$\hat{b}_{\mathbf{k}}^{\dagger}\hat{b}_{\mathbf{k}}^{\dagger}|0\rangle = 0$$

and thus the particles are fermions.

(b) How would the result of your calculation for the Hamiltonian of the preceding problem for the scalar field be modified by using these anticommutation relations? It can be shown formally that there is no viable scalar field theory (the field essentially vanishes) when the field is quantized using anticommutation relations. On the other hand, for a spin-$\frac{1}{2}$ Dirac field, the situation is reversed: commutation relations for the annihilation and creation operators lead to problems—negative energies for free particles and, consequently, a lack of stability—while anticommutation relations produce a Hamiltonian with positive energies.[16]

The next three problems require time-dependent perturbation theory strictly within nonrelativistic quantum mechanics. The electromagnetic field in these problems is to be treated as a classical field. The Hamiltonian in Problems 14.6 and 14.7 is altered by the presence of an electric field **E** by the addition of a term

$$\hat{H}_1 = -\hat{\boldsymbol{\mu}}_e \cdot \mathbf{E}$$

where $\hat{\boldsymbol{\mu}}_e$ is the electric dipole moment operator.

**14.6.** A particle with charge-to-mass ratio $e/m$ in a one-dimensional harmonic oscillator with spring constant $k$ is in the ground state. An oscillating uniform electric field

$$\mathbf{E}(t) = \mathbf{E}_0 \cos \omega t \qquad \omega_0 = \sqrt{\frac{k}{m}}$$

is applied parallel to the motion of the oscillator for $t$ seconds. What is the probability that the particle is excited to the state $|n\rangle$? Evaluate the probability of making a transition when $\omega = \omega_0$, the resonance condition.

**14.7.** A hydrogen atom is placed in a time-dependent homogeneous electric field

$$\mathbf{E}(t) = \mathbf{E}_0 e^{-t/\tau}$$

where $\mathbf{E}_0$ and $\tau$ are constants. At $t = 0$ the atom is in its ground state. Calculate the probability that it will be in a $2p$ state at $t \to \infty$.

**14.8.** A spin-$\frac{1}{2}$ particle is immersed in a constant magnetic field $B_0$ in the $z$ direction and an oscillating magnetic field $B_1 \cos \omega t$ in the $x$ direction. The spin Hamiltonian can be written in the form $\hat{H} = \omega_0 \hat{S}_z + \omega_1 \cos \omega t \hat{S}_x$. See (4.34). Assume $\omega_1 \ll \omega_0$ and treat the time-dependent piece as a perturbation. Calculate the probability that the particle is in the spin-down state at time $t$ if it is in the spin-up state at $t = 0$. Evaluate your result in the resonance region when $\omega$ is near $\omega_0$. Compare your perturbative result with Rabi's formula (4.44). You may wish to review the material on pages 105–106 of Chapter 4, where an approximation made in deriving Rabi's formula is described.

**14.9.** (a) Show for the one-dimensional harmonic oscillator that the position and momentum operators in the Heisenberg picture are given by

$$\hat{x}_H(t) = \hat{x}_H(0) \cos \omega t + \frac{\hat{p}_{x_H}(0)}{m\omega} \sin \omega t$$

$$\hat{p}_{x_H}(t) = \hat{p}_{x_H}(0) \cos \omega t - m\omega \hat{x}_H(0) \sin \omega t$$

---

[16] See R. F. Streater and A. S. Wightman, *PCT, Spin & Statistics, and All That*, W. A. Benjamin, New York, 1964.

respectively, where $\hat{x}_H(0) = \hat{x}$ is the usual position operator in the Schrödinger picture and $\hat{p}_{x_H}(0) = \hat{p}_x$ is the usual momentum operator in the Schrödinger picture. *Suggestion:* Use (14.114) and (14.115) together with (7.8) and (7.9).

   **(b)** Show that the equal-time commutator of the position and momentum operators in the Heisenberg picture is given by

$$\left[\hat{x}_H(t), \; \hat{p}_{x_H}(t)\right] = i\hbar$$

What happens if times for the two operators are different?

**14.10** A spin-$\frac{1}{2}$ system initially in the spin state $|+\mathbf{z}\rangle$ undergoes a magnetic dipole transition, emitting a photon. Use the spin Hamiltonian (14.189) to show that the angular distribution of the photons is isotropic, provided we sum over the probabilites of making transitions to both the $|+\mathbf{z}\rangle$ and $|-\mathbf{z}\rangle$ states.

**14.11.** Show that the number of states, apart from spin, for an *electron* with energy between $E$ and $E + dE$ is given by

$$\frac{V}{(2\pi)^3\hbar^3} \, m_e \, p \, d\Omega \, dE$$

**14.12.** Show that the cross section for the photoelectric effect in which an electron in the ground state of hydrogen is ejected when the atom absorbs a photon is given by

$$\frac{d\sigma}{d\Omega} = 32\alpha \frac{\hbar}{m_e\omega} \frac{k_f(\boldsymbol{\varepsilon} \cdot \mathbf{k}_f)^2}{a_0^5} \frac{1}{[(1/a_0^2) + q^2]^4}$$

where $\mathbf{q} = \mathbf{k}_i - \mathbf{k}_f$. Assume the ejected electron is sufficiently energetic that the wave function for the electron can be taken to be the plane wave

$$\langle\mathbf{r}|\mathbf{k}_f\rangle = \frac{e^{i\mathbf{k}_f\cdot\mathbf{r}}}{\sqrt{V}}$$

where $\mathbf{p}_f = \hbar\mathbf{k}_f$ is the momentum of the electron and $\hbar\mathbf{k}_i$ is the momentum of the incident photon. *Suggestion:* The cross section is the transition rate divided by the incident photon flux, which is equal to $c/V$ in the box normalization used in this chapter.

**14.13.** **(a)** Show that the magnetic dipole Hamiltonian (14.187) yields a zero matrix element for the $3d$ to $1s$ transition in hydrogen. *Suggestion:* Express the angular momentum operator $\hat{\mathbf{L}}$ in terms of $\hat{L}_+$, $\hat{L}_-$, and $\hat{L}_z$.

   **(b)** Show that the electric quadrupole Hamiltonian can yield a nonzero matrix element for this $3d$ to $1s$ transition. *Suggestion:* Use the "trick" (14.164) to express the first term on the right-hand side of (14.184) solely in terms of position operators.

# APPENDIX

# A

# ELECTROMAGNETIC UNITS

Let's start with Maxwell's equations in SI units:

$$\nabla \cdot \mathbf{E} = \frac{\rho}{\varepsilon_0} \tag{A.1}$$

$$\nabla \cdot \mathbf{B} = 0 \tag{A.2}$$

$$\nabla \times \mathbf{E} = -\frac{\partial \mathbf{B}}{\partial t} \tag{A.3}$$

$$\nabla \times \mathbf{B} = \mu_0 \mathbf{j} + \mu_0 \varepsilon_0 \frac{\partial \mathbf{E}}{\partial t} \tag{A.4}$$

In these units, the force on a charged particle $q$ is given by

$$\mathbf{F} = q\mathbf{E} + q\mathbf{v} \times \mathbf{B} \tag{A.5}$$

We can use Gauss's law (A.1) most easily in its integral form

$$\oint \mathbf{E} \cdot d\mathbf{S} = \frac{q_{\text{enclosed}}}{\varepsilon_0} \tag{A.6}$$

to determine the magnitude of the electric field from a point charge $q$. Using a spherical Gaussian surface of radius $r$ centered on the charge, we find

$$E4\pi r^2 = \frac{q}{\varepsilon_0} \tag{A.7}$$

or the familiar

$$E = \frac{q}{4\pi\varepsilon_0 r^2} \tag{A.8}$$

Thus the magnitude of the force $F$ between two charges $q_1$ and $q_2$ separated by a distance $r$ is

$$F = \frac{q_1 q_2}{4\pi\varepsilon_0 r^2} \tag{A.9}$$

which we can express in the form

$$F = k_1 \frac{q_1 q_2}{r^2} \tag{A.10}$$

with the constant $k_1 = 1/4\pi\varepsilon_0$. In SI units, the unit of charge, the coulomb, is defined to be equal to 1 ampere-second, while the unit of current, the ampere, is actually determined by magnetic measurements. This is a natural set of "experimentalist's" units because it is possible to make very accurate current measurements and hence fix the unit of current, and thus the unit of charge, quite precisely. Once we know the unit of charge, the force between two charges is determined. The value of the constant $k_1$ is found experimentally to be roughly $9 \times 10^9$ newton-meter$^2$/coulomb$^2$. See (A.19).

Another system of units that is commonly used for Maxwell's equations is Gaussian units. These units can be described as "theorist's" units, since they are somewhat impractical for use in the laboratory but much more useful than SI units for revealing the true structure and beauty of electricity and magnetism. In addition, they are more commonly used than SI units for describing microscopic phenomena. In these units we begin by first determining the unit of charge, not the unit of current. In Gaussian units we *define* the constant $k_1$ to be unity. Then the force $F$ between two charges is simply given by

$$F = \frac{q_1 q_2}{r^2} \tag{A.11}$$

where the unit of charge is determined by the requirement that two units of charge separated by a distance of 1 centimeter exert a force on each other of 1 dyne. This unit of charge is called a *statcoulomb*. The corresponding electric field produced by charge $q$ is just

$$E = \frac{q}{r^2} \tag{A.12}$$

and consequently [compare (A.8) and (A.1)]

$$\nabla \cdot \mathbf{E} = 4\pi\rho \tag{A.13}$$

in these units.

Let's see what happens to the rest of Maxwell's equations. In SI units, Ampere's law, (A.4), can be expressed in integral form as

$$\oint \mathbf{B} \cdot d\mathbf{l} = \mu_0 \int \mathbf{j} \cdot d\mathbf{S} + \mu_0\varepsilon_0 \frac{d}{dt} \int \mathbf{E} \cdot d\mathbf{S} \tag{A.14}$$

where

$$\int \mathbf{j} \cdot d\mathbf{S} = I_{\text{enclosed}} \tag{A.15}$$

is usual current enclosed by the closed line contour on the left-hand side of (A.14). For a collection of charges that form a current, the magnetic portion of the force (A.5) due to a differential length $dl$ of current is given by $d\mathbf{F} = I\,d\mathbf{l} \times \mathbf{B}$. Since the magnitude of the magnetic field produced by a long current-carrying wire is given by

$$B = \frac{\mu_0 I}{2\pi r} \tag{A.16}$$

in SI units, the magnitude of the force per unit length between two parallel wires separated by a distance $r$ carrying currents $I_1$ and $I_2$ is just

$$\frac{F}{l} = \left(\frac{\mu_0}{4\pi}\right)\frac{2I_1 I_2}{r} \tag{A.17}$$

We can also express this force per unit length in the form

$$\frac{F}{l} = k_2 \frac{2I_1 I_2}{r} \tag{A.18}$$

In SI units it is the constant $k_2$ that is *defined* to be $\mu_0/4\pi = 10^{-7}$ newton/ampere$^2$. This then determines the unit of current by the requirement that two long wires each carrying a current of 1 ampere separated by a distance of 1 meter exert a force per unit length on each other of $2 \times 10^{-7}$ newton/meter. Note that

$$\frac{k_1}{k_2} = \frac{1}{4\pi\varepsilon_0}\left(\frac{4\pi}{\mu_0}\right) = \frac{1}{\varepsilon_0\mu_0} = c^2 \tag{A.19}$$

where $c$ is the speed of light. Thus measuring the speed of light determines experimentally the value of $k_1$ in SI units. In Gaussian units, on the other hand, where $k_1$ is defined, it determines the value of $k_2$.

We still have a little freedom in our choice of units left to play with. In particular, even though the force between the two current-carrying wires is determined, the magnetic field produced by the wires can still be adjusted. We introduce a constant $\lambda$ so that the magnetic portion of the Lorentz force is given by $F = \lambda^{-1} q\mathbf{v} \times \mathbf{B}$, or $dF = \lambda^{-1}I\,d\mathbf{l} \times \mathbf{B}$. Since it is the forces that we measure directly, not the magnetic fields, we can avoid changing any physics by adjusting the value of the magnetic field produced by the wire so that

$$B = \lambda \frac{\mu_0 I}{2\pi r} \tag{A.20}$$

More generally, we need to modify the right-hand side of Ampere's law, which we now express as

$$\nabla \times \mathbf{B} = \lambda\left(4\pi k_2 \mathbf{j} + \frac{1}{c^2}\frac{\partial \mathbf{E}}{\partial t}\right) \tag{A.21}$$

In Gaussian units we use this freedom to choose $\lambda = c$. This choice shows the inherently relativistic nature of magnetic effects [compare the magnetic force $q(\mathbf{v}/c) \times \mathbf{B}$ with the electric force $q\mathbf{E}$] and brings the relativistic nature of electricity and magnetism to the fore with the explicit appearance of the speed of light. It also means that in Gaussian units the electric field $\mathbf{E}$ and magnetic field $\mathbf{B}$ have the same dimensions.[1] Thus in Gaussian units (A.21) becomes

$$\nabla \times \mathbf{B} = \frac{4\pi}{c}\mathbf{j} + \frac{1}{c}\frac{\partial \mathbf{E}}{\partial t} \tag{A.22}$$

Of course, $\mathbf{E}$ and $\mathbf{B}$ having the same dimensions means that Faraday's law, (A.3), must also be adjusted in Gaussian units. We introduce a third constant $k_3$:

$$\nabla \times \mathbf{E} = -k_3 \frac{\partial \mathbf{B}}{\partial t} \tag{A.23}$$

Since the gradient on the left-hand side of (A.23) has the dimensions of 1/length, $k_3$ must have the same dimensions as $1/c$. In fact, in order for electromagnetic fields to propagate at the speed of light, the constant $k_3$ must equal $1/c$. Thus the full Maxwell's equations in Gaussian units are given by

$$\nabla \cdot \mathbf{E} = 4\pi\rho \tag{A.24}$$

$$\nabla \cdot \mathbf{B} = 0 \tag{A.25}$$

$$\nabla \times \mathbf{E} = -\frac{1}{c}\frac{\partial \mathbf{B}}{\partial t} \tag{A.26}$$

$$\nabla \times \mathbf{B} = \frac{4\pi}{c}\mathbf{j} + \frac{1}{c}\frac{\partial \mathbf{E}}{\partial t} \tag{A.27}$$

with the force law

$$\mathbf{F} = q\mathbf{E} + q(\mathbf{v}/c) \times \mathbf{B} \tag{A.28}$$

Trading $\varepsilon_0$ and $\mu_0$ for the explicit appearance of $c$ seems to be a step in the right direction. In fact, these units make more advanced fully covariant presentations of Maxwell's equations in terms of relativistic four-vectors and tensors both straightforward and elegant.[2]

For completeness, we should mention one other system of units that is also attractive to theorists. If you are interested in making the full Maxwell's equations as simple and as elegant as possible, you can eliminate the factors of $4\pi$ that

---

[1] Since the torque on a magnetic moment $\boldsymbol{\mu}$ is given by $\boldsymbol{\mu} \times \mathbf{B}$, scaling up the magnetic field by a factor of $c$ means that the magnetic moment of a current loop picks up a factor of $1/c$ in comparison with its value in SI units, as indicated in (1.1).

[2] A good trick (see Sections 10.2 and 11.6) for evaluating expressions such as energies of the hydrogen atom is to replace $e^2$ with $(e^2/\hbar c)\hbar c$ since the quantity $e^2/\hbar c = \alpha$ ($e^2/4\pi\varepsilon_0\hbar c$ in SI units) is a dimensionless quantity whose value is roughly 1/137. In this way you may never need to recall that the charge on an electron is $4.8 \times 10^{-10}$ statcoulombs in Gaussian units rather that the more familiar $1.6 \times 10^{-19}$ coulombs in SI units.

appear in the charge and current source terms of Gauss's law and Ampere's law at the expense of a slightly more complicated expression for the fields produced by these charges and currents. In this other system of units, known as Heaviside-Lorentz units, we take $k_1 = 1/4\pi$. The corresponding electric field produced by a point charge $q$ is then $E = q/4\pi r^2$. The explicit appearance of these factors of $4\pi$ in the forms for the fields, however, is compensated for by their disappearance in Maxwell's equations. From (A.19) we see that $k_2 = 1/4\pi c^2$, and therefore in Heaviside-Lorentz units Maxwell's equations are given by

$$\nabla \cdot \mathbf{E} = \rho \tag{A.29}$$

$$\nabla \cdot \mathbf{B} = 0 \tag{A.30}$$

$$\nabla \times \mathbf{B} = \frac{\mathbf{j}}{c} + \frac{1}{c}\frac{\partial \mathbf{E}}{\partial t} \tag{A.31}$$

$$\nabla \times \mathbf{E} = -\frac{1}{c}\frac{\partial \mathbf{B}}{\partial t} \tag{A.32}$$

with the force law still given by (A.28). Heaviside-Lorentz units are sometimes referred to as rationalized Gaussian units.

# THE ADDITION
# OF ANGULAR
# MOMENTA

In this appendix we would like to investigate a simple way of adding the angular momenta $j_1$ and $j_2$ together. Our goal is to determine the linear combinations of the basis states for two angular momenta that form eigenstates of total angular momentum. Rather than give a general proof that the values for the total angular momentum $j$ run from $|j_1 - j_2|$ to $j_1 + j_2$ in integral steps, we will take the specific example of adding spin $\frac{1}{2}$ and spin 1 to illustrate a procedure that can be readily extended to the addition of any two angular momenta.

We label in the usual way the two basis states for a spin-$\frac{1}{2}$ particle by $|\frac{1}{2}, \frac{1}{2}\rangle$ and $|\frac{1}{2}, -\frac{1}{2}\rangle$ and the three basis states for a spin-1 particle by $|1, 1\rangle$, $|1, 0\rangle$, and $|1, -1\rangle$. By taking the direct-product of these basis states, we can form six two-particle basis states:

$$|\tfrac{1}{2}, \tfrac{1}{2}\rangle_1 \otimes |1, 1\rangle_2, \qquad |\tfrac{1}{2}, \tfrac{1}{2}\rangle_1 \otimes |1, 0\rangle_2, \qquad |\tfrac{1}{2}, \tfrac{1}{2}\rangle_1 \otimes |1, -1\rangle_2$$

$$|\tfrac{1}{2}, -\tfrac{1}{2}\rangle_1 \otimes |1, 1\rangle_2, \qquad |\tfrac{1}{2}, -\tfrac{1}{2}\rangle_1 \otimes |1, 0\rangle_2, \qquad |\tfrac{1}{2}, -\tfrac{1}{2}\rangle_1 \otimes |1, -1\rangle_2$$

$$\text{(B.1)}$$

We have arbitrarily chosen to call particle 1 the spin-$\frac{1}{2}$ particle and particle 2 the spin-1 particle. As usual, we can drop the direct-product symbol $\otimes$ without generating any confusion.

The two basis states $|\frac{1}{2}, \frac{1}{2}\rangle_1 |1, 1\rangle_2$ and $|\frac{1}{2}, -\frac{1}{2}\rangle_1 |1, -1\rangle_2$ are sometimes referred to as "stretched" configurations. What is special about these configurations is that the $z$ component of the total angular momentum takes on its maximum and minimum values for these two states, respectively. For example, applying

$$\hat{J}_z = \hat{J}_{1z} + \hat{J}_{2z} \tag{B.2}$$

to these kets, we find

$$(\hat{J}_{1z} + \hat{J}_{2z})|\tfrac{1}{2}, \tfrac{1}{2}\rangle_1|1, 1\rangle_2 = \hat{J}_{1z}|\tfrac{1}{2}, \tfrac{1}{2}\rangle_1|1, 1\rangle_2 + \hat{J}_{2z}|\tfrac{1}{2}, \tfrac{1}{2}\rangle_1|1, 1\rangle_2$$

$$= \tfrac{1}{2}\hbar|\tfrac{1}{2}, \tfrac{1}{2}\rangle_1|1, 1\rangle_2 + \hbar|\tfrac{1}{2}, \tfrac{1}{2}\rangle_1|1, 1\rangle_2$$

$$= \tfrac{3}{2}\hbar|\tfrac{1}{2}, \tfrac{1}{2}\rangle_1|1, 1\rangle_2 \tag{B.3}$$

Similarly,

$$(\hat{J}_{1z} + \hat{J}_{2z})|\tfrac{1}{2}, -\tfrac{1}{2}\rangle_1|1, -1\rangle_2 = -\tfrac{3}{2}\hbar|\tfrac{1}{2}, -\tfrac{1}{2}\rangle_1|1, -1\rangle_2 \tag{B.4}$$

Clearly, none of the other four states in our set of basis states (B.1) has these eigenvalues for $\hat{J}_z$.

Since the total angular momentum operator

$$\hat{\mathbf{J}}^2 = (\hat{\mathbf{J}}_1 + \hat{\mathbf{J}}_2)^2 = \hat{\mathbf{J}}_1^2 + \hat{\mathbf{J}}_2^2 + 2\hat{\mathbf{J}}_1 \cdot \hat{\mathbf{J}}_2 \tag{B.5}$$

commutes with the operator $\hat{J}_z$, these two operators must have eigenstates in common. Consequently, the states $|\tfrac{1}{2}, \tfrac{1}{2}\rangle_1|1, 1\rangle_2$ and $|\tfrac{1}{2}, -\tfrac{1}{2}\rangle_1|1, -1\rangle_2$ must each be an eigenstate of the total angular momentum $\hat{\mathbf{J}}^2$. We label these states in general by $|j, m\rangle$. Since $m = \tfrac{3}{2}$ and $m = -\tfrac{3}{2}$ for these two states, respectively, you may be tempted to guess that these are both $j = \tfrac{3}{2}$ states. To verify that this is the case, we express $2\hat{\mathbf{J}}_1 \cdot \hat{\mathbf{J}}_2$ in terms of angular momentum raising and lowering operators, just as we did in (5.10):

$$2\hat{\mathbf{J}}_1 \cdot \hat{\mathbf{J}}_2 = \hat{J}_{1+}\hat{J}_{2-} + \hat{J}_{1-}\hat{J}_{2+} + 2\hat{J}_{1z}\hat{J}_{2z} \tag{B.6}$$

Then we can apply the operator (B.5) to these states. For example,

$$\hat{\mathbf{J}}^2|\tfrac{1}{2}, \tfrac{1}{2}\rangle_1|1, 1\rangle_2 = (\hat{\mathbf{J}}_1^2 + \hat{\mathbf{J}}_2^2 + 2\hat{\mathbf{J}}_1 \cdot \hat{\mathbf{J}}_2)|\tfrac{1}{2}, \tfrac{1}{2}\rangle_1|1, 1\rangle_2$$

$$= \left[\tfrac{1}{2}(\tfrac{1}{2} + 1)\hbar^2 + 1(1 + 1)\hbar^2 + \hat{J}_{1+}\hat{J}_{2-} + \hat{J}_{1-}\hat{J}_{2+} + 2\hat{J}_{1z}\hat{J}_{2z}\right]|\tfrac{1}{2}, \tfrac{1}{2}\rangle_1|1, 1\rangle_2$$

$$= \left[\tfrac{1}{2}(\tfrac{1}{2} + 1) + 1(1 + 1) + 2\tfrac{1}{2}(1)\right]\hbar^2|\tfrac{1}{2}, \tfrac{1}{2}\rangle_1|1, 1\rangle_2$$

$$= \tfrac{15}{4}\hbar^2|\tfrac{1}{2}, \tfrac{1}{2}\rangle_1|1, 1\rangle_2 = \tfrac{3}{2}(\tfrac{3}{2} + 1)\hbar^2|\tfrac{1}{2}, \tfrac{1}{2}\rangle_1|1, 1\rangle_2 \tag{B.7}$$

where we have taken advantage of the fact that the raising operators for each of the particles yields zero when it acts on a state that has the $z$ component of the angular momentum for that particle equal to its maximum value:

$$\hat{J}_{1+}|\tfrac{1}{2}, \tfrac{1}{2}\rangle_1 = 0 \quad \text{and} \quad \hat{J}_{2+}|1, 1\rangle_2 = 0 \tag{B.8}$$

Thus

$$|\tfrac{3}{2}, \tfrac{3}{2}\rangle = |\tfrac{1}{2}, \tfrac{1}{2}\rangle_1|1, 1\rangle_2 \tag{B.9}$$

Similarly, you can verify that

$$\hat{\mathbf{J}}^2|\tfrac{1}{2}, -\tfrac{1}{2}\rangle_1|1, -1\rangle_2 = \tfrac{3}{2}(\tfrac{3}{2} + 1)\hbar^2|\tfrac{1}{2}, -\tfrac{1}{2}\rangle_1|1, -1\rangle_2 \qquad (B.10)$$

or

$$|\tfrac{3}{2}, -\tfrac{3}{2}\rangle = |\tfrac{1}{2}, -\tfrac{1}{2}\rangle_1|1, -1\rangle_2 \qquad (B.11)$$

So far we have found two of the four $j = \tfrac{3}{2}$ states. We can determine the other two by applying the lowering and raising operators for the total angular momentum to the $|\tfrac{3}{2}, \tfrac{3}{2}\rangle$ and $|\tfrac{3}{2}, -\tfrac{3}{2}\rangle$ states, respectively. Using (3.60), we find

$$\hat{J}_-|\tfrac{3}{2}, \tfrac{3}{2}\rangle = \sqrt{\tfrac{3}{2}(\tfrac{3}{2} + 1) - \tfrac{3}{2}(\tfrac{3}{2} - 1)}\hbar|\tfrac{3}{2}, \tfrac{1}{2}\rangle = \sqrt{3}\hbar|\tfrac{3}{2}, \tfrac{1}{2}\rangle \qquad (B.12)$$

Since

$$\hat{J}_- = \hat{J}_{1-} + \hat{J}_{2-} \qquad (B.13)$$

we also know that

$$\hat{J}_-|\tfrac{1}{2}, \tfrac{1}{2}\rangle_1|1, 1\rangle_2 = \hat{J}_{1-}|\tfrac{1}{2}, \tfrac{1}{2}\rangle_1|1, 1\rangle_2 + \hat{J}_{2-}|\tfrac{1}{2}, \tfrac{1}{2}\rangle_1|1, 1\rangle_2$$

$$= \sqrt{\tfrac{1}{2}(\tfrac{1}{2} + 1) - \tfrac{1}{2}(\tfrac{1}{2} - 1)}\hbar|\tfrac{1}{2}, -\tfrac{1}{2}\rangle_1|1, 1\rangle_2 + \sqrt{1(1 + 1) - 1(1 - 1)}\hbar|\tfrac{1}{2}, \tfrac{1}{2}\rangle_1|1, 0\rangle_2 \qquad (B.14)$$

Equating (B.12) and (B.14), we find that

$$|\tfrac{3}{2}, \tfrac{1}{2}\rangle = \tfrac{1}{\sqrt{3}}|\tfrac{1}{2}, -\tfrac{1}{2}\rangle_1|1, 1\rangle_2 + \sqrt{\tfrac{2}{3}}|\tfrac{1}{2}, \tfrac{1}{2}\rangle_1|1, 0\rangle_2 \qquad (B.15)$$

Either by applying the lowering operator again to (B.15) or applying the raising operator to (B.11), we can show that

$$|\tfrac{3}{2}, -\tfrac{1}{2}\rangle = \sqrt{\tfrac{2}{3}}|\tfrac{1}{2}, -\tfrac{1}{2}\rangle_1|1, 0\rangle_2 + \tfrac{1}{\sqrt{3}}|\tfrac{1}{2}, \tfrac{1}{2}\rangle_1|1, -1\rangle_2 \qquad (B.16)$$

Thus we have determined all four of the $j = \tfrac{3}{2}$ states.

Since our basis (B.1) is six dimensional, there are two states left over. These states turn out to be total angular momentum $j = \tfrac{1}{2}$ states. We can generate them by taking advantage of the fact that $\langle \tfrac{1}{2}, \tfrac{1}{2}|\tfrac{3}{2}, \tfrac{1}{2}\rangle = 0$, that is, the amplitude to find a state with $j = \tfrac{3}{2}$, $m = \tfrac{1}{2}$ in the state $j = \tfrac{1}{2}$, $m = \tfrac{1}{2}$ is zero. This is enough information to deduce that

$$|\tfrac{1}{2}, \tfrac{1}{2}\rangle = \sqrt{\tfrac{2}{3}}|\tfrac{1}{2}, -\tfrac{1}{2}\rangle_1|1, 1\rangle_2 - \sqrt{\tfrac{1}{3}}|\tfrac{1}{2}, \tfrac{1}{2}\rangle_1|1, 0\rangle_2 \qquad (B.17)$$

up to an overall phase. Note that only the two basis states $|\tfrac{1}{2}, -\tfrac{1}{2}\rangle_1|1, 1\rangle_2$ and $|\tfrac{1}{2}, \tfrac{1}{2}\rangle_1|1, 0\rangle_2$ are candidates to be involved in this superposition, since they are the only two of the six basis states with the eigenvalue for the $z$ component of the total angular momentum equal to $\tfrac{1}{2}$. Lastly, we can determine the $|\tfrac{1}{2}, -\tfrac{1}{2}\rangle$ state, either by applying the lowering operator $\hat{J}_- = \hat{J}_{1-} + \hat{J}_{2-}$ to the state (B.17) or by choosing the linear combination of the basis states with total $m = -\frac{1}{2}$ that is orthogonal to the state (B.16). In this way we find

$$|\tfrac{1}{2}, -\tfrac{1}{2}\rangle = \sqrt{\tfrac{1}{3}}|\tfrac{1}{2}, -\tfrac{1}{2}\rangle_1|1, 0\rangle_1 - \sqrt{\tfrac{2}{3}}|\tfrac{1}{2}, \tfrac{1}{2}\rangle_1|1, -1\rangle_2 \qquad (B.18)$$

Of course, we haven't really proved that the two states (B.17) and (B.18) are $j = \frac{1}{2}$ states, although this is consistent with the fact that there are two states remaining after we constructed the $j = \frac{3}{2}$ states. Real proof comes from applying $\hat{\mathbf{J}}^2$ to one of them.

In conclusion, note that our initial two-particle basis consisted of six states and that we have now determined the linear combinations of these states that are eigenstates of total angular momentum with $j = \frac{3}{2}$ (four states) and $j = \frac{1}{2}$ (two states). The procedure that we have used to determine these states can be utilized to add together any two angular momenta. For example, if the system consists of two spin-1 particles, the two-particle basis is nine dimensional. The total angular momentum takes on the values 2, 1, and 0. We can start with the stretched configuration $|2, 2\rangle = |1, 1\rangle_1|1, 1\rangle_2$ and apply the lowering operator (B.13) to determine the other four $j = 2$ states. We then use orthogonality relation $\langle 1, 1|2, 1\rangle = 0$ to determine the $|1, 1\rangle$ state and apply the lowering operator to determine the other two $j = 1$ states. Finally, we can take advantage of the orthogonality relations $\langle 0, 0|2, 0\rangle = 0$ and $\langle 0, 0|1, 0\rangle = 0$ to determine the single $j = 0$ state. We need two orthogonality relations in this case because there are three two-particle states, including $|1, 1\rangle_1|1, -1\rangle_2$, $|1, -1\rangle_1|1, 1\rangle_2$, and $|1, 0\rangle_1|1, 0\rangle_2$, that can comprise the $|2, 0\rangle$, $|1, 0\rangle$, and $|0, 0\rangle$ states.

The amplitudes $\langle j, m|(|j_1, m_1\rangle_1|j_2, m_2\rangle_2)$ are known as Clebsch-Gordan coefficients. Although we have called the individual angular momenta spins in this appendix, these angular momenta could be orbital as well as spin angular momentum. Thus (B.17), for example, could result from the determination of the total angular momentum states of a spin-$\frac{1}{2}$ particle that has orbital angular momentum $l = 1$. Compare the specific results of this appendix with the more general results of adding spin $\frac{1}{2}$ and orbital angular momentum $l$ in Section 11.6. These Clebsch-Gordan coefficients are routinely tabulated, so you don't actually need to calculate them each time you need them, once you understand how they are obtained.

# APPENDIX
# C

## DIRAC DELTA FUNCTIONS

The Dirac delta function is actually not a function at all but a "generalized function," or a "distribution," that is defined through the relation

$$\int_{-\infty}^{\infty} dx\, f(x)\delta(x - x_0) = f(x_0) \tag{C.1}$$

for any smooth function $f(x)$. From (C.1) we conclude that

$$\delta(x - x_0) = 0 \qquad x \neq x_0 \tag{C.2}$$

and by setting $f(x) = 1$ in (C.1) that

$$\int_{-\infty}^{\infty} dx\, \delta(x - x_0) = 1 \tag{C.3}$$

Thus the delta function is a "function" that vanishes everywhere except at a single point but nonetheless has unit area. You can think of a delta function as the limit of a sharply peaked function of unit area (see Fig. C.1), as it becomes progressively narrower and higher. In this limit, the function $f(x)$ in the integral in (C.1) can be set equal to its value at $x_0$ since this is the only region in which the integrand is nonzero, and then the constant $f(x_0)$ can be pulled outside the integral.

We can derive a number of properties of delta functions, with the understanding that identities involving delta functions make sense only when the delta

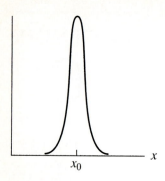

**FIGURE C.1**
A sharply peaked function with unit area. The Dirac delta function arises in the limit that the function becomes infinitesimally narrow and infinitely high.

functions appear within an integral. To derive

$$\delta(ax) = \frac{1}{|a|}\delta(x) \tag{C.4}$$

first consider the result

$$\int_{-\infty}^{\infty} dx\, f(x)\delta(ax) = \frac{1}{a}\int_{-\infty}^{\infty} dy\, f\left(\frac{y}{a}\right)\delta(y) = \frac{1}{a}f(0) \qquad a > 0 \tag{C.5}$$

where in the second step we have made the change of variables $y = ax$. Note that if $a < 0$, this same change of variables would switch the limits of integration, leading to

$$\int_{-\infty}^{\infty} dx\, f(x)\delta(ax) = \frac{1}{a}\int_{\infty}^{-\infty} dy\, f\left(\frac{y}{a}\right)\delta(y) = \frac{1}{-a}\int_{-\infty}^{\infty} dy\, f\left(\frac{y}{a}\right)\delta(y)$$

$$= \frac{1}{-a}f(0) \qquad a < 0 \tag{C.6}$$

These results can be combined together in the form of (C.4). Note that one of the corollaries of (C.4) is

$$\delta(-x) = \delta(x) \tag{C.7}$$

The delta function is an even function. Another relation that follows from (C.4) is

$$\delta(x^2 - a^2) = \frac{1}{2|a|}[\delta(x - a) + \delta(x + a)] \tag{C.8}$$

Since $x^2 - a^2$ vanishes at both $x = a$ and $x = -a$, we can write

$$\delta(x^2 - a^2) = \delta[(x - a)(x + a)]$$

$$= \delta[2a(x - a)] + \delta[-2a(x + a)]$$

$$= \frac{1}{2|a|}[\delta(x - a) + \delta(x + a)] \tag{C.9}$$

More generally, suppose $f(x)$ is a function that has a zero at $x_0$, that is, $f(x_0) = 0$. Expanding $f(x)$ in a Taylor series about $x_0$:

$$f(x) = f(x_0) + \left(\frac{df}{dx}\right)_{x=x_0}(x - x_0) + \cdots = \left(\frac{df}{dx}\right)_{x=x_0}(x - x_0) + \cdots \quad (C.10)$$

and taking advantage of (C.4) again, we obtain

$$\delta[f(x)] = \frac{1}{|df/dx|_{x=x_0}}\delta(x - x_0) \quad (C.11)$$

where we can safely ignore the higher-order terms in the Taylor series because the delta function vanishes everywhere except at $x = x_0$.

When the delta function is multiplied by a smooth function within an integral, we can give meaning to the derivative of a delta function:

$$\int_{-\infty}^{\infty} dx\, f(x)\frac{d}{dx}\delta(x) = f(x)\delta(x)\Big|_{-\infty}^{\infty} - \int_{-\infty}^{\infty} dx\, \frac{df(x)}{dx}\delta(x)$$

$$= -\left[\frac{df(x)}{dx}\right]_{x=0} \quad (C.12)$$

where the second step follows from an integration by parts. Also the integral of a delta function satisfies

$$\int_{-\infty}^{x} dy\, \delta(y - a) = \begin{cases} 0 & x < a \\ 1 & x > a \end{cases}$$

$$\equiv \Theta(x - a) \quad (C.13)$$

where $\Theta(x - a)$ is the standard step function. From this result, we also see that

$$\frac{d}{dx}\Theta(x - a) = \delta(x - a) \quad (C.14)$$

A convenient way to represent a delta function is as the limit of a sequence of regular functions that have unit area but that grow progressively more narrow as some parameter is varied. Some examples:

1. The function $\sin \lambda x/\pi x$ is plotted in Fig. C.2. This function is well behaved for any finite value of $\lambda$. The width of the function is of order $1/\lambda$, since the first zero of the sine function occurs when $\lambda x = \pi$. Moreover, the height of the function at the origin is $\lambda/\pi$. Thus as $\lambda$ increases, the function grows narrower and taller. In fact, the normalization factor of $1/\pi$ has been chosen so that the function has unit area. Therefore, as $\lambda \to \infty$, the function behaves as a delta function:

$$\delta(x) = \lim_{\lambda\to\infty} \frac{1}{\pi}\frac{\sin \lambda x}{x} \quad (C.15)$$

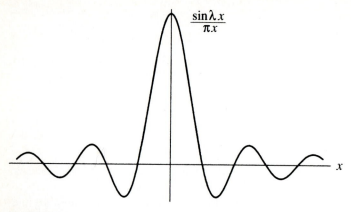

**FIGURE C.2**
The function $(\sin \lambda x)/\pi x$, which represents a delta function in the limit $\lambda \to \infty$.

2. An alternative way of expressing the representation (C.15) of the delta function is especially useful. Since

$$\frac{\sin \lambda x}{x} = \frac{1}{2} \int_{-\lambda}^{\lambda} dk\, e^{ikx} \qquad (C.16)$$

we can write

$$\delta(x) = \frac{1}{2\pi} \int_{-\infty}^{\infty} dk\, e^{ikx} \qquad (C.17)$$

3. Another representation of the delta function is given by

$$\delta(x) = \lim_{\alpha \to \infty} \frac{\alpha}{\sqrt{\pi}} e^{-\alpha x^2} \qquad (C.18)$$

as can be verified by using the results of Appendix D on Gaussian integrals.

4. Finally, you can show that

$$\delta(x) = \lim_{\varepsilon \to 0} \frac{1}{\pi} \frac{\varepsilon}{x^2 + \varepsilon^2} \qquad (C.19)$$

# GAUSSIAN
# INTEGRALS

We first wish to evaluate the integral

$$I(a) = \int_{-\infty}^{\infty} dx \, e^{-ax^2} \tag{D.1}$$

where Re $a > 0$. A useful trick is to consider the integral squared,

$$I^2(a) = \int_{-\infty}^{\infty} dx \, e^{-ax^2} \int_{-\infty}^{\infty} dy \, e^{-ay^2} = \int_{-\infty}^{\infty} dx \int_{-\infty}^{\infty} dy \, e^{-a(x^2+y^2)} \tag{D.2}$$

which can be easily evaluated by switching from Cartesian to polar coordinates:

$$I^2(a) = \int_0^{\infty} r \, dr \int_0^{2\pi} d\theta \, e^{-ar^2} = \frac{\pi}{a} \tag{D.3}$$

Thus

$$I(a) = \int_{-\infty}^{\infty} dx \, e^{-ax^2} = \sqrt{\frac{\pi}{a}} \tag{D.4}$$

What about integrals such as

$$I(a, b) = \int_{-\infty}^{\infty} dx \, e^{-ax^2+bx} \tag{D.5}$$

Here we can convert the integral into one in which we can take advantage of (D.4) by completing the square in the exponent:

$$ax^2 - bx = a\left(x - \frac{b}{2a}\right)^2 - \frac{b^2}{4a} \tag{D.6}$$

Making the change of variables $x' = x - (b/2a)$, we find

$$I(a, b) = e^{b^2/4a} \int_{-\infty}^{\infty} dx'\, e^{-ax'^2} = e^{b^2/4a} \sqrt{\frac{\pi}{a}} \tag{D.7}$$

We can also evaluate integrals of the form

$$I'(a) = \int_{-\infty}^{\infty} dx\, e^{-ax^2} x^2 \tag{D.8}$$

by differentiating (D.1) under the integral sign:

$$I'(a) = -\frac{d}{da} \int_{-\infty}^{\infty} dx\, e^{-ax^2} = -\frac{d}{da}\sqrt{\frac{\pi}{a}} = \frac{1}{2}\sqrt{\frac{\pi}{a^3}} \tag{D.9}$$

This technique can be easily extended. For example,

$$\int_{-\infty}^{\infty} dx\, e^{-ax^2} x^4 = \frac{d^2}{da^2} I(a) \tag{D.10}$$

Finally, we should note that although we have derived (D.4) for Re $a > 0$, we can extend this result to include $a$ in (D.1) being purely imaginary. This is most easily done through contour integration. First consider the closed integral

$$\oint dz\, e^{-az^2} \tag{D.11}$$

in the complex plane for the contour shown in Fig. D.1, with $a$ real and positive. Since the integrand is analytic within the contour, the closed contour integral vanishes:

$$\oint dz\, e^{-az^2} = 0 \tag{D.12}$$

Writing

$$z = re^{i\theta} = r(\cos\theta + i\sin\theta) \tag{D.13}$$

we see on the circular arcs of radius $R$ that the integrand is given by

$$e^{-aR^2(\cos 2\theta + i\sin 2\theta)} \tag{D.14}$$

which goes exponentially to zero as $R \to \infty$ for $0 < \theta < \pi/4$ and $\pi < \theta < 5\pi/4$ since $\cos 2\theta > 0$ for these angles. Thus the contribution of the circular arcs to the contour integral vanishes as $R \to \infty$. We can parametrize the diagonal line

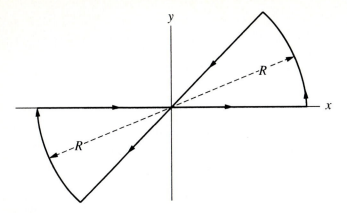

**FIGURE D.1**
A closed contour in the complex $z = x + iy$ plane. The contributions on the circular arcs vanish as $R \rightarrow \infty$.

by $z = r e^{i\pi/4}$, with $r$ running from $\infty$ to $-\infty$. Since $z^2 = r^2 e^{i\pi/2} = i r^2$ and $dz = dr\, e^{i\pi/4}$, we obtain

$$\oint dz\, e^{-az^2} = \int_{-\infty}^{\infty} dx\, e^{-ax^2} + e^{i\pi/4} \int_{\infty}^{-\infty} dr\, e^{-iar^2} = 0 \qquad \text{(D.15)}$$

Consequently

$$\int_{-\infty}^{\infty} dr\, e^{-iar^2} = e^{-i\pi/4} \int_{-\infty}^{\infty} dx\, e^{-ax^2} = e^{-i\pi/4} \sqrt{\frac{\pi}{a}} = \sqrt{\frac{\pi}{ia}} \qquad \text{(D.16)}$$

which is just the same as (D.4) with $a$ replaced by $ia$, provided we take $\sqrt{i} = e^{i\pi/4}$.

# APPENDIX
# E

# THE LAGRANGIAN FOR A CHARGE $q$ IN A MAGNETIC FIELD

How do we handle magnetic fields within the framework of a Lagrangian? Purely electric forces are easy. After all, the electric potential $\varphi(\mathbf{r})$ is introduced in electrostatics as the work done per unit charge to bring the charge to the position $\mathbf{r}$ from some reference point, which is often taken as at infinity. Then the potential energy of a charge $q$ is $V = q\varphi$ and the Lagrangian is given by

$$L = T - V = \tfrac{1}{2}m\mathbf{v}^2 - q\varphi \tag{E.1}$$

In terms of the Cartesian coordinates $x_1 = x$, $x_2 = y$, and $x_3 = z$, the Euler-Lagrange equation of motion

$$\frac{\partial L}{\partial x_i} - \frac{d}{dt}\left(\frac{\partial L}{\partial \dot{x}_i}\right) = 0 \tag{E.2}$$

for the Lagrangian (E.1) is given by

$$-q\frac{\partial \varphi}{\partial x_i} - \frac{d}{dt}m\dot{x}_i = 0 \tag{E.3}$$

This equation of motion is simply

$$m\ddot{x}_i = -q\frac{\partial \varphi}{\partial x_i} \tag{E.4}$$

Since the electric field **E** is given in electrostatics by $\mathbf{E} = -\nabla\varphi$, the equation of motion can be expressed in terms of vectors as the force law $m\mathbf{a} = \mathbf{F} = q\mathbf{E}$.

The full Lorentz force

$$\mathbf{F} = q\mathbf{E} + q(\mathbf{v}/c) \times \mathbf{B} \tag{E.5}$$

includes velocity-dependent magnetic forces, which cannot be obtained from a Lagrangian of the form (E.1) that is just the difference of the kinetic and potential energies. Since the magnetic force always acts at right angles to the velocity, it doesn't change the magnitude of the velocity and thus does no work. However, we can show that the Lagrangian

$$L = \frac{1}{2}m\mathbf{v}^2 - q\varphi + \frac{q}{c}\mathbf{A} \cdot \mathbf{v} \tag{E.6}$$

which differs from (E.1) by the addition of a velocity-dependent term involving the vector potential **A**, yields the Lorentz force (E.5) for the equations of motion. The magnetic field **B** can always be expressed in the form $\mathbf{B} = \nabla \times \mathbf{A}$, since the magnetic field satisfies $\nabla \cdot \mathbf{B} = 0$ and the gradient of a curl vanishes:

$$\nabla \cdot \mathbf{B} = \nabla \cdot (\nabla \times \mathbf{A}) = 0 \tag{E.7}$$

Since for the Lagrangian (E.6)

$$\frac{\partial L}{\partial \dot{x}_i} = m\dot{x}_i + \frac{q}{c}A_i \tag{E.8}$$

the canonical momentum $p_i = \partial L/\partial \dot{x}_i$ is given in vector form by

$$\mathbf{p} = m\mathbf{v} + \frac{q}{c}\mathbf{A} \tag{E.9}$$

In order to evaluate

$$\frac{d}{dt}\left(\frac{\partial L}{\partial \dot{x}_i}\right) = m\ddot{x}_i + \frac{q}{c}\frac{dA_i}{dt} \tag{E.10}$$

notice that $A_i = A_i\big[x(t), y(t), z(t), t\big]$ and therefore

$$\frac{dA_i}{dt} = \frac{\partial A_i}{\partial t} + \sum_{j=1}^{3}\frac{\partial A_i}{\partial x_j}\frac{dx_j}{dt} = \frac{\partial A_i}{\partial t} + \mathbf{v} \cdot \nabla A_i \tag{E.11}$$

Using

$$\frac{\partial L}{\partial x_i} = -q\frac{\partial \varphi}{\partial x_i} + \frac{q}{c}\mathbf{v} \cdot \frac{\partial \mathbf{A}}{\partial x_i} \tag{E.12}$$

(E.2) becomes

$$-q\frac{\partial \varphi}{\partial x_i} + q\frac{\mathbf{v}}{c} \cdot \frac{\partial \mathbf{A}}{\partial x_i} - m\ddot{x}_i - \frac{q}{c}\left(\frac{\partial A_i}{\partial t} + \mathbf{v} \cdot \nabla A_i\right) = 0 \tag{E.13}$$

or

$$m\ddot{x}_i = -q\frac{\partial\varphi}{\partial x_i} + q\frac{\mathbf{v}}{c}\cdot\frac{\partial\mathbf{A}}{\partial x_i} - \frac{q}{c}\left(\frac{\partial A_i}{\partial t} + \mathbf{v}\cdot\nabla A_i\right) \tag{E.14}$$

In vector notation, (E.14) can be expressed in terms of the force $\mathbf{F}$ on the particle as

$$\mathbf{F} = q\left(-\nabla\varphi - \frac{1}{c}\frac{\partial\mathbf{A}}{\partial t}\right) + \frac{q}{c}\left[\nabla(\mathbf{v}\cdot\mathbf{A}) - (\mathbf{v}\cdot\nabla)\mathbf{A}\right] \tag{E.15}$$

or

$$\mathbf{F} = q\mathbf{E} + \frac{q}{c}\mathbf{v}\times(\nabla\times\mathbf{A}) = q\mathbf{E} + \frac{q}{c}\mathbf{v}\times\mathbf{B} \tag{E.16}$$

as desired.

Given the Lagrangian (E.6), we can determine the Hamiltonian in the usual way:

$$H = \sum_{i=1}^{3} p_i\dot{x}_i - L$$

$$= \sum_{i=1}^{3}\left(m\dot{x}_i + \frac{q}{c}A_i\right)\dot{x}_i - \left(\sum_{i=1}^{3}\frac{1}{2}m\dot{x}_i\dot{x}_i - q\varphi + \frac{q}{c}\sum_{i=1}^{3}A_i\dot{x}_i\right)$$

$$= \sum_{i=1}^{3}\frac{1}{2}m\dot{x}_i\dot{x}_i + q\varphi \tag{E.17}$$

At first it appears that the vector potential has disappeared entirely from the Hamiltonian. However, if we express the Hamiltonian in terms of the canonical momentum (E.9), we obtain

$$H = \frac{(\mathbf{p} - q\mathbf{A}/c)^2}{2m} + q\varphi \tag{E.18}$$

This suggests a mnemonic for the way to turn on electromagnetic interactions in terms of the Hamiltonian: take the energy for a free particle of charge $q$

$$E = \frac{\mathbf{p}^2}{2m} \tag{E.19}$$

and make the replacements $\mathbf{p} \to \mathbf{p} - q\mathbf{A}/c$ and $E \to E - q\varphi$ to generate (E.18), with the energy $E$ replaced by the symbol for the Hamiltonian.

# VALUES
# OF PHYSICAL
# CONSTANTS

$1 \, \overset{.}{\text{A}} \equiv 10^{-10}$ m

$1$ fm $\equiv 10^{-15}$ m

$1$ barn $\equiv 10^{-28}$ m$^2$

$1$ dyne $\equiv 10^{-5}$ newton (N)

$1$ gauss(G) $\equiv 10^{-4}$ tesla (T)

$1$ erg $\equiv 10^{-7}$ joule (J)

$1$ eV $= 1.60217733(49) \times 10^{-19}$ J

$\quad = 1.60217733(49) \times 10^{-12}$ erg

$1$ MeV $\equiv 10^6$ eV

$1$ eV$/c^2 = 1.78266270(54) \times 10^{-36}$ kg

$2.99792458 \times 10^9$ esu $= 1$ coulomb (C)

$0°$C $\equiv 273.15$ K

| Quantity | Symbol | Value | Gaussian | SI |
|---|---|---|---|---|
| Speed of light | $c$ | 2.99792458 | $10^{10}$ cm/s | $10^8$ m/s |
| Planck's constant | $h$ | 6.6260755(40) | $10^{-27}$ erg s | $10^{-34}$ J s |
| | $\hbar = h/2\pi$ | 1.05457266(63) | $10^{-27}$ erg s | $10^{-34}$ J s |
| Electron charge | $e$ | 1.60217733(49) | — | $10^{-19}$ C |
| | | 4.8032068(15) | $10^{-10}$ esu | — |
| Electron mass | $m_e$ | 9.1093897(54) | $10^{-28}$ g | $10^{-31}$ kg |
| | | 0.51099906(15) | MeV/$c^2$ | MeV/$c^2$ |

*(Continued)*

*(Continued)*

| Quantity | Symbol | Value | Gaussian | SI |
|---|---|---|---|---|
| Proton mass | $m_p$ | 1.6726231(10) | $10^{-24}$ g | $10^{-27}$ kg |
| | | 938.27231(28) | MeV/$c^2$ | MeV/$c^2$ |
| Neutron mass | $m_n$ | 939.56563(28) | MeV/$c^2$ | MeV/$c^2$ |
| Reciprocal fine-structure constant | $1/\alpha$ | 137.0359895(61) | | |
| Bohr radius ($\hbar/m_e c\alpha$) | $a_0$ | 0.529177249(24) | $10^{-8}$ cm | $10^{-10}$ m |
| Bohr magneton ($e\hbar/2m_e c$) | $\mu_B$ | 5.78838263(52) | $10^{-9}$ eV/G | $10^{-5}$ eV/T |
| Boltzmann constant | $k$ | 1.380658(12) | $10^{-16}$ erg/K | $10^{-23}$ J/K |
| Avogadro's number | $N_0$ | $6.0221367(36) \times 10^{23}$ /mol | | |

Values from "Review of Particle Properties," *Phys. Lett.* **B239** (April 1990).

**1.1.** $1.2 \times 10^3$ G/cm

**2.17.** 0.12

**2.18.** $\sqrt{3}N\hbar/2$

**4.12.** $\sin^4(\omega_0 t/2)$

**6.11.** $\Delta x \Delta p_x = 0.57\hbar$

**6.14.** (a) $\sqrt{30/L^2}$  (b) $960/\pi^6$
    (c) $5\hbar^2/mL^2$

**6.16.** $-\hbar^2\lambda^2/8mb^2$

**7.12.** 0.16

**9.9.** 1.13Å

**9.19.** (d) Prob$(L_x = \hbar) = 0.5$, Prob$(L_x = -\hbar) = 0.5$

**10.3.** 0.24

**10.6.** 0.70

**10.10.** (a) $(2.40)^2\hbar^2/2\mu a^2$, $(5.52)^2\hbar^2/2\mu a^2$, $(8.65)^2\hbar^2/2\mu a^2$
    (b) $(2.40)^2\hbar^2/2\mu a^2$, $(3.83)^2\hbar^2/2\mu a^2$, $(5.14)^2\hbar^2/2\mu a^2$

**11.7.** (b) $E_{1s}^{(1)} = (2/5)(e^2/a_0)(R/a_0)^2$, $E_{2p}^{(1)} = (1/1120)(e^2/a_0)(R/a_0)^4$

**12.5.** $E \le -\hbar^2\lambda^2/4\pi mb^2$

# INDEX

# INDEX